Development of a Field Petrographic Analysis System and its Application to the Study of Socioeconomic Interaction Networks of the Early Harappan Northwestern Indus Valley of Pakistan

Development of a Field Petrographic Analysis System and its Application to the Study of Socioeconomic Interaction Networks of the Early Harappan Northwestern Indus Valley of Pakistan

Graham Mansfield Chandler

BAR International Series 995
2001

British Archaeological Reports are published by John and Erica Hedges Ltd and by Archaeopress

This volume has been published by:
John and Erica Hedges Ltd
British Archaeological Reports
7 Longworth Road
Oxford OX2 6RA
England
Tel/Fax +44(0)1865 511560
E-mail: bar.hedges@lineone.net
www.barhedges.com

Enquiries regarding the submission of manuscripts for future publication may be sent to the above address

BAR S995

Development of a Field Petrographic Analysis System and its Application to the Study of Socioeconomic Interaction Networks of the Early Harappan Northwestern Indus Valley of Pakistan

Volume Editor: Erica Hedges

Printed in England by Biddles Ltd

ISBN 1 84171 196 9

All BAR titles available from:

Hadrian Books
122 Banbury Road
Oxford OX2 7BP
England

The current BAR catalogue with details of all titles in print, prices and means of payment, is available free from Hadrian Books

All volumes are distributed by Hadrian Book Ltd

ABSTRACT

To meet a requirement for conducting petrographic analysis of archaeological ceramics in the field, a light-weight, portable, and easily obtainable kit which permits on-site excavation decisions and allows analysis in countries which restrict or prohibit the export of cultural artefacts was developed. A high-quality field microscope was modified for petrography and tested with a lightweight saw and seven cold-curing resins. The entire kit, containing all equipment and materials necessary to prepare and analyse 100 standard thin sections, weighed about five kilograms. It was used in field applications in the Northwest Frontier Province and the Punjab of Pakistan; and Kilisetepe, Turkey. The data obtained from the Pakistan evaluation were used for the second part of the thesis. This involved preparation and analysis of 175 thin sections of Early Harappan Period (ca 3500 to 2650 BC) ceramics from four sites in Pakistan. Selections were made from seven morphological pottery types common to Early Harappan levels at the sites of Rehman Dheri, Harappa, Tarakai Qila and Lewan. Kiln wasters and/or locally produced pottery were examined in thin section for comparison. Mineral inclusions were identified using standard petrographic techniques, and a textural analysis based on the quartz inclusions was conducted. Size, angularity, sphericity and weathering of inclusions were recorded and examined graphically and statistically and compared with available local and regional geological data. The results were used to examine localised ceramic technologies at the sites and the extent of intersite interaction with respect to ceramic production and technology amongst craftspeople at the sites during the Early Harappan period. Close similarities amongst mineral suites observed at each of the sites and a lack of detailed geological survey data complicated provenance analysis, so the thesis provides an example of petrographic techniques which can be employed in such situations. No evidence of trade in ceramics amongst the sites was revealed. Results suggested that Rehman Dheri and Harappa independently developed fairly sophisticated ceramic industries, as there appeared to be two independent groups of fabric types from the two sites. Tarakai Qila and Lewan fabrics fit with three of the Rehman Dheri types, suggesting less developed ceramic industries which may have been influenced by innovations from Rehman Dheri. The use of data from the portable kit demonstrated its efficacy in field situations.

ACKNOWLEDGEMENTS

This thesis could not have been completed without the immense amount of assistance provided by so many individuals and organisations. Although a PhD is independent research, it cannot be a solo effort. It is impossible to acknowledge the contribution provided by everyone who made it possible, including so many who gave moral support and encouragement when it seemed as if the only way was out, so I hope the following does not leave too many unmentioned.

First and foremost, I would like to express my most sincere gratitude to Dr M. Rafique Mughal, and his predecessor Dr Ahmed Nabi Khan, Directors General of Archaeology and Museums of the Government of Pakistan and their staff for enabling me to use the Early Harappan material from Rehman Dheri, Harappa, Lewan and Tarakai Qila. Their contribution was much more than simple enablement, their warm hospitality, patience and encouragement contributed immensely to my confidence especially at the outset of the programme when problems always seemed insurmountable. In addition, on many occasions, in the absence of the Director General, the Deputy Director Dr Niaz Rasool provided invaluable assistance. In connection with permission to utilise collections from Rehman Dheri, Lewan and Tarakai Qila, equally sincere thanks are due to Dr Farzand Ali Durrani, vice-chancellor of the University of Peshawar; Dr Abdur Rahman, then Chairman of the Department of Archaeology, Dr Farid Khan, Professor of Archaeology and Ihsan Ali, Associate Professor of Archaeology at the University of Peshawar; Drs Richard Meadow, J. Mark Kenoyer and Rita Wright of the Harappa Project for permission to use ceramic samples from Harappa; Dr Ken Thomas of the Institute of Archaeology, University College London and Mr J. Robert Knox of the British Museum for their assistance with the material from Lewan and Tarakai Qila. My stay in Peshawar was enhanced by the generosity and warm hospitality of the staff and MPhil students of the Department of Archaeology at the University of Peshawar.

To my supervisors, Drs Ian Glover and Dafydd Griffiths, goes my heartfelt gratitude. Their kind and generous assistance and encouragement from the outset, from help with the topic selection and objectives of the research right through to preparation for the examination, enabled me to accomplish this dream. I am grateful for their professionalism in allowing me a free rein in the research while providing just the right amount of guidance when it was needed. Their review of earlier drafts of this manuscript provided priceless help.

I am indebted to all of those organisations which provided funding assistance for the research. The Overseas Research Student Award granted by the Committee of Vice Chancellors and Principals of the Universities of the United Kingdom provided substantial help with the tuition fees. Travel expenses for the initial trip to Pakistan were kindly provided by the Gordon Childe Research Fund. Equipment purchase was generously assisted by a grant from the Central Research Fund of the University of London. Travel and accommodation expenses for the main research in Pakistan were most gratefully provided by grants from the Charles Wallace Trust, the University of London Graduate School Research Fund, and the Ancient India and Iran Trust. Without the kind financial assistance of these organisations, the programme could never have been completed.

I could not have made it through without the support of all my friends and family, who encouraged me to do this without wondering about my motives. There are so many others as well. The firm of K.W. Kirk & Sons, unfortunately now in receivership, generously helped with the design and manufacture of the prototype rotating stage for the McArthur Microscope; especially Mr David Toller. Kind assistance was given by Dr Brinley Roberts of the Geological Sciences Department of Birkbeck College who helped with those difficult to identify minerals; Dr M. Javed Khan, Shahab Danishwar and all of the other staff of the Geology Department at the University of Peshawar who provided facilities and assistance with thin section analysis; Dr Eric Robinson of the Department of Geological Sciences at University College London who assisted me in obtaining admission to the library of the Royal Geographical Society of London; laboratory technicians Deborah Lewis, Sandra Bond, and Kevin Reeves for their help with equipment; and fellow PhD and MPhil students in the research room who kept up the morale with stimulating talk and lively debate.

Finally, Beverly was always there with her unfailing encouragement and moral support. The pursuit of my dream generated many difficult times and storms which would have driven others to sheltered ports, but she weathered them all.

If I have forgotten anyone, I apologise. Any mistakes, errors, or omissions in this thesis are solely my own doing. There would have been a lot more without all of those above.

TABLE OF CONTENTS

PART TWO

SOCIOECONOMIC INTERACTION NETWORKS OF THE EARLY HARAPPAN NORTHWESTERN INDUS VALLEY

LIST OF FIGURES

LIST OF TABLES

PART ONE
PETROGRAPHY IN THE FIELD

CHAPTER 1: INTRODUCTION AND BACKGROUND

1.1 INTRODUCTION

"Archaeology is the study of the human cultural and social past whose goals are to narrate the sequent story of that past and to explain the events that composed it." (Willey and Sabloff 1980: 1). The archaeological record cannot be properly interpreted without examination of the human element involved in its formation. As archaeological science becomes ever more rigorous and laboratory-bound, there is a continual risk of losing sight of this human element. Adoption of the rigorous scientific approach in archaeology was causing concern in some circles more than twenty years ago (e.g. Clarke 1973). At that time, scientific archaeology or archaeometry was dominated by scientists. With the possible exception of Carbon-14 dating, it is now archaeologists who are predominantly using scientific instruments (Ehrenreich 1995: 1). This is a very welcome trend which has alleviated much of the concern expressed by Clarke some twenty-two years ago. To preserve the momentum of this movement, a balance needs to be maintained between the purely scientific approach and the hands-on approach of the field archaeologist. This can be accomplished by developing new techniques which place some emphasis on refinement of existing techniques as well as the advancement of new high technology techniques (Tite 1991: 148). Furthermore, development of portable scientific equipment that has direct applicability at the field site brings the two factions closer together. A recent example is development of a portable X-ray spectrometer (Helmig et al. 1989). By placing technological development within social development, such archaeometric studies permit a clearer picture of the presence and nature of culture contact and the possibilities of technical diffusion or parallel development (Ehrenreich 1995: 4).

The technical study of ceramics from the archaeological record is no exception. It is one area of archaeological science that has seen the application of scientific techniques grow by leaps and bounds in popularity over the past thirty years. Techniques developed from the

1940's to 1960's in the other sciences, such as the Scanning Electron Microscope, the Electron Microprobe, Atomic Absorption Spectroscopy, Instrumental Neutron Activation Analysis, X-ray Diffraction, and others, began to be applied to the study of archaeological artefacts. The same period saw relatively little development attention given to existing techniques in the study of artefacts, most notably ceramic petrology. Techniques developed in the geological and sedimentary sciences during the 1930's and 1940's (see Chayes 1956 for descriptions) were borrowed and began to be applied to the study of archaeological ceramics in thin section during the same period. But they were not significantly developed to tailor them to the field of archaeology. Ceramic petrology, and particularly associated fabric analysis, generally produces the sort of data which are more readily interpretable directly in human terms than are lists of elements produced by chemical analysis methods. In fact, with respect to ceramic petrology, the most successful applications to date have been simple ones (Freestone 1991: 399).

Maintaining the balance between the complex and sophisticated scientific techniques and raw field data in the study of ceramics could be enhanced by the development of simple portable scientific equipment. It would put ceramic provenance and technological data squarely in the hands of the field archaeologist and enable him or her to use them to interpret the situation at the site as the excavation progresses. The value here is that it would permit the excavation to be guided so as to optimise interpretation of the human element, thus bringing the application of scientific archaeology directly in contact with field directors and supervisors. Moreover, there are further benefits in taking science applications directly to the field; such as applications on location in countries which ban or severely restrict the export of cultural artefacts. These considerations led to the research and development of a portable system which would provide the capability for preparation and analysis of thin sections of archaeological ceramics on site. See also Johnson (1992) for an early statement on this approach.

Detailed descriptions of petrography can be found in most monographs on the technical analysis of ceramics or stone (e.g. Gribble and Hall 1985; Kerr 1977; Rice 1987: 371-82). Very briefly, the method consists of cutting a small thin slice from a rock or potsherd with a diamond saw, polishing a flat side of it, mounting it on a glass slide, then grinding and polishing it to a thickness of 0.03 mm (30 microns) using an increasingly fine series of abrasive powders in an aqueous slurry on glass plates. The resulting thin section is normally covered with a slip of very thin glass and examined using a polarising light microscope to identify and analyse the mineral inclusions present. In addition, a textural analysis of the fabric can be conducted. This is a statistical analysis of the sizes, shapes, orientation and frequency distributions of the mineral inclusions. Findings from these procedures can characterise a ceramic fabric. If one can account for the alterations to a clay caused by an ancient potter, comparison with local clays and/or a knowledge of the local geology can assist in defining whether the ceramic is of local or non-local manufacture. Ethnographic analyses have shown that 85 % of potters travel less than 7 km to their clay sources (Rice 1987: 116). By comparison, finished pots and ceramic products can be found hundreds of kilometres from their original clay source, especially if transported by water. The best results of a provenance analysis are usually obtained by analysing a representative collection of waster sherds from a known kiln. Thus the characterisation is conducted on fabrics after all alterations of the clay by the potter, and it can be assumed with a high degree of confidence that they are of local manufacture. It eliminates the complication of having to analyse a systematic sampling of local clays, or having to make extensive assumptions regarding the potter's treatment of the clays.

With this sort of information one can systematically determine the manufacturing locations of ceramics, and, through comparisons with findspots of ceramic artefacts, make important inferences about the distribution networks of ceramics or their contents. Moreover, if one examines stylistically and chronologically similar sherds from geographically separated sites that are not traded or transported items, other inferences can be made. If they are seen to be technologically similar, for example, one might suggest that potters were exchanging technological information or innovations. These investigations allow study and examination of the nature and extent of regional socioeconomic interaction in ancient times. Such studies have been successfully employed both in Europe where it is in quite common usage (recent examples include Buko 1984; Echallier 1991; Rigby et al 1989) and in the New World (e.g. Isaacson and Aleto 1989; Jones 1991; Stoltman 1991). The latter is a good example of how ceramic petrography can be used in documenting cultural interaction including the transport of material goods. All too often, indirect evidence (most commonly the analysis of style) in ceramics has been used to infer material exchange. Examining stylistic attributes is valuable as a means of identifying exchanged ideas, but to document the movement of goods it must be done in conjunction with mineralogical, chemical, and geological analyses (Hantman and Plog 1982: 238) in order to minimize the likelihood of tentative inferences being incorrect.

1.2 BACKGROUND

As briefly outlined in the introduction, the preparation of ceramic thin sections for archaeological analysis requires a series of steps. What follows is a more detailed description of these procedures, as an understanding of them is important to appreciating the requirements of the field system under development.

Following the selection of sherds from an archaeological collection (which is in itself a critical process and should be done with some knowledge of sampling theory and a good understanding of the problem to be examined), the first stage is to cut a thin slice of the ceramic material, approximately 1 to 2 mm thick. The cut should preferably be made in a cross section of the pot, to ensure inclusion of details relevant to technological aspects of the manufacturing process such as surface finishes or the concentration of heavier inclusions toward the outside surface which may show that the vessel was wheel-thrown; or to reveal forming technique (Courty and Roux 1995). The cut can be made with a simple hand saw, but this is slow and laborious and results in a poor surface on which to commence polishing. Best results are obtained with a diamond circular blade in a powered saw. Required rim speed for circular saws is normally dependent on the type of diamond blade employed. Water or liquid cooling is needed to prevent overheating and burning the sample and to extend blade life. In most current applications, this is done in the laboratory using saws powered by motors exceeding one-quarter horsepower. In order to keep subsequent grinding to a minimum, the slice should be made as thin as possible without it breaking. It is possible, with practice, to achieve thicknesses of 1.0 to 1.5 mm. In many cases, especially where fabrics are low-fired or coarse and friable, this will not be possible without first consolidating the material using an impregnating or consolidating resin. Details of these are discussed below.

After the cut pieces are allowed to dry, they are consolidated with the resin, if this was not done prior to cutting. Traditionally, Canada Balsam or Petropoxy 154 (a trade name of epoxy resin) is used. The resin is mixed with hardener and/or heated, according to which one is used, and applied to the cut slices of ceramic material. The resin or epoxy absorbed in the samples is then allowed to cure or harden. For both of these commonly used consolidants, this requires a thermostatically controlled heating plate which maintains a constant temperature of 120 degrees Celsius.

Once the resin has cured, the samples are polished to a glossy finish on one side using a graded series of increasingly fine grinding powders. Carborundum or aluminium oxide (alumina) is normally used for this purpose. A sequence of three different grades is normally used, starting with a coarse grade (usually 220 grade), then a medium (400) and finally a fine (1000) grade. The powders are placed on a glass plate and mixed with water to form a concentrated slurry. The ceramic slices are placed face down in the slurry, and ground by hand using a circular motion and slight downward pressure until the desired very smooth finish is obtained on one side. It is important to thoroughly rinse the samples with clean water before moving on to the next finer grade of grinding powder. If this is neglected, particles of the coarser abrasive contaminate the finer powder and scratch the surface of the thin section which will interfere with the identification of minerals and fabric analyses. A very smooth finish is important to ensure a perfectly flush mating with the glass slide in the next step, and to enhance the surface of the material for transmitted light to pass without significant scattering.

The sections are then rinsed with clear water, allowed to dry, and mounted onto clean glass slides (usually 76 x 26 x 1.0-1.2mm thick, low iron clear glass). The same resin or epoxy as was used for consolidation is normally used for mounting. A small amount is spread onto the prepared polished surface, and the sample is placed face down on the glass slide. There should be no bubbles trapped between the sample and the glass slide, as this will interfere with analysis. If bubbles are present, they can be worked out using gentle downward pressure together with a slight rotating motion of the sample. They are then left on the heating plate to cure again. Curing times of various epoxies tested under varying conditions are discussed in Chapter 2. Generally they are of the order of one half-hour to two hours.

The mounted sections are then ready for grinding and polishing to a thickness of 30 microns (0.03 mm)[1]. A properly equipped thin section laboratory will have a powered grinding machine in which the mounted sections can be ground down to about 100 microns or slightly less. There are thin section preparation machines which complete grinding and polishing, but normally it is completed by hand using the procedure outlined above for the initial polishing. Care must be taken to reduce the thickness at a constant rate over the entire viewing area of the section, otherwise a slight wedge-shaped thin section will result and analysis will be difficult or impossible. As the required 30 micron thickness is approached, the section is frequently examined under a polarising microscope to ensure that it is not exceeded. This is accomplished by observing the interference colours of the quartz inclusions usually present in most archaeological ceramics and clays. These inclusions will change from yellow to light grey or white first-order interference colours (Michel-Levy chart) under cross-polarised light when the correct thickness of 30 microns is achieved. It is important that the thickness be as close as possible to 30 microns; as most of the standard mineral identification and analysis techniques are based upon it. The hand polishing should be conducted such that 30 microns is reached when the finest grade of grinding powder is used; this will provide the optimum surface for a smooth bond with the cover slip and ensure surface irregularities will not interfere with mineral analysis.

Finally, a thin glass cover slip is normally applied over the finished surface to protect it from scratches and provide an enhanced surface for mineral identification purposes. They are usually 22 x 40 or 50mm, and of borosilicate glass. Again, the same epoxy is used and it must be ensured that no bubbles are trapped between the sample surface and the cover slip. An alternative covering sometimes used by geologists is transparent cellulose lacquer. Use of the lacquer is less desirable because it tends to scratch easily, and, if applied via a spray, bubbles from the propellant can become trapped upon hardening (Adams et al. 1984: 98). Following time for the resin to cure, the thin sections are ready for examination.

---

[1] There are some advantages in looking at thinner sections. Sedimentologists sometimes prepare 10 micron polished sections so that medium to fine silt inclusions are not under-represented in modal analyses. However, for the standard identification of mineral inclusions, it is necessary that they be 30 microns.

A petrological or polarising microscope is used for examining and analysing the thin sections. It is similar to a normal transmitted light microscope but has some additional features which permit identification of mineral inclusions in a thin section of rock or ceramic. These additional features include a polariser, analyser, rotating stage, Bertrand lens, auxiliary condenser and slots for accessory plates. Figure 1.1 is a schematic diagram of a laboratory model of polarising microscope showing the sequence of these features. Their respective functions are explained through the following brief explanation of the process of analysing thin sections.

The science of analysing minerals in thin section is a highly specialised field in geology, and full details are not provided here. For an undergraduate level of understanding, excellent coverage is given in Kerr (1977), or Gribble and Hall (1985). In addition to allowing identification of different kinds of mineral inclusions, the microscope permits study of their size distributions, concentration, orientation and shapes. Characterisation of these diagnostic features of ceramic fabrics provides valuable information regarding clay sources and tempering materials. Other technological characteristics such as surface treatments and forming techniques can also be inferred.

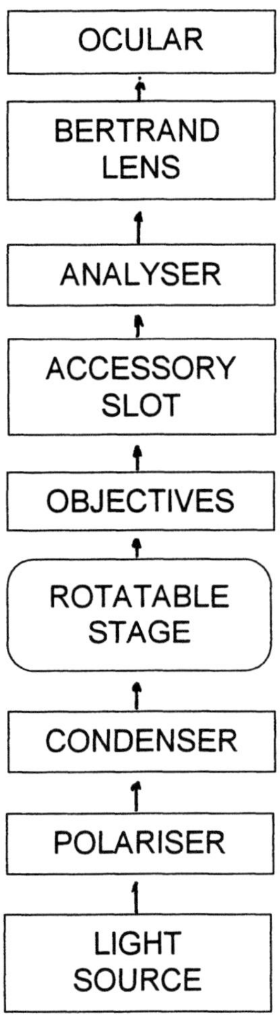

FIGURE 1.1
POLARISING MICROSCOPE SCHEMATIC

In the analysis of minerals in thin section with a polarising microscope, the sequence normally is

1) examination under plane polarised light (PPL);
2) examination under crossed polarisers (XPL); and
3) examination of interference figures and directions in crystals.

Basically, each step allows the researcher to progress a step further in identifying a mineral inclusion. More often than not, the first two steps are the only ones required, especially in situations normally encountered in archaeology. The polariser (see Figure 1.1) produces plane-polarised light which passes through the thin section on the rotatable stage. A rotatable stage is required because in most minerals, rotating the sample under PPL or XPL produces diagnostic colour and pattern changes depending on the angle of the crystal axes with respect to the direction of light polarisation. The polariser and the rotating stage are all that are required for the first step. Under plane-polarised light, relief, colour, cleavage, shape, and texture of mineral inclusions are observed. Crossed polarisers (step 2) are necessary to observe diagnostic interference colours, or to determine extinction angles in minerals such as plagioclase feldspars. Cross-polarisation (or XPL) is accomplished through use of an analyser in addition to the two features described above for plane-polarised observation. As seen in Figure 1.1, the analyser is inserted between the rotating stage and the ocular. It is necessary that either the polariser or the analyser be easily removable and replaceable because during analysis the researcher must constantly alternate between PPL and XPL. The remaining features, Bertrand lens, auxiliary condenser, and accessory plates permit further diagnostic tests. These include examination of interference figures, and determination of fast and slow directions in mineral crystals. These tests are used far less frequently than observation under PPL and XPL, and are usually employed in situations where full identification cannot be made using the first two steps.

Higher technology approaches using chemical analyses of pottery such as Neutron Activation Analysis, Electron Microprobe, Scanning Electron Microscope, and Inductively Coupled Plasma Emission Spectrography are also used for provcnancc studics. These approaches produce lists of elements in the composition and not the minerals as does petrographic analysis. The latter are much more assignable to a source location. The petrological approach offers the possibility to predict the area or region in which a fabric was produced from its geological characteristics. Knowledge of the mineral suite in the ceramic allows some inference to be made of parent rock formations from which they could have been derived. This distinguishes petrographic analyses from chemical analyses, which depend on the analysis of known products from the area of origin (Freestone 1991: 399). Elemental analyses are nevertheless very useful and should be used to complement a petrographic analysis. Neither chemical nor mineralogical characterisation is inherently superior (Neff 1993: 33). Where possible, complementary but independent techniques should be employed based on familiarity and availability of equipment (Kamilli and Steinberg 1985: 314). Elemental data become decisive when there exist considerable differences in clay matrices which cannot be detected by petrographic analyses (Schubert 1986: 177). This was observed in a study by Stoltman et al. (1992) which compared conclusions based on mineralogical examination alone with those based on a chemical analyses alone. They found that the two approaches provided different perspectives on the same data set and were complementary.

A drawback of the elemental analysis methods is that they are expensive and they rely on fixed-base, high technology, specialised equipment not normally accessible to most researchers. Descriptions of the principles, practice and results of instrumental elemental analysis are not included here as the present research is concerned with development and application of petrographic analysis systems. General information on the archaeological applications of these techniques may be found in Rice (1987: 389-404).

1.3 THE NEED FOR FIELD-BASED SYSTEMS

As previously indicated, petrographic analysis is at present almost entirely laboratory-based. The equipment employed in laboratories for both preparation and analysis of ceramic thin sections is by and large much too cumbersome and has not been adapted for field use. The diamond saws and grinders are mostly sturdy floor-standing or bench-mounted models, and the epoxy resins used for consolidation and adhesion require a thermostatically controlled heat source. Polarising microscopes weigh up to 10 kg. There are at least four situations in archaeological research, in particular the technical study of ceramics, which dictate a need for field-based, transportable, systems:

1. <u>Many countries severely restrict or prohibit the export of cultural artefacts</u>. Due to increased instances of illegal sale of antiquities and concerns for the protection of cultural and archaeological heritage, many countries have enacted legislation to prevent cultural artefacts from being exported, even for purposes of scientific research (e.g. Turkey, Egypt, Pakistan). This type of legislation is becoming increasingly common, with the result that archaeological investigation in these countries can become fettered. Research must thus become more locally based, as it is increasingly difficult to apply more advanced and

sophisticated analytical techniques to recovered artefacts. Many of the countries who impose the above export restrictions lack the necessary equipment to conduct the appropriate research in the country of origin; or the foreign-based researcher has no access to it. As a result it has become necessary to extract as much information as possible from archaeological material while in the country of origin. Ceramic petrography as it is presently used allows little opportunity for technical analysis of ceramic artefacts in these situations.

2. On-site ceramic classification schemes will be more accurate if the mineralogical and textural content of the sherds is known. Especially following the opening stages of an excavation, when the classification scheme is being defined, a knowledge of the mineralogical and textural characteristics allows these variables to be used to define fabric groups. Usually, it is necessary to classify fabric type at the initial sorting stage by simple visual examination of a fresh break. This results in very general fabric categories such as 'coarse-grained', 'micaceous', or 'shell-tempered' to name a few seen in the recent literature. These are often not easily replicated by a new or follow-on researcher, and may bear little resemblance to true mineralogical or textural classifications. In many instances the petrological analysis is done months or years after the excavation, necessitating extensive readjustment of a classification scheme after the sorting has been completed once. Johnson (1992) observed this when he petrographically analysed a collection of ceramic material years after the initial fabric classification categories were done in the field. He found a general lack of correlation between the originally defined fabric types and those he later determined via petrographic analysis (*ibid.*: 184).

3. On-site excavation strategies may be optimised if it is known during the course of the excavation that ceramic material can be matched to a particular source location. An archaeological excavation programme is usually an expensive commitment of resources, put into place by difficult-to-obtain funding. It is critical that these resources be efficiently employed to extract the most useful data in the shortest time. The field director must have at his or her disposal ongoing data which can enable on-site decisions to be made, primarily the location and extent of excavation units. Scientific field analysis systems such as the field petrology kit can provide this. For example, if it can be shown at the early stages of a programme that some excavated sherds may be of non-local origin, excavation strategies can be adjusted to emphasise specific areas of the site. Resources could be

directed or concentrated on the areas which yielded the non-local material. Alternatively, if sherds stylistically similar to those from a different known site are encountered, on-site petrographic analysis could be used to confirm whether they are actually of non-local origin. This could prevent a researcher from erroneously assuming they are of non-local origin and guiding excavation strategies in the wrong directions. In this way, resources can be applied in the areas of maximum interest during the course of the excavation instead of the next season, after the laboratory analysis as is often the case at present. It would in many cases result in much more efficient use of funds and minimise instances of misguided excavation strategies.

4. Sampling from a ceramic sherd assemblage may be optimised. In situations where it is necessary to minimise the transport of artefacts, on-site petrographic analysis allows selection of the optimum number of sherds. Transportation of large quantities of ceramic sherds from a remote location can be very expensive and is often inconvenient. Sometimes the excavation or survey is on a one-time basis. On-site petrographic analysis in these situations is an important selection technique. It ensures that only those sherds with particular defined characteristics are identified for transport. Alternatively, only thin sections may need to be transported.

The lack of field portable systems to identify and record mineral inclusions and textures in ceramics has undoubtedly resulted in many instances in the loss of valuable diagnostic information and misguided excavation strategies. Cumbersome and expensive existing laboratory equipment can of course be transported to the field, and has been in many cases. However, transport difficulties and the necessity of having delicate and expensive apparatus such as professional microscopes away from laboratories for months at a time usually makes this option unattractive.

1.4 REQUIREMENTS FOR A FIELD-BASED SYSTEM AND PREVIOUS RESEARCH

Specifications for a complete field kit were drawn up by first defining the considerations necessary. These broadly consisted of :

1. Size and weight: This primary requirement is relatively straightforward and requires little elaboration. The system needed to be very easily portable, especially by commercial

passenger air transportation. It was also preferred that all of the necessary equipment and consumables to prepare and analyse about one hundred thin sections could be carried easily in one case.

2. Ease of use: This was an important requirement also. Because one of the prime objectives of the exercise was to keep the system inexpensive and readily usable, the operator should not be required to possess rare or expensive expertise. Any researcher who has prepared and analysed thin sections in a fixed-base laboratory should be able to do the same in a field situation with the kit.

3. Commercially available components: If the system was to be readily available to most projects and researchers and not fall into disuse, it would not be wise to include components which are available only through specialised suppliers or which have to be manufactured by specialised workshops. It should be able to be assembled using the usual scientific supply houses or locally with a minimum of post-purchase modifications.

With respect to equipment considerations, the process of petrographic thin sectioning can be broken down into four main steps:

- consolidating;
- cutting and grinding;
- polishing; and
- analysis.

Equipment and consumables associated with each of these steps was examined with a view to streamlining, miniaturising, and minimising to achieve the three initial objectives.

Conducted in a field environment, cutting and grinding procedures generally impose the largest weight considerations. Most commercially available powered saws and grinders for ceramics or rocks are designed and developed for use in a laboratory, either fixed to a bench or free floor-standing. In addition, they are usually heavy. This is for two reasons. Firstly, especially with respect to powered grinders, a reasonably powerful electric motor (of at least one-quarter horsepower) is needed in order to remove material efficiently without slowing. Also, grinding wheels are usually of heavy material due to the nature of the abrasives and the requirement for high momentum in the wheel. Secondly, the heavier the weight of the device, the higher its stability. This, especially with respect to saws, allows controlled and precise cuts. Reducing the weight of laboratory saws and grinders has therefore not been a primary concern in their design and development.

It was necessary, as outlined above, to utilise only components and consumables which are generally available through commercial outlets. This eliminated design and manufacture of an entirely new saw and grinder. The major problem rested with the grinder. Grinders are by far the heaviest component in the sequence; yet are important labour- and time-saving devices in laboratory thin section preparation. Because efficient lightweight grinders are not available, a saw with which very thin ceramic slices could be made was necessary. If very thin cross-sections of consolidated material can be made, grinding will be kept to a minimum. When a minimum of grinding is necessary, it can be done by hand using coarse grinding powders or abrasive sheets (see following paragraphs). Therefore, importantly, if an accurate small lightweight saw can be used, the requirement for a grinder can virtually be eliminated. It was realised, however, that use of such a saw would negate the advantages discussed above which are gained by using a heavier model.

Ceramic sherds can also be cut using a hand saw. There are numerous small hand saws which use tile cutting blades which would be efficient in cutting ceramic material. It is, however, laborious and time-consuming; and difficult to obtain accurate, thin, parallel-sided slices from sherds. The result is that extensive grinding becomes necessary, losing the advantage. For preparation of small numbers of thin sections where no powered saw is available, hand cutting may be useful. In addition, it can be used where only polished sections are being prepared for examination under a reflected light microscope.

A requirement for a saw powered by a rechargeable battery was considered. There are instances where a cordless rechargeable saw would be advantageous, especially in areas far away from mains electrical supplies. Although these situations are not particularly rare, electrical mains would still be required; for battery recharging. A prototype of a rechargeable portable saw, designed specifically for thin sectioning in the field, was constructed and tested by Hunt and Griffiths (1989). It was adapted using a Makita Cordless Trimmer and field tested preparing thin sections of andesite at a remote location in the Andes Mountains. It was found that an average of three sections could be cut and ground efficiently before a fresh battery pack was required. Since the researchers were cutting stone, clearly more could be prepared cutting softer materials such as archaeological ceramics. However, in order to prepare 50 to 100 sections, either several fully charged battery packs would be needed, or one would have to return to a mains location to recharge batteries. Provided a large number of sections were not required at a remote site, a rechargeable saw would clearly be most useful, but necessarily heavier than a mains-powered model (due to the battery pack weight). Both mains and rechargeable models were evaluated in the laboratory, while only the mains model was field tested for the present research.

With respect to abrasives, some consideration was given to employing the silicon carbide abrasives fixed to a paper backing, such as the Buehler-Met Metallographic Grinding Paper. It was thought this might offer a more convenient alternative to the loose powders. The coarsest grade, P-120, would in fact be useful for initial coarse grinding to save using excess 220 grade alumina powder. In addition, use of the Metallographic Grinding Paper would probably consume less water which would be an advantage where water is at a premium. Use of the finer grades of papers, however, is not recommended due to possible microscopic damage to the specimen. A powder used in a slurry allows the abrasive to function in a rolling manner and results in less damage to the specimen, especially when working with brittle materials such as archaeological ceramics (Bousfield 1992: 35-7). While the fixed abrasives could be used for initial, coarse grinding, fine alumina powders in a slurry on a glass plate are therefore recommended for final polishing. At least one current researcher has used the silicon carbide abrasive papers for preparation of ceramic samples in the field. Prior (pers. comm.) recently used them for coarse grinding in preparation of polished surfaces for examination via reflected light microscope, and found them generally acceptable.

Overall, there is little scope for reduction in weight or bulk, or improved portability in the area of grinding and polishing powders. The alumina powders are relatively dense. Some weight savings could be achieved using the grinding papers, but there appear to be few other alternatives. A small weight saving can be achieved by using circular glass plates instead of square ones. During hand grinding and polishing, most of the motion is circular, and the corners of the plates are not normally utilised.

Consolidants in regular use for archaeological ceramic petrology require a controlled heat source to use. The two most commonly used, Petropoxy 154 and Canada Balsam, both require a thermostatically controlled thermoplate to maintain a constant temperature of 120 degrees Celsius. The thermoplate uses a heavy iron or steel surface to provide an even heat source, resulting in weights up to 15 kilograms. Local improvisations could be used, but it would be difficult to maintain the constant temperature. To keep weight at a minimum, it was necessary to either locate a very light thermoplate, or test the suitability of using cold curing resins for consolidation and mounting. To eliminate an additional piece of equipment, the latter option was preferred. Criteria for acceptability of cold curing resins were set: These were that a cold curing resin acceptable for field applications should:

- cure in <24 hours at normal room temperatures (approx. 20 deg C);
- function without ancillary equipment e.g. heat source;
- have low viscosity at normal room temperatures;
- have an Index of Refraction very close to 1.54;
- be efficient for both consolidation and adhesion;
- have low shrinkage;
- mix by volume with reasonable latitude;
- be readily available commercially; and
- be safe for commercial air transportation.

The first criterion was considered a reasonable figure for efficient production of thin sections under field conditions, based upon laboratory experience. A researcher may have to make 30 to 100 or more sections for analysis. Sections are normally, for the sake of efficiency, prepared via an 'assembly line' procedure; that is, a batch of 15 to 20 is first consolidated and allowed to cure. When these have cured, they are then polished for mounting. Another batch of epoxy is mixed with hardener, which is used to mount these and also consolidate a new batch of 15 or 20, and so on. Thus, only the required amount of epoxy and hardener is used, which minimises wastage as most epoxies have limited pot lives once mixed. The researcher is thus carrying out the next phase of the preparation while the epoxy is curing in the last completed one. A maximum cure time of 24 hours would provide adequate time for the researcher to complete one phase of the preparation process without being unduly delayed waiting for the resin to cure.

The requirement for the epoxy to function without ancillary equipment needs little elaboration. It would eliminate the necessity of transporting a thermoplate or other item of equipment thus contributing to significant weight and bulk reduction. Low viscosity is another particularly important one; without a low viscosity the epoxy will not penetrate the ceramic microstructure adequately to consolidate it prior to cutting and polishing. This will result in lack of cohesion in the sections, especially in the case of low-fired and friable ceramic materials, and it will crumble or lose inclusions during grinding. A reasonably fast and thorough penetration is an important factor if fast curing times are desired. A slower, more viscous penetrant may harden before it has sufficiently penetrated the microstructure. The viscosity should be low at ambient temperatures or with simple non-controlled heating because there would not be the benefit of a controlled heat source as in the laboratory. An alternative often used in the laboratory is vacuum impregnation (Jim 1985: 598). This option was rejected for field use because it would have introduced an additional piece of equipment into the system.

An Index of Refraction very near 1.54 permits comparison of thin sections prepared using the field epoxy with those prepared using the established laboratory epoxies noted earlier. Using an epoxy with a radically different Index of Refraction could result in misinterpretation of a mineral inclusion in the hands of a researcher who was unaware that the traditional epoxies had not been used, or was accustomed to analysing thin sections consolidated with the traditional epoxies whose Indices of Refraction are 1.54. However, if the Index of Refraction is not 1.54 but its value is nevertheless known, analysis will be relatively unaffected.

The use of one epoxy for both consolidation and adhesion simply minimises the requirement to transport two different chemical substances. It also eliminates potential problems which might arise from the presence of two different Indices of Refraction in the matrix. The requirement that the epoxies be readily commercially available should ensure that the kit remains accessible and viable, that specialised or experimental sources or products would not have to be relied upon. Finally, the epoxy selected must be of sufficiently low hazard that it is air transportable. Many hazardous chemicals are not permitted to be carried in hand or checked baggage on commercial airline flights.

Hunt and Griffiths (1989) used an ultra violet curing adhesive for mounting sections of rock. The method has been developed for use by petrologists for thin section preparation in the laboratory (Yanguas and Paxton 1986: 539). Dymax Light-Weld 304, a resin which cures under ultra violet light, was found to be entirely satisfactory for mounting stone thin sections in the field. It was also considered in the present research for field preparation of archaeological ceramic thin sections. The main drawback in connection with ceramics is in effecting a cure for absorbed epoxy. The ultra violet only cures the epoxy at surface level and does not penetrate into the ceramic where the consolidating epoxy also needs curing. The method would therefore probably be entirely effective for mounting specimens that do not require consolidation. However, in another field application, Johnson (1992: 185) found that Dymax was not effective when used with ceramics for consolidation nor adhesion, under the hot and humid conditions encountered in lowland Thailand in April-May. It is perhaps sensitive to environmental conditions. It was therefore not included in the present testing for field suitability; primarily because it was desired to use a substance that did not require ancillary equipment and because of its lack of effectiveness in curing absorbed epoxy.

The final step to be considered in the process is analysing the prepared thin sections. In many cases, this can be left for completion in the home laboratory, especially where the host country permits export of the prepared sections as scientific samples. This is not a recommended procedure because the correct thickness of a thin section cannot be determined without examination of the quartz inclusions with a polarising microscope. It is also desirable to conduct at least a preliminary on-site observation of the mineral inclusions present to assist in fabric categorisation for sorting purposes. More importantly, it is one of the objectives of the present research to develop a field analytical capability.

Examination of petrological thin sections is always done using a petrographic microscope. As previously indicated, commercially available models are large, heavy, and designed for fixed-base laboratory use. It was therefore necessary to locate a portable one or select and modify an alternative (non-petrographic) transmitted light microscope. Design and manufacture of an entirely new model was not considered, for cost reasons, and because of the need to confine the field kit to commercially available components. A list of criteria was first set up. The main considerations were that it meet the following:

•portability - the microscope must be small enough and light enough to be easily hand carried;

•analytical capability - the microscope must be of adequate quality to perform all of the basic analyses necessary to make decisions in the field; and

•cost and availability - the microscope must be relatively inexpensive and readily available commercially, at least with only minor modifications.

With respect to analytical capability, the minimum requirements were that the microscope be able to be used to perform mineral identifications using both plane-polarised light (PPL) and crossed polarisers (XPL), including measurement of extinction angles in mineral crystals. In addition, it must be capable of measuring size distributions and concentrations of inclusions for modal analysis of fabric inclusions. In order to have the capability to carry out the basic mineral identifications, at least three of the features unique to polarising microscopes were needed. They are a polariser, an analyser, and a rotating stage with 360 degree graduations. It was felt that requiring the capability of studying interference figures and directions in crystals would place too stringent a demand on the portable microscope specifications. Particularly troublesome mineral identifications could be done at the home laboratory. The vast majority of the minerals encountered in archaeological fabrics can be identified using the three basic features. In addition, to measure inclusion sizes, a calibrated eyepiece graticule or scale was needed. To determine concentration of inclusions, an eyepiece grid or stage that moves longitudinally and laterally in calibrated increments was required. This would permit both area and point counting to be performed.

Finally, the field model should be relatively inexpensive compared with laboratory models, and be readily commercially available for use without expensive or extensive modifications. If it was too expensive or necessary to manufacture one from obscurely available components, the microscope would not come into general and regular usage.

Overall, the components should be conveniently containable in a portable case. The case should be rugged enough to withstand field transport and use, compact enough to be carried as hand luggage on commercial aircraft, yet large enough to include all hardware and consumables necessary for preparation and analysis of at least 100 thin sections.

CHAPTER TWO

DEVELOPMENT OF THE FIELD PORTABLE SYSTEM

Once the specifications had been set for the various components as described in the previous chapter, development and testing was begun. The initial step was to locate components approaching the criteria specified, and test them under laboratory conditions to determine any changes or modifications which might be needed to meet the specification for field use. Following this, the kit was ready for testing in a field situation. For discussion purposes, development and testing is broken down into sections corresponding to those used in Chapter One.

2.1 THE SAW

Beginning with the saw, listings of commercial stockists in the UK. were reviewed for suitable lightweight portable models, with the specified criteria in mind. These included Logitech, Evans Lapidary, Buehler, and Hillquist. Numerous efficient diamond lapidary saws suitable for the laboratory were found, but most are much too heavy for field use. Only one suitable lightweight model was found. It is the Evans 4-inch Combi-Saw, manufactured and distributed by Evans Lapidary Machines of Dartford, Kent, UK (addresses of all suppliers used are presented at the end of the section). It is shown in Figure 2.1. Powered by a 240 volt AC motor whose speed can be varied from 0 to 7000 RPM, it weighs about 2 kg complete. The unit accepts various 100mm (4 inch) diameter circular blades, from diamond wafering blades to abrasion wheels. In addition to its light weight, it is relatively compact, measuring 18 x 18 x 9 cm high. The body is manufactured of stainless steel, and a distinct advantage with respect to field use is that it is a very simple and basic design. It has few electronic components or controls, which minimise the potential of breaking down in the field. Its drawbacks are mainly that it lacks an arbour or holding device with which to immobilise the sample during cutting to facilitate maintaining a cut parallel to the plane of the saw blade; and it is underpowered. The low power of the motor is an advantage in keeping overall weight to a minimum but is a disadvantage because it cannot adequately drive a grinding wheel and results in very slow cutting. This problem is somewhat overcome by the high speed of the saw, which provides a light blade with improved momentum. Also, the lack of a material holding attachment causes difficulties in maintaining cuts parallel to the saw blade plane. If the cut is not kept parallel to the saw blade, the blade can bind in the material, with the resultant risk of damaging the saw motor or an expensive diamond cutting blade. It would be possible to design and adapt a holding device to the saw; but this would entail a possibly expensive modification. These drawbacks however did not prove to be insurmountable, and could be overcome using a technique which an operator could easily learn with a small amount of practice.

The Evans saw included a grinding attachment in the standard purchase price. It consisted of a circular attachment similar to the saw blade but with a grinding face on one side. The wheel is faced with an abrasive material which appears not to be designed for grinding any hard materials such as ceramics. Grinding ceramic material was attempted, using a locally-manufactured aluminium platform upon which to rest the material. The material was held squarely against the abrasive wheel using double-sided tape to secure it to a holding block. Two problems arose with this arrangement. Firstly, the saw motor was not powerful enough to efficiently grind off ceramic material; application of even slight pressure slowed the motor and placed it under considerable load. Secondly, the abrasive on the wheel material quickly became clogged with ground ceramics, to the point that it ceased to remove further material until it was cleaned. Moreover, the wheel was difficult to clean. Some thought was then given to utilising a wheel of different material, but the lack of power of the motor would likely have also rendered it ineffective. Removing a given amount of material would in general require a given amount of power. Because of the low power of the motor, other ways of reducing section thicknesses had to be explored.

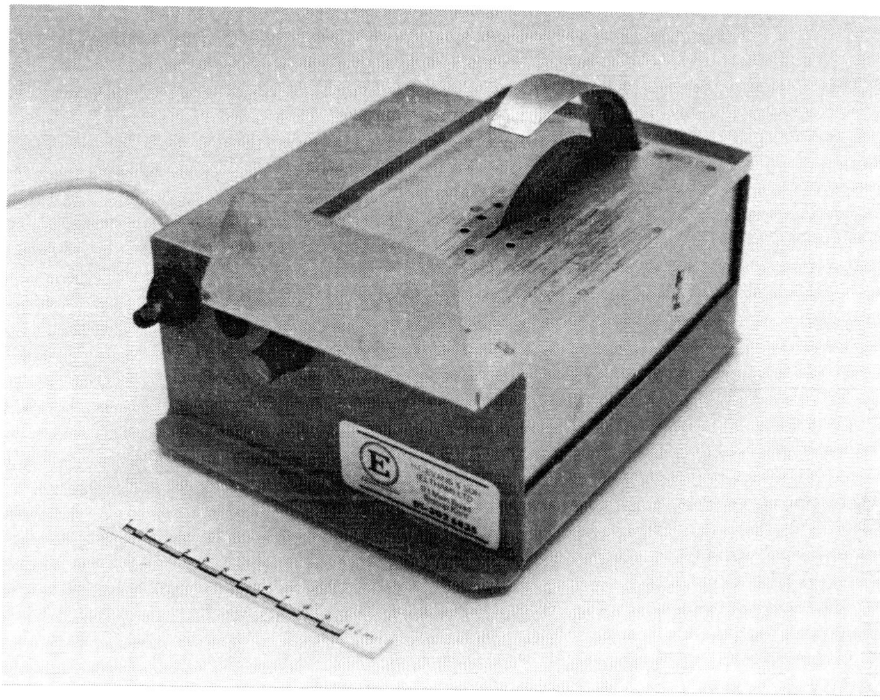

FIGURE 2.1
EVANS 4-INCH COMBI-SAW
Scale is 10 cm.

The Evans saw was evaluated with five different low-concentration diamond blades to select at least one suitable for field use. Main acceptability criteria were a reasonable cutting speed, robust enough to withstand minor misalignments during cutting without damage, ability to function using only water as a lubricant, and a reasonable cost. The blades tested are summarised in Table 2-I. They were selected after consulting manufacturer's recommendations for the field application using the Evans portable saw; with the exception of the Extec XL12235 diamond blade. It is a very high quality and expensive blade designed for low-speed applications. As one was available on loan for testing, it was included in order to provide a wider basis for comparison. All five of the test blades are commercially available in the UK.

TABLE 2-I

DIAMOND SAW BLADES EVALUATED

Blade	**Thickness**	**Diamond concentration**	**Manufacturer**	**Price in UK incl. taxes**
AGST	0.5 mm	100 (low)	Agate and General Stonecutters	£96.00
Extec XL12235	0.3 mm	low	Excel Tehnologies Enfield, CT	£140.00
Meteor	0.3 mm	low	Star Diamond Industries, Carson, CA	£14.00
Famad	0.3 mm	low	Star Diamond Industries, Carson, CA	£18.00
Evans 220 Grit	0.3 mm	low	Evans Lapidary Machines, Kent, UK. (Distributor)	£12.00

Each blade was mounted in the Evans Combi-Saw, and used to cut archaeological ceramic samples. Two medium hard grades of ceramic fabric were used, one coarse and one fine. The sherds were hand-held and fed through the blade at a rate so as not to cause the speed to reduce excessively; i.e. a rate which could be maintained without overloading the saw motor. As well as determining the ease of making a simple straight cut, the ability to cut a slice 1.5 to 2.0 mm thick while maintaining a relatively smooth cut was evaluated. Water was used as a lubricant and coolant as recommended by each of the blade manufacturers. Testing was purely qualitative; and comments were recorded regarding acceptability of each. A summary of the test evaluations is presented in Table 2-II.

TABLE 2-II

SUMMARY OF BLADE EVALUATIONS

AGST	Very slow cutting speed due to blade thickness. Poorly balanced. Moderately expensive. Not recommended for use with the Evans saw for field applications.[1]
Extec XL12235	Fastest cutting speed of all blades tested. Most easily damaged in use, as not designed for high speed applications. Expensive. Not recommended for use with Evans saw for field applications.
Meteor	Moderately fast cutting speed. Rugged; withstands some flexing and binding. Inexpensive. Good value and recommended for field use with the Evans Combi-Saw.
Famad	Moderately fast cutting speed but slower than the Meteor or the Extec. Rugged and inexpensive. Best value of the five tested and recommended for field use with the Evans Combi-Saw.
Evans 220 Grit	Moderately fast cutting speed but slower than the Meteor or the Extec. Rugged and inexpensive. Reasonable value and recommended for field use with the Evans Combi-Saw.

[1] Test blade may not have been representative of production models. See text.

As seen in the summary, the three lower-priced blades, Meteor, Famad and Evans 220 Grit, proved to be the most suitable for field use in the Evans Combi-Saw for the ceramic fabrics tested. Although not the most efficient in terms of cutting speed, all three combine low cost, robustness and durability, and adequate cutting speeds without the need for special maintenance or storage considerations. The three are similar in construction and diamond concentration. The fastest cutting blade was the Extec 12355, which was also the most expensive. As previously mentioned, this blade was designed for low-speed applications, as was indicated by the manufacturer, but still proved to be very efficient at higher rim speeds. Extreme care was required, however, to maintain a parallel cut because the blade can be easily damaged if the sample is slightly turned or twisted during cutting. This problem could be largely overcome with the use of a vice or holding device built into the saw, which the blade is probably designed for. As it is a very expensive blade, it would not be practical to purchase two or three for use as spares in the field. Moreover, it requires regular cleaning and careful storage.

The least effective of the five blades tested was the AGST. Its main drawback appeared to be its thickness, 0.5 mm, which slowed the cutting considerably. Furthermore, the test blade was not well-balanced and the resulting vibrations exacerbated the slow cutting. This blade was also relatively very expensive. However, following testing, the author was informed that the blade tested was an older design and that the present design has largely overcome the problems encountered in testing. Unfortunately, an example of the newer blade was not available for the period of testing.

The prototype rechargeable saw developed and used by Hunt and Griffiths (1989) for use with stone cutting in the field, earlier described, was also briefly evaluated in the laboratory for cutting archaeological ceramics. It is not the intent of the present research to re-test the efficiency of that system, but to determine whether or not it would be suitable for use in the field for ceramics in particular. The blades evaluated above with the Evans Combi-Saw were too large to be accommodated by the arbour designed by Hunt and Griffiths, so the existing blade (a Makita Diamond Saw Blade 724950-8A, 8.65cm in diameter) was used. Again, one coarse and one fine sample of archaeological ceramic fabric was used. The more powerful motor and the gearing which reduces the 8,000 RPM motor speed resulted in a higher torque and much faster cutting speed than had been obtained with the Evans Combi-Saw utilising any of the blades tested. Based on this brief evaluation, it appears that it would be fully acceptable in the field for cutting ceramics. In a field situation without access to electrical mains, this rechargeable model would be very useful provided one had the facilities to transport a sufficient number of freshly charged batteries. Or, alternatively, provision could be made for recharging from an automobile battery.

2.2 CONSOLIDATION AND ADHESION PRODUCTS

Following establishment of the specifications set out in Chapter One, commercial scientific supply houses' product lists were reviewed for available resins which met the basic criteria specified above. Testing was entirely qualitative and restricted to practical considerations.

There are three types of cold curing resins generally in use: epoxy, polyester, and acrylic (Bousfield 1992: 23). Each has its advantages and disadvantages. Epoxy has the lowest viscosity of the three, is the hardest, and has the lowest shrinkage. For this reason, the epoxy type appeared to be the best of the three upon which to concentrate testing. These three advantages are important: a low viscosity assures efficient absorption into the ceramic microstructure for consolidation, hard curing assures a solid end product for grinding and lapping to a polished surface, and a low shrinkage minimises sample

TABLE 2-III

SUMMARY OF COLD CURING RESINS TESTED FOR FIELD SUITABILITY

See text for further details and specifications

Resin	**Type**	**Mixing Ratio**[1]	**Temp. deg.C.**	**Cure Time**	**Consol. Efficiency**	**Comments**
Araldite MY-753[2]	epoxy	10:1	16-18	4 days+	good	too slow to cure
		5:1[3]	16-18	2 days+	good	too slow to cure
		5:1[3]	50	24 hrs+	good	acceptable
Demotec	acrylic	1:2[3]	16-18	20 min	poor	poor consolidation
Epo-Kwick	epoxy	5:1[3]	16-18	4 hrs +	poor	poor consolidation
Epo-Tek 301	epoxy	4:1[3]	16-18	24 hrs+	excellent	good curing and consolidation
		4:1[3]	60	1.5 hr	excellent	excellent curing and consolidation
		2.5:1	16-18	24 hrs+	excellent	good curing and consolidation
		2.5:1	60	1.0 hr	excellent	excellent curing and consolidation
Scandiplast	polyester	2.5:1[3]	16-18	60 min	poor	poor consolidation
Scandiplex	epoxy	2.5:1[3]	16-18	1-7 hr	poor	poor consolidation
		2.5:1[3]	80	1 hr	poor	poor consolidation
Scandiquick	acrylic	1:2[3]	16-18	20 min	poor	poor consolidation

[1] parts resin to hardener by volume

[2] with hardener HY 956

[3] manufacturer's recommended mixing ratios

disturbance due to movements within the impregnant (Jim 1985: 599). With respect to the latter characteristic, caution had to be exercised with quick-setting resins such as the acrylics, because contraction can also increase with reduced setting times (Bousfield 1992: 17).

Table 2-III lists seven cold curing resins selected for laboratory testing. These are the resins whose written specifications generally met those set out in Chapter One and the foregoing before being subjected to testing. The list is not considered exhaustive; it is possible that there exist others which were not revealed in the search. However, the search did consist of consulting laboratories and technicians involved in making thin sections, and the major manufacturers and distributors of mounting systems and supplies. Also, the search was limited to UK. availability; further sources may be available in other countries.

For each of the resins shown, the first test was to determine each of their absorption or consolidating efficiencies when applied to archaeological ceramics. No published information on these characteristics was available. Each resin was mixed according to the manufacturer's specifications; with the exception that all measuring and mixing ratios were done by volume and not by weight as was recommended by some of the manufacturers. The reason for this exception was to minimise the amount of equipment required, and that one would not want to transport a delicate weighscale to an archaeological site. Each was thoroughly mixed in a 10ml disposable plastic cup for 2-3 minutes to ensure effective combination of the resin and its hardener. Two thin (2-3 mm in thickness) cuttings of ceramics, one coarse and one very fine grained fabric, were prepared on a diamond saw. These were either immersed in the mixed resin, or applied to the surface of the cuttings in the manner normally used for consolidation of ceramics in preparation for thin sectioning; or both. Each was then allowed to cure at 'room' temperature (16 to 18 degrees C in the Institute of Archaeology thin sectioning laboratory) until completely hard. Where curing times at laboratory temperatures were in excess of the desired maximum of 24 hours, the experiment on that resin at that temperature was discontinued. Mixing ratios were then altered or higher temperatures were applied and the process repeated. As previously discussed, it was desired that auxiliary heating should not be required for effective curing;

in order to minimise the amount of field equipment needed. Nevertheless, in some cases testing was conducted at higher temperatures, because often in the field it is possible to increase temperatures slightly using simple room heaters or placing samples in the sun. Following curing, samples were cut through in cross section with a diamond saw and examined under a reflected light microscope at 40X magnification to determine the depth of penetration of the epoxy into the ceramic material. If the depth of penetration was less than about one millimetre, the resin was rejected as unsuitable for use in consolidating archaeological ceramics, and further testing was not conducted. At depths less than this, difficulties would be likely to be encountered in obtaining a slice that would hold together during cutting. Qualitative aspects of the testing included recording comments with respect to ease of mixing, whether special conditions e.g. ventilation would be required, etc.

Results of the testing are summarised in Table 2-III. Of the seven tested, only two proved to be acceptable for field use: Araldite MY 753 used with Hardener HY 956, and Epo-Tek 301. The remainder failed to penetrate far enough into the ceramic material to effectively consolidate it and were therefore rejected. In most cases, penetration of the resin into each of the ceramic sections was immeasurably small, and clearly ineffective for consolidation. The majority of the cold curing resins on the present market are intended for embedding metallurgical samples for polishing; applications which do not require the lower viscosities necessary for consolidating porous materials such as archaeological ceramics. This is especially true of Scandiplast, Scandiquick, and Scandiplex.

With respect to curing times at 'room' temperatures (16-18 degrees C in the laboratory), in most cases it was observed that curing occurred generally in accordance with each resin's specification when in its mixing vessel, but not when spread thinly on a ceramic sample or glass slide. When spread thinly, the resins remained sticky to the touch long after the specified curing period had passed. In the polymerisation process, there is heat generated which produces the solid plastic. In some of the fast-setting resins this can be as high as 100 degrees C (Bousfield 1992: 21). For this release of thermal energy to be effective in raising the temperature and so accelerating the polymerisation reactions, it is necessary that the resin mixture be placed in a mixing cup or vessel which allows a critical mass of mixed resin to be concentrated in order to contain the heat and allow it to build up. When mixed resins were spread onto a ceramic sample, therefore, this rise in temperature could not be attained; this may have contributed to the lack of effective curing in these instances. In addition, the low temperature of the laboratory may have been a factor.

In later applications using Epo-Tek 301 in the hotter seasons of Pakistan and Turkey (see field testing sections following), resins spread thinly on ceramic cuttings cured readily at ambient temperatures in much shorter time periods than those attained at the laboratory room temperatures. Also, Johnson (1992: 185) found that Scandiplast cured satisfactorily in conditions encountered in lowland Thailand in April-May, although it was ineffective as a consolidating agent. Of the two acceptable epoxies, Araldite and Epo-Tek 301, the latter proved to be the most effective. The Epo-Tek 301 had a lower viscosity at room temperatures, while the Araldite was most effective for consolidation when heated to about 40 to 50 degrees C. However, both possessed acceptable absorption characteristics in ceramics sections, and, assuming a simple local heat source was available, the Araldite would probably be equally as effective in the field. Araldite is hygroscopic, however (this was not evaluated), and therefore should be further evaluated before attempting to use it in the field, particularly under humid conditions. Because of this, and because Epo-Tek 301 had superior consolidation characteristics at ambient temperatures, it is the recommended resin out of the seven tested.

Epo-Tek 301 was found to cure faster when mixed in the ratio of 2.5:1 rather than the manufacturer's recommended ratio of 4.0:1. At 60° C, curing times improved from 1.5 to 1.0 hours with the change in mixing ratio. The effect of the changed ratio on the Refractive Index of the resulting polymer is not known. Refractive Indices are known to change with many factors, e.g. ageing (Roff and Scott 1971, cited in Horie 1987: 27).

After Araldite MY 753 and Epo-Tek 301 were determined to be most suitable for field use with respect to consolidation characteristics, they were further evaluated preparing and analysing thin sections in the normal manner. The remainder were deemed unsuitable in these aspects, and eliminated from further evaluation. Archaeological ceramic samples were consolidated using each of the two acceptable resins and polished on glass plates using the standard three grades of alumina polishing powders. It was found that samples consolidated using those resins were equally as workable as those prepared using the traditional resins, Petropoxy and Canada Balsam. Required levels of polishing were accomplished with relative ease. The prepared sections were then mounted on glass slides using the same resin as was used to consolidate them. The thin section preparation process was continued with grinding and polishing the mounted sections, and then mounting the cover slips. It was found that all of these steps could be accomplished with the same ease as with using the traditional resins.

2.3 THE MICROSCOPE

Development in respect of the final step in the thin sectioning process, that of analysis, is now considered. The first two steps enable the sections to be made in the field, which in some cases will suffice. In the majority of cases, however, it is desirable that analysis take place both in the field and in the home laboratory, as was the case in the present research.

As indicated earlier, inclusion of an analytical capability for the field demands a field-portable polarising microscope. Field-portable polarising microscopes are not readily found on the commercial market. A search was conducted which revealed there are none which meet the specifications, at least in marketed form. The search consisted of a review of scientific supply houses and optical shops, and contact with other universities involved in ceramic and geological petrography. One portable model was used by the Open University several years ago. It was based on the McArthur design (see description following) and was intended for use by students at home to gain an appreciation of petrography. Apparently it was supplied as part of the study package and used with thin sections that were also provided. In order to keep costs very low, because one had to be supplied to each student, the optics of the microscope were below professional standards. As a tool to introduce home students to the techniques of petrographic analysis it offered an excellent compromise. It was examined and evaluated for the present research and found to be wholly inadequate for detailed professional petrographic analyses.

Only two portable transmitted light microscopes which could be satisfactorily modified to meet the specifications set out earlier were available: the Kirk Technology McArthur

Biological Microscope, and the Swift FM-31 Field Microscope. They are shown in Figures 2.2 and 2.3 respectively. Both models possess excellent optics and are of good quality manufacture. Additionally, both were found to be relatively easily convertible to a polarising capability.

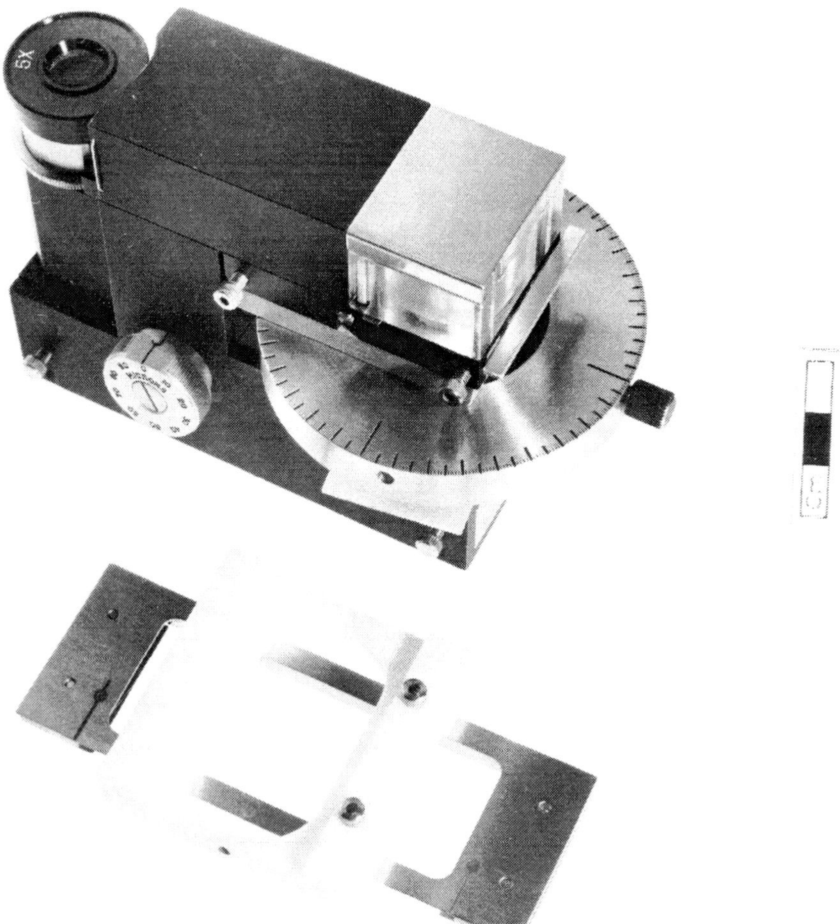

FIGURE 2.2

THE McARTHUR MICROSCOPE

With rotating stage and polariser modifications installed. Gliding stage is in the foreground.

Recalling the discussion from Chapter One, the additional features of a polarising microscope over those of a regular transmitted light microscope include in the main a polariser, an analyser, a rotating stage, a Bertrand lens, an auxiliary condenser, and slots for accessory plates. Modification of a (portable or laboratory) regular transmitted light microscope to include all of these features would have incurred substantial redesign and concomitant development and manufacturing costs. Examination of the structure and configuration of both of the field microscopes showed that installation of certain of the above features would entail significantly greater costs and difficulties than others. Consequently, in view of keeping modifications to a minimum, a compromise was necessary. In analysis of minerals in thin section, the usual sequence is to examine the slide first in plane polarised light (PPL) and in crossed polars (XPL). The majority of identifications required in a field situation can be accomplished by examining the thin section while rotating it on the microscope stage under each of these light conditions. Modifications to a transmitted light microscope necessary to permit this are restricted to installation of a polariser, an analyser, and a rotating stage. The remaining features can be omitted while still retaining the capability to identify a majority of minerals. For particularly troublesome identifications, thin sections can usually be re-examined later in a fully-equipped laboratory without serious loss of diagnostic capacity in the field.

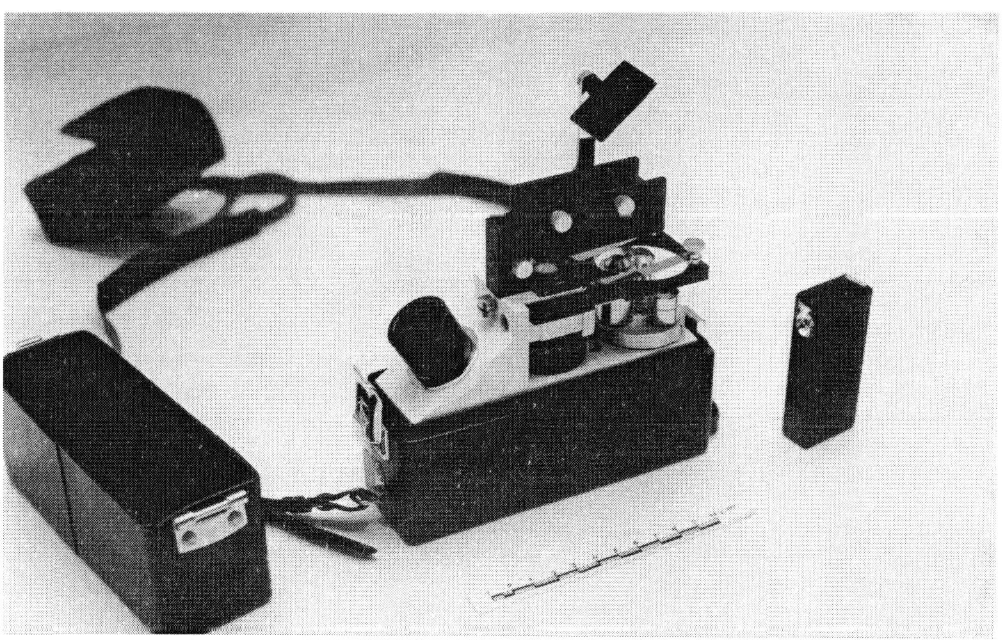

FIGURE 2.3

THE SWIFT FM-31 FIELD MICROSCOPE

Both the McArthur and the Swift microscope were considered for modification. The McArthur is the more expensive, at £492 for the basic model, compared with £322 for the Swift Model FM-31 (1994 prices). The McArthur is the more compact and lighter of the two, and it had more optional accessories available which would permit use of the microscope in other field applications, such as are discussed below. They offer similar magnification ranges of about 40x to 1000x, although the upper end of that range is not normally required for mineral identification. Both offer excellent optics. Design and installation of a rotating stage modification on the Swift would have entailed a more substantial modification, as there is no natural mounting place for it, as is the case with the McArthur. A rotating stage could be installed on the stage mount of the Swift using existing mounting holes, but it would have to be a completely new design to allow for maintaining the slide centrally under the objective as it is rotated under examination. This is due to a major difference in the design of the plate utilised for changing objectives; the McArthur uses a sliding plate (described below), and the Swift uses a rotating barrel similar to the laboratory models. The design used on the McArthur provides an ideal base upon which to mount a rotating stage. A final consideration in the selection of the McArthur over the Swift was the nature of the artificial light unit. The McArthur lighting unit utilises a halogen bulb which provides an intense white light, and the Swift uses a normal incandescent bulb. The use of as white a light as possible provides easier identification of minerals (B. Roberts, pers. comm.). While the Swift could be modified to include a halogen source, it would have entailed yet a further modification. The McArthur microscope shown in Figure 2.2 has the rotating stage modification installed.

The Mc Arthur unit tested was obtained from the manufacturer, K.W.Kirk and Sons Ltd of Milton, Cambridge. Unfortunately, since the date of purchase in November 1993, the firm has gone into receivership. At the time of writing, the winding up of the affairs of the company has not yet been completed. The McArthur microscope's main distributor, Agar Scientific Limited (see Table IV at the end of this section for address) has indicated that the license for manufacture is to be sold as a company asset and that the purchaser of the license and patent will be resuming manufacture in the near future. Agar plans to continue to be the main distributor, so future availability should not become a problem.

The McArthur Biological microscope was originally designed by John McArthur in 1932 for performing medical diagnoses under unfavourable conditions (Watt 1993: 27). It is a transmitted light microscope which achieves its compactness by folding the optical path into a smaller space, as shown schematically in Figure 2.4. The Swift FM-31 uses the same principle. A traditional simple mirror, or an optional integral light source powered by two AA batteries and featuring a halogen bulb, provides illumination. This light travels downward through the adjustable iris and the condenser, and passes through the slide mounted on the stage, then on through one of the objectives, deflected by two mirrors in the optical light tube and finally upwards through the ocular. Three high quality objectives are mounted on a sliding plate under the stage and are available in magnifications of 10x, 40x, and 100x. Coupled with the standard 10x ocular, this provides overall magnifications of 100x, 400x, and 1000x. Optional 5x or easily obtainable other sizes can extend or modify this range. The unit is focused with knurled knobs situated on either side of the housing.

In microscopes, the initial magnification (the magnification provided by the objective) equals the optical tube length divided by the focal length of the objective (Kerr 1977: 28). Because the optical tube length has been shortened to achieve the compactness of the McArthur relative to laboratory models, shorter focal length objectives are employed to maintain comparable magnification levels. This unfortunately results in shorter working distances with the McArthur, but because the sections are thin, this does not present a problem in application to field petrography.

The basic McArthur microscope weighs 630 grams and measures 10 cm x 9 cm x 5 cm. At the time of purchase, optional extras included the light unit described earlier, a gliding stage (as seen in the foreground of Figure 2.2) for easy and accurate movement of slides when the rotating stage is not being used, a camera attachment, a tripod, and a protective carrying case. The gliding stage is a very useful addition for other archaeological applications in the field such as use-wear analysis using acetate peel replicas (Chandler 1994: 45-6) or examination of imbedded botanical samples.

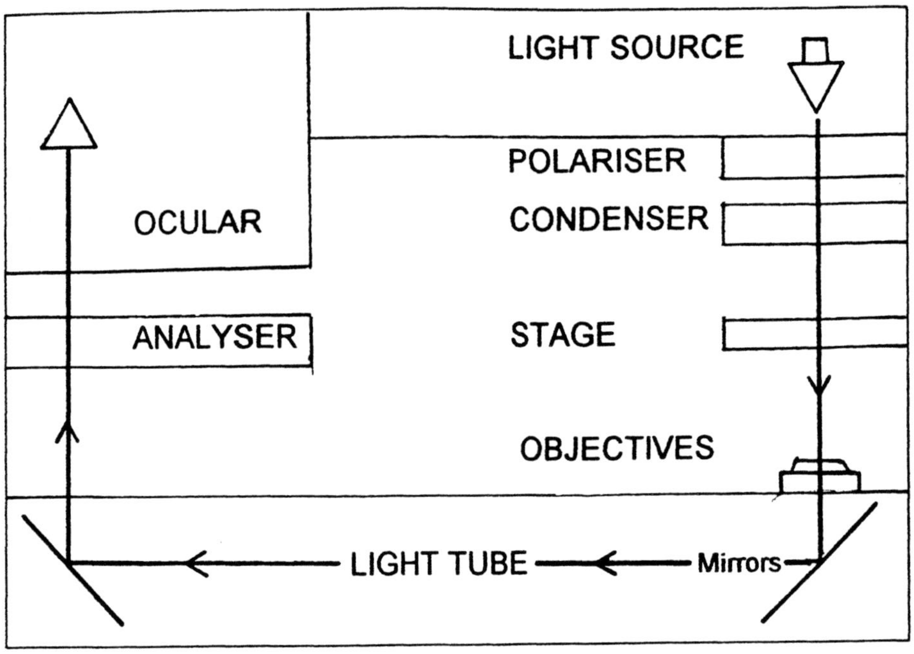

FIGURE 2.4

SCHEMATIC REPRESENTATION OF THE McARTHUR MICROSCOPE

As modified for petrographic field use

The McArthur model was modified to include a polariser, analyser, and rotating stage. Addition of the polariser presented no serious problem; it easily fits into the filter slot on the microscope condenser arm. High transmission Polaroid sheets were cut to fit into the aforementioned slot, ensuring that the axis of polarisation was parallel with the condenser arm. A second piece of high transmission Polaroid sheet was cut to fit inside the ocular housing to serve as the analyser. It was inserted inside at the base of the ocular drawtube below the ocular. This position enabled rotation of the ocular drawtube to a position such that the Polaroid analyser axis was perpendicular to the polarisation of the light from the polariser, to achieve crossed polars. The analyser was held in place with a simple neoprene O-ring whose outside diameter (23 mm) was slightly larger than the inside diameter of the ocular drawtube. This provided the right amount of tension against the inside of the drawtube to hold the analyser in place but still allowed it to be removed or changed if required. It held the analyser in place so that crossed polars could be achieved at an easily identifiable position, i.e. when the flat part of the drawtube was flush with the microscope body. Minor realignments, if required, could be accomplished via rotation of the drawtube as described above. This ensured that unnecessary time was not consumed in searching for the correct position each time the researcher changed from PPL to XPL.

A prototype rotating stage was developed based upon an earlier model designed by Dr. Dafydd Griffiths of the Institute of Archaeology, University College London. That original design was necessarily limited to lower magnifications due to the aforementioned restricted working distances imposed by the microscope design. It was effective in mineral identifications at magnifications using the 10x objective, but when using objectives greater than 10x the thin section interfered with the objective. This restricted identification of mineral inclusions. Moreover, it did not allow accurate area or point counts to be conducted, thus limiting fabric analyses. The new prototype was manufactured in collaboration with the microscope manufacturer Kirk Technology, and is shown in Figure 2.5. It is made from a single piece of aluminium and engraved with 360 degree graduations, in increments of one degree. An index marker attaches to the microscope body to provide a point against which to read stage positions when measuring extinction angles. The stage mounts on the condenser housing. To install it, it is necessary to first remove the lamp housing, and the condenser. The standard stage must also be removed first, by removing the two screws on each side which hold the two sections of the stage in place. This became necessary to eliminate the above mentioned physical interference problems between the slide and the objectives. The rotating stage is then set underneath the condenser arm and held centrally by re-installing the condenser. It is secured in place against the condenser using a set screw with a knurled knob for finger tightening. This arrangement could be improved upon in future production models by installing two set screws and using recessed slotted head screws in place of the single screw with knurled knob. This would alleviate stress concentrated at a single point on the condenser where the screw contacts it, and in addition allow more freedom of rotation of the stage. In the present prototype, the knurled knob protrudes such that it interferes with the microscope housing and prevents full rotation of the stage. A clip on the underside of the stage supports the slide under examination while facilitating manipulation for examination. In the prototype stage, this clip is manufactured of one piece spring steel which holds the slide tightly in place, but scratches it slightly when it is being manipulated on the stage. As the clip holds the slide very tightly, it also makes it more difficult to move around easily while conducting an analysis. Future production models would benefit from using a two-pronged clip manufactured from softer, less binding material; for example, Nylotron.

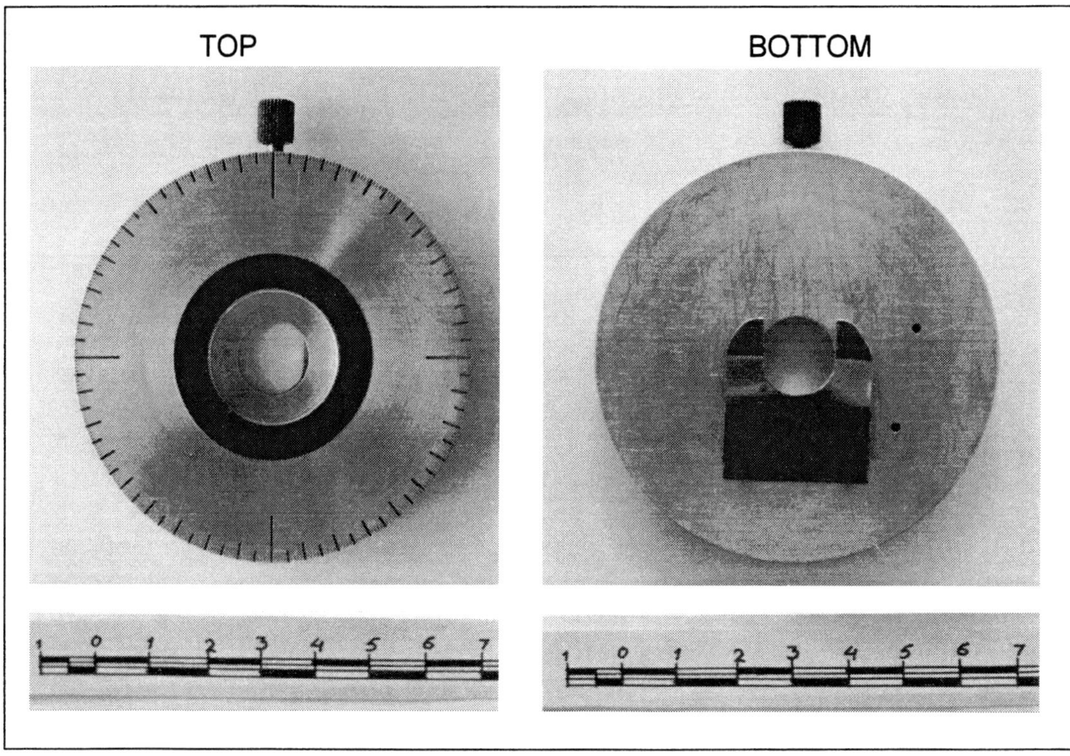

FIGURE 2.5

ROTATING STAGE FOR THE McARTHUR MICROSCOPE

Bottom view shows clip for holding slide. Scales are in cm.

A final modification was the installation of an eyepiece graticule in the ocular. This is necessary for conducting fabric analysis in the field; it enables measuring diameters or lengths of inclusions and determining their density per unit area. The graticule chosen was a 10 x 10 grid[1] which fills the entire viewing area and installed in a 7x focusing ocular, both obtained from Graticules Limited (address is given in Table 2-IV at the end of this section). The focusing feature permits the graticule to be maintained in focus at all times, so that it appears on the same viewing plane as the thin section image. It appears as if laid on the surface of the slide image. The selection of a 7x focusing ocular was only opportunistic; a 5x was not available from Graticules Limited at the time. The graticule was calibrated using a stage micrometer obtained from Agar Scientific, at each magnification. A grid was chosen over a simple crosshair image so that point counting as

[1] Grid square of side length 1.0 millimetre.

well as area counting could be done. Usually, point counting is accomplished through the use of a travelling stage, which moves the slide under observation in pre-determined increments. The inclusion, void, or matrix under the crosshair at each increment is recorded. As a travelling stage would have been quite difficult and expensive to adapt to a small microscope, the alternative is to use an eyepiece grid. With a grid, the nature of each inclusion under each intersection of the grid is recorded. Area counting, to determine densities of each inclusion type, is also easily done using a grid. All particles enclosed within the perimeter of the grid are measured and counted. These methods are outlined in detail in Chapter 5.

To determine acceptability of the modified McArthur microscope, a series of thin sections of ceramics and a stage micrometer were examined, and compared with the same items on a full-size laboratory polarising microscope. The comparison model was an Olympus Model BH-2 laboratory microscope. There was seen to be very little difference in the quality of the respective images. The available camera attachment was also evaluated. Thin sections of archaeological ceramics were photographed and compared with the same images on the laboratory model. Thin sections typically have a dark background against smaller, brighter inclusions. The artificial lighting available on the McArthur microscope is not as intense as the laboratory models for photomicrography. This results in some difficulties unless a camera with automatic exposure is used. Observations of relief, pleochroism, cleavage and weathering of inclusions appear to be unaffected, and the minimal distortion ensures that measurement of extinction angles will be sufficiently accurate for diagnostic purposes.

Finally, the system thus developed had to be contained in a suitable carrying case. It was desired that all of the hardware and consumables required be contained in one case. A suitable aluminium case was procured from Jessop's Camera Supplies (address is given in Table 2-IV at the end of this section). Figure 2.6 illustrates the case and all of the supplies and equipment needed for the preparation and analysis of 100 thin sections of archaeological ceramics in the field. Figure 2.7 shows the same kit packed with the hardware and supplies. Complete as described, it weighed under five kilograms. Table 2-IV lists the contents, and suppliers and costs (1994) of each item.

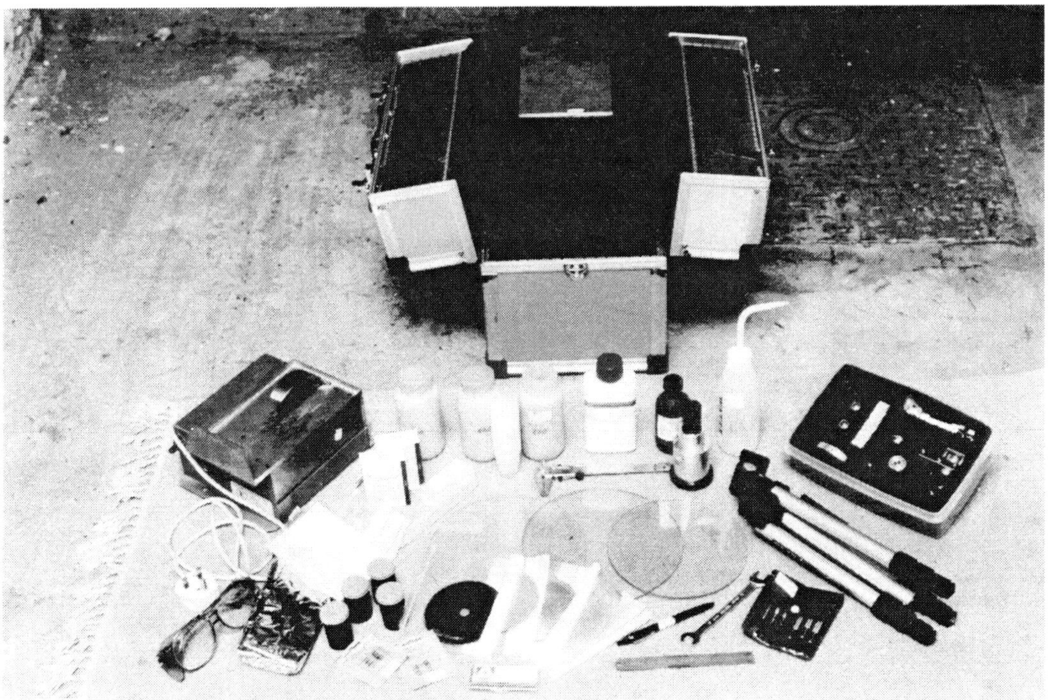

FIGURE 2.6

FIELD PETROGRAPHY KIT WITH CONTENTS

McArthur microscope and accessories are in container on the right

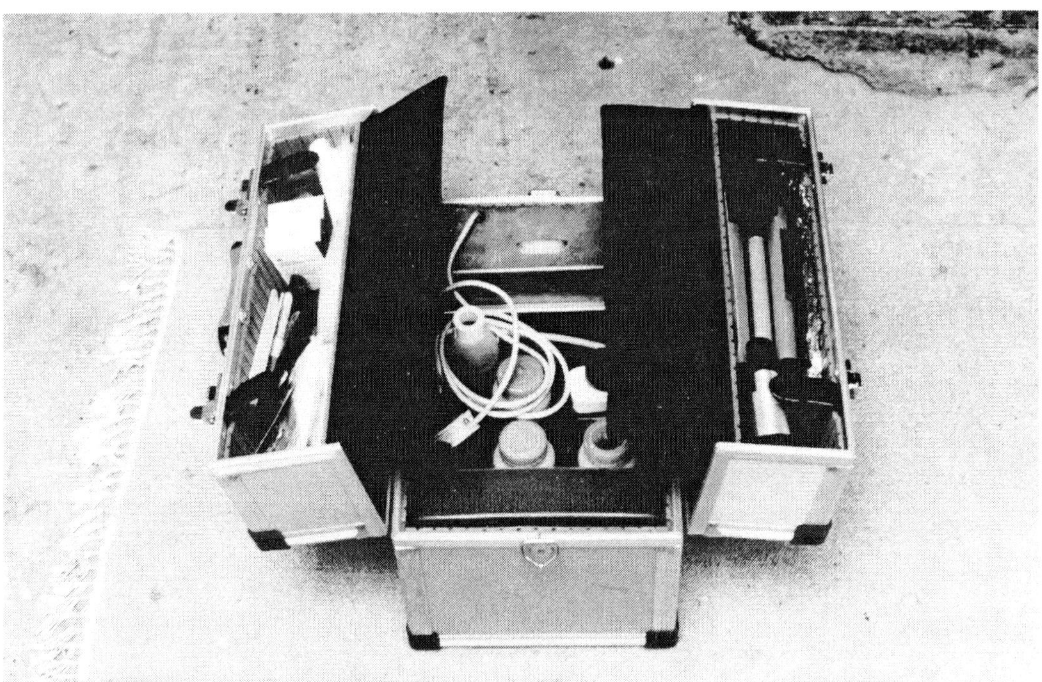

FIGURE 2.7

FIELD PETROGRAPHY KIT PACKED FOR TRANSPORT

TABLE 2-IV

FIELD KIT COMPONENTS

Qty	Item	Supplier (see list)	Cost (1994)[1]
1	Evans 4-inch Combi-Saw	1	£291.00
3	4-inch Famad diamond blades	2	£36.00
1	half-inch spanner for saw arbour	3	£7.00
5	spare 0.7 ampere fuses for saw	3	£2.00
1 pr	safety glasses	4	£7.50
1	McArthur biological microscope	5 (also 4)	£590.00
1	rotating stage modification	5	£120.00 (est.)[2]
2	extra halogen light bulbs	5	£28.20
10	AA batteries for light unit	4	£12.00
1	camera attachment	5	£88.00
1	high transmission polarising sheet	4	£8.00
3	spare 23mm dia. Neoprene O-rings	3	£1.50
1	focusing ocular with graticule	6	£71.60
2	Allen keys for light unit (1 spare)	3	£2.50
1	instrument screwdriver set	4	£16.20
1	tripod	7	£29.00
2	250ml Epo-Tek 301 epoxy	8	£50.50 (both)
2	50ml Epo-Tek 301 hardener	8	see line above
25	10ml plastic mixing cups	4	£3.00
100	glass slides	4	£15.00 (est.)[3]
100	glass cover slips	4	£10.00 (est.)
10m	aluminium foil	locally procured	
120	slide labels	4	£4.00
1	diamond marker pen	4	£5.50
3	500ml containers alumina powders	9	£4.00 (est.)
3	8-inch diameter glass plates	locally procured	
1	water squirt bottle	4	£2.00
1	aluminium case	7	£54.00
[1] Includes Value Added Tax of 17.5%		**TOTAL**	**£1,458.50**
[2] Rotating stage prototype was donated by K.W. Kirk & Sons (see [5] next page; text)			
[3] Remainder of estimated cost items were procured from laboratory stock			

List of Suppliers

1. Evans Lapidary Machines
   54 Beesfield Lane
   Farningham, Dartford
   Kent DA4 0BZ

2. Malvern Lapidary
   39 Broadlands Drive
   Malvern WR14 1PW

   Manufacturer: Star Diamond Industries
   17022 Montanero Avenue
   Carson, California 90746

3. Buck and Ryan Hardware
   Tottenham Court Road
   London WC1 (or any standard hardware shop)

4. Agar Scientific Ltd
   66a Cambridge Road
   Stansted, Essex CM24 8DA

5. K.W. Kirk and Sons Ltd (now in receivership - see text)
   Microscopes and accessories soon available from supplier 4 above.

6. Graticules Limited
   Morley Road
   Tonbridge, Kent TN9 1RN

7. Jessop of Leicester Limited
   Jessop House
   Scudamore Road
   Leicester LE3 1TZ

   In London:
   Photo and Video Centre
   67/69 New Oxford Street
   London WC1A 1D

8. Promatech Ltd
   Unit 1, The Elliott Centre
   20 Elliott Road
   Cirencester, Gloucester GL7 1YS

9. Washington Mills ElectroMinerals Ltd
   Mosley Road
   Trafford Park
   Manchester M17 1NR

CHAPTER THREE

FIELD TESTING, SUMMARY AND CONCLUSIONS

3.1 FIELD TESTING: PAKISTAN

The assembled system was first taken to Pakistan to use in the investigation of socioeconomic interaction networks of the Early Harappan Northwestern Indus Valley, details of which form Part Two of this thesis. It was packed and carried easily as hand luggage aboard international and domestic commercial airline flights to Karachi and Peshawar, respectively. At Peshawar, it was set up in a small room in a residence rented as accommodation by the author. The trip was undertaken from February to May 1994, during which time the weather ranged from rainy, wet, and 12 degrees C to sunny, dry, and 43 degrees C. This range of conditions provided a solid test of the susceptibility of the epoxy resin to some of the extremes in temperature and humidity that can be encountered in archaeological field work.

240V 50Hz electrical power was available, albeit intermittently. Also, the facility had a source of running water adjacent to the room being used as a laboratory which proved convenient for washing the sections between different grades of alumina powder. The Evans Combi-Saw was set up on a small bench and the microscope in another, cleaner, room. Figure 3.1 shows the microscope in use at the location.

Following selection of a suitable collection of sherds from each location studied (details are provided in Part Two), they were washed and properly recorded and photographed. They were then cut into very thin sections using the Evans saw and one of the recommended less costly, low concentration diamond blades, i.e. the Evans 220 Grit, the Meteor, or the Famad, all of which functioned adequately in the previous testing in the laboratory. The AGSC blade was also again evaluated under these conditions but it was very slow cutting, out of balance, and caused the saw to overheat due to the extra material that the 0.5mm blade thickness had to remove. Another problem experienced in using the Evans saw was that if the motor was overloaded even slightly by forcing the material into the cutting blade such that the motor speed reduced, the motor would cut out after 5 or 6 sections had been cut. It would then not re-start for around 30 minutes, and on two occasions it caused the internal saw fuse to blow. The problem was experienced using any one of the diamond blades. This was probably due to overheating; however, it had not been experienced in the laboratory trials in London. The problem could be overcome by feeding the material into the blade very slowly, ensuring that the motor speed did not reduce. This resulted in some very slow, but effective, cutting speeds. When a large number of cuts had to be made, therefore, some patience was needed.

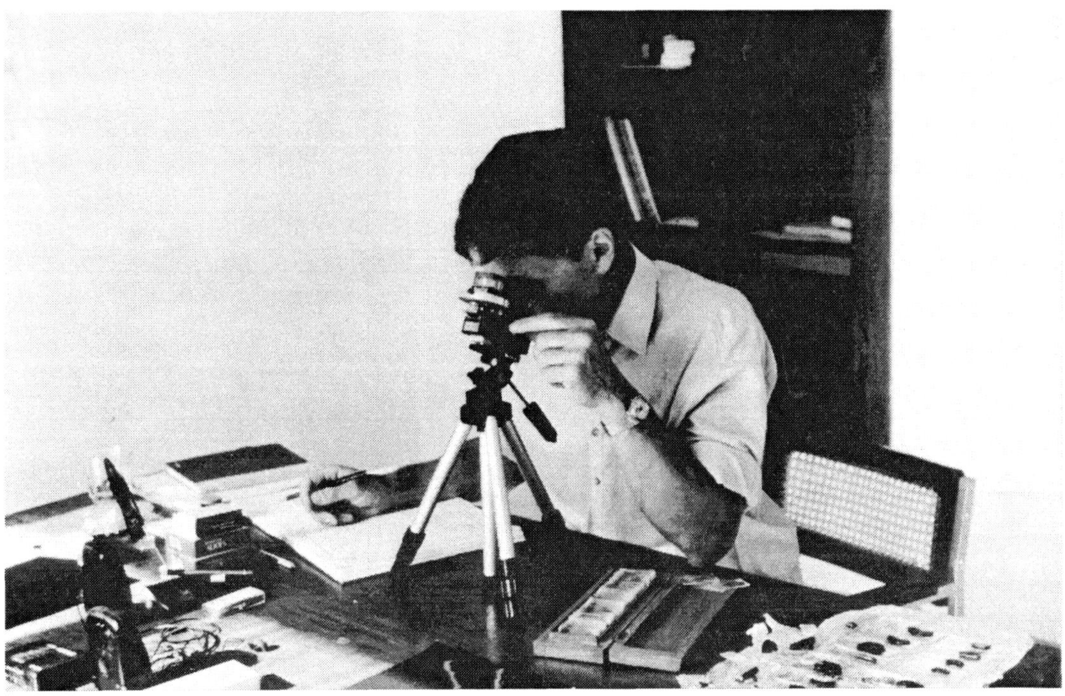

FIGURE 3.1
McARTHUR MICROSCOPE IN USE AT PESHAWAR

The most significant problem, although not insurmountable, was the curing of the Epo-Tek 301 epoxy in the earlier part of the season when the weather was wetter and colder. The mixture was prepared in a 2.5:1 resin to hardener ratio as had been determined to be the most effective in the laboratory testing. The first time this was attempted, the weather was very humid and rainy, and the temperature was about 14 degrees C. The preparation room was not heated. Under these conditions, the mixed epoxy did not cure even when left for 3-4 days. It is not known if this was in part due to the wet conditions[1], to the colder temperatures, or both. Temperatures this low were not experienced in the laboratory in London. Because these curing times were unacceptable, heating the two constituents before mixing was attempted. Both ingredients were heated to approximately 40 degrees C on a natural gas room heater, then mixed and applied to the samples. This reduced its viscosity and consequently improved its absorption. Following application of the mixed resin, the samples were placed on aluminium foil on the same heater and allowed to cure. Under these conditions, the Epo-Tek cured in about 3 hours. As the season became sunnier and warmer, another technique to speed the curing process was discovered which proved quite successful. After consolidation using the auxiliary heater, when the samples were mounted on glass slides, they were placed in direct sunlight to cure. Although air temperatures were only in the low 20's C, curing was effected in around 2 hours. It is not known if the curing was speeded by ultra-violet light in the sunlight, or if it was due to direct radiant heating by the sun. In any event, the technique proved effective on many occasions.

Toward the hotter part of the season, ambient temperatures approached 42 degrees C. Under these temperatures it was found that the epoxy, when used at a ratio of 2.5:1, cured in only 1.5 to 2.0 hours, even when spread very thinly on slides. On one occasion at these temperatures, mounted sections were placed in the direct sunlight. The result was formation of very small bubbles of gas which became trapped in the epoxy between the coverslip and the thin section of ceramic. These bubbles unfortunately remained as the epoxy cured. Therefore, placing samples in the sun to cure should not be undertaken at temperatures this high. It appears that it assists in curing at temperatures up to the mid-20's C, but at higher temperatures the advantage is lost.

The sections prepared as above were hand-ground and polished, as no powered grinder was available. In order to remove as much material as possible at the outset, the 220 grade alumina powder was used. It was found that if the alumina powder was changed to a fresh amount after grinding 3 or 4 sections, material removal was more efficient.

[1] Many epoxy consolidants are hygroscopic (see Jim 1985: 598).

However, it was noted that had a larger supply of 220 grade alumina powder been available, it could have been changed more often, which would have speeded the process. Because of the risk of consuming all of the coarse grinding powder before sufficient sections had been prepared, changes were limited to every 4 sections. Efficiency was noticeably reduced after 2 slides, and times required to reduce the section from the initial saw cut thickness of about 2mm down to 40 or 50 microns (when the 400 grade powder is used) lengthened from 5 or 6 minutes to about 10 or 12 minutes of laborious hand grinding per section. Initial slices were cut as thinly as possible so that the amount of material which had to be removed was kept to a minimum.

Once the thin sections had been prepared as above, they were analysed using the McArthur microscope as modified and discussed in Chapter Two. Because the completed thin sections were permitted to be brought out of the country, it was not necessary to complete a full mineralogical and textural analysis on site. However, it was desirable to carry out the analysis as far as possible such that a reasonable field evaluation could be made on the microscope which would be useful for future studies.

The microscope was mounted on the short table tripod, as shown in Figure 3.1. The first examination of the thin sections was during the final polishing. As the required thickness of 30 microns was approached, the slides were examined under crossed polars (XPL) to observe the interference colour of the quartz inclusions. No difficulties were encountered in this operation, but it had to be ensured that the sections were dry before mounting them on the rotating stage because excess water interfered with the viewing. Also, any moisture in the light unit may promote corrosion. Following mounting of the coverslips, more detailed analysis was conducted with the microscope. This included identification of the main mineral groups in each and ensuring that there was a sufficient amount of data to complete the investigation contemplated. It was found that the McArthur microscope was as effective in this task as in the laboratory evaluation. No new problems were encountered with the field application. It was noted however, because longer periods of time were spent viewing thin section through the microscope, that eye fatigue can become a problem with prolonged use. Although eye fatigue is also a problem with laboratory models, it is likely to be more of a problem with the McArthur because of the small size of the ocular, and because it is a monocular instrument. A binocular attachment was advertised as being available from the manufacturer, Kirk Technology, but at the time of evaluation none were available to be tested.

Overall, in the Pakistan application, the field kit generally performed as expected and no serious problems were encountered. A total of 170 thin sections were prepared at Peshawar. Recommendations are summarised after the following section.

3.2 FIELD TESTING: TURKEY

The second field application of the kit was in Turkey. As with Pakistan, Turkey also disallows the export of cultural artefacts, which includes ceramic sherds. The same equipment and supplies were taken as with the Pakistan evaluation, with the exception that, based the on the experience gained at Peshawar, an additional 250ml of 220 grade alumina powder was included, but the AGSC diamond saw blade was abandoned.

The evaluation in Turkey was part of the investigations at Kilisetepe (originally named the Maltepe Archaeological Project), a rescue project on the Göksu River near Müt, in southwestern Turkey. Its duration was late July to mid-September 1994, and climatic conditions were much different than those encountered during the earlier part of the Pakistan project. Temperatures reached a high of 47 degrees C in August and daytime highs were routinely above 36 degrees C. The humidity was extremely low, as the site was located just on the inland side of the south coastal mountain range in the south. As with the Pakistan evaluation, a simple room was available to be used as a crude laboratory. The building did not have running water, but an unlimited supply was available from a standpipe outside. A 240V 50Hz power supply was available.

No further problems were experienced with the Evans saw. The technique of feeding material to the blade very slowly ensured that the motor did not cut out under load. For most of the cutting, the Famad diamond blade was used.

Environmental conditions at Kilisetepe proved to be the best of all those tested for the Epo-Tek 301 epoxy. It remained at a very low viscosity, and usually cured in about 1.5 hours when left at ambient temperatures. On no occasion was auxiliary heating necessary to speed curing; nor was it necessary to place samples in sunlight. A 2.5:1 mixing ratio was again used, with excellent results. Best results were obtained when it was used mid-day to late afternoon when the temperature was the hottest. Evidently Epo-Tek 301 epoxy is sensitive to climatic conditions, and appears to function best under hot, dry conditions.

When hand grinding the sections to 30 microns, the process was speeded significantly by using a fresh batch of 220 grade alumina grinding powder approximately every two sections. When this was done, each section took about 5 minutes of hand grinding to reduce the thickness from 2mm down to about 50 microns. However, this consumed approximately 500ml of 220 grade grinding powder in preparing 60 sections, a significantly higher rate than that experienced in the Pakistan evaluation. The advantage gained was a 50 to 100% improvement in time taken for each section preparation. Three separate buckets of water were used for rinsing the slides, one for each grade of grinding powder. It was found that using buckets was not as efficient as running water, because in order to avoid contamination, it was necessary to change the water often.

Follow-on analysis was similar to the Pakistan evaluation using the McArthur microscope. It was used in the same configuration. No new problems were encountered.

Overall, the two field applications of the system revealed some problems not encountered in the laboratory testing. Those identified during the inital field testing in Pakistan were largely overcome in the second evaluation. The field trials demonstrated its flexibility and usefulness as a field analysis tool. Recommendations for field use are summarised in the following section.

3.3 RECOMMENDATIONS FOR THE FIELD

As a result of the laboratory and practical field evaluations, some recommendations can be made. There are two general recommendations. Firstly, and most importantly, adequate practice should be made with the materials and components before using them in the field. It is very important that the researcher or technician become familiar with all materials and equipment prior to field work. This will substantially reduce the potential for anything to go wrong far away from sources of material or advice. Secondly, unless one is completely familiar with the local markets, it should not be assumed that materials or components can be procured near the site location. Making such assumptions often results in extensive running around or delays while an item is ordered or shipped in; or having to utilise materials of inferior quality. This is especially true of materials such as glass or saw blades, which are often of inferior quality. For both of the field evaluations, to ensure sufficiently flat glass, high quality plates were obtained before leaving. The field kit should be prepared with all materials and components necessary, with only possible minor exceptions such as electrical mains plugs to conform to local standards which are usually easier to obtain in the host country.

With respect to sawing the sherds, one smooth cut should first be made to provide a flat edge. This should then be consolidated and allowed to harden. This will permit the second cut to be made very thin, and substantially reduce the amount of grinding required. The procedure ought to be practised prior to the field application, as it can take some experience to perfect. If it can be done efficiently on location, the minimum amount of epoxy and coarse grinding powders will be needed. As all coarse grinding is done by hand, which is a very laborious and time-consuming process, every effort needs to be made to minimise it.

If possible, a facility with running water available should be used. This ensures complete rinsing between grades of grinding powders and will reduce the possibility of contaminating the finer grinding powders and scratching the surface of the sections. It is realised, however, that in many cases this will be a luxury; indeed in some cases water may be a luxury. The alternative is to use buckets of water, as was done in the field evaluation in Turkey. If the water supply is not limited, it is recommended that three separate buckets be used; one for each grade of grinding powder. Otherwise, particles of coarse alumina which remain in suspension in the rinse water may be picked up on the slide and transferred to the next finer grade of powder. If water is in short supply, a recommendation is to use just one bucket, and complete a set of slides, say 5 or 6 at a time, on one grade of powder before going on to the next finer grade. This would conserve water while still minimising the risk of coarse powder contamination.

When hand grinding the sections from 2mm down to about 50 microns, i.e. when the maximum amount of material had to be removed, it was found that the 220 grit alumina powder became dull and lost its cutting efficiency after about two slides. This was especially true of those sections with a high cured resin content. To overcome this, and substantially reduce fatigue, it is recommended that the powder be changed every two slides or so. Therefore, a larger amount of the 220 grade alumina powder should be taken, or the coarse metallographic grinding papers should be used. It is estimated that an additional 250ml of 220 grade alumina, or a total of 750ml, be made available if one is expecting to make about 100 thin sections. Although this adds weight to the kit, it is well worth the investment in reduction of time and energy in preparing the sections.

With respect to the epoxies, use of Epo-Tek 301 is highly recommended, at least for the climatic conditions of the areas where testing was conducted. Based on the laboratory testing, it is expected that it would also be the best in other conditions. However, if very humid conditions are to be encountered in the field, it is recommended that some testing be first conducted in an environmental chamber if possible. If the thin sections will be prepared at temperatures less than about 25 deg C, a source of auxiliary heating should be considered to keep curing times to a minimum. The testing showed that constancy of temperature was not critical; any device that can provide temperatures around 30 to 40 deg C will suffice. The temperature should not be above this however, as it may result in bubbles forming under the slide or cover glass. Alternatively, thin sections can be placed in direct sunlight to aid in curing. For ambient temperatures above 30 deg C, no auxiliary heating should be used; nor is sunlight curing advised. For best curing times, it is recommended that the ratio of epoxy to hardener be 2.5:1. Above all, as recommended in the initial paragraph of this section, it is important that one is familiar with the epoxy's characteristics under varying temperature conditions. Moreover, further initial experimentation should be done on site, using discarded sherds, to determine the epoxy's response to local conditions before commencing any work on research material. All conditions cannot be simulated in the laboratory before departure. Determining absorption and curing characteristics under the local conditions in advance will ensure that cuts made on rare ceramic material are kept to a minimum. Often, it is the larger, more complete vessels, or the most decorated, that are selected for thin sectioning, and in these cases saw cuts need to be kept very small and few in number. Finally, in respect of the epoxies, a copy of the specifications of the Epo-Tek 301 should be carried when boarding commercial airliners. Although it was not experienced by the author, inspectors may disallow on board any chemical substances that cannot be shown to be non-hazardous for flight.

When viewing and analysing slides, the use of small table tripod is recommended. The microscope is very compact, and if used directly on a tabletop, fatigue quickly becomes a problem. The tripod may then also be used for photographing the sherds, an important first step before sawing them into sections. A tripod with extendible legs could also double for other uses at the site.

Apart from the foregoing considerations, preparation and analysis of thin sections in the field is much the same as in the laboratory. While conditions are often not as clean or comfortable as the laboratory, one can usually overcome these and improvise where necessary.

3.4 SUMMARY AND CONCLUSIONS

At the outset, a need for closer liaison between field and scientific archaeology was recognised in order to maintain the human element in the discipline. Simpler, more accessible, and more portable science-based analysis systems were identified as potential connectors between the two. The foregoing development of a field-portable petrographic analysis system has been largely successful in helping to bridge this gap.

To make science-based archaeology closer and more useful to the field archaeologist, a need for systems that could alleviate these shortcomings was identified. In particular, the ability to conduct on-site ceramic petrographic analysis was seen to fill this gap with respect to the study of ceramic provenance and technology. It also filled a need for conducting petrographic analysis in locations where samples may not be exportable; and allowed on-site decision making.

Generalised parameters for the requirements of a field-based petrographic analysis system were defined to be portability, ease of use, and ready commercial availability of components. A step-wise description and review of the preparation of ceramic thin sections and their analysis as it has traditionally been carried out was done. This identified main obstacle areas to conducting petrography in the field. Major obstacles were in the grinding process, the need for a constant heat source for effecting resin cures, and the heavy, bulky nature of laboratory petrographic microscopes.

Each of these was studied with a view to elimination or reduction in weight and size. With respect to cutting and grinding ceramic samples, it was found that careful and diligent use of a lightweight portable saw could produce very thin slices of properly consolidated ceramic material that need little follow-on grinding. Relatively inexpensive but rugged diamond blades may be used to provide a section with smooth cuts from pre-consolidated ceramics. These could then be ground to the required thickness by hand using 220 alumina grinding powder or coarse grit metallographic grinding papers. These steps could eliminate the need for a heavy power grinder. For consolidation and adhesion of the ceramics to glass slides and coverslips, use of cold mounting resins could eliminate the requirement for a heavy thermostatically-controlled heating plate. Finally, development of a compact field-portable petrographic microscope with most of the capabilities of a laboratory model would significantly reduce the weight and bulk problems associated with standard laboratory models. If these could be made acceptable for field use, a fully portable system could be realised.

Specifications that would ensure an acceptable and accurate standard of thin section preparation and analysis were then set for each. As one of the original system specifications was that all components and consumables be readily available commercial products, standard stockists and suppliers' offerings were reviewed for acceptability. This was followed by laboratory testing, and finally actual field testing and use.

One suitable portable saw was located, and tested on archaeological ceramics using five different diamond blades. Although somewhat underpowered, the Evans 4-inch Combi-Saw was found to be entirely adequate when used with the more inexpensive diamond blades such as the Famad, using a technique easily mastered with practice. Seven cold-curing resins were evaluated under varying combinations of mixture ratios and curing temperatures. Two of them, Araldite and Epo-Tek 301, were found to be acceptable for field use. Both have excellent consolidation and curing characteristics, but Epo-Tek was the more effective of the two at lower ambient temperatures.

There was found to be no field-portable petrographic microscopes commercially available, but two compact transmitted light microscopes were located that could be readily and inexpensively modified. The McArthur biological microscope and the Swift FM-31 were examined and both found to be adaptable to the addition of a polariser, an analyser, and a rotating stage. These modifications were carried out on the McArthur microscope and the unit tested. It was found that, despite the fact that they represent only some of the features of full-sized laboratory petrographic microscopes, these three basic modifications permitted the McArthur microscope to be used to identify most minerals at the level required for field work. In addition, an eyepiece graticule installed in the ocular permitted fabric analyses to be conducted in the field. Although not chosen for testing because it was slightly heavier and less adaptable to a rotating stage modification, the Swift FM-31 could also be made entirely adequate.

A standard aluminium photographic equipment carrying case was adapted to house this relatively light weight combination of materials and components. It was carried in commercial airline hand luggage and used on location in Pakistan and Turkey, two countries which prohibit the export of archaeological artefacts. The system was set up as a crude laboratory in basic rooms, one with running water, and one without. Early in the season in northwestern Pakistan, at temperatures around 12 deg C in damp conditions, it was found that the Epo-Tek 301 epoxy was very slow to cure. The problem was solved by applying heat from a basic gas room heater. This, apart from having to use the coarse alumina powder very sparingly, was the only difficulty encountered in the field testing. In the hotter seasons in both countries, when daytime temperatures were in the range 38 to 45 deg C, the epoxy cured within one hour without auxiliary heating applied.

Overall, the foregoing development, testing and evaluation of the field petrography kit was generally successful. It may be concluded that the original problem, that of lack of a complete system capable of allowing both preparation and analysis of archaeological ceramics in thin section away from the laboratory, has been largely overcome. The kit developed proved this in two independent applications miles away from a home laboratory. It is relatively inexpensive, components are available commercially, and it is very portable. The main drawback at the present time is the non-availability of the McArthur Biological microscope; however, it is expected that manufacture will be resumed in the near future. If not, the alternative field microscope earlier discussed, the Swift Model FM-31, could quite easily be modified to perform analyses.

PART TWO

SOCIOECONOMIC INTERACTION NETWORKS OF THE EARLY HARAPPAN NORTHWESTERN INDUS VALLEY

CHAPTER FOUR: INTRODUCTION AND THEORETICAL CONSIDERATIONS

4.1 INTRODUCTION

The Harappan, or Indus Civilisation, is an advanced and complex urban civilisation which flourished in the Indus River valley region of present-day Pakistan, northwestern India, and part of Afghanistan during the period ca. 2650 to 1850 BC. Generally accepted as the third earliest of the world's civilisations behind the Mesopotamian and the Egyptian, it has been much less studied than the other two. It meets the classic "ten rather abstract criteria" as defined by Childe that qualify it as an urban civilisation (Childe 1950: 9). One of those criteria, a writing system, has been a source of lack of progress in studying the nature of the civilisation, because it has not yet been fully deciphered. Although much progress has been made (see Parpola 1994 for a recent summary), a complete decipherment has eluded researchers so far. The archaeological record has therefore been the primary source of information to date. Numerous puzzles exist, one of the major ones being the rapid rate of the civilisation's increase in complexity from a village farming and pastoral economy to a full-fledged urbanised state, much unlike the rise of the Mesopotamian or Egyptian urban states. In contrast to the Indus, the Mesopotamian civilisation developed over the course of roughly three millennia (Kohl 1987: 15). The Indus civilisation's rapid transition to urbanisation of only 100-150 years presents archaeologists with a challenge to the 'uniformitarian' model of cultural change. Rather than emerging from a long, slow period of gradual and constant cultural growth and modification leading to a slowly emerging pattern of urbanisation and social complexity, the Harappan Civilisation seems to have resulted from a very short period of transformation, a "veritable revolution within the urbanisation process" (Possehl 1990: 261). To shed light on the nature of this change, much more research is needed in order to define aspects of Harappan life leading up to and during the transition to an urban society.

A fuller picture would assist archaeologists and social theorists in explaining the Harappan urbanisation phenomenon. One of these aspects is the nature of socioeconomic interaction between sites during the formative years.

A fully urbanised society such as the Indus Civilisation must accommodate full-time specialist craftsmen and mechanisms for distribution of produced surpluses (Childe 1950: 11). In order for this to be accomplished over an area as vast as that covered by the Harappan Civilisation, some sort of structured communication and exchange systems and processes needed to be in place. It is well known amongst Harappan scholars that stylistic and morphological similarities in ceramics existed geographically and temporally throughout the Mature Harappan and, regionally, throughout the Early Harappan period (see descriptions in Chapter Five). This alone, however, tells us little about the nature and extent of the interaction networks that were in existence to maintain this commonality and uniformity of style. Was each production area acting independently to produce pottery with similar visual characteristics, or was pottery being produced, transported and traded purely as a commercial activity? Or, something in between? Examination of these questions can provide us with a picture of the type of intersite socioeconomic activity that was present, and help to define the complexity of the social structure of the time. A study to determine the extent to which pottery was being transported or the extent that technological information and innovations were being exchanged amongst sites during the Early Harappan period would provide some of the necessary insight. This thesis undertakes to examine that, by selecting samples of Early Harappan ceramics from four Indus valley sites, and comparing and contrasting their respective mineralogical and textural compositions.

This chapter provides the background and discusses some theoretical considerations and models regarding the nature and understanding of socioeconomic interactions in prehistory in a general sense, and how their study can help to address the problem of examining how the Indus Civilisation rose to a complex level in such a short period of time. It is shown how petrographic data derived from archaeological ceramics can be applied and used to address the problem. Chapter Five then presents a description of the Indus Valley and its environs both present-day and during the Harappan period, and a description of the

Harappan Civilisation and associated research. Descriptions of the four sites under study are provided, followed by a description of the Early Harappan ceramic typologies. Chapter Six describes the methodology employed, from sample selection to methods of analysis of the ceramic thin sections. Chapter Seven presents and discusses the results of the analyses and their implications. Chapter Eight then presents a summary of the thesis and conclusions to be drawn from the research, together with suggestions for further research.

4.2 SOCIOECONOMIC INTERACTIONS IN PREHISTORY

Because standard styles of ceramic decoration and morphology occur throughout the geographical extent of the Harappan Civilisation, our starting point is to ask what this means. What can it tell us about the nature of Harappan social and economic organisation and complexity during the earlier periods where it is first observed? The presence of stylistic and morphological similarities demonstrates that there was some form of communication interaction. Obviously, a first explanation could be that the ceramics were produced in a relatively few large production centres and, through a highly organised and elaborate distribution system, re-distributed throughout the entire region. Or, the groups may have simply produced pottery of a similar style locally to show their affiliation with a broader-ranging culture.

A considerable amount of research and study has been conducted on the subject of prehistoric exchange. This research appeared to reach its peak in the late 1970's and early 1980's. In many of these studies, particularly in the case of ceramics, indirect evidence has been used to infer physical exchange of goods. For example, Jacobsen (1984) used ethnographic observations, Cordell and Plog (1979) used the presence of stylistic similarities to infer the exchange of goods. There are numerous other studies of these types seen in the literature. It began to be realised that the spatial distributions of stylistic characteristics that were assumed in some studies to be caused by exchange also could have been produced by a number of other processes.

The reliance on exchange to explain patterns in the distribution of specific styles is in part a reaction to the lack of recognition of this phenomenon in many of the classic stylistic studies of the 1960's (Hantman and Plog 1982: 238). According to Hantman and Plog, these analyses were characteristically based on a social learning-enculturation model of the factors underlying stylistic behaviour. In this formulation, artefactual style was viewed as a passive epiphenomenon reflecting the intensity of social interaction between spatially defined social units. Interaction in these studies was a poorly defined term, although it most often referred to reconstructed patterns of kinship and residential descent systems. Style, particularly on pottery vessels, was viewed as learned behaviour that was largely unintegrated with other aspects of prehistoric social systems.

An alternative model to explain stylistic distributions was proposed by Wobst (1977), described in Hantman and Plog (1982: 239). He considers style as functioning to maintain social systems, and also as an indicator of social interaction, but not restricted by the concept of style as solely a learned behaviour. In Wobst's model, style is interpreted as a dynamic variable, serving specifically in the processes of information exchange and boundary maintenance. This information-exchange model posits that zones of interacting individuals can be defined by an analysis of stylistic distributions, and that the recognition of variability of these zones reflects directly on patterns of change in regional social organisation (Wilmsen 1973: 24). Closely akin in the study of cultural diffusion processes to the concept of the *interaction sphere* of Caldwell (1965), it has been adopted by many archaeologists. The communication of information presented in this model does not, however, always flow freely, and there are many socioeconomic factors that can affect the regional pattern of information flow, as delineated by Hantman and Plog (1982: 239). Population size and density, variation in settlement patterns, the degree of population dispersal, and, most importantly, the degree of social differentiation within a region will affect information exchange. In a socially- or spatially-differentiated society, information would be expected to be processed in a hierarchical fashion. There are many aspects of social differentiation that may influence the distribution of styles including social status, political affiliations, entrepreneurial behaviour, and economic specialisation. Geographical or physical barriers can also inhibit or foster the exchange of information and stylistic factors. Any of these can serve to accelerate information sharing between some subsets of the population, while deterring the process in others.

Hantman and Plog (1982) therefore expand on the Wobst model by incorporating some of these factors that can produce differential patterns of information exchange. By defining style as 'a correlated cluster of attributes', they argue that systematic change in combinations of these attributes may be treated as stylistic *innovations*. They propose that analysing the distributional patterns of ceramic styles as regional forms of the communication and adoption of innovations can provide a satisfactory framework to understand distributions not explained by commodity exchange alone (*ibid.* 240). However, most archaeological studies of the adoption of innovation have not recognised the way in which social factors influence the spatial communication of innovations. Two fundamental models used to describe information diffusion are (*a*) the contagious wave model (Hagerstrand 1967) and (*b*) the hierarchical diffusion model (Hudson 1969). The contagious wave model focuses primarily on household adoption of innovations that diffuse among households or individuals and are probably accepted by all members that have 'certain characteristics' (Pedersen 1970: 205). This model fails to recognise the potential effects of social differentiation and complex political and economic systems on the spatial pattern of information communication. Hantman and Plog (1982: 241) comment that Hudson (1969) provided a model which largely overcomes this problem and does take into account socially and spatially differentiated societies. Hudson's model states that, rather than simply spreading out across a region (as would happen with the contagious wave model, in the absence of socially-induced barriers), goods and ideas move downward through a central-place system with a rank-size hierarchical structure. This model focuses on entrepreneurial innovation (Pedersen 1970: 205). Hantman and Plog state that in innovation diffusion of this type, the innovation is likely to occur first in the settlement that has the greatest amount of intraregional and interregional interaction, usually the largest settlement or political centre. Hagerstrand's (1966) model appears to straddle both and models by combining the hierarchical diffusion approach with the wave effect approach to produce a comprehensive synthesis of how information is communicated throughout a regional system. Information is spread initially through the urban hierarchy, followed by centres next in rank, then finally breaking up and being replaced by the wave effect.

There are many socioeconomic factors that can create both hierarchical and nonhierarchical patterns of information diffusion. One that can affect the distribution of ceramic styles is the division of social groups into varying status levels (Hantman and Plog 1982: 242). Status and prestige are cognitive constructs that, though based on material economic advantage, are often signalled through aspects of art and other symbolic systems. Numerous researchers have observed correlations between groups of ceramic artefacts and social status (e.g. Miller 1980; Rubertone 1978). This concept of ceramic style as symbol of socioeconomic status can be extended to a concept of ceramic style as functioning to designate belonging to a particular group, apart from its function of designating an individual's status. Thus it can be conceived to exist as an indicator in less stratified societies as well as those with sharp social status differences. Style here would identify with a particular culture group rather than a particular status group. In these instances, stylistic similarity is not a result of the intensity of interaction, but rather it functions as a symbol of affiliation within a regional political system. Moreover, certain symbols may act regionally as both symbols of alliance and of status, as in the case of the Olmec (Flannery 1967).

The information exchange model, as traditionally applied in nonstratified societies, relies on the concept of stylistic similarity as a function primarily of the size of territorial mate and material exchange systems. In more complex systems where production is impacted by political determinants, ceramics and the adoption of stylistic innovations will often play an increasingly political role (Hantman and Plog 1982: 243). Where the production of ceramics is a function of regional political competition, the distribution of regional ceramic styles should be expected to vary primarily in accordance with change in the size of polities. In these systems, it should be expected also that alliances between political leaders would be signalled stylistically.

Variation in the mode of ceramic production can also dramatically affect the distribution of ceramic styles. Focusing on patterns of distribution as well as consumption, a convenient classification system divides production modes into six categories: household production, household industry, individual industry, workshop industry, village industry, and large-scale industry (Van der Leeuw 1977). These, respectively, are marketed for: own use, group use, regional use, village use, wide-regional use, and regional and export. The distribution of styles produced within these levels of productive modes should be expected to vary accordingly. For instance, where production is specialised and distribution widespread, a pattern of regional homogeneity would be expected in ceramic style. Conversely, household production and local distribution may produce patterns of extreme stylistic diversity (Hantman and Plog 1982: 245). Although other social factors as discussed above may ultimately affect these distributions, identifying the level of production and distribution helps to establish a baseline from which to evaluate these factors.

4.3 USE OF PETROGRAPHIC DATA

A traditional way of determining the geographical source of pottery production has been to note spatial occurrence of pottery styles, then apply the 'criterion of abundance' concept which places the locus of manufacture in the zone where the style most frequently occurs (Rice 1987: 177; Rapp 1985). However, this concept can only be effective in situations where one can assume that all (or a representative selection) of the possible sites containing the ware have been fully or representatively excavated, and that pottery production is in fact being transported or traded to neighbouring sites. Moreover, it assumes that commonality in pottery styles implies a commonality of manufacture location, which of course is not always the case. More effective methods use mineralogical and chemical data to apply the 'provenience postulate', which states that differences between sources of raw materials can be recognised analytically and positional variations will be greater between than within sources (Rice 1987: 413). The two methods involve different data sources and analyses. Mineralogical or petrological analysis identifies mineral inclusions in ceramic pastes while chemical analyses provide lists of elements present. Neither is inherently superior. Researchers have used either, or both in combination, most successfully.

Basic descriptions of the various chemical methods can be found in Rice (1987: 371-404), Orton et al. (1993: 144-151), Kamilli and Steinberg (1985), or Rapp (1985). The major advantage of the chemical methods is that of repeatability. Because the methods are fully instrumental, the subjective component is virtually eliminated. Variability due to differences in researcher or technician interpretation or judgement is minimised and this allows more meaningful inter-laboratory comparisons to be made. The chemical methods also have a distinct advantage for the analysis of clay sources. Because clay particles are much too small to be identified with a petrographic microscope, and the clay minerals are altered during the vitrification process, chemical methods are the most often used to identify clay types. This has been effectively demonstrated by Topping and Mackenzie (1988) and Adan-Bayewitz and Perlman (1985). However, one must be careful in these cases to ensure that the clays analysed have been modified in an identical way to that used for the ancient ceramic under consideration because these alterations have been shown to have a significant influence on chemical composition (Kilikoglou et al 1988). Clays as used by potters are by nature not homogeneous and levels of elements can be altered by human intervention as well as by post-depositional processes. For example, as pointed out by Neff et al. (1988: 160), the addition of temper may enrich some elemental concentrations while diluting others, and that the magnitude of enrichment or dilution is not likely to be uniform over all elements (see also Bishop 1980; Bishop, Rands et al. 1982; Bishop, Stein et al. 1982; and Neff et al. 1989). All too often, researchers have embraced the modern chemical methods while overlooking the immense value of accompanying it with a petrographic analysis. Elemental data only become decisive when there exist considerable differences in the elemental composition of the clay matrix which cannot be detected by petrographic analysis (Schubert 1986). Most importantly, chemical analyses alone can tend to ignore or drift away from the all-important human element involved in the pot's manufacture, or so fail to contribute to attempts to understand 'real-life' pottery-making behaviour. It is most unlikely that ancient potters chose a particular clay combination because it contained certain concentrations of rare elements. A further major disadvantage of the chemical methods is their restricted availability and the cost of the facilities necessary; this was outlined in Part One.

Numerous research projects on the study of ancient trade and commerce have been successfully conducted using petrographic data alone. Some of the more recent include Jones (1986), who conducted an extensive petrological analysis of lowland Maya wares to study temper types in use; Isaacson and Aleto (1989) demonstrated the value of petrographic analysis in conducting provenance studies of ceramics from Ecuador; Stoltman (1991) used the method as a technique for documenting cultural interaction in the Mississippian; Gerrard and Gutierrez (1991) defined intra- and inter-kiln variations in Spanish medieval pottery; and Donahue et al (1990) examined sources of temper in the islands of the northern Lesser Antilles. Numerous others may be found in the literature. These and other researchers have demonstrated its value as a convenient, low-cost, accessible tool for the study of interaction networks. It is not the purpose here to provide details of the methodology nor underlying theories; much of this has been discussed in Part One.

Petrological analysis involves making thin sections of the pottery, identifying the mineral inclusions, and matching them to a geological formation source. It functions best in regions where results of extensive geological mapping are available to provide details of mineralogy and geological formations to match those inclusions observed in the archaeological ceramic under study, and where geological formations and mineralogy vary significantly within local regions. However, this is not always the case. For example, vast areas of the Northwest Frontier Province of Pakistan where three of the archaeological sites of the present study are located suffer from a lack of detailed geological studies. Also, the Bannu Basin and the Gomal Plain, two of the geological formations containing sites of the present study, show limited mineralogical differences (see Chapter 5). Ideally, a systematic sampling of the local clays, if properly done, will provide the best comparative data. But because clays, and the inclusions which characterise their sources, can vary widely within a region, and ancient clay sources can become exhausted or buried, e.g. by river course changes, this can be not only a laborious and time-consuming process but may even be misleading. For the same reasons, it is not good practice to rely on non-systematic, opportunistic local clay sampling as is often done in provenance studies. Studies done in England often suffered from lack of geological diversity in some regions. It largely because of the latter case that textural analysis was introduced into the study of archaeological ceramics in Britain in the early 1960's (Peacock 1971). The petrographic method therefore can entail more than simple mineral identifications and matching to a source lithology to determine provenance. It can be applied to the study of fabric textures, i.e. an examination of the shapes and size frequencies of inclusions present. These methods are discussed later in the section. It can also be used to study the technology used by ancient potters, from the type of temper chosen to determination of firing temperature ranges. Before discussing its use for deriving technological information, some discussion follows on the use of thin-section petrography for textural analysis.

Before embarking on a discussion of the use of petrography for textural analysis, it is important to understand temper. The inclusions which are studied can very simply be classed into two sources: naturally occurring, or added by the potter. The term 'temper' in archaeology is generally used to denote the addition of coarse components to a clay paste by the potter to modify the properties of the clay (Rice 1987: 406). But much of what is called temper in the archaeological literature may constitute inclusions naturally present in the original clay material, not added independently at all. Indeed, the potter may have selected a particular clay source because of its natural inclusions. The distinction between natural and added temper is difficult to make, particularly if the inclusions are those that often occur naturally in clays such as quartz, calcite, mica, and sometimes shell. "The greatest problem in distinguishing added temper from natural inclusions is the presence of quartz sand. Typically the shape of the grains is the basis for such distinctions. Pronounced angularity of the grains is interpreted as meaning a rock was crushed and hence that the quartz was added, whereas rounded grains are taken as evidence of alteration during transport as part of a sedimentary clay deposit." (Rice 1987: 410). Stoltman (1991), Lombard (1987) and other researchers have tended to use this method of distinction in granulometric analysis. However, it is not always a logical assumption that highly angular inclusions, especially of quartz, seldom occur naturally in deposits such as clay banks. Where quartz occurs in primary clay deposits, it may be angular. Where it occurs in secondary clay deposits, the assumption would generally be a safer one.

The implications of this assumption are that very angular quartz is always temper. However, of course the converse is not always true. In other words, one cannot always assume that the lack of crushed quartz indicates a lack of added sand or quartz temper. If a potter had access to a satisfactory natural sand deposit (as is often the case near river clay deposits), he or she would naturally choose to exploit this source rather than go through the time-consuming and laborious process of crushing rock for temper. It is true however that the more angular the sand-size quartz inclusions are, the stronger the interlocking frameworks it will provide as a temper (Rice 1987: 73-4); and therefore potters would probably seek to include angular or subangular quartz. If natural sources possessing acceptable angularity are accessible they would likely be chosen over crushing. Aeolian and riverine sand deposits are generally much more subangular than freshly-crushed material as a result of the transportation process. It is often the case that natural sands, aeolian, lacustrine or river, have acceptable angularity, and are exploited for use as temper. They are used either in their naturally occurring grain size distribution or sieved to obtain the desired temper grain size range. Both have been ethnographically recorded in the areas of northwest Pakistan of the present study. In workshops observed in Karmathu, north of Bannu district, local potters use sieved aeolian sand deposits for temper (Rye and Evans 1976: 39). In Swat, to the north of the study area, potters have been observed using sieved sand obtained from river deposits (*ibid.*: 28); in Dera Ismail Khan and in Panjab, near the site of Harappa, naturally-occurring inclusions are left in the clays because they produce the desired characteristics (*ibid.*: 50-51,55). Use of natural river sand deposits as temper has also been observed at Shaikhan Dheri (Husain 1992: 172). There are therefore a variety of situations, at least with respect to the present research, to be considered as well as the assumptions generally made by past researchers that only very angular inclusions should be considered as temper (in the case of quartz).

Where naturally occurring quartz sand deposits are being exploited for temper, it is much more difficult to distinguish between the natural and potter-added inclusions in thin section. Naturally occurring sands have usually suffered some erosion through transportation or chemical weathering and are not as angular as a newly-crushed quartz. If there is no strongly angular quartz fraction present, but there is a determinable size distribution or a bimodal size distribution in the quartz inclusions, it may well be assumed that part of that quartz fraction could be added temper[1]. In these cases where the very

[1] Naturally occurring quartz inclusions in primary clay deposits may be angular. These can become mixed in a paste preparation with less angular river or lacustrine sands added as temper. This would also result in a bimodal angularity distribution.

angular fraction is not present, it may be assumed that temper, if added, was from local sources. It has been shown that in addition to clay sourcing, potters seldom travel more than 6 km to sand temper sources (Arnold 1980: 149; cited in Rice 1987: 116); if the sands are further away than this, it would likely be more efficient to crush some local rock.

Additionally, one can make a case that it does not matter whether or not the quartz fraction was intentionally added. A potter often prepares clay by removing some inclusions (e.g. by levigation); and by adding some (e.g. temper). These processes are designed to produce the desired characteristics in the clay body. We can therefore assume the inclusions *that are present* were intentional, in that the potter was satisfied, before forming the vessels, that all necessary inclusions were there regardless of how they came to be there. This is so for purposes of provenance studies at least. To make behavioural inferences regarding clay and temper sourcing based on this line of reasoning might require a different approach. Therefore, an analysis of the quartz inclusions in each thin section can be extremely useful for provenance studies even if one cannot distinguish natural inclusions from those intentionally added by the potter. It can be used to characterise a ceamic fabric.

If one makes the assumption that all quartz inclusions, both naturally and intentionally added, are from local sources, and that quartz particles can be provenanced as to type of source environment by their characteristics, then an analysis of them can be helpful in differentiating pottery sources. The approach has been used only rarely in archaeological provenance studies, perhaps because it is primarily a method used by geologists and sedimentologists and so requires a reasonable understanding of sedimentology. A recent innovative attempt in archaeological application was conducted by Mason (1995). He (or she) demonstrated that provenance of Islamic stonepaste ceramics of the Near East could be inferred by analysing the degree of cloudiness of quartz as caused by fluid inclusions. Lombard (1987) used composition of sand temper in sherds to establish probable geographic location of pottery manufacturing region, by comparing them to samples of sands collected from the same region. He (or she) however made the assumption that the only temper added was the coarse gneissic fraction present (*ibid.:* 105); and that all remaining, more rounded quartz inclusions were naturally present in the original clays; this made it easier to match them to local clay deposits. This is a risky assumption because it assumes that the potter did not add any natural sands as temper nor did the potter alter the clay via levigation. Dickinson and Shutler (1971) studied the nature of temper sands of Pacific island pottery sherds and inferred their origins from a knowledge of the varying composition of sands based on bedrock geology of some of the islands. Again, the above assumptions had to be made regarding the potter's preparation of the clay mixture.

Extensive research has been carried out by sedimentologists that can be used to advantage in archaeology to assist in sourcing of quartz sand tempers. The basis of the research is that particulate matter, including quartz, is subject to abrasion and weathering from various influences in their polycyclic history (Selley 1988). This affects their size, shape and other characteristics. While one must be very careful in inferring the depositional environment of ancient sediments from their granulometry, it has been shown that size distribution characteristics can be used to make such inferences. Studies of the characteristic parameters of the distributions of sand particles of a sediment assortment has been shown to be revealing as to the nature of the depositional environment. The earlier studies by Friedman (1961: 524; 1967) demonstrated that river sands and dune sands generally have positively-skewed distributions because river water normally contains a fairly high concentration of suspended clay and silt (Blatt et al. 1980: 73). Beach sands tend to be better sorted than river sands because of winnowing by wind and wave action, and thus tend to have lower numerical standard deviation values. Inman, however, in an earlier study (1952: 126), stated that many size-frequency distribution curves of sediments approach normal distributions, and considered this an advantage in that it allowed the properties of the distribution to be described in terms of deviations or anomalies from the normal. Folk and Ward (1957: 25) showed that extreme high or low values of kurtosis in sand particle-size distributions imply that part of the sediment achieved its sorting elsewhere in a high-energy environment, and that it was transported essentially unmodified into another environment where it was mixed with another type of material. In this case the two distributions will retain their original characteristics, i.e. strongly bimodal. Buller and McManus (1972) examined grain-size data from non-biogenic sediments from a variety of environments and found also that sorting is a function of depositional environment. They found that quiet-water sediments exhibited sorting relatively constant over a wider range of median grain sizes than other environments; that beaches represent environments of maximum energy; and that shallow marine deposits fall somewhere in between and so experience a wider spectrum of sorting gradients (*ibid.:* 16). These observations have been corroborated by Friedman (1979: 5-6). These sorts of studies are useful in archaeological temper provenance studies only if the behavioural aspect of potters is taken into account. That is, ethnographic studies of present day potters in the area have shown that potters employing river sand tempers sometimes sieve the sand before mixing it into the clay in order that only the larger size fraction is used as temper (see previous). This action would confound use of naturally occurring size distributions to infer depositional source. However, if, on the basis of the subangular nature of the larger quartz inclusions, it appears that naturally-occurring sand sources are being used (i.e. not crushed), it can be inferred by their distribution in the ceramic whether or not they were sieved before use.

In addition to the above effects on the sorting, or size frequency distribution, of quartz particles, much can be inferred from their angularity and sphericity. Quartz grains become weathered according to their depositional and transport history as discussed above; it has also been shown that they can become significantly weathered in postdepositional environments (Blatt et al. 1980: 246; Crook 1968). Pye and Mazzello (1994: 506) carried out studies which demonstrated that this can occur by chemical processes in periods as short as a few thousand years after deposition, via silica dissolution and grain etching. These sorts of influences, occurring at different rates in different postdepositional environments, can result in different degrees of weathering. In the case of the present research, this was explored because of the wide possible differences in their respective transportational and depositional environments and that a few thousand years in different environments could have produced distinctively different weathering effects. Most specifically, those which weathered on the Gomal Plain around the site of Rehman Dheri would probably have spent more time and distance in a fluviatile environment than those from the Bannu Basin; due to their respective distances from their probable original mountain sources.

The weathering influences so far discussed have related only to their effect on quartz particles and inclusions. Other commonly occurring inclusions in clays such as biotite, muscovite, and feldspars are even more susceptible to the influences of weathering, both in transport and in postdepositional situations (Blatt et al. 1980: 250-65). Because of the differences in geological history between the Gomal Plain and the Bannu Basin (see Chapter 6), it was expected that differences could exist in the weathering history of these non-quartz inclusions that could aid in distinguishing clay sources of the two areas. Size distribution analysis is rather more difficult and much less reliable as applied to these inclusions. The micas, for example, typically occur both in detrital form and in larger particles. If the larger laths are randomly oriented in a thin section, the probability that they would all be cut so as to make a repeatable long-axis measurement possible is quite minuscule. The detrital micas are too small to be accurately measured using an optical microscope; moreover, if many of the smaller particles are less than 30 microns, it is pointless to even attempt a size distribution analysis using a 30 micron thin section because many of them would be hidden within the thickness and escape measurement. (The diagnostic value of the detrital micas however is not lost in the present research. Their influence can be measured through the birefringence of the clay matrix as seen in thin section. This is discussed in Chapter 6.) The larger mica particles, those greater than 30 microns, can be observed easily in thin section and a quantitative value placed on their degree of weathering. This factor can be of diagnostic value in differentiating clay sources.

Finally, feldspars are very subject to the influences of weathering. This mineral is constantly being eroded and broken down both during transport and *in situ*. The eventual residua of feldspars in fact are clay minerals. For example, the clay mineral kaolinite is formed from the degradation of K-feldspars through diagenetic processes, especially in marine environments (Millot 1970: 240; Chamley 1989: 29). The degree of weathering in feldspars can then often be used as a tool to differentiate sources.

The foregoing considerations were discussed to provide an appreciation of the choice of methodology used in the provenance analysis. A further aspect of archaeological thin section analysis involves research into the technological attributes of pottery-making. Some of these are:

1. determination of the volume of tempers utilised for certain fabric types;
2. determination of the preferred temper grain size;
3. observation of slips and surface finishes;

4. determination of firing temperatures; and
5. forming techniques.

Study of these attributes of pottery-making, and making comparisons of technology between sites or areas of origin can provide clues as to the nature and intensity of intersite or inter-regional interaction networks. As outlined in Section 4.2, socioeconomic interaction in prehistory can be at varying levels: exchange of stylistic information, exchange of technological and craft specialisation information (or innovations), or exchange at an economic level of actual products. Even if provenance analysis shows that actual trade may not have been occurring, study of the technological attributes from thin sections of pottery from sites of different regions can provide clues as to the nature of interaction. For instance, if standardisation is seen within technological parameters from site to site, it might be inferred that potters were exchanging technological information and innovations. This could show that the observed similarities in style and decoration were more than only indicators of cultural identity; that there existed networks for exchange of information and innovations as well. Standardisation (technological or compositional) in ceramics is often used as evidence for specialised craft production as being a key factor in the political economy of complex societies (e.g. Clark and Parry 1990: 290). Standardisation is a relative concept that can only be identified through comparison of two or more artefact assemblages with differing degrees of homogeneity (Costin 1991: 35; cited in Blackman et al 1993: 61). Variations in homogeneity, or the lack of them, from site to site, can be used to make some inferences regarding transfer of technology between regions.

Different types of data acquisition and presentation are required for the two purposes of provenance analysis and evidence of technological similarities. For provenance analysis, an examination of the size distribution and morphological characteristics of the inclusions is revealing as it can serve to characterise the source of these inclusions. On the other hand, in order to observe and analyse technological attributes, it is revealing to know, in addition to the characteristics of the distributions and morphologies of the inclusions, how these distributions relate, in terms of volume, to the clay matrix. That is, it is important to know the concentrations or densities of the various sizes of inclusions in the clay matrix; either as a percentage of the whole or in absolute terms of numbers per unit volume. As previously mentioned, the potter most probably subtracted or added inclusions according to volumes relative to the volume of paste. An accurate measure of these densities from site to site will reveal any standardisation present in the paste mixes.

There is much debate in the archaeological and sedimentological literature on the relative advantages and disadvantages of different methods of textural analysis, in particular the method of counting to be employed. The basics are that it is required to obtain a particle count that is representative of the particle distribution of the entire ceramic vessel. The simplest approach is to just measure the smallest and the largest grain in each thin section (Orton et al 1993: 141). Another simple method which is used often is to report just the average grain size or to estimate sizes and distributions using visual charts of known or prepared standards (e.g. Matthew et al. 1991). These approaches are generally more expedient and provide an estimation of grain size distributions, but they lack the precision and rigour of actually counting and measuring the grains, and examining their size distribution characteristics. The latter is the more accurate approach, as the data are amenable to statistical study and analysis. Importantly, it permits technologically important information such as bimodality in size distributions to be detected by examination of the data in graphical form (Freestone 1991: 406). It offers varying degrees of accuracy depending on the selection of sampling method. There are basically four sampling strategies: simple random, stratified random, systematic, or clustered (Shennan 1988: 314-320). If the population has an inherent patterning, a systematic sample will provide the best estimation of the actual population.

Point counting is a systematic sampling procedure by which thin-section constituents are estimated from a series of observations made at fixed intervals across the entire area of the thin section. Either a mechanically-driven stage which moves at pre-set intervals or an eyepiece graticule is normally used for the purpose. A statistically valid number of points is counted (see Van der Plas and Tobi 1965; Chayes 1965), each is classified according to its properties depending on the nature of the investigation, and usually barcharts, pie charts, or ternary diagrams are used for presentation of the relative concentrations of the various inclusions. Thus classification into fabric types is made possible and comparative groupings facilitated.

For the study of grain size distribution alone (i.e. irrespective of its density with respect to the clay matrix), various counting methods are in use. This commonly entails a choice of four methods of selecting points to count: point-, line-, ribbon-, or area-counting (Middleton et al. 1985: 66). Figure 4.1 illustrates the four methods. In point-counting, selection of inclusions is made systematically by measuring and counting only those which fall under the intersection of grid lines. A variation of the point-counting method was used by Streeten (1982), in the case where there was a low density of inclusions in the sherds. This was the 'nearest grain' method whereby a grain is measured for every grid-point, by taking the grain nearest the grid-point. This method effectively abandons the use of a standardised grid (Middleton et al. 1985: 71). It is not clear why the area count method was not used in this case, however, as it is much more applicable to the case of sherds with

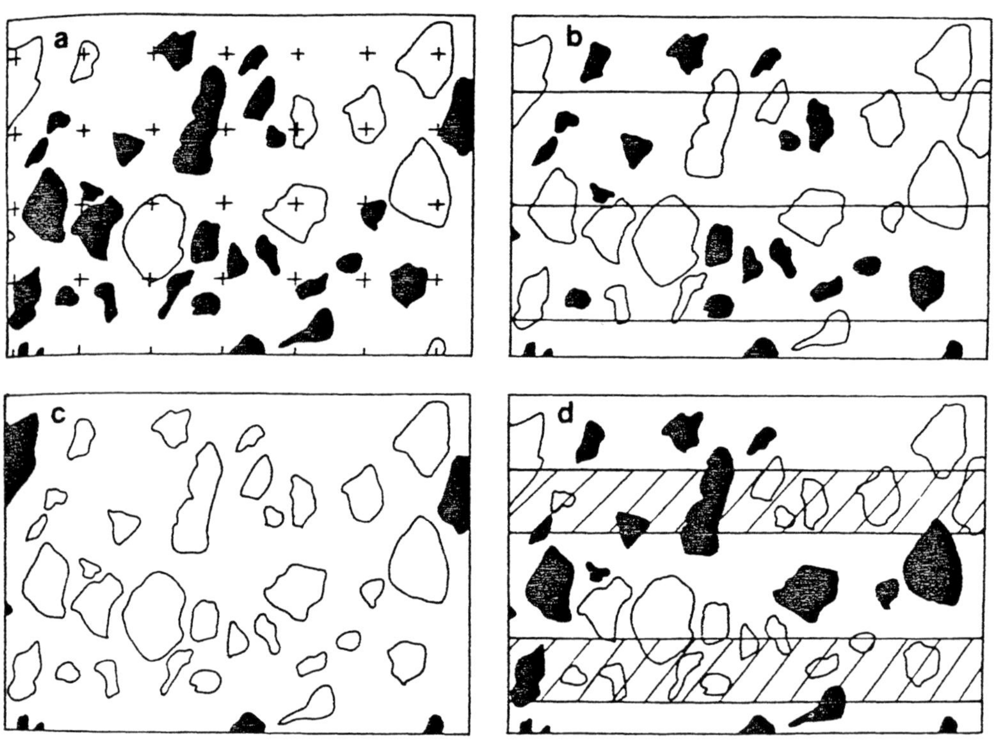

FIGURE 4.1

METHODS OF GRAIN SELECTION FOR MODAL ANALYSIS

a= point, b= line, c= area, d= ribbon. Black grains are those not counted. (Middleton et al. 1985)

low densities of inclusions. In line counting, all those which are contacted by the line are counted. In ribbon counting, all those which fall within a pre-determined band or ribbon are included; and in area counting, all inclusions falling within a defined area of the thin section are counted. It has been generally found that area and ribbon counting provide the most accurate estimation of the actual distribution (Middleton et al. 1985: 68, 71; Betts 1982). Some criticisms against point-counting are presented in the following paragraph. The use of area counting makes it possible to deduce the volumetric proportion of inclusions as well as determining their size distributions. This is done by counting the inclusions in a known area; then using the absolute number per square unit of area (usually millimetres) as a direct function of number per unit volume of paste. While this does not directly provide the number per unit volume, comparison of two or more fabrics on the basis of number per unit area can provide a direct comparison. It also relates more to the 'human factor' in ceramic textural studies, i.e. the potter adds a pre-determined amount of a particular size distribution of temper usually based on paste volume. Point-counting produces a similar concept; in that it is actually an area measurement, producing figures that represent what area of the slide is covered by particular categories. However, because of the superior accuracy of area-counting over point-counting (it reduces the bias against the smaller inclusions--see following paragraph), volumetric-related results derived from area counts/area are more indicative of the actual distribution. Moreover, for size distribution analysis area-counting allows the counting of a fewer number of points for the same accuracy because only inclusions are counted, and not matrix (see discussion in Section 6.4).

Shortcomings of the point- and line-counting methods were pointed out early in the sedimentological literature (e.g. Smithson 1939: 350), and Van der Plas advocated ribbon sampling because "the use of line and point counting methods in granulometric analysis leads to results in which the influence of the sampling method is so strong that the practical value of these techniques is doubtful." (1962: 145). The main reason for these criticisms of point- and line-counting methods is that (see Figure 4.1) the probability of a particle being counted decreases in indirect proportion to its size. Mathematically expressed, as particle size approaches zero, the probability of it not being counted approaches 1.0. One would expect therefore that the utility of the area-counting method relative to the others would increase as the density of smaller particles increases relative to the larger ones.

For percent counts with respect to the density within the clay matrix, a variation on point counting is best used; here, clay matrix is counted as a point. This method is well-described and championed by Stoltman (1989, 1991) who used it successfully to document cultural interaction in the Middle-Mississippian. The method is in common use among geologists for estimating rock composition in thin section. Other applications to archaeology have included Braun (1982), Hantman and Plog (1982), Plog (1980) and Tankersley and Meinhart (1982).

Slip and surface finishes can be observed and analysed in thin section. Often, one can measure thickness of applied slips for inter-vessel comparison. Or, the nature of inclusions in the slip can be determined.

Petrographic thin section data has been used by some researchers to estimate ranges of firing temperatures of ancient ceramics. Temperatures above about 1,000 degrees C cause defects such as exfoliation (separation along cleavage planes) within plagioclase feldspars and biotite (Kapur et al. 1992; Sakarya et al. 1990). Porat et al. (1991) report estimation of firing temperatures of Early Bronze Age Egyptian pottery by observing mineralogical changes which took place under firing. They observed decomposition of calcite and reaction of the products with the clay minerals between 750 and 850 degrees C which caused vitrification of the matrix (but see Feathers 1989 following). Also, oxidation of green hornblende to a deep red oxyhornblende occurred between 800 and 850 degrees. They reported that firing temperatures of 850 to 950 degrees can be indicated by completely decomposed limestone and a highly vitrified, milky matrix. Feathers (1989: 581) reports that rate and duration of firing are as important as maximum temperature in determining what physical changes occur during firing. For example, with a low rate of heating, calcite decomposition can begin at 700 degrees C; at a faster rate it may not begin until 800 degrees C. Changes in some amphiboles can sometimes be used as a firing temperature indicator, but they can occur as low as 600 degrees C (Middlemost 1985); too low to be a reliable indicator of terracotta temperatures. Rice (1987: 431) reports that green hornblendes become brown at temperatures of about 650 degrees. See also further discussion in Section 7.1.1.

Finally, clues as to forming techniques employed can be inferred from thin section observations (Rice 1987: 380). Often, interfaces between coils can be noted where the coil construction method has been used (Courty and Roux 1995). These researchers recently conducted some experimental archaeology investigations which they claimed can show that the observation of surface features alone cannot always be relied upon to infer that pots were wheel-thrown. Thin section examination of their experimental sherds compared with excavated material (including some Harappan material) showed that a combination of coiling and wheel-finishing could have been employed. Also, characteristic alignment of inclusions in the fabric observed in thin section can reveal directions of stress deformation and thus an indication of forming techniques. Further discussion is presented in Section 7.1.1 under 'Fabric', where observations made in the Rehman Dheri sherds are discussed.

CHAPTER FIVE: BACKGROUND AND DESCRIPTIONS

5.1 THE INDUS VALLEY AND ITS ENVIRONS

5.1.1 Geography

The region encompassing the extent of the Harappan civilisation and its progenitors is located in the South Asian continental area. The following geographic description is drawn from Ahmad (1969), except where noted otherwise. This area includes part of a system of fold mountains that stretch from the Mediterranean to the Pacific coast. The Himalayas and their offshoots are part of this system of mountains which dominate the continent of Asia and historically have provided a formidable barrier to communication between Central Asia and South and Southeast Asia. To the south of this complex of fold mountains lay another block of ancient, hard rock which once extended from Brazil to Australia, known as Gondwanaland. Although still disputed by some (e.g. Crawford 1979), it is generally accepted that this once formed part of what is now known as the Antarctic peninsula. Between this northern and southern block, there was an ancient shallow sea during the Palaeocene Epoch. The fold mountains which now dominate the centre of Asia emerged from this sea. The sandy material brought down by the rivers accumulated on the bottom in layers which, when compressed by the weight of sediment, became sandstone. In other parts of the shallow sea, the shells and bones of sea animals accumulated and limestone was formed. In this way a thick layer of sedimentary rock accumulated between the two continental blocks of ancient, hard rock. As the earth contracted upon cooling, these two blocks were drawn together and the sedimentary rock between was compressed and distorted to form the mountain folds. Over the millennia the exposed rocks have weathered and deposited the resulting material on the sunken plateau known as the Indo-Gangetic plain, which now has alluvial deposits over 2,000 metres thick. The area is still tectonically unstable and has experienced a number of earthquakes.

The present country of Pakistan spans these regions of fold mountains and recent deposits. About forty percent of the country consists of a plain which is drained by the river Indus and its tributaries, with mountains on the north and west and a plateau in the south west. The country can be divided into six major physical regions:

1. The Northern Mountains
2. The Western Bordering Mountains
3. The Salt Range and the Potwar Plateau
4. The Upper Indus Plain
5. The Lower Indus Plain
6. The Baluchistan Plateau

1. The Northern Mountains consist of the Himalayas and the Trans-Himalayan Ranges. The Himalayas are divided into four ranges: The Outermost, or sub-Himalayas consisting of low hills known also as the Siwaliks; the Outer, or Lesser Himalayas further to the north which includes the Pir Panjal with altitudes averaging about 5,000 metres; the Central, or Great Snowy Himalayas averaging 6,000 to 7,000 metres with some of the highest peaks in the world; and the Ladakh Range, also called the Inner Himalayas. The Trans-Himalayan Ranges lie to the north of the Himalayas and include the Karakorum in Northern Kashmir and Gilgit, and the Hindu Kush range in the Northwest of the country which runs into neighbouring Afghanistan.

The importance of these Northern ranges is that they have historically formed a great barrier to movement from Central Asia into the plains of present-day Pakistan. This physical isolation is broken by only a few passes. They also intercept the winds from the Arabian Sea and the Bay of Bengal to cause and define patterns of rainfall. This rainfall, combined with the melting snows from the ranges in the spring and summer, feeds the river systems of the area including the great system of the Indus.

2. The Western Bordering Mountains bound the Indus Plain to the west. Regionally the whole area falls into three main divisions (Spate et al. 1972: 480): the hills of the Northwest Frontier; the sub-montane Indus region, including the Vale of Peshawar and the Bannu Plain (Basin), the Potwar Plateau and the Salt Range; and Baluchistan. Politically, it lies within the Peshawar and Dera Ismail Khan Divisions (Northwest Frontier Province), and the Quetta and Kalat Divisions (Baluchistan Province).

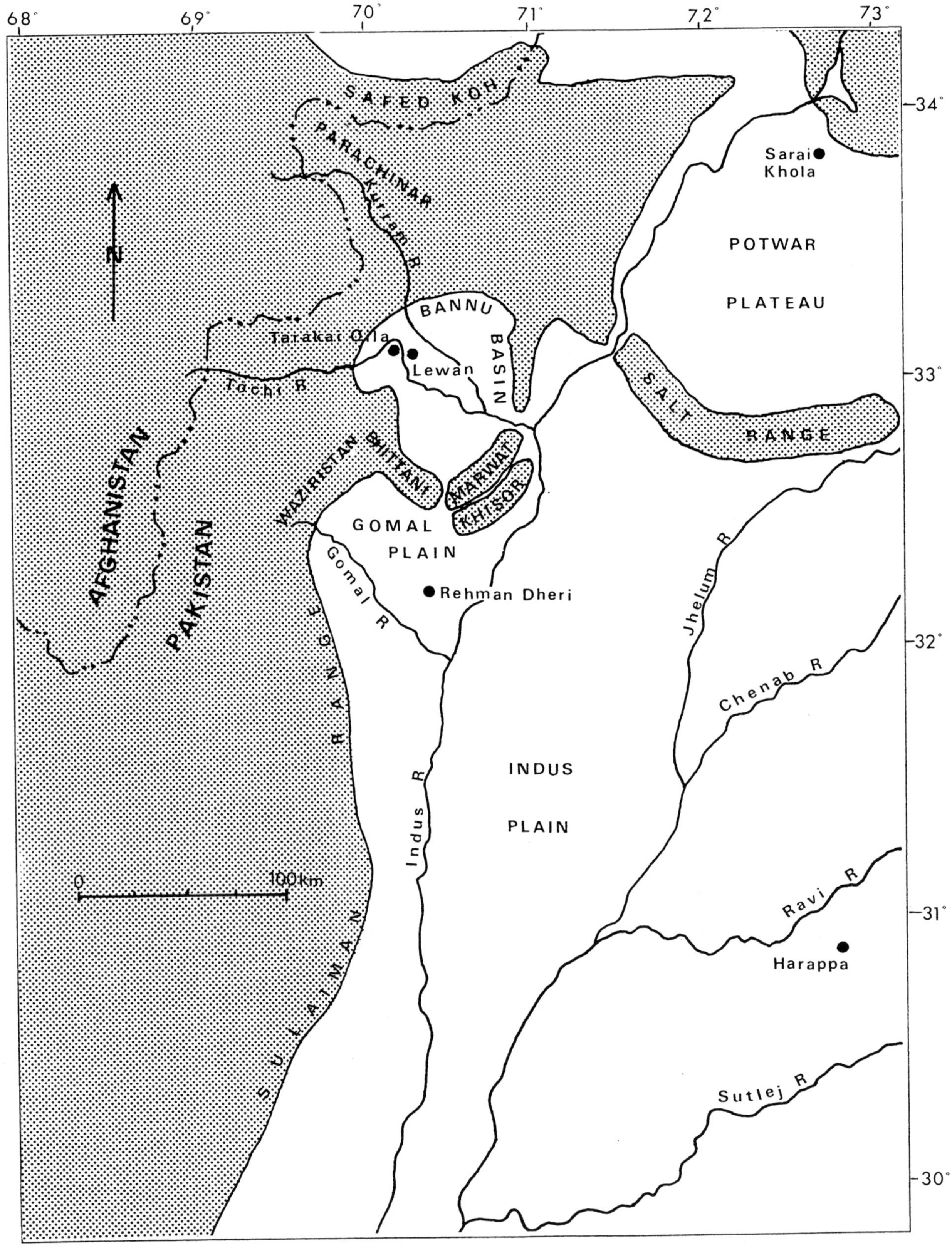

FIGURE 5.1
PHYSICAL MAP OF PAKISTAN STUDY AREA
Shaded areas indicate main mountain regions

Three minor ranges run south from the Hindu Kush to the Kabul river, and between these ranges flow the Swat, the Panjkora and the Chitral-Kunar rivers. South of the Kabul river lies the Khyber Pass which connects Peshawar with Kabul. Also south of the Kabul river is the Safed Koh Range which runs east and west, and has an average height of 4,000 metres. To the south of this range is the river Kurram which cuts through the Bannu Basin near the sites of Lewan and Tarakai Qila (see Figure 5.1). The Kurram Pass provides a route into Afghanistan. Between the Kurram and the Gomal rivers is the Waziristan hill area. The Gomal rises in Afghanistan and joins the Indus near Dera Ismail Khan, the closest large city to the major Early Harappan site of Rehman Dheri, which is of major interest to this research. The Gomal Pass in its valley has traditionally been used by the Powindahs, a nomadic tribe, for movement between present-day Afghanistan and Pakistan (Durrani 1988: 7). Joining Bannu, in the Bannu Basin, and Dera Ismail Khan, in the Gomal Plain, is the Pezu Gap, which nearly connects the Bhittanni and Marwat Ranges (Spate et al. 1972: 490). This gap probably provided communication between the two areas in prehistory (Kenoyer 1991: 30). South of the Gomal river, the Sulaiman mountains run southwards for about 500 kilometres to connect with the Bugti and Marri hills which run from the southeast to the northwest (Scholz 1983: 13). The Bolan river is the main one of this region along the valley of which lies the Bolan Pass. The western boundary of the Indus plain is completed by the low Kirthar hills which are drained in the south by the Hab and Lyari rivers flowing into the Arabian Sea near Karachi.

Unlike the Himalayas, most of these ranges lie outside the path of the monsoons (see Climate this chapter) and so the rainfall is low and they are almost bare of vegetation. The river beds are usually dry but occasional heavy rains in the mountains cause destructive torrents and floods.

3. The Salt Range east of the Indus starts in the Jogi Tilla and Bakrala ridges near Jhelum and runs southwest to the north of the river Jhelum then turns northwest to cross the Indus near Kalabagh. West of the Indus, it continues south into the districts of Bannu and Dera Ismail Khan. It consists of two parallel ridges with an elevated fertile plateau intersected by ravines and peaks in between. This area presents the most complete geological sequence in the world from the Cambrian to the Pleistocene.

North of the Salt Range lies the Potwar Plateau, varying from 300 to 600 m above sea level. A large part of it has been dissected and eroded by running water resulting in ridges and troughs, and basin plains. These exposures have provided a unique opportunity to observe extensive fluvial successions which has greatly assisted research into floodplain aggradation of the area (Willis and Behrensmeyer 1994: 63). Remnants of the ice age here consist of residual hills and hillocks of glacial debris.

4. The Upper Indus Plain is situated south of the Himalayas and is drained by the Indus and its northern tributaries: the Sutlej, Beas, Ravi, Chenab and Jhelum rivers. It is made up largely of fertile alluvium deposited by these rivers, the major part of which constitutes the rich, agricultural province of the Punjab. The past two centuries have seen the extensive construction of canals and irrigation systems to develop the region towards its full potential. The broad river valleys of the upper Indus Plain have been formed by the erosive action of the rivers which meander over a large area. The active flood plains of the rivers are typically inundated seasonally by summer floods. In many parts, embankments have been constructed to contain these floods. The deposits are composed of the youngest alluvium, predominantly coarse in texture and finely stratified. Away from the rivers and floodplains the land rises to the old flood plains which the rivers have now abandoned through their meandering (the meander flood plains). The old flood plains contain the oldest alluvium. The sub-montane plain area west of the Indus is known as the Derajat or the Sulaiman piedmont and is divided into the districts of Dera Ghazi Khan and Dera Ismail Khan.

North of Dera Ismail Khan, two areas of lowland reach into the western mountains. These are the plains of Bannu and Peshawar. The former, containing the sites of Lewan and Tarakai Qila, is an alluvial basin drained by the Kurram and Tochi rivers. The bordering mountains rise abruptly from the basin edge and completely isolate it from the Punjab plains. Its only easy access is by the Pezu gap between the Bhittani and Marwat Ranges. The basin shows the normal depositional sequence from coarse textured alluvial fans at the base of the mountains to moderately fine and fine deposits in the lowest part of the basin.

5. The Lower Indus Plains are mostly defined by the present province of Sind. In this area, the Indus is at its largest; in flood season, it is several kilometres wide. It is an

aggrading river, due to the heavy silt load it carries. Frequently changing course over the last millennia, it has formed the great Sind plain by this alluvial and meandering action. An excellent recent study of this action is presented by Flam (1993).

6. The Baluchistan Plateau, which is more than 700 metres above sea level, lies to the west of the Sulaiman and Kirthar Mountains. The Toba Kakar and Chagai ranges in the north separate it from Afghanistan. The coastal Makran range skirts the south.

The foregoing has been a general description of the main geographical features of the overall area, to provide an appreciation of the setting of the regions under the present study. Where required, more specific information pertinent to the local geographies of the site areas is presented in Section 5.3. below.

5.1.2. Climate

Present

The following description is drawn from Ahmad (1969) except where other authors are cited. The climate of Pakistan is, on the whole, arid and extreme or continental. The rainfall is generally low, with hot summers and quite cold winters. In the latter part of the summer season monsoon rains are general throughout the country. The principal rainy season lasts from July to September.

The country can be divided into four main temperature regions:

1. The northern and northwestern mountains which experience very cold winters with frequent frosts and heavy falls of snow, and have mild summers.
2. The upper Indus plains to the south of region (1) have a typically continental type of climate. Summers have high temperatures due to the low altitude and lack of cooling breezes from the sea. This area is very dry and dust-blown. June is the hottest month in this region, when temperatures seldom fall below 28 degrees Celsius, and highs are generally in the mid-40's.
3. The lower Indus plain near the sea does not experience such extremes of temperature, but the proximity of the sea produces high levels of humidity which makes the heat

oppressive in summer. Typical high temperatures in this region range from 35 degrees Celsius in the summer to the low 20's in the winter.

4. The Baluchistan plateau has conditions similar to the northern region but with higher summer and winter temperatures.

Pakistan is dominated by monsoon winds. Monsoons are seasonal winds blowing from sea to land in the summer and from land to sea in the winter. The summer monsoons enter Pakistan from the Arabian Sea in the south and as they blow into the hot dry land of the Indus valley to the north, cause rain on the Himalayan slopes and the western mountains. This summer monsoon is critical in the non-irrigated agricultural areas of the country, however it often causes disastrous flooding. In autumn the summer monsoon gradually retreats and is replaced by the winter monsoons blowing from the land. These are light, dry winds extending down the Indus plain.

The Indus area is generally arid with low rainfall. Levels of rainfall decrease from the Himalayas southwards, ranging from a mean annual of 80 cm at Sialkot at the foot of the Himalayas to 49 cm at Lahore to only 20 cm at Karachi. Half to three-quarters of the annual rainfall occurs during the summer monsoon. An exception to this is in the western part of the plain, which receives slightly more rain in the winter monsoon season, e.g. Peshawar, which receives an annual rainfall of 34 cm, receives about 18 cm during the winter monsoon and 16 cm during the summer monsoon.

Palaeoclimate

Early archaeological researchers of the Indus valley were under the impression that there had been a significant decrease in rainfall in the area since prehistoric times (Raikes and Dyson 1961: 265). This was based on the original observations of Sir Aurel Stein (1931: cited in Raikes and Dyson 1961) who came to this conclusion after observing certain archaeological evidence. The evidence was based primarily upon three factors: (1) the presence of *gabarbands* or stone-faced dams presumably to control surface water (see also Harvey et al. 1993) (2) the location and number of prehistoric mounds in contrast to present-day settlements, and (3) the depth of the mounds and sharply defined cultural areas, indicative of cultural stability and lack of movement. These observations also supported the interpretations of subsequent researchers (Marshall 1931: 2; Vats 1940: 4;

Piggott 1950: 134-135; Wheeler 1953: 5-6). This established view was then brought into question by Raikes and Dyson (1961), who closely examined and re-interpreted these lines of evidence. They found that in almost every case the evidence - hydrographical, zoological, botanical, archaeological, and architectural - failed to substantiate the hypothesis that climatic change of any major proportion has occurred. Coincidentally, Fairservis examined the same evidence independently and reached a similar conclusion (Fairservis 1961; cited in Misra 1984: 461).

Later studies summed up the geomorphic and geological evidence. Goudie (1973) based his evidence on sand dune accumulations which are indications of arid periods because of the requirement of low precipitation for their formation. His study of sand dunes in areas of higher present rainfall in Rajasthan purported to infer that arid conditions in the area came to an end between 9250 and 5000 BP. The latter date (Verstappen 1970) was somewhat tenuous, as it was based on existence of sand dunes; and not stabilisation of dunes which would be a better indicator.

Palynological studies have confirmed arid conditions existed before 10,000 BP. In 1971, Singh revived the theory of a more wet climate following this from his interpretation of palynological data from three salt lakes and one freshwater lake in Rajasthan (Singh 1971 and Singh et al. 1974, cited in Misra 1984: 461). Hashimi and Nair (1980) drew inferences based on differences in abundance of carbonate sediments and differences in size of quartz grains from the inner and outer continental shelf of western India, to show that peninsular India had a semi-arid climate in the late Pleistocene. Later research by the same authors (Hashimi and Nair 1986) concentrated on the distribution of feldspars in the same continental shelf deposits to corroborate their earlier work. Since feldspar is an allogenic mineral derived from the weathering of rocks on adjacent land areas, its presence in continental shelf deposits reflects the variation in weathering rate and thus is a sensitive climatic indicator. In a given basin of deposition, an increase in feldspar content would indicate climatic aridity. This work pointed to the prevalence of a semi-arid climate 11,000 years ago.

Overall, these studies have suggested that approximately 10,000 to 11,000 years ago, the climate of south Asia changed from arid to semi-arid. These conditions have changed little over that period and are much the same as those prevailing in the area today.

The foregoing discussion on the present climate and palaeoclimates has covered in general the overall Indus area. In the regions of the present study, specifically the Bannu Basin, the Gomal Plain, and the western Punjab, some microclimates deviate from the general descriptions. These are covered below in Section 5.3.

5.1.3 Geology

Generally speaking, the Indian subcontinent comprises two main physiographic or geological parts known as the Peninsular and the extra-Peninsular regions (Hunting Survey Corporation 1960: 1):

1. Peninsular India is a great triangular region of relatively low relief, which includes the Western and Eastern Ghats, the Deccan plateau, the Malwa plateau, and numerous low hill ranges. Its rocks are mainly Precambrian to Cretaceous but include Cenozoic strata in small areas around the periphery of the main massif. The Precambrian portion contains two main series of rocks, the older consisting of metamorphic and igneous rocks referred to as the Archean series. The younger, or Late Precambrian series is generally less metamorphosed and contains sedimentary and igneous rocks; it is known as the Purana series. The Palaeozoic rocks are represented by the Vindhyan and Kurnool systems, which are largely correlative, and by the lower part of the Gondwana system.

2. The extra-peninsular region is comprised of two sub-regions (Hunting Survey Corporation 1960: 2, Wadia 1966: 1). These are as follows:

(a) The mountainous region of the Himalayas which borders India to the west, north and east; and

(b) The great Indo-Gangetic plain of the Punjab and Bengal, which separates the folded mountains from the peninsular region.

The geology of the Himalayas falls into three broad stratigraphical zones or belts. They are (Wadia 1966: 10):

1. The Northern, or Tibetan Zone, lying behind the line of highest elevation. This zone is composed of a continuous series of highly fossiliferous marine sedimentary rocks, ranging in age from the earliest Palaeozoic to the Eocene age.

2. The Central, or Himalayan Zone, comprising most of the Lesser or Middle Himalayas together with the Great Himalaya. It is mostly composed of crystalline and metamorphic rocks - granites, gneisses and schists, with unfossiliferous sedimentary rocks of very ancient (Purana) age.

3. The Outer, or Sub-Himalayan Zone, corresponding to the Siwalak ranges, and composed almost entirely of Tertiary, and principally of Upper Tertiary, sedimentary river deposits.

The depositional history during the Palaeocene of the latter area consists of three depositional provinces of the Indus Basin to the east and two depositional provinces in the Baluchistan Basin to the west. The three provinces of the Indus Basin, which is of current interest, are called the Potwar-Kohat, the Sulaiman, and the Kirthar. Lithofacies contour maps of Palaeocene rocks reveal high clastic and high sandstone-shale ratios along the eastern side of the Indus Basin, generally parallel to the Precambrian Indian Shield. These high ratios suggest that the main influx of sediments into the Indus Basin was from the east (Meissner and Rahman 1973: 4). There evidently were areas of calm clear waters along the western side of the basin, where organic life thrived and limestone and calcareous sandstone and shale were deposited.

Two of the areas of prime geological interest in the present study are shown in Figure 5.1. These are the Bannu and Dera Ismail Khan quadrangles, between latitude 31 and 33 degrees N and longitude 70 and 71 degrees E. Detailed surveys of these quadrangles were conducted by the U.S. Geological Survey and the Geological Survey of Pakistan in the 1960's to examine exploitable hydrocarbon potential (Hemphill and Kidwai 1973). Immediately to the north of these lie the Parachinar quadrangle, from latitude 33 degrees to 34 degrees N. This area was part of the same survey, but contains less field observations (Meissner et al. 1975). This was due to the fact that approximately 80 percent of the quadrangle is within tribal areas where it is difficult or impossible to obtain permission to enter. As a result, the majority of it is based on aerial surveys and so it contains much less mineralogical detail.

These surveys have provided the only complete descriptions of the geology and have proven very useful for the present study. In the broader scheme, these studies are important because they are strategically located between the Salt Range-Potwar Plateau region where some of the earliest stratigraphic studies in Pakistan were made (Wynne 1878 and Waagen 1889, 1891; cited in Hemphill and Kidwai 1973: 4) and the Baluchistan-Sind region where comprehensive reconnaissance geologic and stratigraphic studies were carried out by Hunting Survey Corporation Ltd (1960).

The Khisor, Marwat, Bhittanni (these three forming the Trans-Indus range), and Sulaiman Ranges, as well as the highlands of Waziristan form a nearly continuous mountain system between these two regions. Uplift is still an active process in the Trans-Indus range (Khan 1987: 139). Rocks of late Tertiary age are nearly unbroken in their outcrop for more than 300 miles from the Potwar Plateau to the foothills of the southern Sulaiman Range and beyond. The Gomal Plain area, where the site of Rehman Dheri is situated, consists of piedmont deposits of coarse detrital material derived from the adjacent highlands (U.S. Geological Survey and Geological Survey of Pakistan 1964). If local clays were utilised by Early Harappan potters at the sites of Lewan and Tarakai Qila in the Bannu Basin and Rehman Dheri in the Gomal Plain, detrital material derived from adjacent mountains and highlands would be expected to be found in them. The sites are situated in areas of Quaternary sedimentary rocks covered by widespread unconsolidated surficial deposits. The whole area between the Safed Koh and the Tochi River largely consists of Eocene nummulitic limestones and Siwalik sandstones (Spate et al. 1972: 490). The Bannu Basin is a distinct geomorphologic unit that is almost completely surrounded by folded and faulted Plio-Pleistocene rocks of the Siwalik Series, and is bounded to the west by a north-south wrench fault (Rendell 1981: 219). The adjacent areas from which detritals could have been derived include, for the most part, the Bhittanni and Marwat Ranges separating the Bannu and Gomal areas, the Waziristan highlands to the west, and the Sulaiman Range to the west and extending south towards the Baluchistan plateau. From the northwest, the Kurram river drains the entire Parachinar area before entering the

Bannu Basin and much of the detrital material to be found in the clays there would be expected to have been derived from the formations in the drainage area. Some detrital materials found in the Gomal Plain could additionally have been derived from formations in the Khisor Range, which parallels the Marwat Range to the southeast.

The Bhittanni and Marwat Ranges consist of Upper Tertiary sedimentary rocks containing thick clastic deposits mostly of continental origin. On the southward side of the Marwat range and in the adjacent Khisore range, extensive silica sand deposits are found in the basal part of the Datta formation near the towns of Pezu and Paniala. Analysis of these has shown some iron oxide content but they are commercially exploitable (Alauddin et al. 1993; Hussain 1975). Some muscovite and epidote has been found there (Mallett 1887: 94, 97). The adjacent Khisor Range is composed of Palaeozoic, Mesozoic, and Upper Tertiary sedimentary rocks, underlain by the Salt Range Formation of Cambrian or Precambrian age. An interesting geological feature amidst the Plio-Pleistocene sequence exposed in the Bhittanni range is the Bain diamictite, originally termed the Bain Boulder-bed (Morris 1938). The outcrops and lithology of this unit are unique in the area, and contrast with the sandstone/siltstone of the underlying and overlying fluvial deposits (Khan et al. 1985: 53). It has remnants throughout the Bannu Basin. Common sedimentary rock fragments include quarzites, graywackes, shales and mudstones, gypsum and anhydrite, and limestone. Among the igneous rock fragments are granite, diorite, gabbro, dolerite, basalt and pumice fragments. Metamorphic rock fragments are gneiss, mica schist, serpentine, green and black slates and jasper. The lithology and sedimentology of this unusual formation are compatible with a volcanic lahar origin, probably the Dasht-i-Nawar caldera to the WNW in Afghanistan (*ibid.:* 61).

The Waziristan highlands are composed of Mesozoic sedimentary rocks, mostly of marine origin; Lower Tertiary sedimentary rocks, mostly of marine or deltaic origin; Upper Tertiary sedimentary rocks, mostly of continental origin; and Lower Tertiary and Upper Cretaceous intrusive rocks consisting of mafic and ultramafic rocks and metamorphosed sediments in central Waziristan. To the northwest, where the Bhittanni Range meets the Waziristan highlands, deposits consist of the Chaudwan Formation (limestone, quartzite and sandstone) and the Nagri and Chinji Formations (sandstone, claystone, siltstone and the Bain diamictite which includes smaller fragments of sedimentary, igneous and metamorphic rocks). The Sulaiman Range, bordering the west and running approximately north-south, is composed mainly of Upper and Lower Tertiary sedimentary rocks of both continental and marine or deltaic origin. Further to the west, not covered by the Hemphill and Kidwai study of 1973, rocks of the amphibolite facies achieve widespread distribution in eastern Afghanistan and northern Pakistan. In eastern Afghanistan, the amphibolite facies of the Nuristan Complex consist of biotite, garnet-biotite and garnet-biotite-sillimanite gneisses; garnet-biotite, staurolite-garnet-biotite and biotite schists, quartzites and quartzitic gneisses, marbles, amphibolites and pyroxene-amphibole gneisses (Pervil'ev 1982: 260).

The Parachinar area to the northwest consists of intensely folded and faulted sedimentary rocks which are partially metamorphosed and mineralised (Meissner et al. 1975: 3). The northern part of the region is underlain by Jurassic, Cretaceous, and Palaeocene limestone, sandstone, and shale. To the south, Eocene, Miocene, Pliocene rocks occupy most of the lower tracts. The Eocene rocks consist of limestone, clay, shale, sandstone, gypsum, and rock salt. The Miocene and Pliocene sequence consist of sandstone, shale, siltstone, conglomerate and cobble beds. The Jurassic system formations include beds of modular marl, calcareous shale, and quartzose, ferruginous, and calcareous sandstone. The limestones are fine to coarse crystalline and contain scattered, sandy, oolitic, and fossil-shell beds. The Cretaceous system quartzites include mafic igneous intrusive rocks, sandstones and shales contain belemnites and ammonites, and limestones contain thick beds of shale and marl and shale (*ibid.*: 4).

The Bannu Basin owes its origin to the folding and upwarping of Siwalik deposits during the Late Pleistocene. The anticlinal ridges thus formed effectively sealed the area off from both the Gomal Basin to the south and the Indus floodplain to the east. The Late Pleistocene and Holocene sedimentary history of the basin has been one of almost continuous deposition as the large (approximately 5,000 sq. km) structural depression gradually filled with sediments supplied by rivers and torrents that drained into it (Rendell 1986: 1). The depth of sediments is enormous, having been estimated to exceed 900m (Morris 1938). The total catchment area of the basin is about 15,000 sq. km and the basin sediments are therefore derived locally from the largely uncemented Siwaliks or from older rocks in the upper catchments of the Kurram, Tochi, Bharan, Khaisora, Shaktu and Kashu Algad rivers and torrents. Finer grained sediments have been subject to redistribution by wind action. Several different types of deposit may be recognised within the basin. Fragments of older terrace have been identified along the western edge of the basin (Hemphill and Kidwai 1973), while a sequence of cut terrace is present in the Pezu Gap. A series of low-angle boulder fans and gravel spreads mark the mouths of the Khaisora and Shaktu torrents, but much of the basin is covered by alluvial sands and silts and by wind-blown sands (Rendell 1986: 3).

In marked contrast to the Gomal area immediately to the south, the Bannu basin does not possess any alluvial fans. The rate of sedimentation within the basin would have had the effect of 'drowning' any alluvial fans, had they existed. The extent to which the Bannu basin has filled up with sediments when compared to the Gomal system is attested by the relative basin levels on either side of the Pezu Gap where the level of the Bannu basin is 104m higher than the Gomal area (Rendell 1986: 5).

The foregoing demonstrates the range and characteristics of minerals and rocks one would expect to find as inclusions of detrital materials in clays of the Bannu and Gomal regions. Additionally, it points out that difficulties could be expected in attempting to differentiate clay sources from the two regions based on their respective mineral suites, because minerals from the two areas would be expected to be similar. To a degree, this was proven to be the case in the present study (see Chapter 7).

Similar difficulties would not be expected in discriminating between the clays from Harappa and those from the Bannu and Gomal areas. The Indo-Gangetic Plain is comprised partly of the Indus Plain. Except for the hills of Khairpur, the Indus Plain is devoid of major relief and is essentially a broad, shallow basin bordered on the east by the ranges of Baluchistan and Western Sind (Hunting Survey Corporation 1960: 3) The central part is built up with floodplain material of the Indus River, its surface marked by meander scars and old river channels. The western part of the basin is a piedmont plain that slopes gently up to the mountains and is surfaced with outwash material derived from them as described above. The Indus River derives most of its water from the Himalaya and Hindu Kush ranges of northern Pakistan, India, and Afghanistan. Most streams are comparatively small and are intermittent owing to the low annual precipitation.

The Harappa area consists mainly of stream and braided-stream deposits of sediments from the uplift of the Himalayas (U.S. Geological Survey and Geological Survey of Pakistan 1964). The adjacent Ravi River, one of the main tributaries of the Indus, is a meandering river of low velocity which carries suspended silt and clay through gently sloping floodplains (Pendall and Amundson 1990a: 304). The lower reaches of the Ravi near Harappa, where the river joins with the Chenab, Pleistocene-age alluvium is overlain by more recent alluvium of Holocene age. Reworked sediments from glacial loess at the base of mountains 200 km northeast of Harappa are thought to be the source of the fine-grained alluvium of the Holocene (Mian and Syal 1986; cited in Pendall and Amundson 1990a: 302). The older alluvium contains prominent concentrations of calcium carbonate (Amundson and Pendall 1991: 14). However, subsequent alluvial activity has created less distinct Holocene deposits and terraces which lie between the Pleistocene terraces and the currently active floodplains. They exhibit evidence of several depositional cycles, and pedogenic carbonate is Stage 1 or absent (Sehgal and Stoops 1972; cited in Pendall and Amundson 1990a: 304). Mineralogy of all three alluvia is dominated by quartz, feldspars, mica, and carbonate (Pendall and Amundson 1990b).

5.2 THE HARAPPAN CIVILISATION AND ITS ANTECEDENTS

The Harappan Civilisation was known in the western world as early as the mid-1800's, but its significance appears not to have become evident until the early 1920's when the discovery and initial excavations of Mohenjo-Daro were first published (Marshall 1924). Since then, a considerable number of researchers have undertaken excavations and study and our understanding of the nature of the civilisation has been greatly enhanced. What follows in the subsequent sections is a three-phase outline summary of the present state of knowledge about it. Much of the basic information has been gleaned from two works which are generally considered authoritative compendia on the Harappan Civilisation: Allchin and Allchin (1982) and Shaffer (1992). Shaffer's article, although published in 1992, contains only material up to 1985. Where other researchers and sources are used, they are cited.

Shaffer uses the nomenclature and categorisation system popular in North America of dividing cultures and their sequences into cultural *traditions* which are further subdivided into *eras* and *phases.* The system was defined by Willey and Phillips (1958: 37). The central concept of the tradition facilitates a stylistic grouping of diverse archaeological assemblages into a single analytical unit, while limiting the need for establishing the precise nature of cultural and chronological relationships that link assemblages. The concept does however imply that such relationships exist. Shaffer defines three cultural traditions in the Harappan: those of Baluchistan, Helmand, and the Indus Valley. Because it is the region of interest of the present research, this thesis concentrates on the Indus Valley Tradition. Traditions are divided into eras which group archaeological units, or phases, that share a few, very general cultural characteristics which do not have fixed boundaries in time or space. Finally, eras may have one or more phases. A phase represents "an archaeological unit possessing traits sufficiently characteristic to distinguish it from all other units similarly conceived..." (Willey and Phillips 1958: 22). At present, a diagnostic ceramic style located at one or more sites is the major characteristic of a phase (Shaffer and Thapar 1992: 247). A final concept used by Shaffer, which is particularly important to the present study, is that of the *interaction system* (Binford 1965). Interaction systems refer to various avenues of communication existing within and between social groups which may crosscut both traditions and phases. This communication is reflected in a widespread distribution of similar cultural traits and artefacts, intrusive and indigenous within a relatively brief time period.

The presently known extent of the Early and Mature Harappan Civilisation is shown in Figure 5.2. The civilisation at its peak encompassed over one million square kilometres of the northwestern region of the Indian subcontinent, and was mostly centred around the Indus River system drainage together with coastal areas bordering the Indian Ocean.

5.2.1 Before 3500 B.C.

The Indus Valley Tradition dates from pre-6000 BC. The Early Food-Producing Era is mainly defined by the Mehrgarh Phase. Period IA at the type-site of Mehrgarh is dated by the excavators to the beginning of the seventh millennium (Jarrige 1986) and was aceramic. The first occupation contains fireplaces and evidence for both domesticated and wild barley along with domesticated wheat. The animal economy is initially dominated by about twelve species of wild 'big game' (Meadow 1984: 35), indicating that the early population exploited the Kachi Plain and hills surrounding it. By the end of the aceramic period, the faunal assemblage consisted almost entirely of domesticated sheep, goat and cattle (Meadow 1984; 1989). Except for the local exploitation of riverine and marine resources, the importance of hunting and gathering therefore declined through time. Sedentary settlements consisted of simple, multi-roomed buildings with space between them which was used for domestic purposes and burials (Lechavallier and Quivron 1981: 72-85). Some differential social status is indicated by funerary items found with the burials (Samzun and Sellier 1985), and the existence of long distance trade networks or interaction with other groups is suggested by the presence of marine shells, lapis lazuli and turquoise, items which are not available locally. Radiocarbon dates from the aceramic period range between 9385 ± 120 bp and 5530 ± 180 bp; and average to a calibrated date of 5225 BC. These dates confidently place the beginning of the food-producing economy within the seventh millennium, and offers a very strong case for an indigenous process of domestication within the greater Indian subcontinent (Possehl 1990: 265). Continuity and change are marked at the site with the introduction of soft, buff, chaff-tempered pottery in the following period, ca 5500 BC. Many essential traits of the Indus Valley Tradition become established during this phase. The most important of these were a food-producing economy, sedentary villages with mud-brick architecture, and the development of crafts. Moreover, the use of imported materials may show limited interaction with other social groups.

5.2.2 The Early Harappan Period

The Early Harappan Period was initially defined by Mughal (1970; 1990) in order to recognise the existence of continuity between what had been termed the pre-Harappan, and the Harappan. His earlier work (1970) had shown that most of the characteristics that define the Harappan period, now called the Mature Harappan, were already largely evident during the period 3500 to 2500 BC; just prior to the urbanisation period. Other prominent

Harappan researchers, most notably Possehl (1990), use the terms Pre-urban Phase and Urban Phase to describe Mughal's Early and Mature Harappan periods. Both sets of

FIGURE 5.2

EXTENT OF THE EARLY AND MATURE HARAPPAN CIVILISATION

(Adapted from Mughal 1990: 188-9)

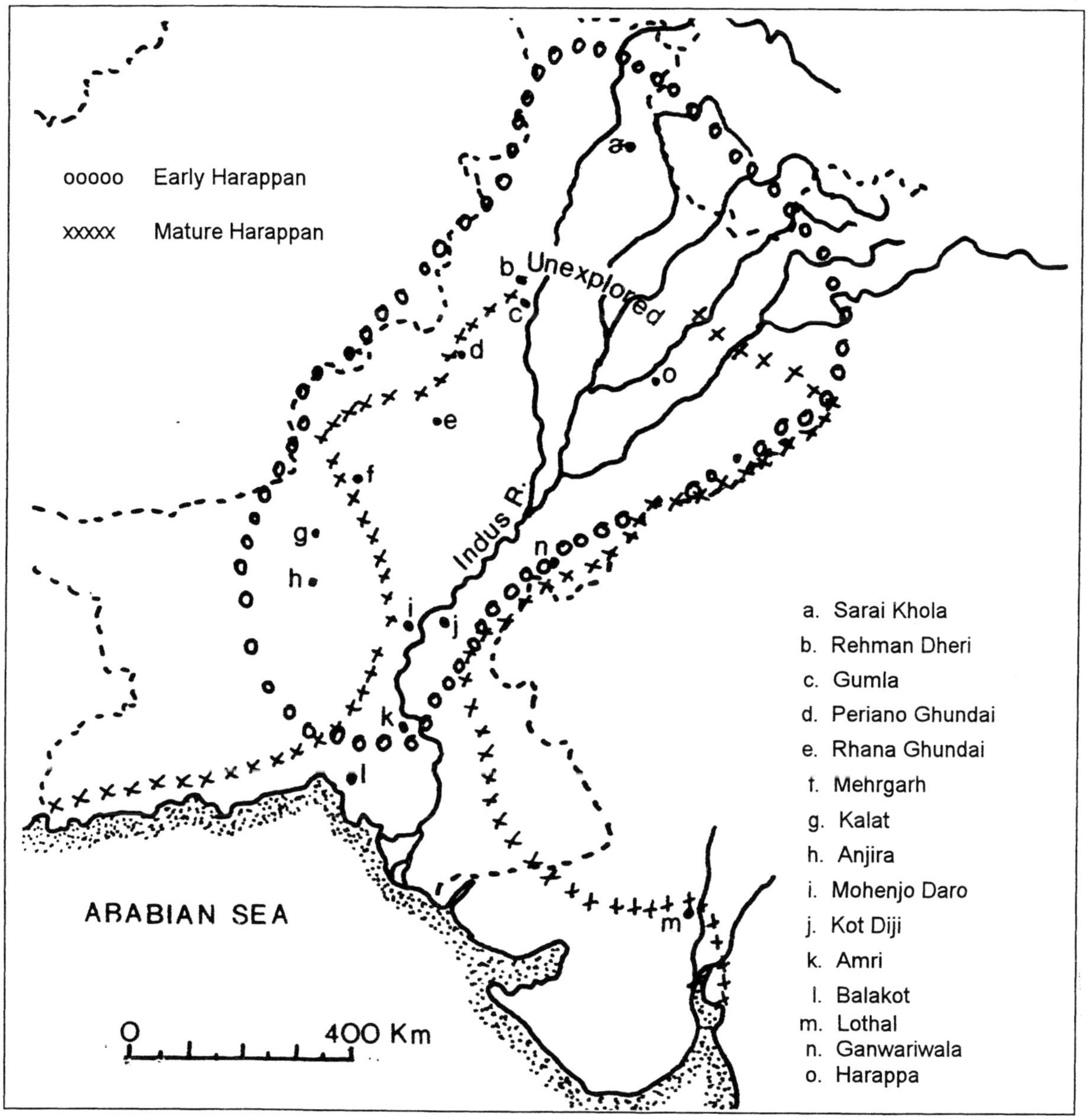

nomenclature tend to emphasise that there was a continuity of sorts throughout; Mughal's describes a smoother transition, but Possehl's depicts an abrupt change to urbanisation which occurred at the end of the Early Harappan. The present thesis will use Mughal's terminology, which is in more common usage.

Shaffer describes the period 4000 BC to about 2350 BC by four phases (he defines a gap in sites between 5000 and 4000 BC): the Balakot, Amri, Hakra and Kot Diji (1992: 426). The four overlap somewhat as they are defined as gradual changes, and they occur in different geographical areas. The Harappan (or Mature Harappan) commences at 2650 BC in most of the Indus area, overlapping the end of the Kot Diji Phase by 300 years in Shaffer's interpretation. Defining each of the phases and how they differ is not within the scope of this chapter; nor is it necessary for an understanding of the research and its interpretation. The salient features of the Early Harappan period are described; and then the last of Shaffer's phases, the Kot Dijian, is covered in more detail because it constitutes the phase and chronological period of the material under study from Rehman Dheri, Lewan, Tarakai Qila and Harappa.

The Early Harappan period demonstrates continuity from the beginnings of the village farming settlements described from the early levels at Mehrgarh, to the Mature Harappan. Some of this continuity is structured by the environment and ecology. The patterns of food production, including the plants and animals utilised, as well as some aspects of architecture remain relatively little changed over the period. Other characteristics, not dependent on environment or ecology, such as worked stone, terracottas and some shell- and bead-making, show continuity also (Mughal 1990: 185).

The Early Harappan period, on the other hand, also saw the emergence of several of the features of the urban Mature Harappan. There appear at present to be two schools of thought on this; those of the two prominent researchers noted above, i.e. Mughal and Possehl. Their positions are illustrated by their terminologies as noted above. On the one hand, Mughal asserts the 'continuity position' where most of the features of urbanisation were present in the initial stages of the Early Harappan period, while Possehl takes the 'discontinuity' view that some of them did not become evident until the very end of the Early Harappan period, if at all. As the features that are differently viewed by the two are somewhat central to the present research, the background to them is discussed below.

Following the initial excavations of the major Harappan sites of Mohenjo-Daro (Marshall 1931, Mackay 1937) and Harappa (Vats 1940, Wheeler 1947), it was hypothesised that the Indus Civilisation had diffused from elsewhere, perhaps Mesopotamia, in its fully-developed form. In fact, the objective of much of the early research was to trace where it diffused from (e.g. Stein 1931: 184) and influenced the choice of many of the site areas that were excavated. This research orientation persisted until well into the 1950's and 1960's.

Beginning in the 1950's, interest was beginning to be generated in searching for evidence of an indigenous start to the Harappan civilisation. Much of the research was conducted in the western borderlands with Afghanistan and Iran. Fairservis (1956, 1959, 1961) conducted several surveys and limited excavations in Quetta, Pishin, Chagai, northern Kalat, Zhob and Loralai. Beatrice de Cardi conducted excavations in the Surab and Kalat regions (1965, 1983), and Raikes (1968) explored sites in the Kalat district. Dales (1962) explored the Makran coast to the south, and excavated at Balakot (1979). Further excavations were carried out to the western borderlands in the Quetta area where significant evidence of indigenous development were unearthed (Jarrige 1979, 1981, 1989).

The new evidence from these surveys and excavations drastically changed perceptions of cultural developments leading up to the Mature Harappan. A number of new and important syntheses emerged in the mid to late-1960's which expounded these. Dales' (1965) was one of the earliest. Dales outlined six phases of development, each highlighting social, political and economic developments in Afghanistan and Baluchistan. He suggested that these developments provided the antecedents to the rise of the Indus civilisation. Rafique Mughal (1970, 1990) reviewed all of the excavation and survey data of the early occupations to date, in the Greater Indus Valley and northern Baluchistan, which demonstrated a very high degree of continuity between developments during the period 3,000 to 2,600 BC, and the nature of the fully developed civilisation after 2,600 BC. He concluded that these indigenous developments were the roots of the Indus civilisation. He popularised the use of the term Early Harappan to describe the period ca 3,500 BC. to 2,500 BC. in the Greater Indus Valley. This new terminology reflected the trend in viewing the civilisation as a continuous evolution of cultural characteristics gradually leading up to the Mature Harappan.

Our knowledge of the Early Harappan culture of the Greater Indus Valley has improved vastly since Mughal's pioneering work in 1970. Much further research has since been conducted (Mughal 1990 provides an excellent summary). Connections have been demonstrated for example in the western borderlands with the Iranian Plateau (Wright 1984). However, the cultural development of the civilisation still contains numerous puzzles with respect to the nature of its rise in the late 4th and early 3rd millennium BC., which seem to challenge the 'uniformitarian' model of cultural change (Possehl 1990). It contains identified continuities and discontinuities with respect to its development towards a fully urbanised civilisation. Continuities that link the Pre-Urban and Urban phases (as defined by Possehl 1990) are evidenced in items relating to subsistence and expressions having to do with style. Discontinuities between the two relate to those features of a sociocultural system that make it distinctively urban in character such as social stratification and craft specialisation (*ibid.*: 269-70).

Shaffer (1992) notes that during the Early Harappan period, there was a distinct homogeneity of style, especially in the ceramic assemblages, over the entire distribution area. This was manifested in mass produced ceramics in standardised shapes, which appeared to be distributed in wider interaction spheres. For example, Rehman Dheri appears to have served as a regional production and distribution centre (Durrani 1988). Similarly, evidence from Mehrgarh III (ca 4,000 to 3,400 BC) showed that ceramics began to be mass produced in standardised shapes and may have been distributed in much wider interaction spheres (Jarrige 1979). These and most other instances of evidence of interaction are based on style considerations alone, as very little direct evidence of product distribution exists in the literature. An exception is Wright (1984) who conducted a chemical anlaysis (neutron activation) and petrographic study of two classes of ceramics from Iran and the Indo-Iranian borderlands to the south of the present study area. Her conclusions (mostly based upon the chemical analysis) suggested that distributional inferences based on stylistic grounds alone had been misleading; that actual product distribution had a much lesser extent. Another example, but from the Mature Harappan period, comes from Blackman and Vidale (1992), who carried out a neutron activation analysis of stoneware bangles from the sites of Harappa and Mohenjo-Daro. Their study provided direct evidence of intersite movement of manufactured stoneware bangles between Harappa and Mohenjo-Daro during the Mature Harappan Period. A point to be made here is that it is necessary to demonstrate the nature of exchange or movement of goods or information in order to show the extent of economic interaction between regions or sites. The aforementioned homogeneity in style present in the Early Harappan assemblages would have required extensive communication and interaction networks. However, one can say relatively little about the nature and extent of these networks unless details of any actual product or information exchange can be known with a reasonable degree of certainty.

Ceramic and artefactual links provide further evidence of the nature of interaction networks during the Early Harappan period. In the area of interest to the present study, the Indus Sub-Montane Tract is defined by Lahiri (1992: 30) for interaction and trade purposes as including parts of the Salt Range and, west of the Indus, the large sections of hills such as the Kohat enclosing the Bannu lowland as well as the Waziristan hills to the west of Derajat. The Derajat, which includes the Gomal Plain, is an integral part of the main Indus valley (see also Section 5.1 above for geographic and geological descriptions of these areas). The Bannu Basin shared ceramic and artefactual links with the Derajat region, the Kot Dijian horizons of upper Sind and the Kaachi plain. A frieze of four fishes showing both eyes, as if fossilised, on a bowl from Lewan exhibits a striking resemblance to one found at Rehman Dheri (Durrani 1988: 202; Knox 1986: 215-6). Also, as discussed in Section 5.4 on Early Harappan ceramics below, other strong links are seen in the ceramics of Lewan, Tarakai Qila, and Rehman Dheri. Broad similarities in morphology, surface treatments and decorations are seen.

The bead-making industries evidenced at Rehman Dheri (Durrani et al. 1994) showed intensive activity and technologies which were seen to be very similar to those at Lewan (Allchin and Rendell 1986: 81-7). The microblades, microdrills, and other tools used in bead manufacture showed that technologies appear to have been shared between the sites. Moreover, many of the raw materials used, such as agate, carnelian, lapis lazuli, quartzite, and steatite are common at both sites. Less extensive excavations conducted at Tarakai Qila, 14.5 km from Lewan, revealed evidence of a bead-making industry also (Allchin and

Knox 1981: 249). Finally, the terracotta human figurines found at Rehman Dheri "...exhibit numerous parallels with the figurines found in the nearby site of Gumla, as well as with figurines from Lewan in the Bannu basin. Some of the types may also be found in the Panjab, in the lower levels of Sarai Khola, Jalilpur, and Harappa." (Flavin et al. 1994: 83).

Much of the present research in the Indus Valley is focusing upon examination of the discontinuities as defined by Possehl (1990). At Nausharo, the French Archaeological Mission is including in its analyses technical and provenance studies of the ceramics of the Early Harappan period to determine the nature and intensity of the apparent interactions in the Kachi Plain area currently shown by stylistic similarities only (S. Mery, pers. comm.). The present work of the Harappa Archaeological Project is examining the transition in detail between levels 1 and 2, which is the chronological period relating to the Early/Mature Harappan transition (J. Kenoyer, pers. comm.).

5.3 SITE DESCRIPTIONS

5.3.1. Rehman Dheri

The following description of the site of Rehman Dheri has been gleaned from the first excavation report (Durrani 1988) except where other sources are cited. Durrani describes the initial excavation work which was undertaken in the seasons 1976 to 1980. The second published excavation was conducted in 1991, to continue the work at the site and surrounding areas (Durrani et al. 1991). At least two other excavations have been conducted at the site, by the University of Peshawar, but these have not been published.

A recent photograph of the site is shown in Figure 5.3. Rehman Dheri was one of a number of archaeological sites discovered in 1971 by Dr. A.H. Dani, who was then conducting the excavations at nearby Gumla (Dani 1971). The site takes its name from the village of Rehman, 5 km. away. The site was originally noted for its roughly rectangular shape incorporating what appeared to be well-designed streets and city planning (see Durrani 1988: Plate VA). Also, its size was impressive, at about 550 m. by 370 m. It is situated about twenty km. from the present banks of the Indus River. Its present location is about 3.5 km. from the main metalled road between the town of Dera Ismail Khan and, to the north, Pezu, very near to the old banks of the Indus River as shown by the present western limit of terrace deposits (Hemphill and Kidwai 1973: Plate 1). It is noted also that this limit of terrace deposits skirts the southern tip of the Khisore Range to the north. Detritals from these deposits could have found their way into the clay deposits utilised by the Early Harappan potters of Rehman Dheri.

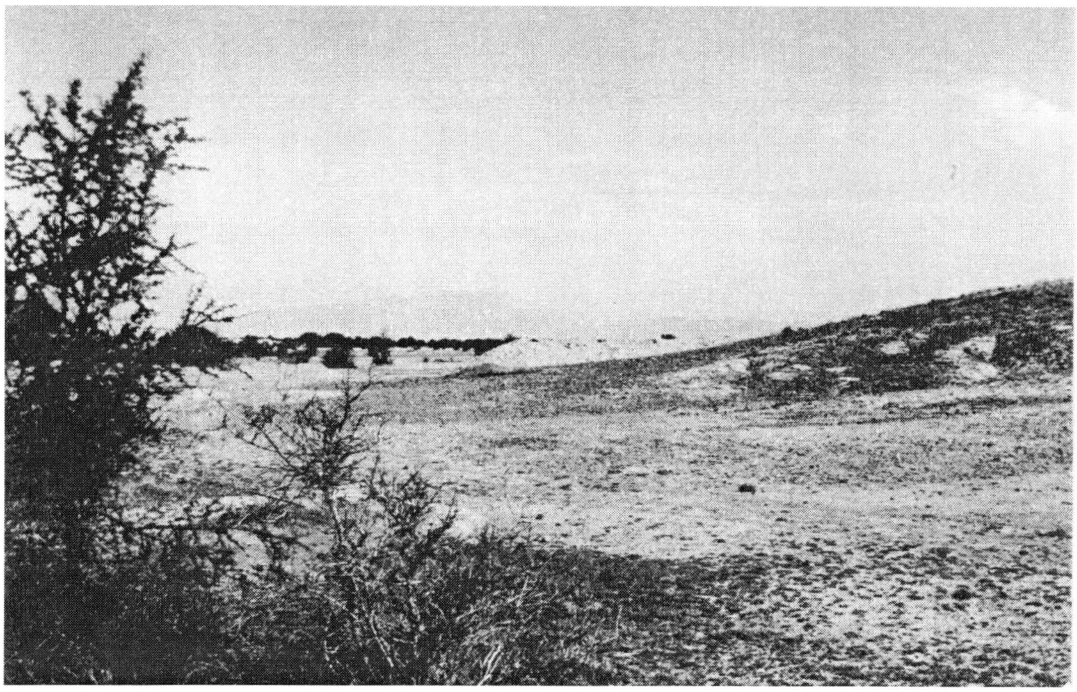

FIGURE 5.3
SITE OF REHMAN DHERI

The site is located in the lower basin formed by the piedmont flood plain which slopes gradually from the Sulaiman Range towards the Indus River. The plain around the site is of undifferentiated surficial deposits, with boulder fans further to the west approaching the foothills of the Sulaiman. The plain has a sub-tropical, continental arid climate. Owing to an annual rainfall of less than 10 inches and a high rate of evaporation, vegetation is sparse and restricted to stands of Lana trees and saline bushes. This plain is exposed to much sheet flooding and is irrigated with *barani daggar* methods. Used during periods of heavy rain, this method consists of embanking the three sides of a field downstream of the sheet flooding which arrests the water flow and concentrates it in the field. It is then allowed to permeate the field which is then ploughed and sown. More recently, the introduction of tube wells has made the area more amenable to cropping.

Rehman Dheri is extremely rich in ceramic waste on the surface. Although not excavated, extensive kiln works are evident on the west side of the site, where the prevailing winds impinge on the edge of the mound. Kiln wall material can be observed eroding through the surface. A dish, thought to be for holding paint, and a terracotta stopper have been found there (Durrani et al. 1991: 67, 78). Additionally, numerous wasters are seen in the immediate area of the kilns. Some of these were collected by the author for thin-sectioning to observe inclusions present in samples that may be assumed to be of local clays. Bones, ash, and charcoal can be easily collected on the surface, as can beads, lapis lazuli, bangles, terracotta cart frames, painted and unpainted pottery and bone tools. The layout is extremely interesting. It appears from the aerial photograph, earlier referred to, to have been organised in a rectangular fashion, with a northwest to southeast orientation. What appears to be a main street, oriented approximately NW-SE, cuts the settlement in two approximate halves, and side lanes meet the main street from either side, somewhat reminiscent of the major Mature Harappan site of Mohenjo-Daro. Aerial photographs also show a possible enclosure wall around the whole settlement with possible towers or bastions at regular intervals. Although it has not been excavated, the wall is also evidenced by regular, straight surface discolorations particularly visible in the early morning. Some degree of occupation appears to have been present to the south, outside the wall. The site appears to have been a major centre during its height in the Early and Mature Harappan, with extensive craft manufacturing.

The site plan of Rehman Dheri is shown at Figure 5.4. A different plan was used for subsequent excavations (Durrani et al. 1991), but the original is shown here because it indicates the areas from which excavated material was sourced for the present research. It is seen that four main excavational units were explored: A, B, C, and D. Units A, B, and C were located at the high points of the mound and excavated to virgin soil, and Unit D on the Main Street. Unit D was not completed. It was intended to explore the relationship of house units on either side of the street to the main street.

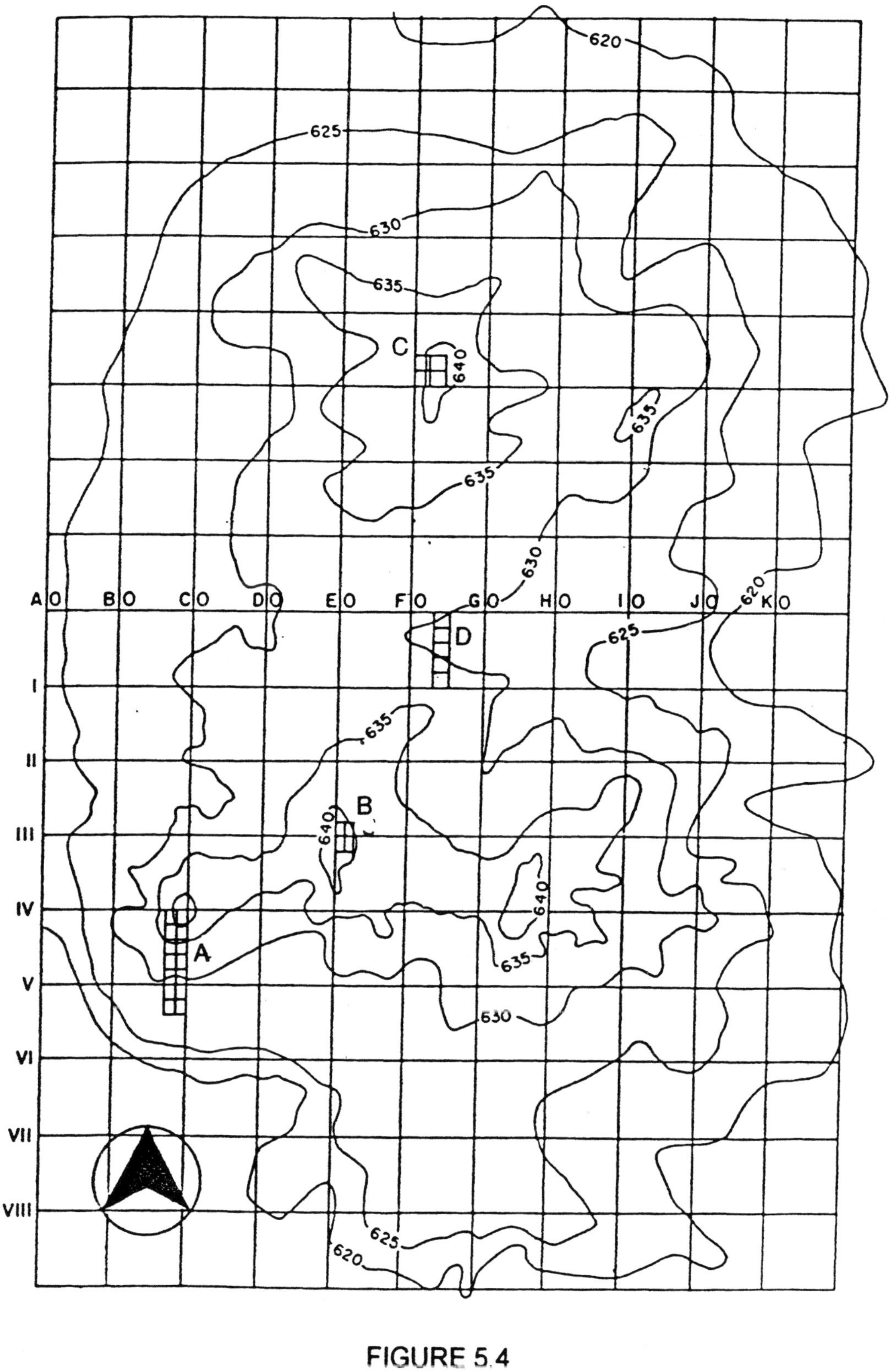

FIGURE 5.4

SITE PLAN OF REHMAN DHERI

(After Durrani 1988: 133)

0 100 200 300

FEET

The initial excavations at the site enabled a sequence of three main periods to be discerned:

Period RHD I:

Radiocarbon dated to ca 3340 to 2850 B.C., this was identified as the earliest period of occupation at the site. Durrani (1988) divided this period into two sub-periods, RHD IA (3340 to 3100 B.C.) and RHD IB (3100 to 2850 B.C.). For purposes of the present research, these are condensed into a single discussion. Architectural features for the whole period included structures of packed mud, brick-shaped clay slabs and clay blocks. Evidence of several rooms was identified but the limited extent of horizontal excavations did not permit definition of the number and nature of rooms in each house unit. Walls appeared to run perpendicular to the city wall. A circular grain silo, with a diameter of 135 cm, was located in one room. Other architectural features include circular ovens about 30 cm in diameter, open on one side. These were in addition to a larger, rectangular, fireplace in the rooms. Ceramic finds are covered in the next section. Plant remains included evidence of charred grain, predominantly wheat and barley, from the grain silos. Faunal remains were buffalo, cattle, sheep, goat, and fish bones. A significant small find, from Period IB, was a unique ivory seal measuring 3 x 2.7 x .02 cm. (*sic*, Durrani 1988*:* 28) displaying extraordinary craftsmanship. It depicted two mountain goats with wavy horns, and two scorpions flanked by a frog on the other side. It is significant in that it shows some continuity with Mature Harappan ideas and designs.

Period RHD II

This period was dated in the range 2850 to 2500 B.C. Architectural features representing a period of platform building and levelling which sealed all lower layers defined the break with Period I. Construction methods show continuity with the earlier period. Features representing domestic activities continued to be rectangular hearths and circular ovens. However, one notable change is in the grain storage techniques, which changed to the use of large storage jars half buried into the habitation floors. Ceramics of the period are treated in Section 5.4 below. Plant remains were unchanged, and faunal remains consisted of cattle, sheep, goats, fish, and birds.

Period RHD III

Also divided into two sub-periods, this period is dated 2500 B.C. to 1900 B.C., contemporary with the Mature Harappan Period. Ceramic styles continued relatively unchanged at Rehman Dheri, and the site appeared not to have developed all of the stylistic changes seen at other sites of the Mature Harappan. An analysis of this is given by Chaolong (1990). It represents a third major phase and involved platform building and levelling which sealed all previous layers. It had no significant structural remains due to its proximity to the surface and related erosion. Some rectangular rooms with small cubicles were identified. Grain silos were not present but large storage jars for grain and water storage continued. Rectangular hearths, open on one side, continued as did the circular ovens but rectangular fireplaces with a central pit as in RHD I, did not occur. Twin hearths were first seen as a major feature of Period IIIB. Walls continued to run parallel and perpendicular to the city wall as in earlier phases. A major feature of this period is the provision of narrow alleys between different house units. Again, ceramics are covered in a subsequent section. Plant and faunal remains continued the same as earlier periods, with the exception that a variety of wild grasses and agricultural weeds were identified.

With respect to external trade and transportation routes, the site of Rehman Dheri lies on a probable ancient trade route running approximately north-south between Mohenjo-Daro and the far northern site of Shortugai, during the Mature Harappan and possibly the Early Harappan periods (Kenoyer 1991: 30). That it was likely a trade route in the Early Harappan is reinforced by the artefactual commonalities and industrial/technological similarities seen between sites of the Bannu Basin, Rehman Dheri, and Harappa as discussed earlier in the chapter.

5.3.2. Harappa

The site of Harappa is the most-excavated of all of the Indus Civilisation sites (Possehl 1991: 6) and as such needs little introduction. Full details of the site excavations are far outside the scope of the present research and are not necessarily required for an appreciation of the technical analysis of the ceramics. Numerous summaries have been provided by the excavation team, and these should be referred to for a full picture of the site (e.g. Meadow 1991). First reported as a site of archaeological significance by Charles

Masson (Masson 1844: 452), the most extensive excavations were conducted by M.S. Vats in 1926 and between 1934-1940 (Vats 1940). R.E.M. Wheeler conducted important excavations in the cemetery and on the Citadel Mound in 1946 (Wheeler 1947). Most recently, it has been excavated each season since 1986 by the inter-disciplinary team currently under the direction of Dr. Richard Meadow.

Harappa lies in District Sahiwal, Punjab, about 175 km west of the present city of Lahore. It is situated some 12 km from the present Ravi river, which still runs approximately the same course as during the early occupations at the site (Amundson and Pendall 1991: 16). However, there is a conspicuous tributary channel that often carries water during monsoon season. It is not known when the main river bifurcated from this channel, but the channel may actually have been the main river during the occupation period of the site. The earliest occupation has been radiocarbon dated to about 3200 B.C. (Period I). Period II dates range from 2470 to 2405 B.C. and overlap with the dates of the following Period III, which range from 2913 to 2047 B.C. (Kenoyer 1991: 38). It has been shown to contain several phases of full urban and post-urban occupation, including major architecture, and is one of the largest and most advanced of all known Harappan sites. Major architecture includes large public structures that have open access, large open courtyards, and so-called 'granaries'. Important features include carefully designed and well-maintained drainage systems, wells and bathing platforms, and sump pits. Harappa is one of only three cities of the Indus Civilisation so far excavated that show a 'typical' Harappan city layout, namely an isolated western citadel and a lower city whose architectural features are oriented with the cardinal directions with thoroughfares running north-south at intervals. The others are Mohenjo-Daro and Kalibangan (Jansen 1993: 46). A massive wall surrounds the site. Habitation structures follow a similar pattern to those described for the site of Rehman Dheri above.

Primary occupation levels of the Early Harappan period include traces of mud-brick walls, hearths, ceramics similar to those found at Rehman Dheri, fired bangles, blades from dark grey-black chert, beads and human figurines all too numerous to elaborate on in this report. Evidence of extensive ceramic production comes from the presence of a kiln with many vitrified sherds and kiln wasters that indicate long-term ceramic production. The ceramic corpus from the site is further discussed in the section following site descriptions.

The topography around Harappa is relatively flat and subdued, and the climate of the region is semi-arid to arid and hot, with summer monsoon rains the predominant form of precipitation. Mean annual temperature is about 26 degrees C. Natural vegetation is semi-arid to arid, subtropical thorn forest (Dakshini 1985; cited in Pendall and Amundson 1990a: 302). The natural vegetation has been severely altered by grazing and cultivation, brought on by an extensive canal network constructed by the British in the 1920's.

5.3.3. Lewan

The site of Lewan is located in the Bannu Basin about 12 km from the present town of Bannu. It lies on a fragment of degraded river terrace measuring approximately 450 by 325m and was probably originally associated with the Tochi river, which has since shifted its course (Allchin and Allchin 1992: 526). With the exception of a gravel lens 2.4m below surface in trench LWN IV, the Lewan terrace is composed of fine-grained overbank deposits, probably laid down by the Tochi river (Allchin and Rendell 1986: 35). The surface of the site is covered with stones, partially due to its nature as a stone tool working site, and partially due to significant deflation which has occurred since abandonment. The site appears to have been the location of an extensive stone-working industry, including microlithic; and also of a prominent bead-working industry (Allchin and Rendell 1986; Allchin 1992). The pottery from Lewan was dated by the excavators by its similarity to the sequence from Rehman Dheri. It was observed that all of the pottery from Lewan belonged to a single main craft tradition, and "had a certain homogeneity which immediately discloses that it could not belong to a later period" (Allchin and Knox 1986: 103).

Excavated during 1977-8, the site covers about 12 hectares. Eight separate trenches were opened, revealing two houses and nine pits. Most of the pottery used for the present research was recovered from the house areas. Pottery descriptions are presented in Section 5.4 below. Sherds litter the surface, and the presence of kiln wasters at the site attests to a high probability that some of the pottery was manufactured at the site. A sample of these was collected by the author for comparative material.

The climate of the Bannu basin is sem-arid, with little or no rainfall in the period from October to February, but with occasional periods of heavy rainfall during the summer months. The rainfall distribution is highly variable both from month to month and from year to year (Rendell 1986: 5). As a result, the discharge regime of the Kurram and Tochi rivers is torrential with the maximum flows in the summer months. The area between the rivers is now intensively irrigated.

The vegetation of the Bannu Basin is highly degraded, due to pressure from the human and animal population, but falls within the tropical thorn forest zone (as defined by Beg 1975; cited in Thomas 1986: 19). As a result of these pressures, the natural vegetation is now restricted to only a few areas which are either inaccessible, or protected. The natural vegetation includes common plants tolerant of saline areas and sandy areas of tropical plains and tropical inland thorn forests. Due to a combination of environmental conditions, intensive agriculture, competition with man's domestic animals, and hunting, the vertebrate fauna of the basin is not rich (Thomas 1983: 132).

5.3.4. Tarakai Qila

The second of the Bannu Basin sites, the material of which is under investigation in the present research, Tarakai Qila is some 14.5 km to the northwest of Lewan. It was contemporary with the later levels of Lewan (Allchin and Allchin 1986: 200). It is on the right bank of the Tochi river some 11 km west of Bannu City and the upper surface is about 10m above the present bed of the river. It is located on a mound of about 400m by 300m but appears to have been larger in the past. Total thickness of occupation horizons exceed 6m. It was excavated in 1978-9. Six C14 dates from excavated levels give a time span ranging from c. 2800 to c. 2040 B.C., calculated according to the MASCA calibration (Khan 1986: 194). These dates correspond to later stages of the Early Harappan to the Mature Harappan.

No complete house or room plans were uncovered, but extensive stone foundations were excavated, and associated with the Kot Dijian period. The occupation was apparently continuous as no breaks in the sequence were recognised apart from one or two levels of minor burning (Allchin and Knox 1981: 246). The house remains included some hearths, floors, pits, and rubbish deposits which permitted recovery of some bioarchaeological remains. These included charred remains of barley, wheat and lentil (Thomas 1983: 139). Animal remains were dominated by domesticated cattle, sheep, goat and water buffalo. The cattle and water buffalo are only represented by adult individuals; this led the researcher to infer that their importance was for non-meat purposes such as milk production or traction. By contrast, some 30 per cent of the sheep and goats were of juveniles, probably indicating their exploitation for meat. Remains of wild animals were few in number. Exploitation of river resources is indicated by remains of catfish and shells of bivalve molluscs. Overall, the whole assemblage of domesticated animals and plants from Tarakai Qila suggests an economy founded on pastoral and arable elements, with marked seasonality in the latter.

Compared with Lewan, there was a paucity of material culture revealed by the excavations (Allchin and Knox 1981: 249). Apart from the pottery, which is described in the following section, most of the finds were of a strictly utilitarian nature. There were small numbers of beads, including some of lapis lazuli, and toy cart wheels of terracotta. Important finds included four small stone seals. Terracotta goddesses, such as were found at the early levels of Lewan, were not found at Tarakai Qila. The stone tool industry, although not as extensive as that found at Lewan, appeared to be fairly well developed (Allchin 1981: 234). It was noted that many of the lunates were found to have gloss extending over part of the surface only and that they appeared to have been hafted diagonally, leading the researcher to speculate that they had been employed embedded in a matrix so as to form a composite saw-like cutting edge. This, together with finds of saddle querns used for grinding cereals and other seeds, and the presence of wheat, barley and lentils further support the hypothesis that Tarakai Qila was very much an agricultural community.

5.4 CERAMIC TYPOLOGIES OF THE EARLY HARAPPAN PERIOD

Because of the similarities in the ceramic typologies of the Early Harappan at the four sites from which material is being investigated for the present research, they are being treated in this separate section. Indeed, it is due to these similarities occurring at geographically separated sites that the research was commenced. Because the larger part of the sample selection was made from Rehman Dheri, the typology employed by Durrani (1988) is used for all four collections. The ceramics from the Early Harappan period at the site of Sarai Khola, near the present city of Taxila, have close affinities (Mughal and Haleem 1972). These are mentioned here because it was intended at the outset of the research to include material from that site for comparison with the others. Future research into the networks of interaction of the Early Harappan should incorporate this comparison (see Chapter 8).

Durrani (1988: 31-2) divided the pottery into three wares:

1. Grey ware: Examples of these were extremely rare and confined to the top levels of period RHD III. Only six examples were recovered in the excavation. As none of these was sampled for the present research, their descriptions are omitted here.
2. Red ware: This ware was predominant. It had a great deal of variation in types, texture, and decoration, which are outlined following.
3. Buff ware: A lesser number of these occurred.

The typology used by Durrani (1988) first groups vessels on the basis of morphology, and subdivides them using rim forms and decoration/surface treatment. The same approach is currently used for classifying the ceramic sequences at Harappa (Jenkins 1994: 320). The categories as they are defined are polythetic; that is, a vessel 'may be assigned to a type if it exhibits a large number, but not all, of the defining attributes of the type' (Adams and Adams 1991: 355). Durrani (1988) first divided the sherds and vessels into two basic morphological classes: jars and bowls. These two classes were then further broken down, morphologically, into three types of jars and six types of bowls as follows:

<u>Jars</u>: Types AI, AII, and AIII

<u>Bowls</u>: Types BI, BII, BIII, BIV, BV, and BVI. These bowl types are based on the geometric height of the vessel walls.

These morphological types are described in the following paragraphs. For purposes of the present research, decoration and surface treatment were not considered as a variable (see Chapter 6 for discussion). However, they were recorded for the sherds that were studied. The system used at Harappa also did not use secondary characteristics such as surface treatments as defining criteria, however, all are recorded because the 'presence or absence of various attributes and changes in their frequency are useful in defining to which period a given excavation unit should be assigned' (Jenkins 1994: 320). At Sarai Khola, wheel-made vessels were categorised as to type by both morphology and surface treatment (Mughal and Haleem 1972: 45-110). While all of the same types were present, the system does not lend itself efficiently to direct one-to-one cross-referencing to the other sites' defined types. The following description includes discussion on variations observed in the morphologies with respect to rim type, etc. and some limited discussions on surface treatment where it is deemed important to the present research.

Type AI Jar

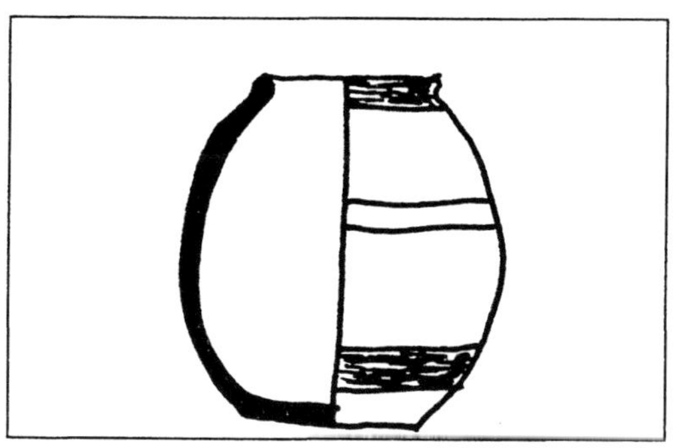

The type AI jar is illustrated above. It is described as ovoid in cross section (Durrani 1988: 32, examples p 150), and is further subdivided into rim types: (a) with plain, everted, or flaring rims; and (b) with flanged rims. These categories are then broken down into 17 varieties, based primarily upon a broad diversity of surface decoration. Surface treatments range from slipped to unslipped, incised, combed, and roughened. Painted designs include horizontal red, white, and black bands; geometric, floral, or zoomorphic motifs; and panels formed by white and cream bands. Similar surface treatments were recorded for the current excavations at Harappa (Jenkins 1994: 321-24). Bases were round or flat, and not used for typology type. This jar type is very common in the Early Harappan, and is found in the early and transitional levels at Harappa, where is it termed a 'jarpot' (Jenkins 1994: 322-3; rim and base forms illustrated p 318), Lewan (Knox 1986: 209-51), and Tarakai Qila (Allchin and Knox 1981: 249). At Sarai Khola, it is classed as a Type VII, VIII, or IX, depending on surface treatment and rim type (Mughal and Haleem 1972: 60-70).

Type AII Jar

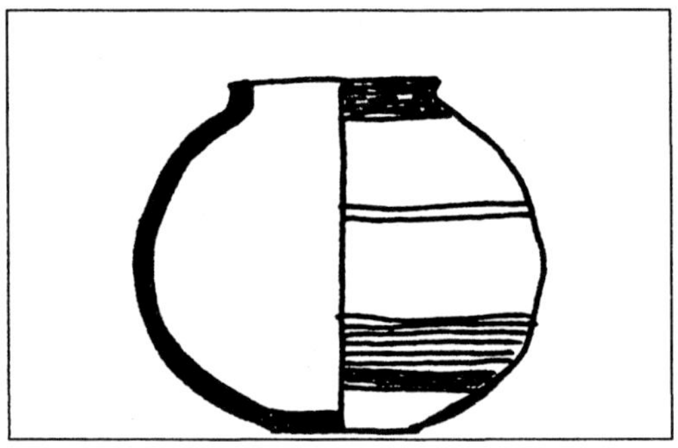

Type AII is described as a spherical jar (Durrani 1988: 32). As with the AI, these are subdivided on the basis of rim type: (a) plain, everted, or flaring rim; and (b) flanged rim. The same 17 surface decoration categories are used, and these are also found at Harappa, where it is also termed a 'jarpot' (Jenkins 1994: 322-3; rim types described p 318), Lewan (Knox 1986: 209) and Tarakai Qila (Allchin and Knox 1981: 289). At Sarai Khola, it has been termed Type VII, VIII, or IX depending on surface decoration and rim type (Mughal and Haleem 1972: 60-70).

Type AIII Jar

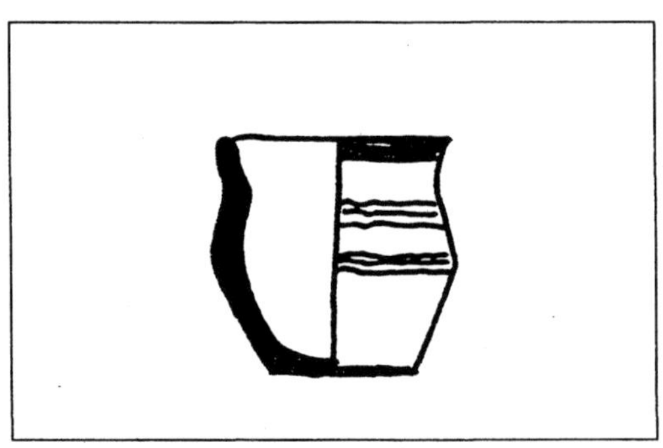

Type AIII jar is described morphologically as inverted, truncated, conical (Duranni 1988: 32). They are not broken down further. These have not been reported from recent excavations at the early and transitional levels of Harappa, but may have been found. No identical types were reported from Lewan, although similar morphologies were found, termed a 'wide mouthed bowl with slightly waisted sides' (Knox 1986: 209). Possibly due to incomplete publishing, Type AIII was not reported from Tarakai Qila. It was not reported from Sarai Khola (Mughal and Haleem 1972). This type was not sampled for the present research.

Type BI Bowl

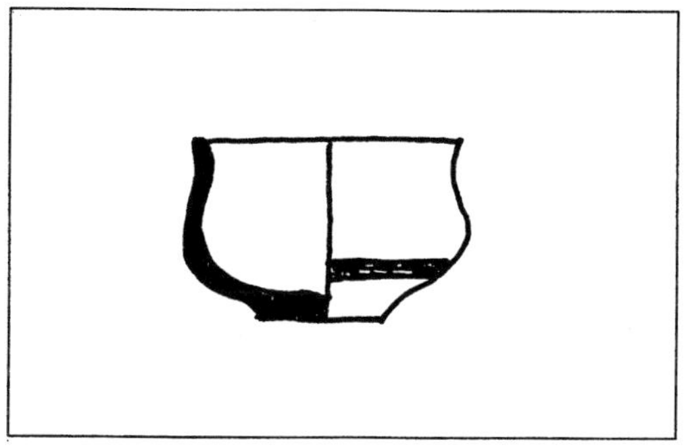

The type BI is termed a horizontal, cylindrical bowl (Duranni 1988: 33). Rim types, not used to further sub-classify, are plain or flaring with some vertical examples. Walls are carinated with the angle usually formed in the lower third of the vessel. Bases are predominantly ring. The BI type is broken down into 5 varieties according to surface decoration: plain or unslipped; burnished; horizontal painted bands or incised lines over red slip; simple black or white painted geometric designs over horizontal bands; and complex geometric, floral or zoomorphic motifs in panels formed by horizontal bands. Type BI is commonly found in the Early Harappan and is present in the early and transitional levels at Harappa (Jenkins 1994: 322; surface decorations p 321), at Lewan (Knox 1986: 220, 228, 238), and Tarakai Qila (personal examination of collection by author 1994). At Sarai Khola, these are classed as Type XI carinated bowls (Mughal and Haleem 1972: 72).

Type BII Bowl

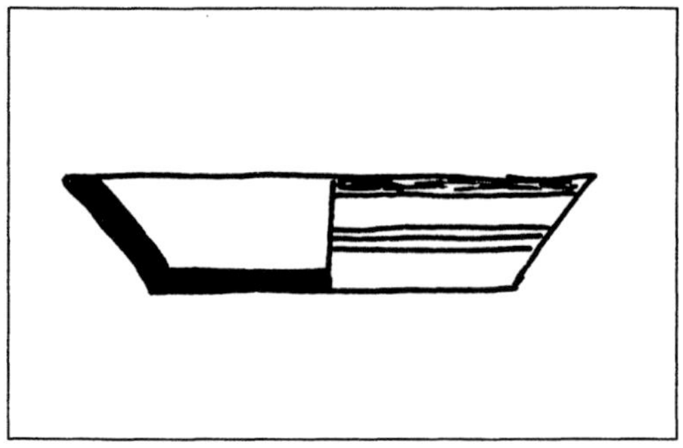

Type BII is termed an inverted, truncated, conical bowl (Duranni 1988: 34). Predominant rim types are plain or flaring. Walls are predominantly concave or straight. All bases are flat. Subdivided into 3 varieties according to surface decoration: plain, slipped or unslipped; horizontal painted bands on interior and exterior; and isolated painted floral and zoomorphic motifs over horizontal bands. Found in early and transitional levels at Harappa (Jenkins, pers. comm.), at Lewan (Knox 1986: 216, 220-1, 232-4, 240), and Tarakai Qila (personal examination of collection by author 1994). They are termed Type XII pans or dishes at Sarai Khola (Mughal and Haleem 1972: 73).

Type BIII Bowl

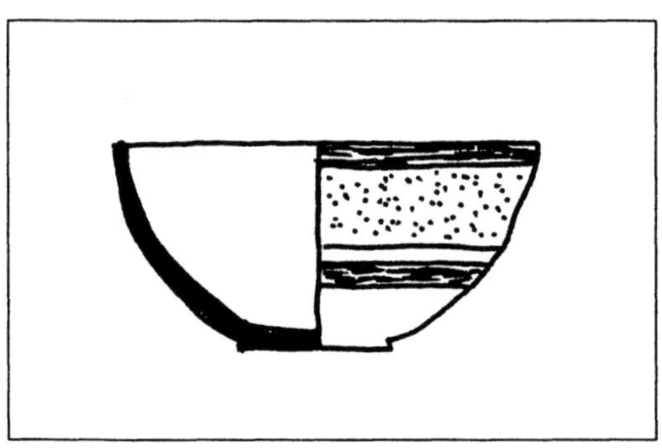

Type BIII is termed a semispherical bowl (Duranni 1988: 34). These are broken down into two sub-types: (a) convex-sided; and (b) carinated walls with everted rims. The BIII (a) rims are plain, sloping rims predominantly and bases are mostly flat, or ring. They are subdivided into 3 varieties according to surface decoration. The BIII (b) rims are mostly everted, walls are mildly carinated, and bases are primarily ring type. They are categorised into 3 surface decoration types, identical to those above for Type BII. They are also found at Harappa (Jenkins 1994: 323), Lewan (Knox 1986: 228-31, 240, 244, 252), and Tarakai

Qila (personal examination of collection by author 1994). They were reported as Type X bowls at Sarai Khola (Mughal and Haleem 1972: 71).

Type BIV Bowl

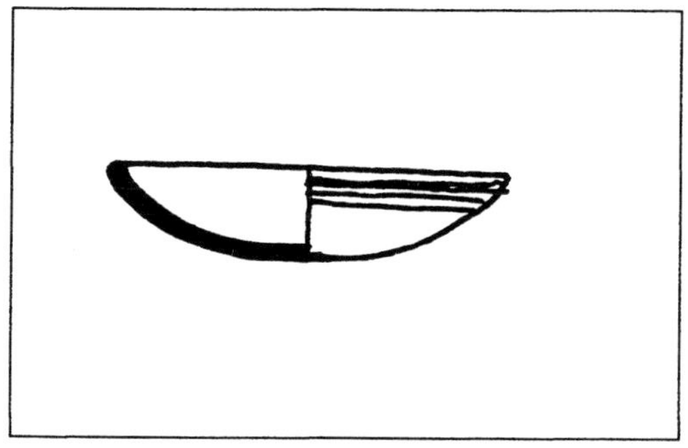

Type BIV is described as a quadrispherical bowl (Duranni 1988: 35). Two sub-types are classed according to wall morphology: (a) with convex walls; and (b) with carinated walls. Type BIV(a) has predominantly plain, vertical rims, convex walls, and round or disc bases. Type BIV(b) has predominantly everted rims, mild to sharply carinated walls, and mostly disc bases. Three varieties of surface decoration are described: plain, no decoration, with or without slip; simple geometric decoration; and complex geometric, floral, or zoomorphic designs painted in black or red over white or cream panels. Found at Harappa (Jenkins 1994: 323), Lewan (Knox 1983: 230, 234), and Tarakai Qila (personal examination of collection by author 1994). At Sarai Khola, they were termed Type XV plates or dishes (Mughal and Haleem 1972: 76).

Type BV Bowl on Stand

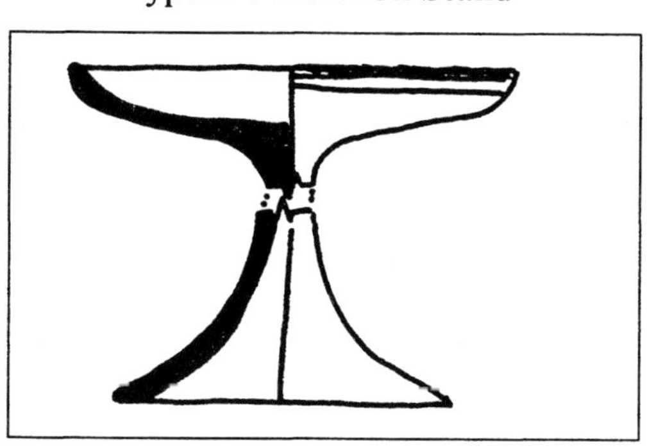

Type BV is described as bowl-on-stand (Duranni 1988: 35). The stand and base are one piece, but are usually manufactured as individual sections. Three sub-types are described:

(a) semispherical bowls on prominent, hollow, concave-sided stands; (b) deeper semispherical bowls with convex sides and externally-thickened rims on low hollow, concave-sided stems; and (c) shallow or quadrispherical bowls on medium or tall stems, usually referred to as 'dish-on-stand' in South Asian archaeology. Decoration is not used to sub-classify. Common throughout the Early to Mature Harappan, they are found in the early to transitional levels at Harappa (Jenkins 1994: 323), Lewan (Knox 1986: 215-6, 230, 240, 250), and Tarakai Qila (personal examination of collection by author 1994). They are found at Sarai Khola and termed Type XIV offering stands (Mughal and Haleem 1972: 75).

Type BVI Bowl

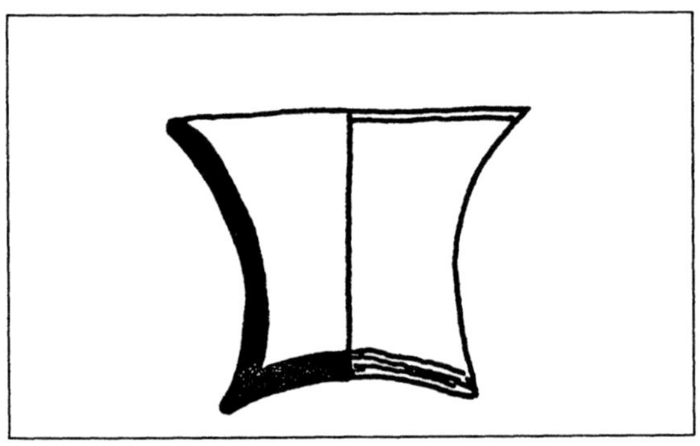

Type BVI is termed an inverse, trapezoidal bowl (Duranni 1988: 36). It is not sub-classified. Rims are plain or flaring, walls are concave, and bases are concave. They may have been found at Harappa, but were not reported from the current excavations as a common type (Jenkins 1994). They were not reported from Lewan (Knox 1986) or Sarai Khola (Mughal and Haleem 1972), nor observed by the author in the collection from Tarakai Qila. This type was not sampled for the present research.

A further type which was sampled in the present research was termed CII by Durrani (1988: 36, 111). This was a term given to polychrome examples, usually body sherds, regardless of parent vessel morphology.

CHAPTER SIX

METHODOLOGY

6.1 Selection of sherds for analysis

In selecting the sample of sherds for thin-sectioning, ceramics from the site of Rehman Dheri were used as the basis for typology. As outlined earlier, Rehman Dheri provided the largest sample of the four sites examined, it appears to have had a major ceramic production industry, and it contains a wide variety of decorative types which also appear at many other Early Harappan sites. It also has many of the distinctive features of Harappan cities, and indications that it could have been a centre for the production and trade in ceramics and other crafts (see Chapter 5). The original reason for selection of this site as the basis was that if ceramic products, or their contents, were being traded, Rehman Dheri appeared to be the most likely centre for the region. It is located on many of the early trade routes, both to the north and south, and to the west through the Gomal Pass. The site has been excavated to sterile levels in more than one area, and so a complete chronological sequence was available. This permitted a single-site study of technological developments and changes in ceramic crafts from the Early Harappan period through to the Mature Harappan. Moreover, the excavated material was securely stratified as to levels, carbon dates were available, and sequences were such that they could be directly compared to levels of similar date at the other sites. The largest portion of the sherds sampled therefore came from this site.

Table 6.1 summarises the number of samples, listed by morphological type and site. Selection was generally simple random sampling from morphological type groups in each of the site collections. The random sampling was not made with the intention of reflecting the ratios of types found in the parent population; it was not the aim for example to determine the percentage of AI jars or BIII bowls in the entire sherd population of each site. The objective was simply to provide a representative sample of each of the most common morphological types found at each of the three chronologically defined levels at Rehman Dheri and select similar samples from the other sites. In this way, fabrics and technologies could be compared from site to site on a common basis and inferences made with respect to provenance and exchange of technological information.

TABLE 6.I

SHERD SELECTION SUMMARY

Number of sherds selected by site and morphological type

For type descriptions, see Section 5.4. Site legend: RHD = Rehman Dheri; HRP = Harappa; LWN = Lewan; TRQ = Tarakai Qila

TYPE	RHD	HRP	LWN	TRQ	TOTAL
AI Jar	7	4	5	5	21
AII Jar	10	5	3	4	22
BI Bowl	12	5	4	2	23
BII Bowl	3	1	0	4	8
BIII Bowl	6	4	5	5	20
BIV Bowl	7	3	3	1	14
BV Dish on Stand	6	1	0	7[1]	14
CII Polychrome	5	0	0	0	5
Wasters	11	3[2]	6	0	20
Bangles	0	18	0	0	18
TOTALS	67	44	26	28	165

[1] Most of these were rim sherds, therefore it could not be certain that the parent vessel was type BV. Based upon rim morphology, some could be type BIV bowl sherds.

[2] Pottery made from local clays was examined in lieu of wasters

The selection from the Rehman Dheri collection was made so as to provide a representative sample, commensurate with time and scope constraints, of each of the common morphological types. In addition, the sample was to provide chronological representation from Period I to Period III; with a distinct concentration on Periods I and II (the Early Harappan levels). Selection of some samples from the Mature Harappan levels would provide a basic look at any changes in fabric or other characteristics between Periods I and III. The concentration on Early Harappan levels was designed to provide intersite comparisons with Harappa and the Bannu sites for that period. The primary investigation of the thesis is to examine intersite interactions during the Early Harappan period. The selection made from the Rehman Dheri collection is shown schematically in Figure 6.1.

There were four excavation units at Rehman Dheri (see Figure 5.4). Material from BIV (Unit A) was chosen, because it was excavated down to sterile levels and contained a complete sequence from Period IA to IIIB. A single unit was chosen so as to eliminate the influences of possible variation with respect to horizontal spacing at the site[1]. Only sherds that could be identified as to parent vessel type were selected, so that possible correlations between fabric types and morphological types could be examined. One exception to this was in the case of Type CII (see Chapter 5). Three of these were chosen for analysis to ascertain if they had any significant differences from the others. Otherwise, surface decoration was not a factor in sample selection. Properly sampling each of the decorative type groups would have resulted in much too large a sample to be contained within the time constraints and scope of the present research.

There were not sufficient samples of certain of the morphological types available in the Rehman Dheri collection that were properly labelled as to provenance. It was important that each of the sherds selected had a known provenance before they could be used in the analysis. In order to examine for changes occurring from the Early to the Mature Harappan periods, dating of each sherd is critical. The selection process took considerable time as a result of the state of storage of the sherds at the University of Peshawar in anticipation of moving them to the new museum storage areas.

Figure 6.1 shows that the majority of samples (73 %) were chosen from Periods I and II, the Early Harappan period at Rehman Dheri. In addition to the excavated sherds selected for analysis, the site was visited and 12 waster sherds collected from the surface, in the

[1] In doing this, it is realised that if such a variation were found to exist, a bias would have been imposed on the present selection. The possibility of this variation should be investigated as a matter of urgency, as is recommended in Chapter 8. Limited time and sample availability precluded the broader sample selection for the present research.

vicinity of the unexcavated kilns described in the previous chapter. A full listing of the sherds and their metric descriptions is included with data at Appendix A. Figures 6.2 to 6.11 illustrate the entire sample selected from the Rehman Dheri collection for thin-sectioning.

Period	**AI Jar**	**AII Jar**	**BI Bowl**	**BII Bowl**	**BIII Bowl**	**BIV Bowl**	**BV Dish**	**CII Poly.**	**Totals**
IIIB (Upper)	2	2	3	0	1	3	2	0	13
IIIA	0	0	0	0	0	0	0	2	2
II	1	2	3	1	1	1	2	0	11
IB	1	3	3	0	3	2	2	2	16
IA (Lower)	3	3	3	2	1	1	0	1	14
Totals	7	10	12	3	6	7	6	5	56

FIGURE 6.1

SCHEMATIC REPRESENTATION OF EXCAVATED SHERDS SAMPLED FROM REHMAN DHERI

Number of sherds selected by type and period

For an explanation of periods, see Section 5.3; type descriptions are given in Section 5.4

In the case of the Lewan material, the excavated sherds had been stored in thin cotton sacks with only a label inside the bag identifying the contents as to provenance. These sacks had been stored in a basement area of the new University of Peshawar museum building which was under construction during 1994. Unfortunately, many of these bags were in a deteriorated condition when placed there, had broken open following a flood in the storage area, and become intermixed. The majority of the sherds were therefore no longer identifiable as to provenance and could not be used for the present research. However, some of the sherds had originally been labelled directly on the artefact, and these were the only ones chosen for examination. This factor restricted the selection to few examples of certain morphologies. A total of 22 sherds was selected from the Lewan material. In addition, the author visited the site and collected 6 ceramic wasters from the surface in order to characterise the local clays. Figures 6.12 to 6.17 show the selection of sherds made from Lewan, and Figure 6.18 illustrates the wasters collected from the surface.

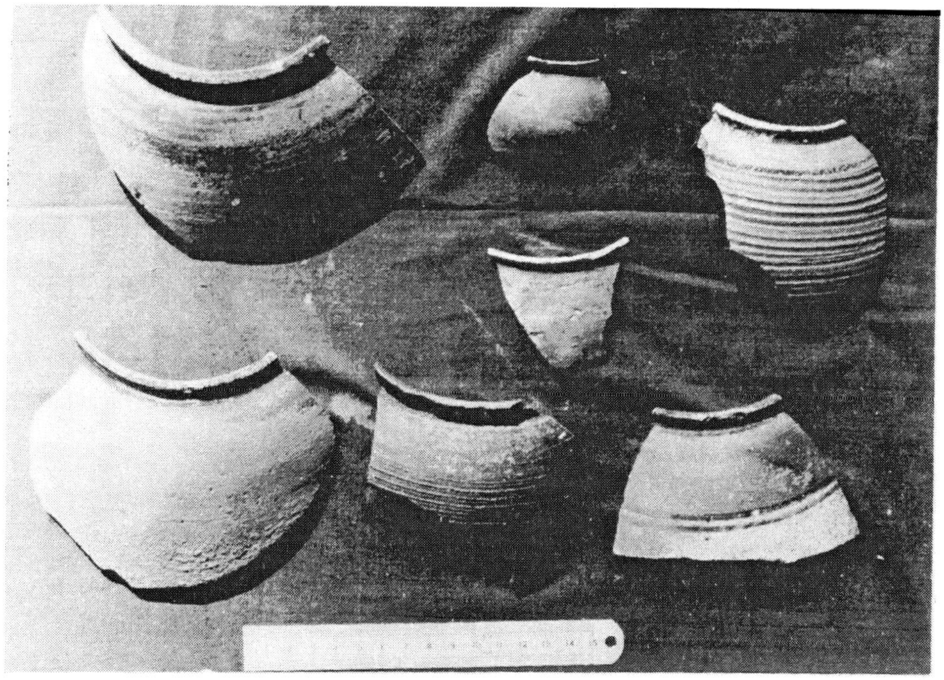

FIGURE 6.2

REHMAN DHERI TYPE AI JAR SHERDS

Clockwise from top left: RHD 36, 41, 37, 39, 35, 40. Centre: RHD 38.

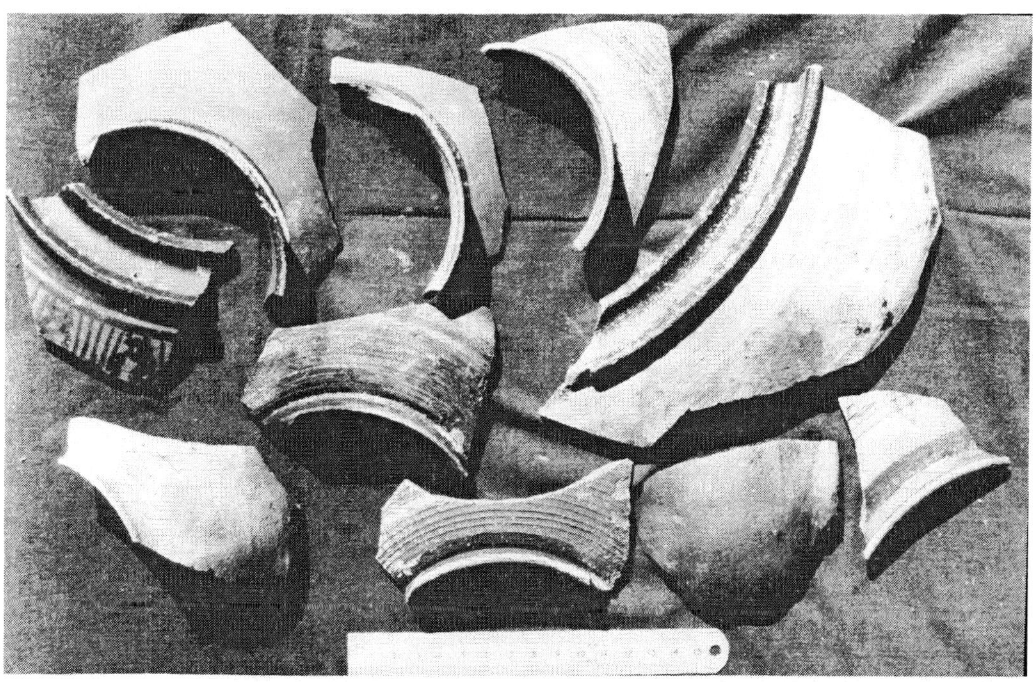

FIGURE 6.3

REHMAN DHERI TYPE AII JAR SHERDS

Clockwise from top left: RHD 49, 56, 53, 47, 55, 54, 50, 52, 48. Centre: RHD 51.

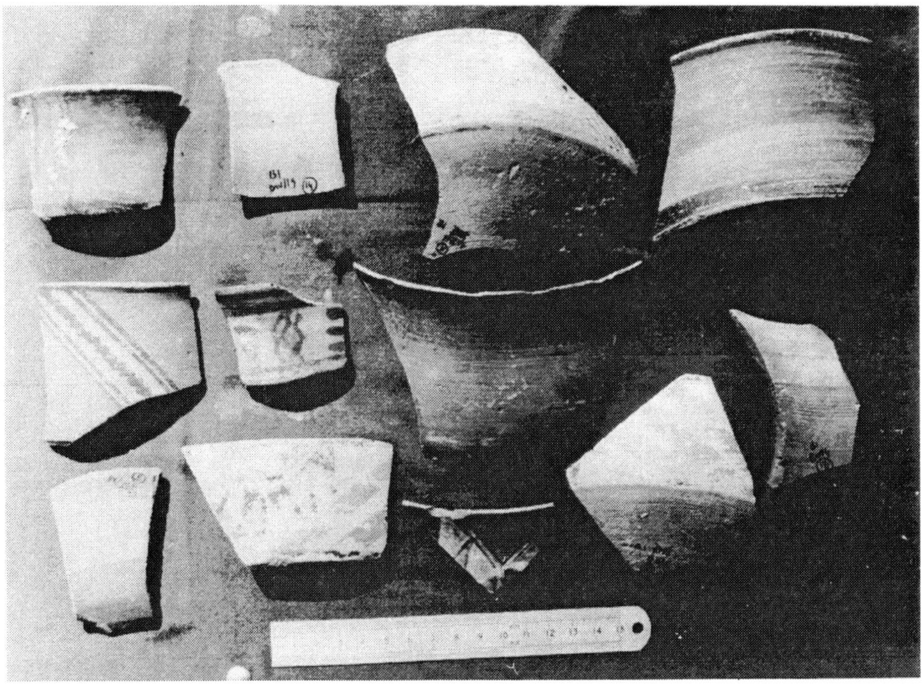

FIGURE 6.4

REHMAN DHERI TYPE BI BOWL SHERDS

Top row: RHD 12, 7, 8, 15. Middle row: RHD 9, 10, 18, 13. Bottom row: RHD 12, 16, 11, 17.

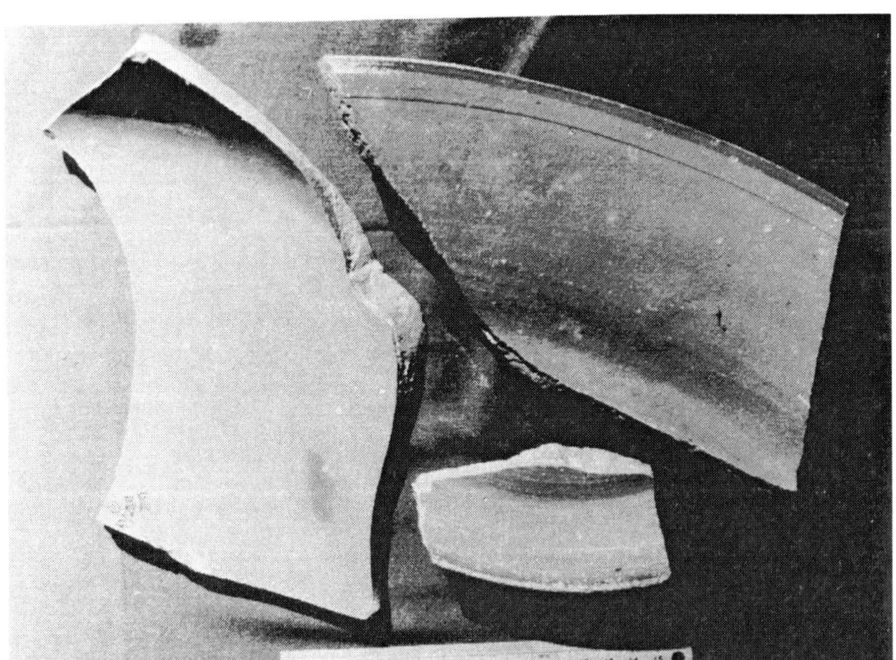

FIGURE 6.5

REHMAN DHERI TYPE BII BOWL SHERDS

Clockwise from left: RHD 25, 26, 27.

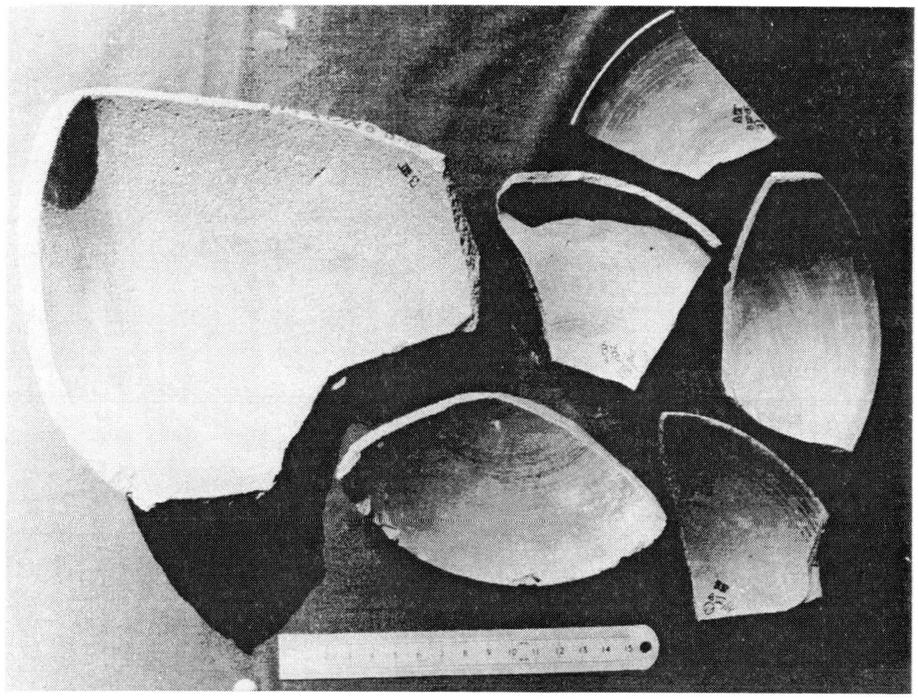

FIGURE 6.6
REHMAN DHERI TYPE BIII BOWL SHERDS
Clockwise from left: RHD 6, 2, 5, 1, 3, 4.

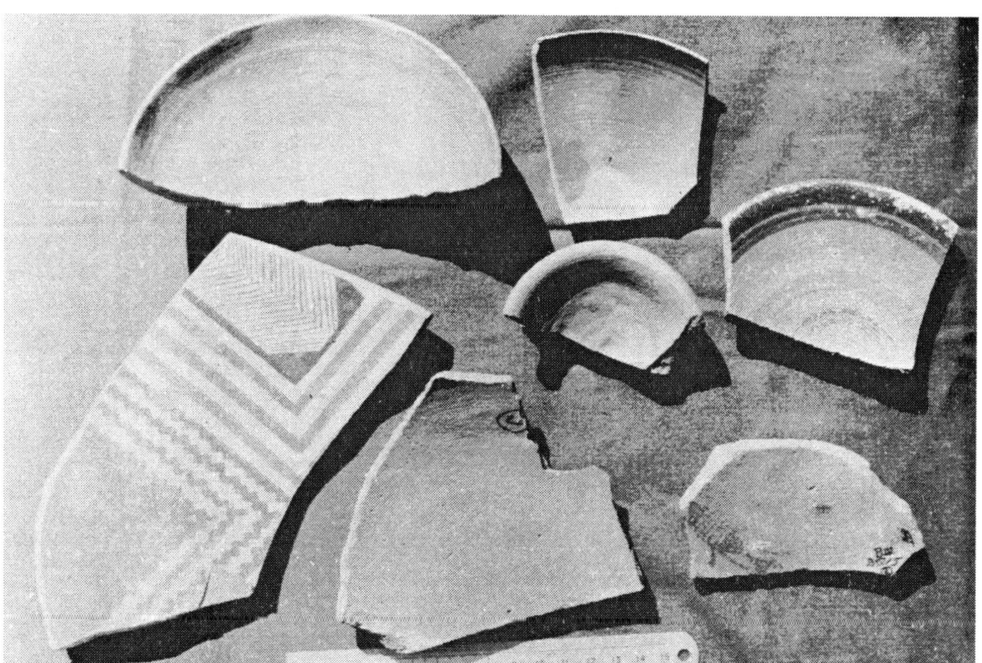

FIGURE 6.7
REHMAN DHERI TYPE BIV BOWL SHERDS
Clockwise from top left: RHD 31, 30, 28, 32, 33, 29. Centre: RHD 34.

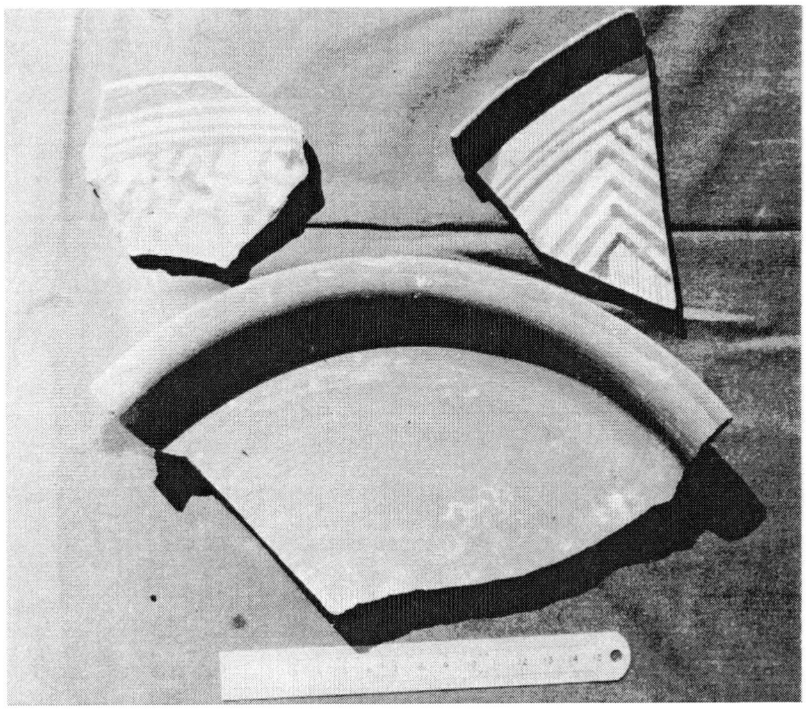

FIGURE 6.8
REHMAN DHERI TYPE BV DISH ON STAND SHERDS
Top row: RHD 22, 23. Bottom: RHD 24.

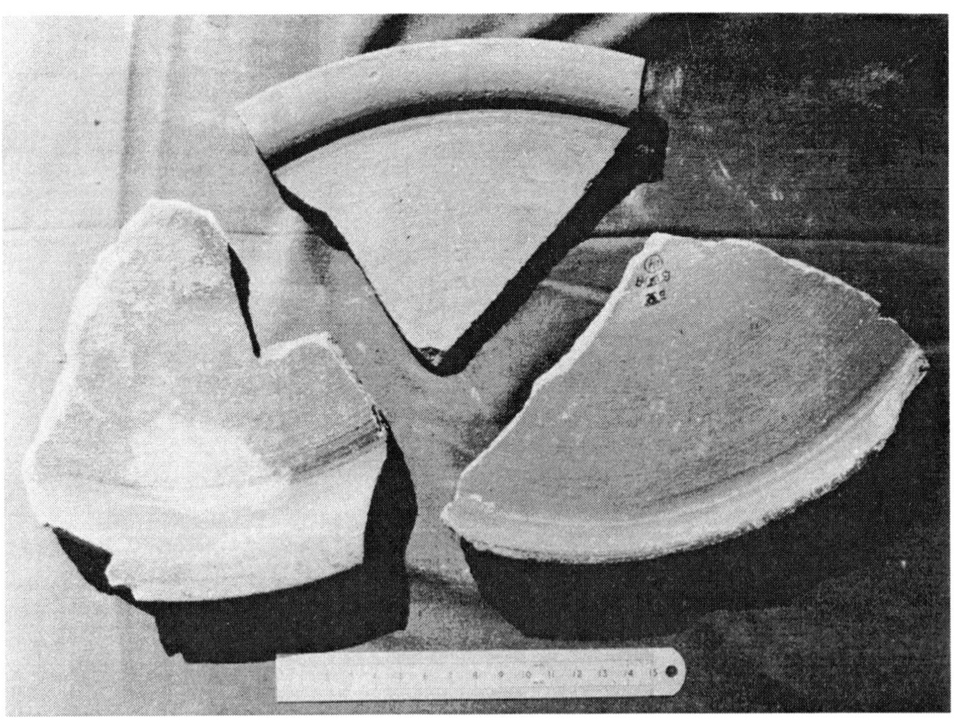

FIGURE 6.9
REHMAN DHERI TYPE BV DISH ON STAND SHERDS
From left: RHD 19, 20, 21.

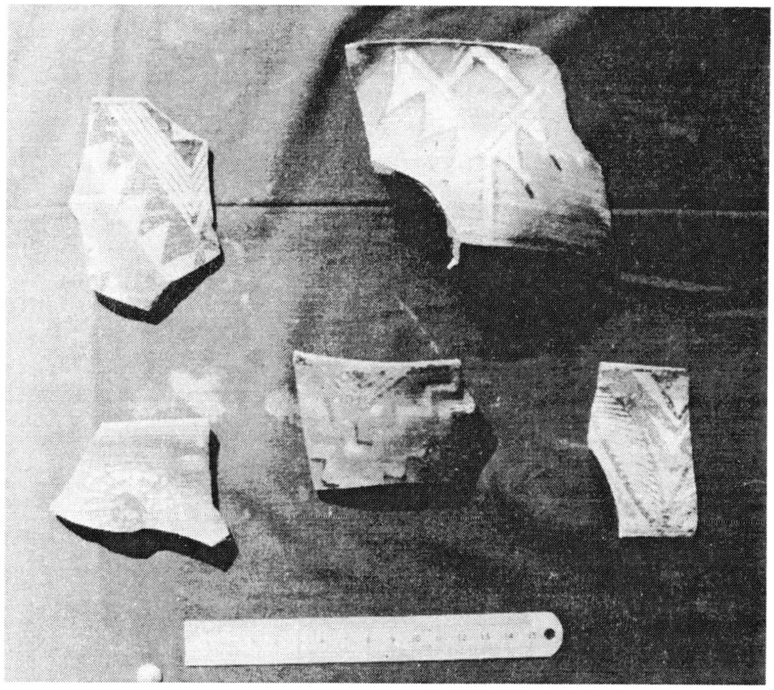

FIGURE 6.10

REHMAN DHERI TYPE CII POLYCHROME SHERDS

Clockwise from top left: RHD 44, 46, 45, 42, 43.

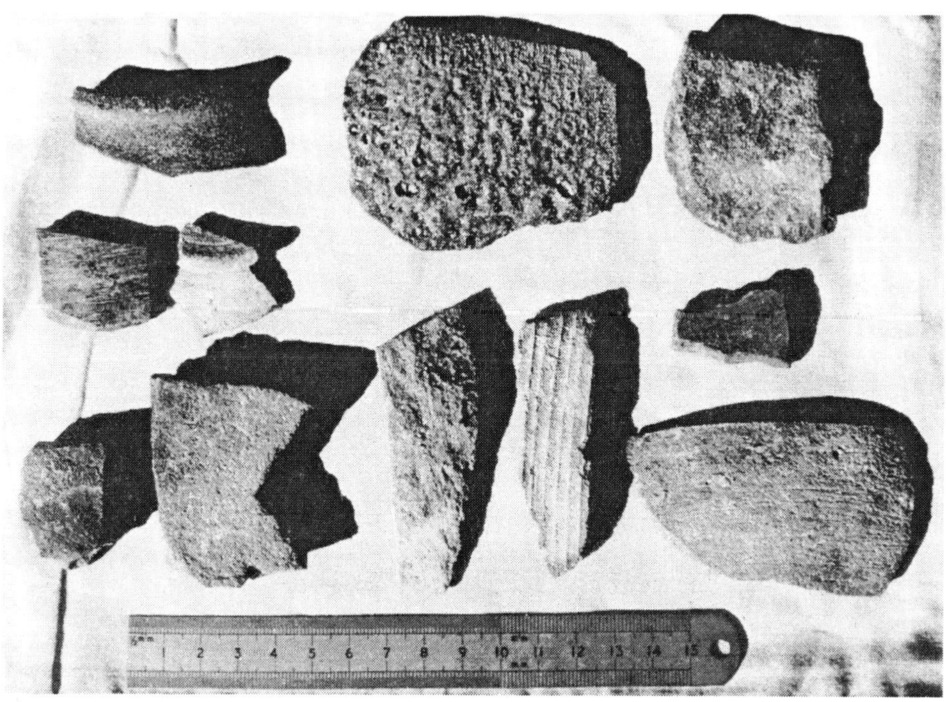

FIGURE 6.11

REHMAN DHERI SURFACE WASTER SHERDS FROM KILN AREA

Top row: RHD 62, 69, 65. Middle row: RHD 65, 63, 71. Bottom row: RHD 64, 61, 67, 66, 60.

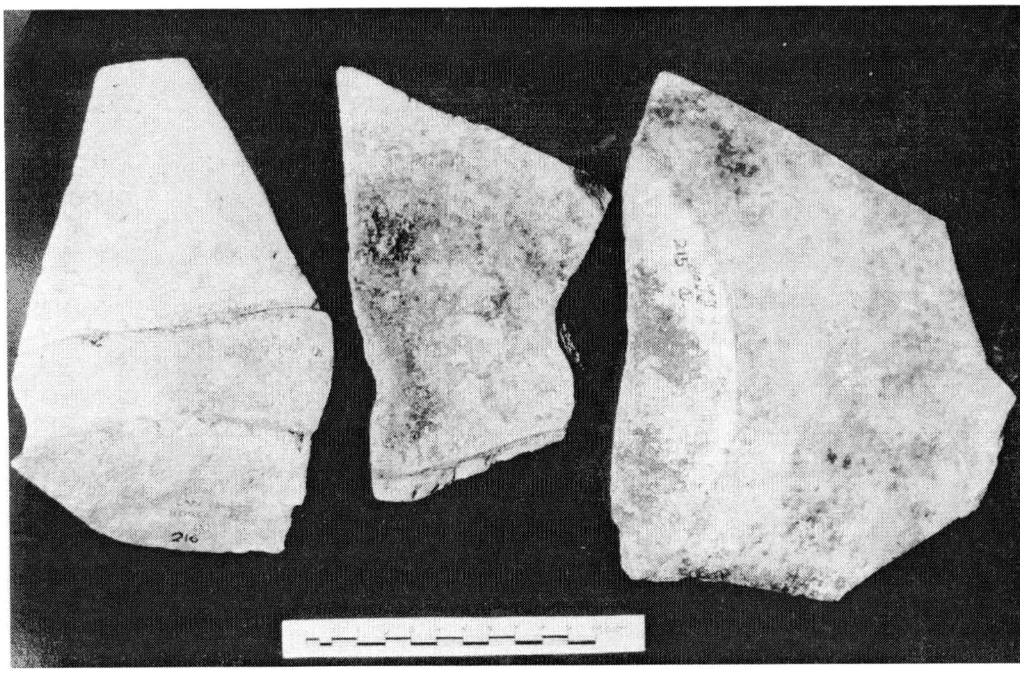

FIGURE 6.12

LEWAN TYPES AI AND AII JAR SHERDS

From left: LWN 216, 219, 215.

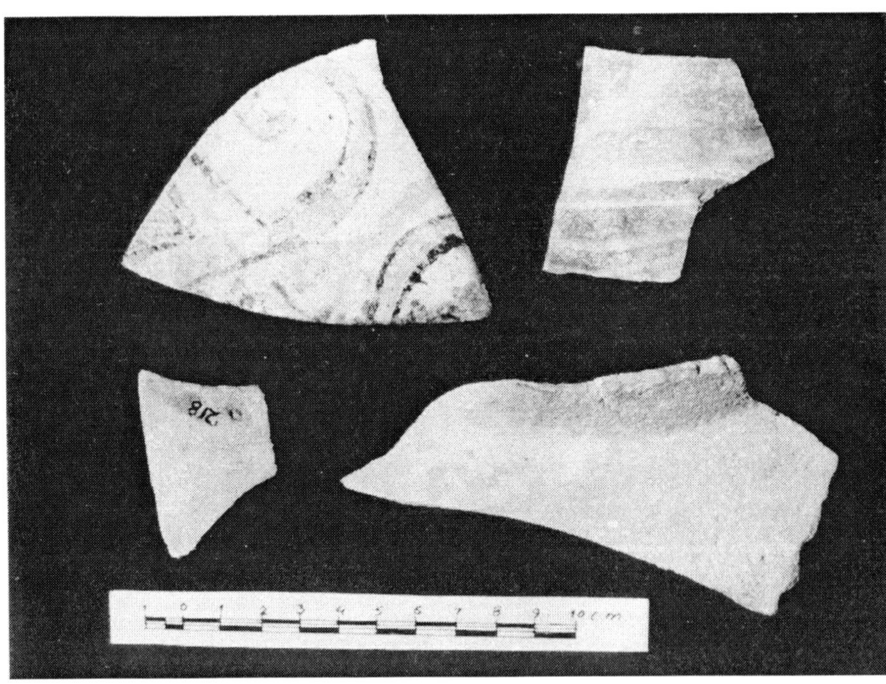

FIGURE 6.13

LEWAN TYPES AI AND AII JAR SHERDS

Clockwise from top left: LWN 210, 209, 207, 218.

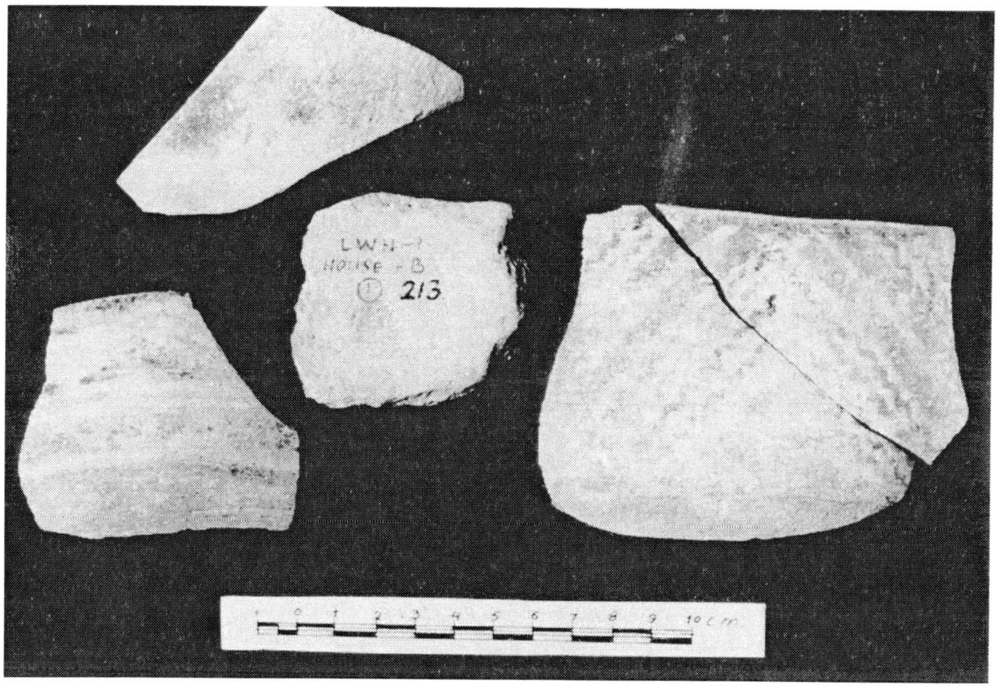

FIGURE 6.14
LEWAN TYPE BI BOWL SHERDS
Top: LWN 201. Bottom, from left: LWN 211, 213, 200.

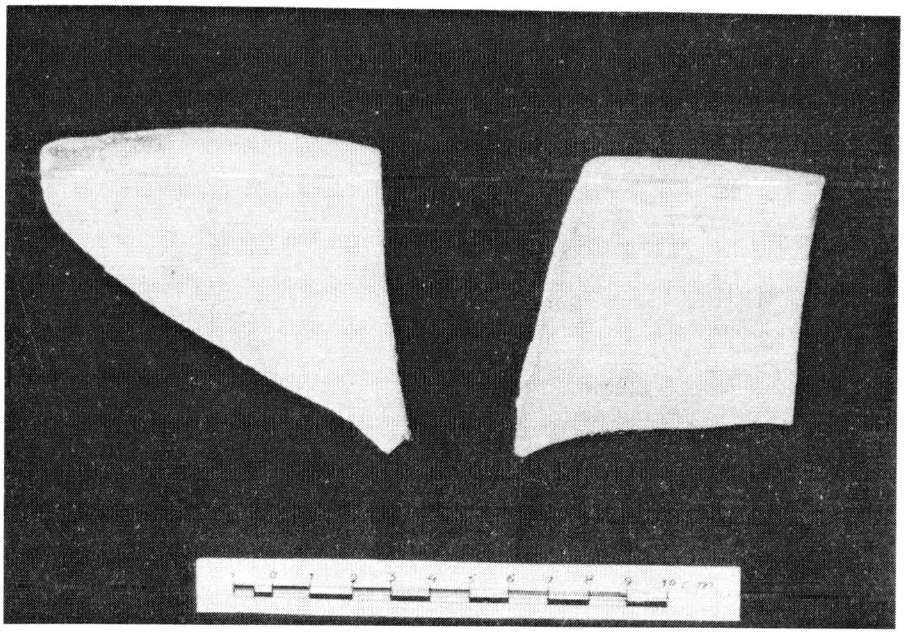

FIGURE 6.15
LEWAN TYPE BII BOWL SHERDS
From left: LWN 202, 206.

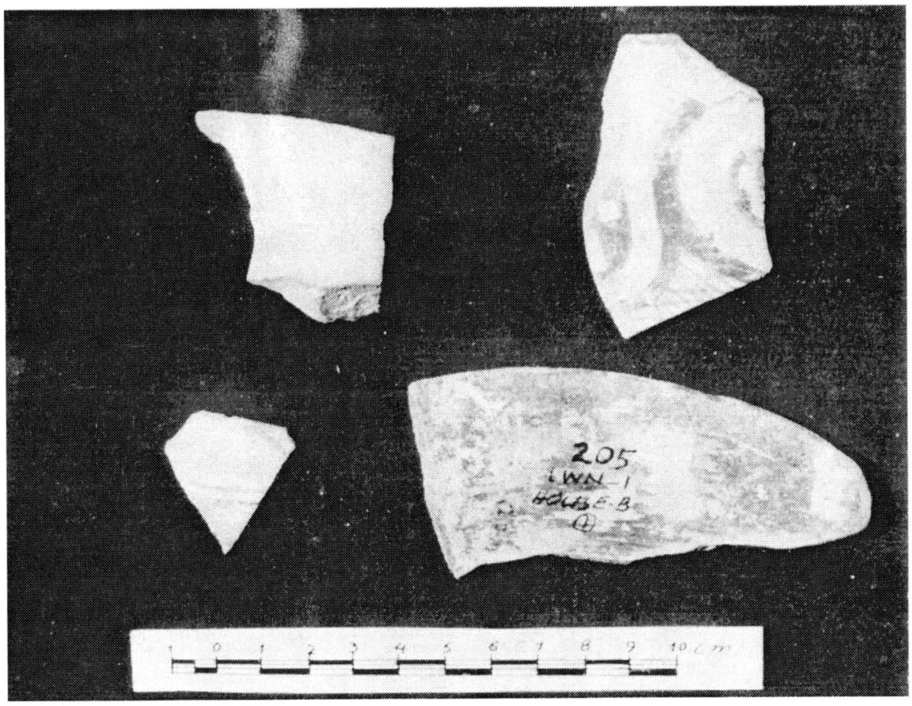

FIGURE 6.16
LEWAN TYPE BIII BOWL SHERDS
Top: LWN 208, 204. Bottom: 217, 205.

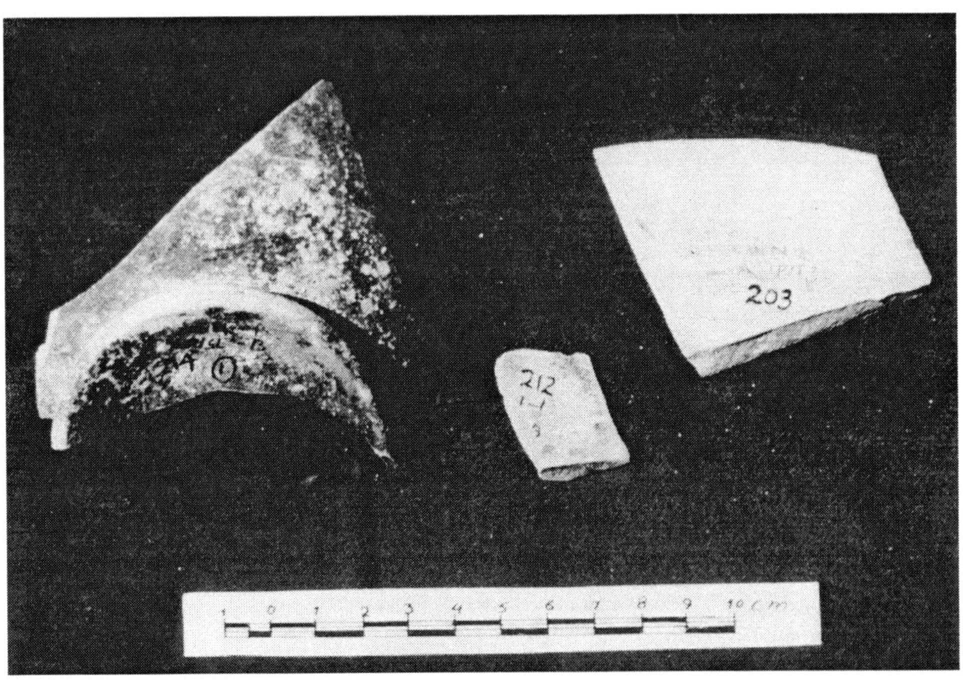

FIGURE 6.17
LEWAN TYPE BIV BOWL SHERDS
From left: LWN 214, 212, 203.

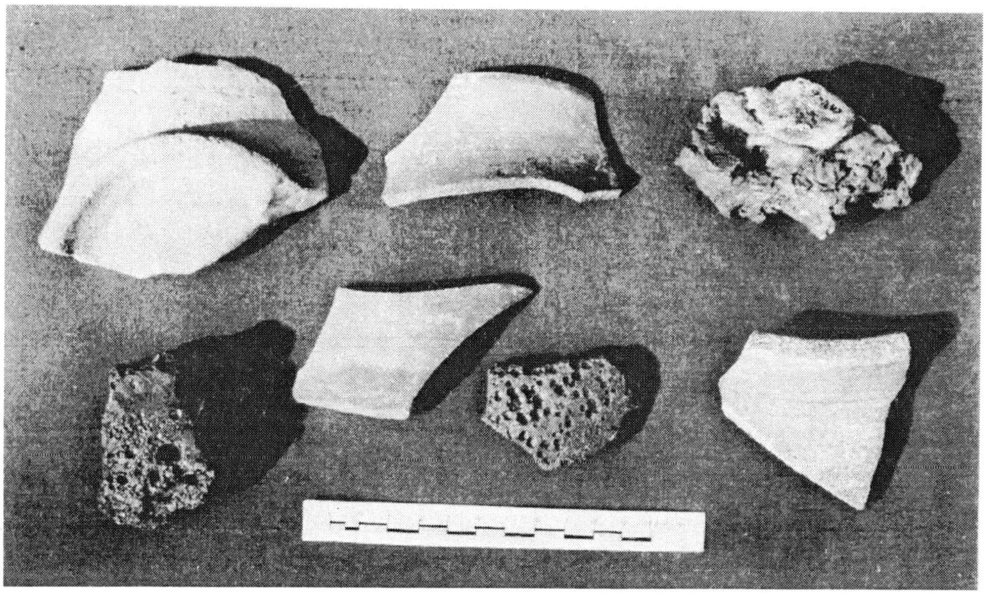

FIGURE 6.18

LEWAN WASTER SHERDS

Top: LWN 251, 249, 253. Bottom: LWN 252, 249 (same as top row), 250, 248.

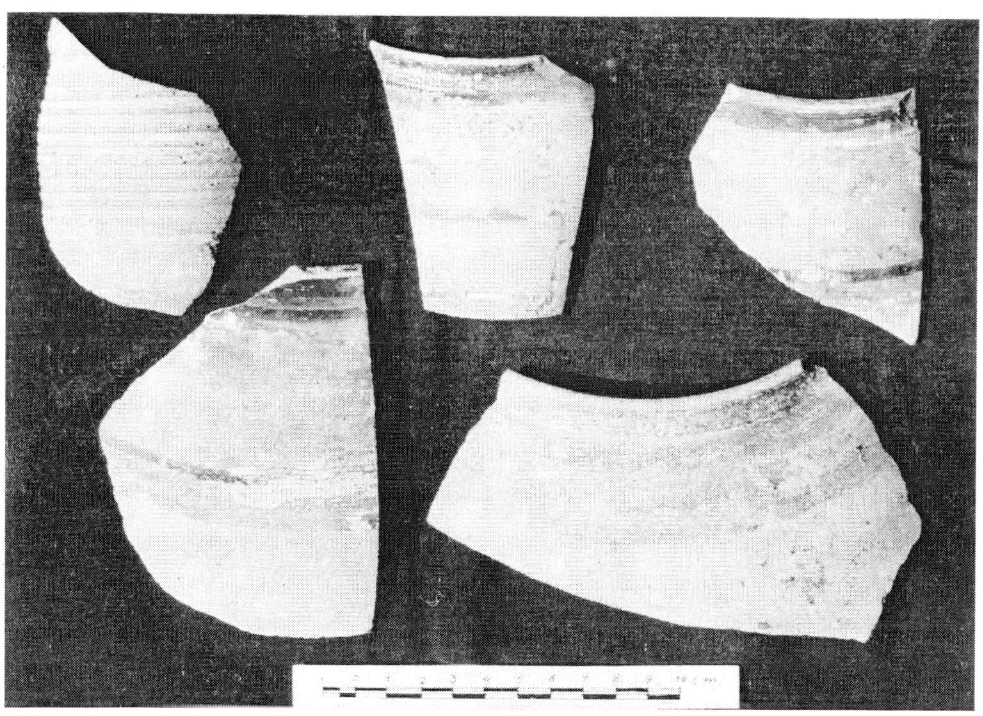

FIGURE 6.19

TARAKAI QILA TYPE AI JAR SHERDS

Top: TRQ 223, 224, 222. Bottom: 221, 220.

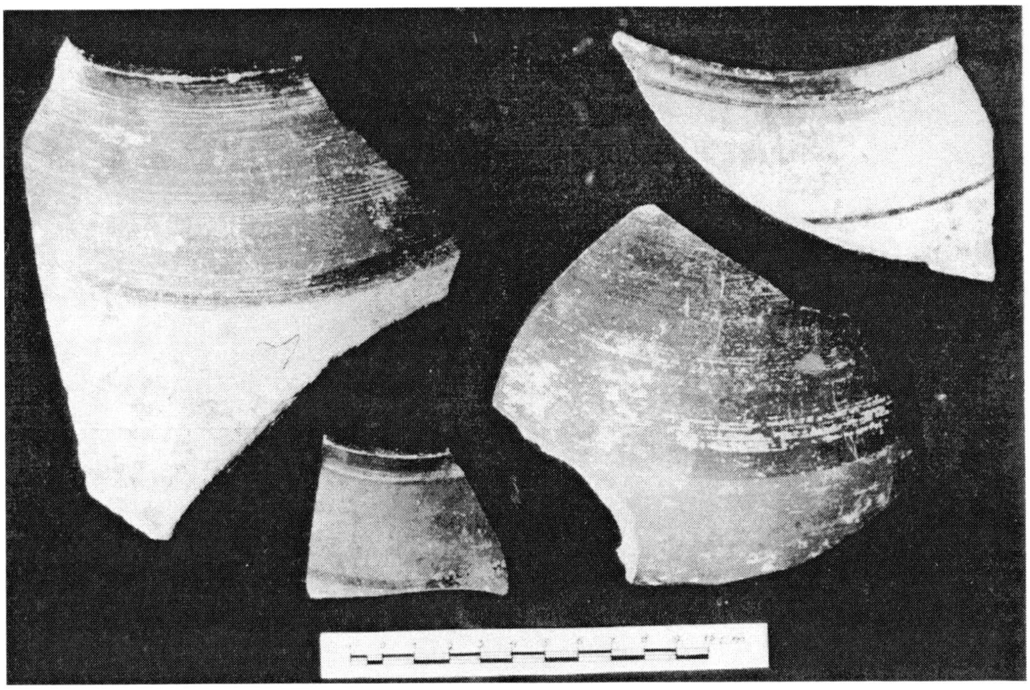

FIGURE 6.20
TARAKAI QILA TYPE AII JAR SHERDS
Top: TRQ 228, 227. Bottom: TRQ 226, 225.

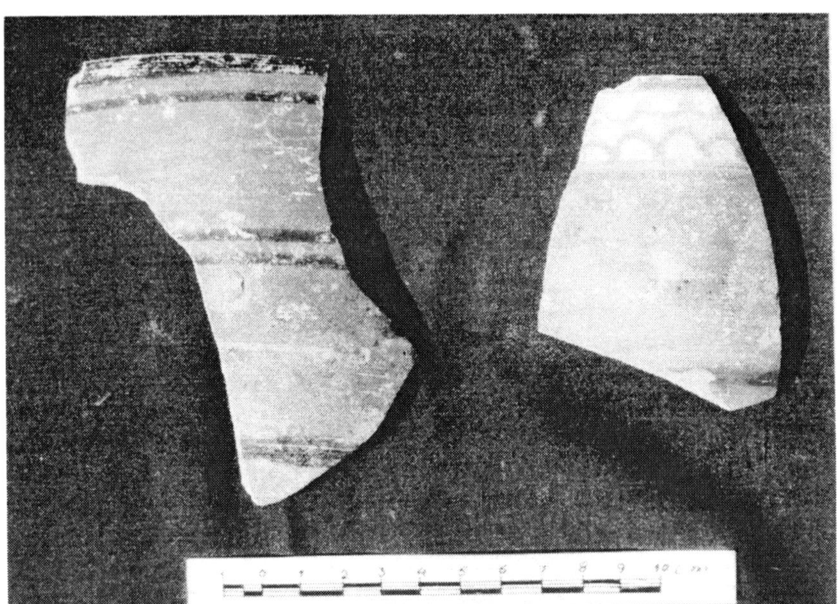

FIGURE 6.21
TARAKAI QILA TYPE BI BOWL SHERDS
From left: TRQ 230, 229.

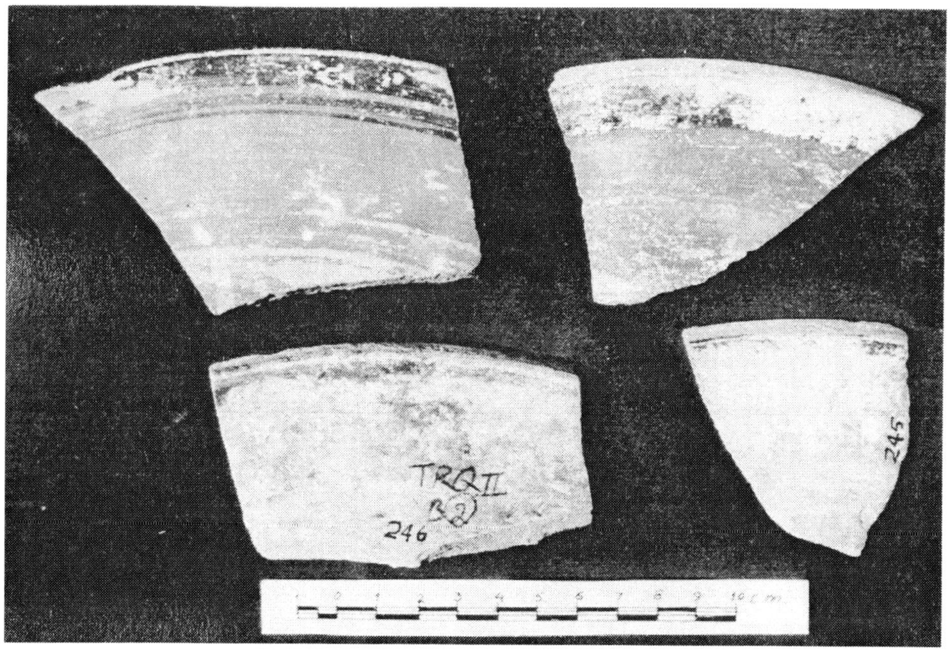

FIGURE 6.22
TARAKAI QILA TYPE BII BOWL SHERDS
Top: TRQ 247, 237. Bottom: TRQ 246, 245.

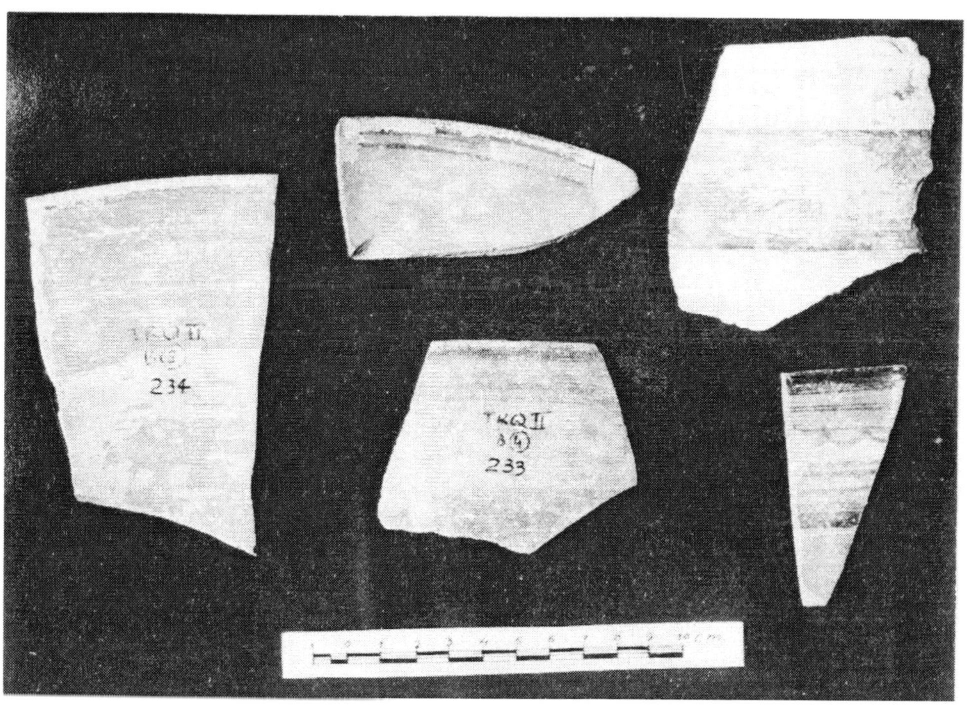

FIGURE 6.23
TARAKAI QILA TYPE BIII BOWL SHERDS
Clockwise from left: TRQ 234, 232, 235, 231, 233.

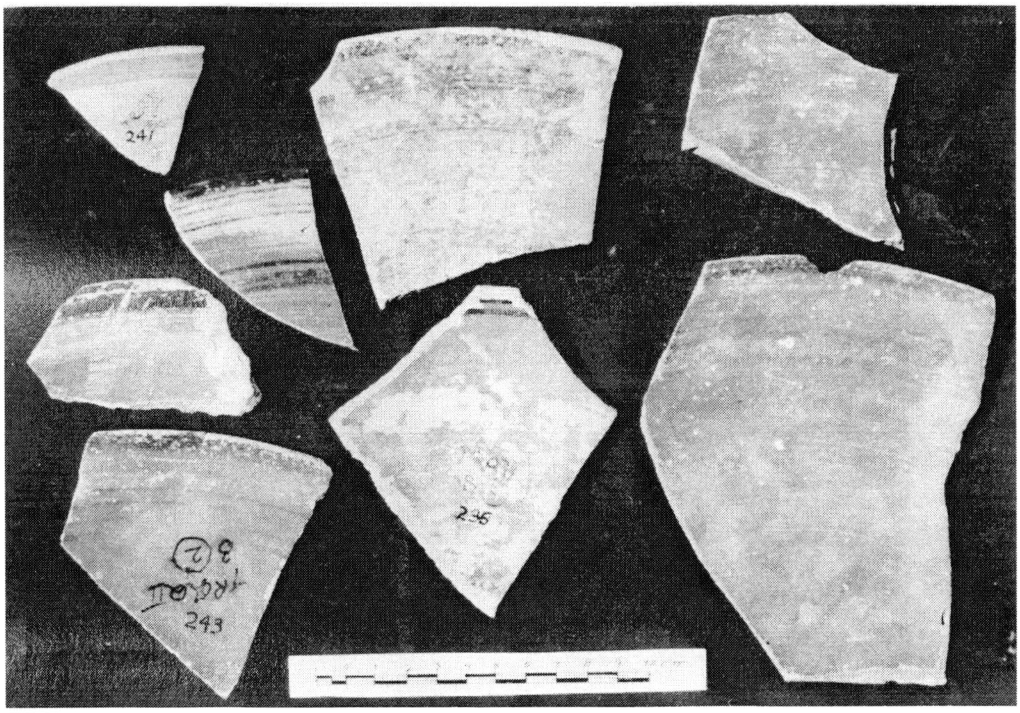

FIGURE 6.24

TARAKAI QILA TYPE BIV BOWL or BV DISH ON STAND SHERDS[1]

Top: TRQ 241, 239, 240, 242. Bottom: TRQ 244, 243, 236, 238.

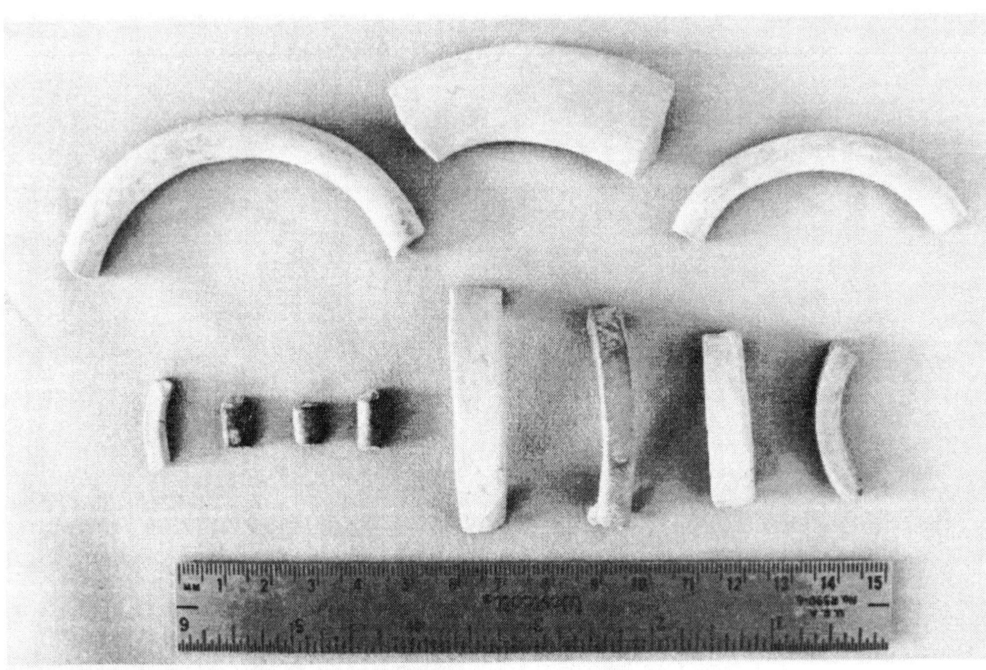

FIGURE 6.25

HARAPPA SINGLE-PIECE BANGLE FRAGMENTS

Top row: HRP 137, 136, 135. Bottom row: HRP 130, 131, 133, 132, 138, 141, 140, 139.

[1] See footnote 1, Table 6.1.

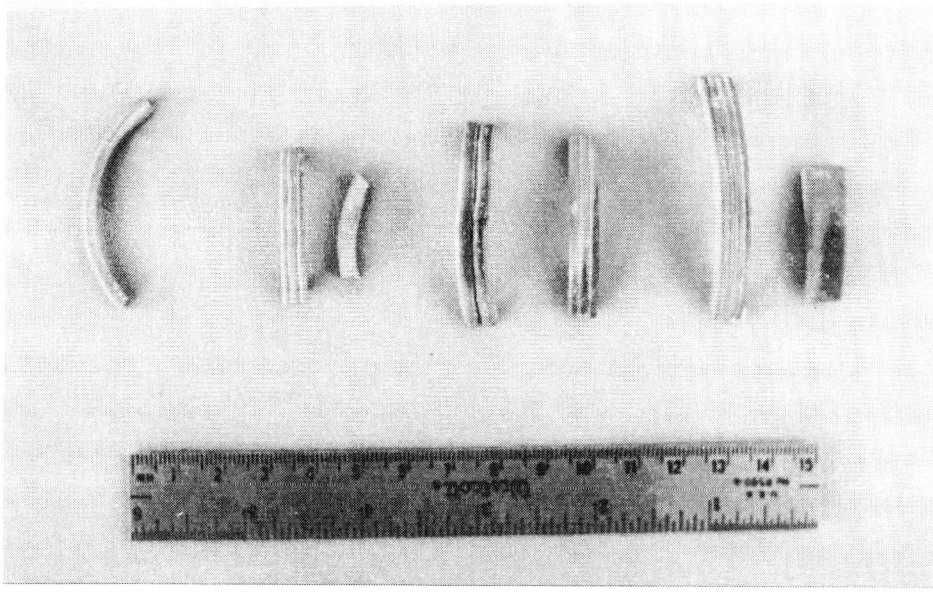

FIGURE 6.26

HARAPPA MULTI-PIECE BANGLE FRAGMENTS

From left: HRP 134, 146, 147, 143, 142, 144, 145 (single-piece).

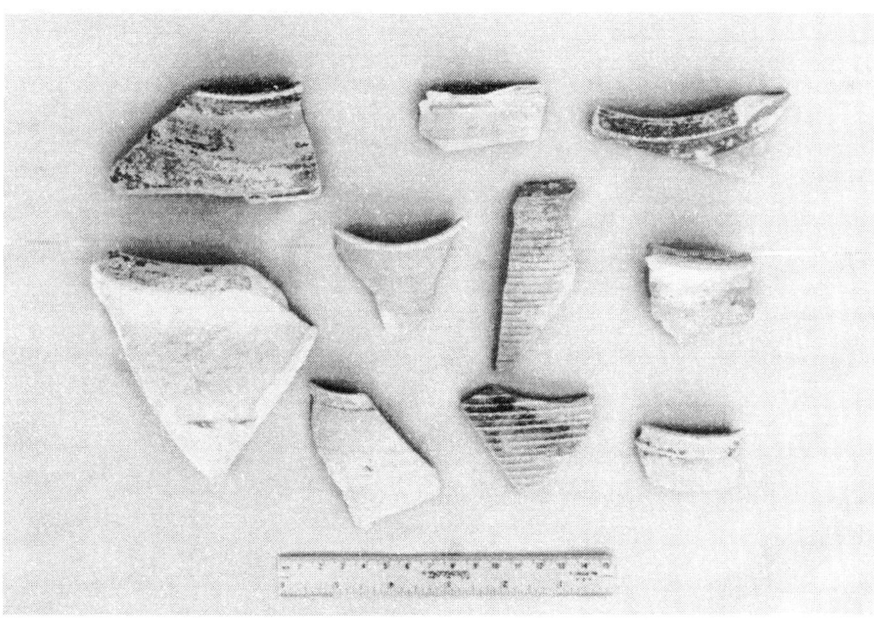

FIGURE 6.27

HARAPPA TYPE AI AND AII JAR SHERDS

Top: HRP 107, 105, 104. Middle: HRP 121, 100, 103, 108. Bottom: HRP 101, 102, 106.

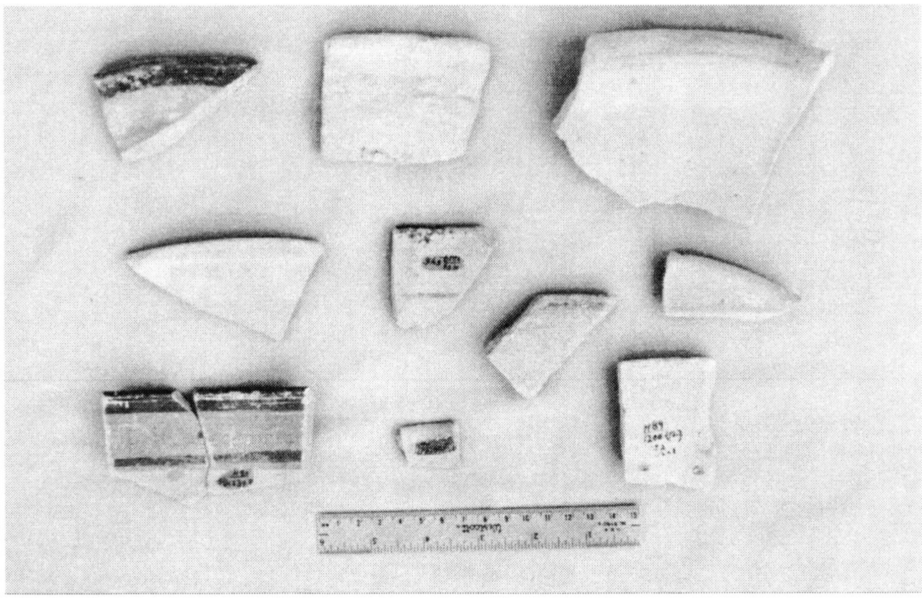

FIGURE 6.28
HARAPPA TYPE BI to BIV BOWL, and BV DISH ON STAND SHERDS
Top: HRP 122, 118, 119. Middle: HRP 114, 116, 113, 120. Bottom: HRP 117, 109, 110.

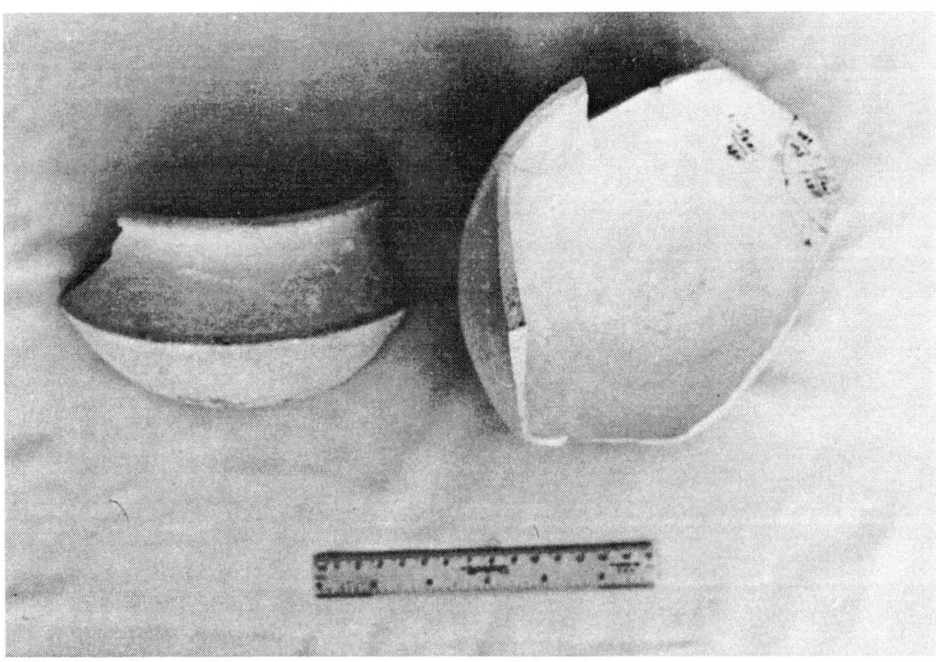

FIGURE 6.29
HARAPPA TYPE BI BOWL SHERDS
From left: HRP 111, 112.

Much of the material from Tarakai Qila had also, unfortunately, been stored in the same area as the Lewan material and suffered the same fate. In addition, a considerable number of the sherds had been selected for publication by the excavators and these had been placed in a storage area without any attached identification as to provenance. During the long delay in preparing the publication, the sherds and their unattached labels became accidentally mixed. These sherds, which had been selected as some of the best examples of each type, therefore could not be used for the present research. Nonetheless, sufficient material was located to provide a (minimal) data base from the site. A total of 28 sherds was selected from provenanced items in the remaining collection. Again, where possible, sherds identifiable as to parent vessel type were chosen. Figures 6.19 to 6.24 show the selection made for analysis from Tarakai Qila. No waster sherds were found with the Tarakai Qila sherd collection; also none were reported in the original excavation report (Allchin and Knox 1981). Moreover, the site was inaccessible due to tribal conflicts in the area during the period that the author was collecting sample sherds, so it could not be visited. Because of its proximity to the site of Lewan however, it was expected that natural inclusions in local clays would be similar to those from that site.

From the site of Harappa, a total of 19 sherds and 20 bangles were selected for thin sectioning. The bangles were chosen because, at the outset of the research, it was intended that these artefacts be compared amongst sites in a similar fashion to the pottery samples. However, too many delays were experienced in obtaining bangles from Rehman Dheri for comparison; and those which had been excavated from the two Bannu sites unfortunately appeared to have been lost. As the bangle samples had already been obtained from the Harappa Project, it was decided to thin section them in order to provide a comparison of their fabrics with those of the pottery. Also, examination of the bangle sections could shed some light on the technology of bangle manufacture in the Early Harappan. The pottery sherds were also selected so as to provide an example of each morphological type. As often happens when one is working to restricted time limits, there is not sufficient time to locate an ideal sampling for thin sectioning. Such was the case at Harappa; hence the limited number in the sample. The bangles which were thin-sectioned are shown in Figures 6.25 and 6.26, and the pottery samples selected are illustrated in Figures 6.27 to 6.29.

6.2 Sample preparation

Each of the sherd samples was photographed and a record made of the parent vessel type where applicable, the wall thickness, rim diameter where applicable, artefact number, colour, and surface decoration if any. Thin sections were then made of each using the portable field kit as described in detail in Part One.

The sections were cut perpendicular to the wall surfaces (and parallel to the wall cross-section). This was to ensure that any differences in particle density that may exist between the inner and outer surfaces of the walls could be observed and a representative count obtained. Also, any technological information such as observation of details characteristic of vessel forming technique, and any surface finish could be properly observed. Cuts were made through a cross-section of the rim, in the case of rim sherds. At least three slices of each sherd were made, in order to allow for spoiled thin sections, and to permit instruction of the technique to students at the University of Peshawar.

6.3 Data acquisition

Using standard petrographic techniques, each of the inclusions was identified by scanning the entire surface of the section under the petrographic microscope. For assistance with the mineral identifications, standard reference works were regularly consulted (Adams et al. 1984; MacKenzie and Adams 1994; MacKenzie and Guildford 1980; Rutley 1988; and Yardley et al. 1990). Particularly troublesome identifications were also kindly assisted by Dr Brinley Roberts of Birkbeck College Dept of Geology. For the quartz inclusions, a modal analysis was additionally carried out. This is described later in the section. For the minerals present in lesser amounts, a record was made of their size, angularity, sphericity, weathering, concentration, type and opacity (this variable is described below), according to that shown in Table 6-II. By applying a measurement scale to some of these variables, especially concentration, it was possible to make meaningful comparisons from site to site. Areas or sites exhibiting similar mineral suites could then possibly be differentiated by observing varying concentrations. Applying a numerical scale to concentration also permits average and mean concentrations to be derived, and a measure of confidence to be placed on differences observed from site to site.

The 'micaceousness' and birefringence of each fabric was estimated on a scale of 1 to 4, according to whether it was subjectively considered to be low or high, respectively. 'Micaceousness' refers to the concentration of very fine detrital micas (usually muscovite) visible in the fabric, and birefringence refers to the changes in transmitted light observed when the stage is rotated under crossed polars. It permitted recording of a further two variables that could be employed to classify or describe fabric types. Data processing is described in Section 6.4.

TABLE 6-II

MINERAL CHARACTERISTICS RECORDED

('x' denotes variable recorded, blank denotes not recorded)

Ang = angularity   Sph = sphericity   Weath = weathering
Conc = concentration   Opac = opacity under crossed polars

MINERALS	Size	Ang	Sph	Weath	Conc	Type	Opac
Plagioclase Feldspar	x	x	x		x		
Microcline Feldspar	x	x	x		x		
Biotite	x			x	x		
Muscovite	x			x	x		
Amphibole					x		
Carbonate					x	x	x
Quartz[1]	x[1]	x	x		x[1]		
Pyroxene					x		
Epidote					x		
Schist					x	x	
Other					x	x	

[1] Quartz data are analysed separately via modal analysis (refer to text)

The following variables were recorded on a scale of 1 to 4:

Size	1= <50 microns	Angularity	1= rounded
	2= 50-150 microns		2= subrounded
	3= 150-250 microns		3= subangular
	4= >250 microns		4= angular

<u>Sphericity</u>	1= low sphericity	<u>Weathering</u>	1= fresh
	2= moderate		2= slightly weathered
	3= semi-spherical		3= moderately weathered
	4= high sphericity		4= highly weathered

<u>Concentration</u> 1= absent
2= trace (estimated <0.5% of thin section area; 1 or 2 inclusions)
3= rare (estimated 0.5 to 2.0% of thin section area; 2 to 5 inclusions)
4= common (estimated 2.0 to 15.0% of thin section area; >5 inclusions)

Sizes were determined using the eyepiece graticule, which was calibrated as described below. With the exception of quartz, which was accorded a much more in-depth analysis with respect to size, the means of each of the variables were determined. These means were then averaged and a measure of the standard error of the mean obtained through statistical formulae, as described in Section 6.4 below. This permitted means to be compared from site to site as outlined above, or between morphological types, and allowed inter-site or inter-type comparisons to be made. Angularity and sphericity were estimated using a modification of scales similar to those presented in Adams et al. (1984: 3; citing Pettijohn et al. 1973).

Weathering of inclusions was estimated more subjectively. It was primarily under consideration as applied to the muscovite and biotite inclusions in order to provide another measure of possible intersite differences in the absence of significant variations in their respective mineral suites. In many cases the mica inclusions were too small for a weathering condition to be determined, but in the majority an accurate estimate could be made. The more advanced weathering features observed in the micas included slightly frayed ends (these were classed as category 3) extending to actual exfoliation separation along the boundaries of the mica plates (category 4).

In the case of the carbonates, under 'Type', a numbering system of 1-3 was used. These are not values but used as indicators of type. The number 1 was used to indicate bioclast. Bioclast is used here as a generic term to label skeletal particles which are the remains (complete or fragmented) of the hard parts of carbonate-secreting organisms (Adams et al. 1984: 39). Where bioclasts were observed, which were few, an attempt to identify the species was made. For this, Bignot (1985) was consulted. Identification proved somewhat difficult, because in order to observe details of microfossils, micropaleontologists usually prepare thin sections thicker than those prepared for mineral identification purposes. They are usually ground to a thickness of 30-50 microns, compared with the mineralogical standard of 30 microns which was used for the sections prepared for the current study (Bignot 1985: 9). Thinner sections tend to obscure details of the microfossils. Type category 2 were, for the most part, inclusions composed of microcrystalline calcite (micrite) or sparite in varying amounts. These were further sub-classified as to their type of opacity. Although it is somewhat unusual to classify carbonate inclusions in this manner, it was done because they appeared to fall into four quite distinct categories, according to the pattern of their opacity under crossed polars. Most of them appeared to be made up of at least two distinct components, i.e. a nucleus, and a surrounding cortex. These two components exhibited independently varying opacities. Therefore in light of detecting possible differences in these carbonate inclusions with respect to area or site, they were recorded according to the following: 1= cortex + nucleus opaque; 2= cortex opaque and nucleus micritic; 3= cortex micritic and nucleus opaque; and 4= cortex + nucleus micritic. These were recorded in the last column of the mineral data block of the data worksheets (refer to Figure 6.30 and its discussion in Section 6.4 below). Finally, with respect to 'Type', category 3 was used to indicate a type of inclusion that was present in many of the sections and could be described as rounded, non-pleochroic, deep orange-brown in plane polarised light, the same colour under crossed polars, and not birefringent. This type of inclusion was also noted to be more prevalent from some areas than others. Further, more detailed discussion of these carbonate inclusions is presented in Chapter 7; the purpose here is primarily to explain the method of recording the observations.

The category 'Other' included, for the most part, argillaceous inclusions, as defined and classified by Whitbread (1986). Argillaceous inclusions were grog; clay pellets (natural); or clay temper (added by potter). Each of these values was recorded on a data sheet and then entered into the computer database for analysis and data presentation (described in Section 6.4). Concentrations or sizes of quartz inclusions were not recorded on this scale because a much more comprehensive measurement of this mineral was conducted through the point- and area-counting procedure described below. Angularity and sphericity of quartz often showed a different value for large and for smaller inclusions in the same section; in these cases, an annotation was made on the data sheet entry to this effect.

Initially, the extinction angle of the plagioclase feldspar inclusions was measured, using the technique outlined in Rutley (1988: 114-5). It was intended that this be used to determine their composition between albite and anorthite (*ibid.:* 417), thus providing a characterisation which might have been useful in determining provenance. The standard method is to determine the extinction angle of at least five plagioclase inclusions, and the maximum obtained is taken as the extinction angle for the sample. This was initially done; however, it soon became evident that the concentrations of plagioclase feldspars in the thin sections were not large enough to permit five particles to be measured in each. Consequently these measurements were abandoned. Using just one extinction angle measurement as representative of the whole section would have given an erroneous indication. As indicated in Chapter 8, it is recommended that chemical analysis be conducted on these inclusions to investigate provenance.

In addition to the mineral identifications, a modal analysis of inclusion size was conducted of each section. Both point- and area-counting methods were used. Point-counting was used to evaluate the size distribution characteristics of all inclusions as a group, and area-counting was used to evaluate the size distribution characteristics of the quartz inclusions only. For both, the eyepiece graticule was used, at 40x. Prior to commencing either type of count, the graticule was calibrated for 40x magnification. This was accomplished by placing a stage micrometer (from Graticules Ltd; 100 x 0.01mm increments = 1mm) on the stage with the graticule installed in the eyepiece. By aligning the linear axis of the micrometer image with one of the axes of the graticule, a measurement of the length of each of the increments of the graticule and the length of a side of the entire graticule square area was determined. This enabled measurement of each inclusion, and the area of the graticule. For inclusion with microphotographs prepared for presentation, a microphotograph of the stage micrometer was taken at each power used. By reproducing the micrometer photograph at the same magnification as the thin section photographs, a reader can measure inclusion sizes from the microphotographs.

For the point-count analysis, four locations on each section were chosen: one near the top or outer surface of the sample; one near the bottom or inner surface; and two at random within the thickness of the section. At each location, the eyepiece graticule was used to obtain 100 points. This provided a reasonable sample size of 400 points for each thin section (see discussion below with respect to the statistical validity of using this number so as to obtain an estimate of the population distribution of all inclusions in the section, together with a confidence interval). For each of the 100 intersections of the graticule, a record was made as to one of three categories: matrix, sand, or silt. Matrix simply means that the intersection fell on a point which was not an inclusion or a void; and could be classed as part of the clay matrix. The categories sand/silt were used as size categories; with no inference as to nature of mineral. These refer to the standard Udden-Wentworth scale for classifying sizes of inclusions, where sand-size inclusions are those greater than 62.5 microns (0.0625mm) and silt-size are less than 62.5 microns. The only exception to recording the size of the inclusion was if the intersection fell on a void; these were not counted for purposes of the compositional analysis, as it was desired to characterise the fabric according to relative concentrations of each of the three classes only. Voids do not add to this description of the original fabric mixture before firing. An inclusion was counted if any part of it fell under the intersection of the graticule. If two or more intersections fell on one particle, it was counted each time. This double-counting is at variance with point counting as used for size distribution analysis, where counting the same inclusion twice would obviously skew results. For these cases it has been standard practice to use a graticule spacing which is at least as large as the largest inclusion diameter. However, the data derived from the point-counting described only the area of the slide that was covered by each category and not numbers of each size range (as was done with the quartz inclusions described below). In order to describe such area-related data it is necessary to count whatever falls under each intersection. Count data were recorded on the data sheet for each section and entered into the computer database for processing and analysis. It was used to produce a percentage compositional chart as part of the fabric description of each section. Data processing and analysis are described in Section 6.4.

To describe the size distribution of the quartz inclusions in each section, an area-count of each inclusion was conducted. Data obtained from point counts can also be used to define size distributions, but because some of the size ranges are represented by such a small percentage of the total number of points counted, to achieve a high enough confidence level would have entailed measuring over 1,000 points in each slide (see Section 6.4). Many of the slides' sizes would clearly not have permitted this, because the point-counting method does not count each and every inclusion in the graticule area as does area-counting. Moreover, counting such a large number would be prohibitively laborious.

For the area count of quartz inclusions, the image of the graticule area was first placed in a fairly central location on the slide, about midway between the inner and outer surface of the ceramic cross-section. All quartz inclusions that fell fully within the area were counted. For each inclusion, measurement was made of its longest axis. If it was less than about 30 microns, it was not counted because many inclusions of this size would have been hidden within the thickness of the section and the count obtained would have been misleading if taken as representative of the whole population of inclusions that size. In some cases, a judgement had to be made as to the sharpness of the boundary of the inclusion. If the boundary was diffuse or merging for those grains slightly smaller than 30 microns, an estimate was made as to whether or not it was 30 microns. This was done because a particle may well have been 30 microns, but it might have been fully or partly contained within the thickness of the section. In these cases, the boundary of the particle is usually indistinct or diffuse because part of the transmitted light is passing through a layer of the matrix. A reasonably accurate estimate can nevertheless be made. For those quartz inclusions less than 30 microns, boundaries can become so indistinct as to render this technique of little use. These were not recorded.

It was found that the most efficient method at each location of the graticule was to break the area down into strips defined by rows of grid intersections. The whole area of the graticule was thusly divided into ten strips. Each strip was scanned, initially only counting those quartz particles which fell into the 30-50 micron size category. By scanning all ten strips of the graticule area, a total for that size category was recorded. Where a grain fell into more than one strip, it was only counted once, and omitted from the count for the other strips into which it fell. The process was then repeated for each size range, up to those >350 microns in size. The method especially facilitated counting of the smaller inclusions where counts were high and it would otherwise have been easy to lose one's place on the graticule area.

A slight difficulty was encountered where a quartz inclusion was in extinction under crossed polars at the stage position used for counting. Because the inclusion was in extinction, it could have been overlooked and not counted. In order to include these for measurement, it was necessary to rotate the stage about 5 degrees so that the inclusion in question would come out of extinction. The problem with doing this was in ensuring that the stage was returned to its original position to resume the count. The rotating stage of the laboratory petrographic microscope has a detent position in which it can be set at the beginning of a count. When it was necessary to deviate from position, it could be easily regained to resume the count. For the McArthur microscope however, this was not possible. When using the McArthur microscope, the position of the stage was noted before a count was commenced; this ensured that the same position could be returned to. The problem was not encountered in quartz inclusions where undulose extinction was present, which was a common occurrence, because part of the inclusion was always visible regardless of the position of the rotating stage.

No distinction between monocrystalline and polycrystalline quartz was made in the measuring and counting. Both were present in all of the sections observed (see Chapter 7). It is likely that characterisation or provenancing of fabrics could have been assisted by recording these separately and observing any distinguishing differences in ratios or size distributions between sites. The option was seriously considered; but, in light of the length of time involved, it was rejected. Each section took approximately 45 minutes to complete the counting process, once the learning curve was complete. To count a statistically valid number of each of the two groups would have doubled this observation time. Additionally, only those quartz inclusions that were purely quartz were included for the area count. Many of the sections contained quartz inclusions which themselves contained other minerals such as micas, epidote, amphibole, or feldspars. These were generally classed as quartz schist or gneiss, according to texture; in those instances where quartz was the dominant mineral.

All measured and counted quartz inclusions greater than 30 microns were placed into one of six size intervals: 30-50, 50-125, 125-200, 200-275, 275-350, or > 350 microns. The total number of inclusions that fell into each of the six categories was recorded on the data sheet. If the total number of inclusions did not exceed 250 in the first area counted, the graticule was moved to a new location on the section and the process repeated. In each instance, all grains that fell within the area were counted. This was continued until at least 250 grains were counted in total for each section. In the majority of cases, over 300 grains were counted, and in many, over 500. This number was chosen in order to ensure that a statistically valid number of points were counted such that the smaller percentages were adequately represented at the level of two standard deviations (see Section 6.4). Only those inclusions that fell fully within the graticule area were counted; those that had any part of their perimeter outside were not counted. A small number of the sections were too small or contained too few quartz inclusions to attain this level of confidence, e.g. a maximum of only 180-200 grains were available to be counted. In these cases, as many points as possible were included. The corresponding confidence interval was given by the process outlined in Section 6.4. Six of the samples did not even contain 150 usable counts; in these cases, counts were not made, and they were not included in the statistical base. All numbers were entered into the computer database for processing, analysis, and data presentation, described in the following section.

6.4 Data processing and analysis

All data recorded were transcribed from the hand-recorded sheets into a Lotus 123R5 database on a Zenith Data Systems Z-Star 433VL computer. This enabled the data to be organised onto computer datasheets, analysed and placed in suitable form for analysis and presentation. A single Lotus 123 worksheet was used for each thin section, and each site was made a separate file. A sample of a worksheet is provided at Figure 6.30.

For each worksheet, organisation of data was as follows[1] (refer to Figure 6.30): Firstly, the top left block of data contained the thin section number, in this case RHD 23, indicating it was from the site of Rehman Dheri and was numbered 23 in the series. This is followed in the same area by sherd identification or artefact number, period and level (provenance), morphological type, rim diameter in centimetres, thickness in millimetres, and finally the area of measurement of the thin section, in square millimetres (sqmm), used for the area count of quartz inclusions.

The block of data immediately to the right of this contains all of the quartz size distribution data. The top row labels each size range column, in microns, finishing with 'numpts', which is the total number of quartz inclusions counted for that section. The second row contains the absolute number in each size range counted, ending with the total in the final column. The remaining rows in this data block are values calculated from the absolute numbers in each size range by using formulae entered into each cell. For example, the third row is the calculated percentage of the total in each size range. It is calculated by applying a simple subroutine in the 1-2-3 software using the absolute number in each size range divided by the total number of points and multiplying by 100.

[1] A key to the worksheet data is also provided in summary on the first page of Appendix A.

FIGURE 6.30

SAMPLE OF DATA RECORD SHEET

See next page for a key to the abbreviations used; and the text for details

	period	level	morph type	rim dia	thick	sqmm
RHD 23						
B IV/19	2	11	BV	38	7	25

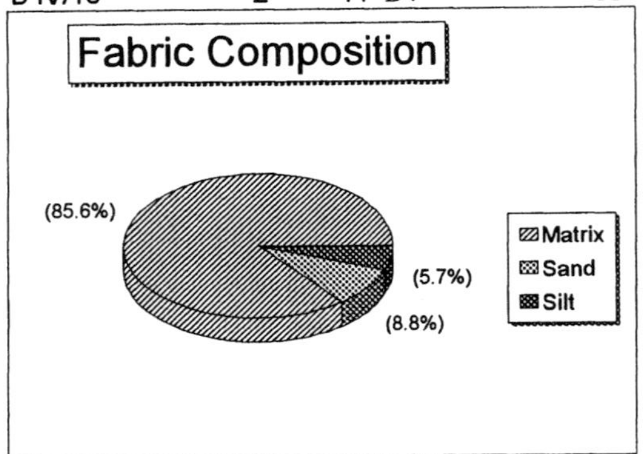

Qz Size	< 50	50-125	125-200	200-275	275-350	> 350	numpts
Number	114	72	64	15	1	1	267
%	42.70	26.97	23.97	5.62	0.37	0.37	
sigma	3.03	2.72	2.61	1.41	0.37	0.37	
2psigma	48.75	32.40	29.20	8.44	1.12	1.12	
2msigma	36.64	21.53	18.74	2.80	-0.37	-0.37	
Conc	4.56	2.88	2.56	0.60	0.04	0.04	
Fabric	Matrix	Sand	Silt	Micac	Biref	Colour	
	332	34	22	3	3	medbrn	388

Minerals	Size	Ang	Sph	Wth	Conc	Type	CarOpa
PFs	2				2		
MFs					1		
Bio	3			3	2		
Mus	2			3	4		
Amp					1		
Car	2				3	2,3	2
Quartz		2	3			m,p	
Pyrox					1		
Epidote					1		
Schist					2		
Other							

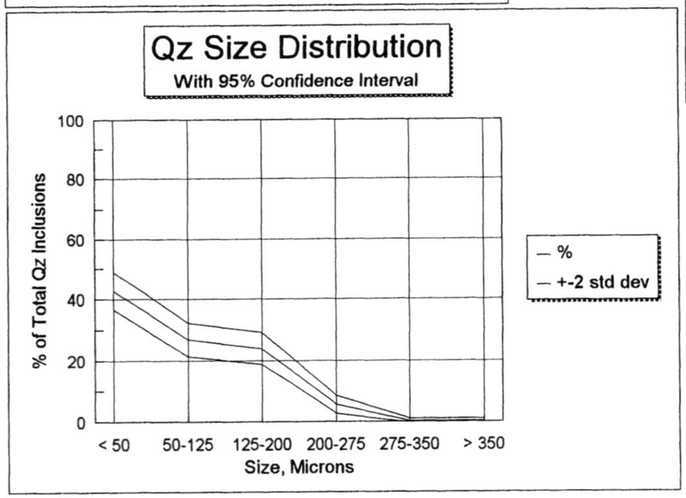

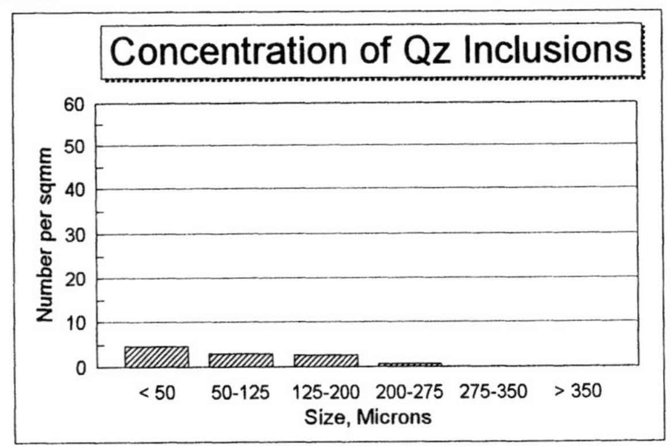

%matrix	85.57
%sand	8.76
%silt	5.67
conc<50	4.56
conc>50	6.12
%<50	42.70
%>50	57.30

KEY TO ABBREVIATIONS USED IN WORKSHEETS AND FIGURE 6.30

(in order of appearance, by row)

For a detailed description, refer to Sections 6.3 and 6.4

morph type	= parent vessel morphological type (described in Section 5.4)
rim dia	= for rim sherds, rim diameter in cm
thick	= average wall thickness, or, for bangles, cross-section dimensions
sqmm	= area of thin section used for area count, in mm^2
Qz size	= refers to size range of quartz inclusions, in microns, next 6 columns
numpts	= total number of points counted for area count
%	= percentage of total counts occurring in each size range
sigma	= one standard deviation, σ
2psigma	= $\% + 2\sigma$
2msigma	= $\% - 2\sigma$
Conc	= concentration, number of each size range, per mm^2
Micac	= micaceousness of matrix in thin section (1 = low, 4 = high)
Biref	= birefringence of matrix in thin section (1 = low, 4 = high)
Colour	= ltbrn (light brown), medbrn (medium brown), dkbrn (dark brown)
Ang	= angularity (1= low, 4 = high)
Sph	= sphericity (1 = low, 4 = high)
Wth	= weathering (1 = low, 4 = high)
Conc	= concentration (other than quartz, 1=absent, 2=trace, 3=rare, 4=common)
CarOpa	= carbonate opacity (refer to section 6.3 for explanation)
PFs	= plagioclase feldspar
MFs	= microcline feldspar
Bio	= biotite
Mus	= muscovite
Amp	= amphibole
Car	= carbonate
m, p	= for quartz, monocrystalline or polycrystalline, respectively
Pyrox	= pyroxene
ind	= indeterminate
arf	= argillaceous rock fragment
vrf	= volcanic rock fragment
conc<50	= concentration of quartz inclusions <50 microns in size, per mm^2
conc>50	= concentration of quartz inclusions >50 microns in size, per mm^2
%<50	= quartz inclusions <50 microns in size as percentage of total quartz
%>50	= quartz inclusions >50 microns in size as percentage of total quartz

The next three rows, 'sigma', 'p2sigma', and 'm2sigma' calculate a confidence level for the representation obtained by taking a sample of the parent population of quartz inclusions. The value for 'sigma', or one standard deviation σ, is obtained from the formula

$$\sigma = \sqrt{\frac{p(100-p)}{n}}$$ (Van der Plas and Tobi 1965: 87)

where: p = the real content of a mineral in percent by volume,
n = the total number of points counted.

This means that the estimate will lie between $p+\sigma$ and $p-\sigma$ in about 68 out of 100 cases, between $p+2\sigma$ and $p-2\sigma$ in about 95 out of 100 cases, and between $p+3\sigma$ and $p-3\sigma$ in about 99.7 out of 100 cases. The above formula was entered into each cell in the row below the percentage previously calculated. The values of each percentage and the total number of points counted were used to calculate σ values corresponding to each size interval. It is noted that instead of the real content the estimated content is used. It can be shown that this does not cause seriously different results (Van der Plas and Tobi 1965: 88). The confidence level of 95 percent was chosen, i.e. plus or minus 2σ, as acceptable for the present study. These values are shown in the next two rows. They mean that, using the percentages calculated in the worksheet shown at Figure 6.30 as an example, we can be 95% confident that the concentration of quartz inclusions 30 to 50 microns in the sherd sampled as RHD 23 lies between 36.64% and 48.75%. If a narrower range at the confidence level of 95% was desired, a larger number of points would have to be counted. This approach provides a lesser degree of confidence in relative terms, however, for the size ranges at the lower concentrations, e.g. in this example, for those in the range 50-125 microns. The result here shows that we can be 95% confident that the concentration in this size range lies between about 21.5% and 32.4%. On the other hand, at the 1σ level, we can be about 68% confident that the concentration of quartz inclusions of size 50-125 microns lies in the range 24.2% and 29.7%. These values may appear at first to be somewhat broad for accurate comparisons of fabrics between different sites or types, or for accurately defining percentage concentrations in fabric types. However, to improve these confidence intervals significantly would have required that at least double the number of points be counted. This would have entailed approximately double the hours of manual counting on the microscope, which would have been unnecessary. It is time-wasting to count more than the number of inclusions necessary to achieve a desired confidence level. The present example, Figure 6.30, is one of the lower numbers of points counted; most are in excess of 300, which provides a narrower range for the confidence interval. The levels chosen proved to be adequate for the purposes and inferences to be made (see Chapter 7).

The final row of this data block contains the calculated concentrations or densities of each size interval in grains per square millimetre (mm^2). These were calculated by first taking the absolute number of quartz inclusions counted in each size interval and dividing it by the size of the area counted, i.e. 25 mm^2 in the case of RHD 23 (obtained from the block 'sqmm').

The 'Fabric' data field, seen on the two rows immediately below the quartz data block, contains data relevant to each fabric. The aforementioned point count results are entered here for each of the three categories matrix, sand or silt. The next two columns recorded the micaceousness ('Micac') and birefringence ('Biref') observed for each fabric. This is followed in the next column by the colour of the fabric as observed in thin section. Finally, the end column here provides the total number of points counted for the point count conducted. Confidence intervals were not calculated for the counts made in this case. This was because the data were used only for an approximate comparative estimate of the relative concentrations of each category and for plotting the fabric composition chart.

Below this data is the mineral data block. Mineral abbreviations used are as follows:

PFs = plagioclase feldspar	Car = carbonate
MFs = microcline feldspar	Pyrox = pyroxene
Bio = biotite	Amp = amphibole
Mus = muscovite	

This block lists all of the parameters recorded relevant to each mineral: size, angularity ('Ang'), sphericity ('Sph'), weathering ('Wth'), concentration ('Conc'), type, and carbonate opacity ('CarOpa'). An explanation and scale used for each of these was presented above in Section 6.3. For quartz type, 'm' and 'p' refer to microcrystalline and polycrystalline, respectively. These values were used to determine the means and standard errors of the means in each category. The method for this is explained in a later paragraph. In the category 'Other', abbreviations were used as follows: argillaceous inclusion ('arg'); grog (no abbreviation); clay (no abbreviation); volcanic rock fragment ('vrf'). Where cells are blank, these parameters were not recorded for the corresponding mineral.

To the right of the quartz concentration chart, concentration data are summarised. The first three entries, % matrix, % sand, and % silt, are self-explanatory and are those contained in the 'Fabric' data block converted into percentages. The next two show concentrations of those quartz inclusions <50 microns and the total of those >50 microns. Finally, the lower two entries show these as percentages. All of these were separately placed on the data sheet so that the program could read the worksheets in 3-D ranges for summarising and averaging textural data for the whole site, and facilitating fabric classifications. These processes are discussed below.

Charts were then produced by the Lotus 123R5 program. The three charts derived for each thin section were: quartz size distribution, concentration of quartz inclusions, and fabric composition. This selection was produced because their combination enables the textural nature of the fabric to be determined at a glance. The visual impact permits an immediate comparison of any two or more of the thin sections without having to wade through numbers and attempt to visualise distributions. Importantly, in case of the quartz size distribution chart, it allowed identification of the modality which showed the possibility of potter manipulation of the raw clay via levigation or temper addition. This was particularly useful because in most cases the larger particles were of similar angularity and sphericity as the smaller ones, resulting in little discrimination for identifying tempers. It is realised that these quartz size data should properly be presented as histograms, but the line graphs provided a stronger visual tool for defining fabric types. Lotus 123 macro subroutines were recorded and used to produce each graph. This feature provided a convenient and time-saving method for producing charts. For the quartz size distribution chart, values from the % row, the 'p2sigma' and the 'm2sigma' rows in the quartz size distribution data block were plotted together on one graph for each section. This provided a presentation which showed the size distribution together with an immediate visual identification of the confidence interval.

The quartz size distribution provides a ready presentation of the relative densities of the different size ranges present in the thin sections, but it does not give an indication of individual densities. To provide density information, a second graph of quartz size data was produced for each section. This histogram shows the concentration of each size group of quartz inclusions. It is plotted using the concentration densities of the last row of the quartz data block, and presents quartz concentration by size range in terms of number per square millimetre of thin section surface area. It provides at a glance the densities of each size range, and is very useful to use as a ready comparison between fabric types or as a defining factor for fabric types. The same ordinate scale (0 to 60 per sqmm) was used in each to facilitate comparisons amongst sherds. Finally, a pie chart of fabric composition data was produced using data from the point count. This presents the composition (all inclusions) of each sherd in terms of percentage of the three categories: matrix, sand or silt.

The foregoing provided summaries of raw data for each thin section. These are included as Appendices A to C, and in this form can be used by future researchers as a baseline for building a regional expansion. Further data analysis was conducted using the Lotus 123R5 database. 3-D ranges are ranges of data spanning a number of different worksheets contained in the database. The normal two dimensions are row and column in a single worksheet. The third permits a single cell or collection of cells in two dimensions to become a range in three dimensions. For example, to obtain the average concentration of plagioclase feldspars in all of the sherds examined from the site of Rehman Dheri, a range spanning the cell where the concentration value was recorded over all of the worksheets for that site is taken for analysis. This selects that value for all sheets so that the programme could be instructed to perform a statistical analysis over that range to provide a measure of the average concentrations of plagioclase feldspars in Rehman Dheri pottery, together with other statistical information. Similar analyses can be performed for other variables and functions. By performing analyses for individual sites, and then combining them, site comparison charts could be prepared for each variable. This process was carried out to determine the average values for each of the variables recorded in the minerals data block, using the pre-recorded @ functions of the programme. The resulting summary of combined data was placed in a new worksheet at the end of each site file, then copied onto a new file which combined the individual site data. A similar process enabled a comparison of variables between chronological levels at Rehman Dheri.

In addition, for the values of mineral concentrations, the standard error of the mean was calculated, using the programme available with Lotus 123R5. This gave a measure of the level of dispersion in the values recorded for each parameter (Shennan 1988: 302-3). For purposes of this standard error of the mean calculation, the estimates made of each of the concentrations in each thin section was treated as a measure of the mean concentration for that section. Charts were then prepared of each of these average values for each of the sites, using bangles as a separate category in the case of Harappa. This provided charts of each with five categories, and permitted a visual comparison of the variations observed site-to-site. For example, one chart would include the average concentration of plagioclase feldspars for all five categories as bars on a barchart. This allowed immediate comparison of the different concentrations of plagioclase observed amongst the sites and site areas.

Plots were prepared of the relative occurrences of concentrations of quartz inclusions in the 30-50 micron size range and for the >50 micron size range for each site, again treating the data from the Harappa bangle selection as a separate category. This was done by programming the Lotus 123R5 to provide the number of samples in each that fell into pre-determined concentration intervals. The ranges were set at intervals of 10 inclusions per square millimetre, and spanned the range 0 to 70 per square millimetre, for the inclusions <50 microns; and intervals of two inclusions per square millimetre, from 0 to 12 per square millimetre, for inclusions >50 microns. The programme simply determined how many sections fell into each range. When plotted, these charts were used to show the mean and distribution of each category. Again, this permitted intersite comparisons; by allowing examination of differences in means and distribution characteristics such as skewness and kurtosis.

A further analysis was made of the quartz size distribution data from the >50 micron range. Four of the five size categories contained therein, i.e. 50-125, 125-200, 200-275, and 275-350 microns were subjected to the same analysis as in the foregoing paragraph. The fifth, >350 microns, was not considered because the very low number of occurrences in this size range did not allow any meaningful site-to-site differences to be observed. It was used however for other purposes such as examination of technological data and grouping into fabric classifications. The number of occurrences of concentrations of each of the four size ranges was totalled by site, using increments of one per square millimetre from 1 to 7. The number falling into each of these increments was totalled and plotted; one chart for each size range. This produced four graphs for each site (and four for the Harappa bangle sample). These were then examined for differences in distribution characteristics.

From the fabric composition data, or the results of the point count, similar summaries were made of each category. Concentrations of silt-size and of sand-size inclusions were prepared on a chart and examined as to distribution characteristics. This was accomplished by creating percentage intervals of five percent and determining the number of occurrences in each. It produced a further two charts per site (and two for the Harappa bangle sample). Plotting these provided a visual indication of the distribution mean, skewness and kurtosis. These were compared site-to-site.

The results of the mineralogical examination and the size data charts produced were visually examined and compared, to form groupings which will be termed 'fabric types'[1]. Variations in mineral concentrations, percentage of silt- and sand-size inclusions, and characteristics of the quartz size distribution data were considered as the basis of the fabric groupings. The purpose of the fabric groupings was to first define a group of fabrics manufactured at Rehman Dheri; and then, using these as 'master fabrics', compare them to fabrics produced at the other sites. In this way, it was thought, a determination could be made as to whether or not technological innovations from a main production site were being communicated to the smaller sites, primarily Lewan and Tarakai Qila. It was expected that Harappa, being a major site of its own accord, may or may not have had similar fabrics as Rehman Dheri under this assumption. This was discussed in Section 4.2. Chapter 7 discusses the analysis and interpretation.

[1] The term 'fabric type' makes an implicit assumption that the potter(s) intended to create them. This of course may not have been the case at all . It is only a value created by the researcher. Further discussion is contained in Chapter 7.

All of the foregoing data were used to define the characteristics of fabrics from each of the sites and the bangles from Harappa; for provenance studies. Variations in concentrations of each of the minerals observed were subjected to a one-way analysis of variance test to determine the degree of confidence which could be placed on the hypothesis that they came from different populations. This was conducted for each of the mineral categories that were measured on the 1 to 4 scale discussed earlier in the chapter. These findings were used to make inferences regarding provenance of sherds.

In addition to facilitating fabric groupings, examination of the size distribution characteristics of the quartz inclusions allowed some assessment of ceramic technologies employed by potters at the respective sites. In particular, it permitted the nature of temper usage to be assessed. Inferences could be made about the amounts of temper utilised, the sources, and the type.

CHAPTER SEVEN

ANALYSIS AND DISCUSSION

This chapter presents the main analysis and discussion of the data. It is divided into two principal sections: individual site analysis; and intersite analysis. Firstly, data from Rehman Dheri is presented and discussed. This site had the largest sample; and, because the samples were selected to represent the same types from all three periods, it provided the opportunity to examine variations in ceramic technology through time as well as from type to type. This is followed by similar analyses of other site data. Somewhat less space is devoted to the other individual sites, as this sort of representative sampling was not available from all of the sites. Following the studies of the ceramics from individual sites, the second section compares them from site to site, assessing the degree of compositional and technological similarity or dissimilarity. From this examination, inferences and suggestions are made with respect to the level of interaction that was present amongst the four sites.

7.1 INDIVIDUAL SITE ANALYSES

7.1.1. Rehman Dheri

Raw data for sherds sampled from Rehman Dheri are presented in Appendix A. As described in the last chapter, these data sheets provide provenance and metric information; quartz size distribution and fabric textural data; mineral concentrations and characteristics; and graphs/charts of these to facilitate visual comparisons. A key detailing the abbreviations used in the sheets is provided on the page immediately before Appendix A. Detailed explanations of the data are presented in Sections 6.3 and 6.4. One individual sheet is presented for each thin section analysed. There are three cases where the sample was too small or of too poor a quality to complete a useful analysis with respect to textural data; these pages present only provenance, metric and mineralogical information.

Mineralogy

Mean concentrations of the minerals observed in thin sections from Rehman Dheri are summarised in Table 7-I. The table uses the author's scale described in Section 6.3, where

1= absent	2= trace
3= rare	4= common

Data shown in table 7-I were obtained by taking the means of each of the individual mineral concentrations recorded on this scale. Standard error of the mean is as described by Shennan (1988: 302-3), and the value recorded for each mineral concentration is used as the individual mean of each thin section. This provides a good estimate which is particularly useful for making intersite comparisons where mineral suites are undifferentiated as to type but not as to concentrations, as is discussed in Section 7.2. The data represents all sherds from the site, both excavated and waster sherds.

TABLE 7-I

REHMAN DHERI MINERAL CONCENTRATION DATA

◆ Excludes quartz ◆ see text for explanation of concentration scale used ◆

Mineral	**Concentration**	**S.E.Mean**[1]
Plagioclase Feldspar	2.73	0.13
Microcline Feldspar	1.19	0.05
Biotite	2.38	0.10
Muscovite	3.63	0.08
Amphibole	2.24	0.13
Carbonate	3.14	0.10
Pyroxene	1.35	0.06
Epidote	1.60	0.06
Schist	2.16	0.11

[1] Standard error of the mean

Quartz concentration data have not been included in the table, as they are treated separately via the area count data, which is discussed in full under 'Fabric' in the following section. Quartz was the most abundant mineral present in all of the sections. Both monocrystalline and polycrystalline grains were present, in an estimated ratio of 80:20. The polycrystalline grain boundaries were generally sutured, which is characteristic of quartz from a metamorphic source (Adams et al. 1984: 5). Some grains were composed of microcrystalline quartz. Instances of sheared or stretched metamorphic quartz were observed in those polycrystalline grains with sutured boundaries. Many of both types exhibited undulose extinction, defined as the grain extinguishing over a range of at least 5 degrees on rotation of the stage. This is characteristic of quartz grains from both igneous and metamorphic sources. Further physical description and discussion of the quartz inclusions, including angularity and sphericity, is presented under 'Fabric' following.

Feldspar inclusions were generally of similar angularity and sphericity as the quartz inclusions. They were mostly fresh to moderately weathered plagioclase. Microcline feldspars were relatively rare, and only occurred in some of the samples. Where they were present, it was in trace concentrations, i.e. no more than one occurrence per thin section. On the other hand, the plagioclase inclusions generally showed two or three, and occasionally four or five occurrences in an individual sample.

Concentration of muscovite averaged 3.63, or relatively 'common', while biotite concentration was somewhat less, with an average of 2.37, or between 'trace' and 'rare'. It was noted that in most of the sections there appeared to be two separate size distributions (subjectively observed only) in most of the micas, especially muscovite. One consisted of the very fine detrital particles which are too small to be measured with the petrographic microscope, and a larger, measurable, population consisting of laths. At the outset of the analysis, an attempt was made to measure the size distribution of the larger population, as it was thought this may have provided some clues for regional or site differentiation. This was abandoned part way through the Rehman Dheri analysis because it became evident that the two populations were not separable enough to warrant continuing. The very small end (<50 microns in length) of the larger lath population tended to grade into the fine, detrital, non-measurable material of the smaller population. It was felt that by ignoring a major part of the mica inclusions, and deriving inferences about mica in general, misleading inferences could have resulted from the data. Estimation of the concentration of the very fine, detrital mica particles was made by assigning a value from 1 to 4 to the micaceousness and birefringence of the fabric under cross-polarised light. These results are discussed under 'Fabric' below.

Weathering of the mica inclusions was assessed by examining the larger laths. Both biotite and muscovite were generally quite weathered in the Rehman Dheri sherds, with an average of 2.98 for biotite and 2.74 for muscovite. Using the scale of 1 to 4 as outlined in Section 6.3 (1 = fresh, 4 = heavily weathered), this means that they exhibited moderate deterioration to some defoliation and minor separation, including frayed ends on the larger laths.

Mean concentration of amphiboles was 2.24. Hornblendes were the most common. They were brown, and generally very small, usually 50-125 microns in their longest axis. Cross sections were rhombic, and they exhibited second-order interference colours. Although not measured as with the mica inclusions, weathering was not common in the amphiboles, as they are not as subject to the effects of weathering.

Carbonate inclusions in the Rehman Dheri thin sections were generally quite variable and difficult to assess. They were recorded in three categories: bioclasts, oolitic, or a third category which is undefined. The oolitic type grains were the most common, occurring in 82% of the sherds. They are termed oolitic, which is used here as a descriptive term only; not intended to infer anything about the origin of the particles. It could not be determined whether the grains were actually concretionary bodies. On the other hand, few exhibited any internal structure, therefore they may be false oolites such as fecal pellets (Pettijohn 1975: 84). The lack of internal structure is also possibly due, in some instances, to the inclusion being plucked out of the thin section during polishing. A typical grain is shown in Figure 7.1. Most were peloid-shaped, and had two distinctive parts: a nucleus, and a cortex. Either, or both, of these parts usually appeared to be micritized, i.e. altered to microcrystalline calcite via micritization. In some instances, the nucleus was composed of sparite (sparry calcite, defined as calcite crystals larger than about 5 microns). Often, the cortex or rim showed layering, both with and without the presence of micrite. This is seen in the lower inclusion of Figure 7.1. As described in Chapter 6, the nature of their opacities was recorded, according to which of their two parts were micritized. This is recorded on each thin section data sheet in Appendix A, using the legend shown on the first page, and explained in Chapter 6. Alone, these data reveal little about the provenance or technology of the pottery from the site, but in comparison with corresponding data from other sites may be revealing as to source differentiation. A cautionary note should be added here. The mineral which appears to be micrite may possibly be dolomite, as the two are often not easily distinguished by their optical properties alone, without first staining the thin sections. Staining was not possible after they had been prepared and coverslipped in Pakistan. Both are present in older limestones. When examined by a carbonate specialist, the mineral properties were consistent with those exhibited by micrite (C. Hayward, pers. comm.). The calcareous nature of the Rehman Dheri pottery was observed in a preliminary macroscopic and X-ray emission spectroscopy study reported by Durrani and Wright (1992: 149-50).

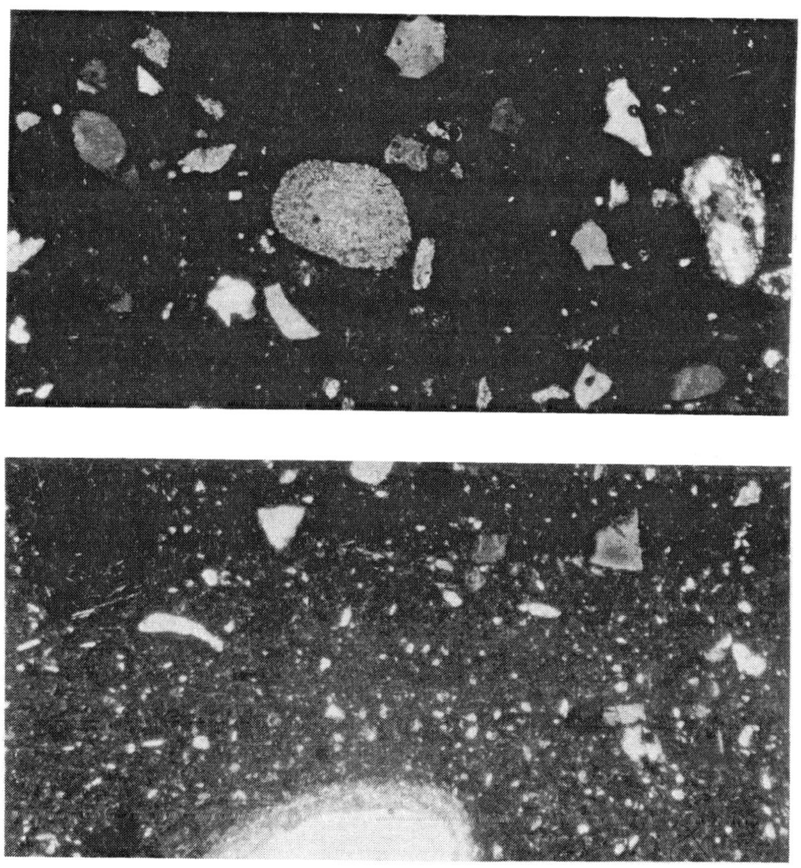

FIGURE 7.1

CARBONATE INCLUSIONS

Crossed polars, 40X, field size 1.0 x 2.0 mm. Upper: RHD 16, 'oolitic' type inclusion in centre. Lower: LWN 206, 'oolitic' type lower centre, also volcanic rock fragment upper left.

A possibility that cannot be ruled out is the use of limestone as temper; although sufficient carbonates are usually naturally present in clays such that their addition is not needed (Rice 1987: 407). Based upon the known abundance of limestone in the source areas of the local clay inclusions, it is most probable however that the limestone inclusions are natural.

In most of the samples, the oolitic type carbonates showed evidence of varying degrees of decomposition, i.e. only remnants, in the fired ceramic. This was particularly evident in most of the waster sherds from the site, where in three of the samples the carbonates were almost completely decomposed. As discussed in Chapter 4, decomposition of calcite occurs at 750 to 850 degrees C, and a firing temperature 850 to 950 degrees or more is indicated by completely decomposed limestone (Porat et al. 1991). According to Feathers (1989: 581), the decomposition can commence at temperatures as low as 700 degrees with a low rate of heating. These observations provide a useful indication of firing temperatures of the Rehman Dheri ceramics. Based upon decomposition of calcite and limestone therefore, firing temperatures could have been as low as 700 degrees C. They probably averaged around 850 degrees, based on the transitory nature of the decomposition levels observed; most had at least some decomposition, while others were at a more advanced level. The more advanced state of decomposition seen in some of the wasters suggests that higher temperatures were experienced in those sherds. This was corroborated by the blistered and spalled state of their surfaces (see Figure 6.11), suggesting firing temperatures above 850 degrees C. Calcareous clays fired to 850 degrees C and above are known to spall upon cooling, due to the CaO (quicklime) picking up moisture from the air to form slaked lime ($Ca[OH]_2$) and releasing heat. The accompanying volume expansion sets up stresses in the clay body which cause cracking and spalling (Rice 1987: 98). If the surfaces of calcite-tempered vessels show conical spalls, it can be inferred that the firing temperature was between 750 and 900 degrees C (Magetti 1982: 128). The absence of exfoliation along cleavage planes of plagioclase feldspars and biotites suggests temperatures did not exceed 1000 degrees C.

Bioclasts were present only in trace concentrations, and in about 9% of the sherds. An example of one type observed in the Rehman Dheri material is shown in Figure 7.2. Because only two of these were observed, it is difficult to identify the species. They are composed of micrite, and appear to be of a type similar to colonial coral (Adams et al. 1984: 61).

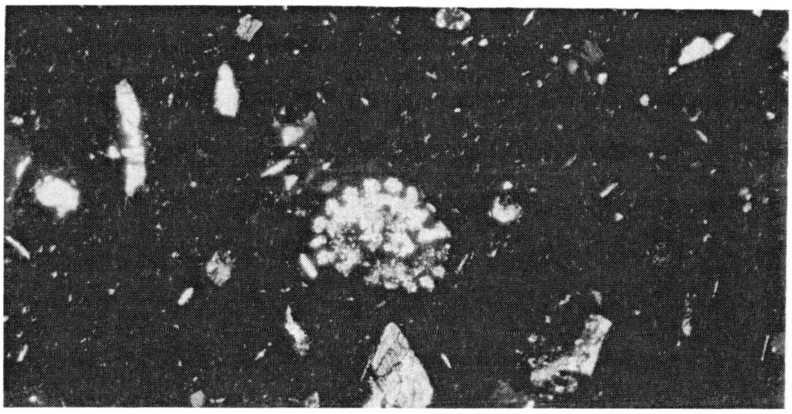

FIGURE 7.2
BIOCLAST INCLUSION FROM REHMAN DHERI
RHD 5. Crossed polars, 40X, Field size 1.0 x 2.0 mm

The third class of carbonates observed were not identified, and are placed in the category of carbonates rather sceptically. They were present in 21% of the sherds sampled. Two examples are shown in Figure 7.1, in the upper right of the lower microphotograph. They averaged 50-125 microns in their longest measurement with the occasional grain as large as 300 microns. They were not pleochroic, and were orange-brown to brown with a plain smooth texture under PPL. The colour was almost identical under XPL, they were not birefringent, and exhibited no crystal structure. Boundaries were sharp. When present, they were always in association with the oolitic type carbonates discussed above. They were recorded primarily to document their presence and provide comparison with inclusions from the other sites.

Minerals observed in the epidote group were mostly clinozoisite. They generally exhibited high relief and some anomalous blue birefringence, with bright first-order interference colours. They were usually small, in the range 50 - 125 microns.

Schists were generally biotite or epidote schists, with occasional inclusions of amphibole, clinozoisite or plagioclase feldspar within the schist grain. Most of those classified as schist were done so to indicate that the grains were schistose in texture; however, some were more gneissic. For comparison purposes, all were classed as schist, and the major mineral recorded. Generally, the major mineral in the schistose and gneissic grains was polycrystalline or microcrystalline quartz.

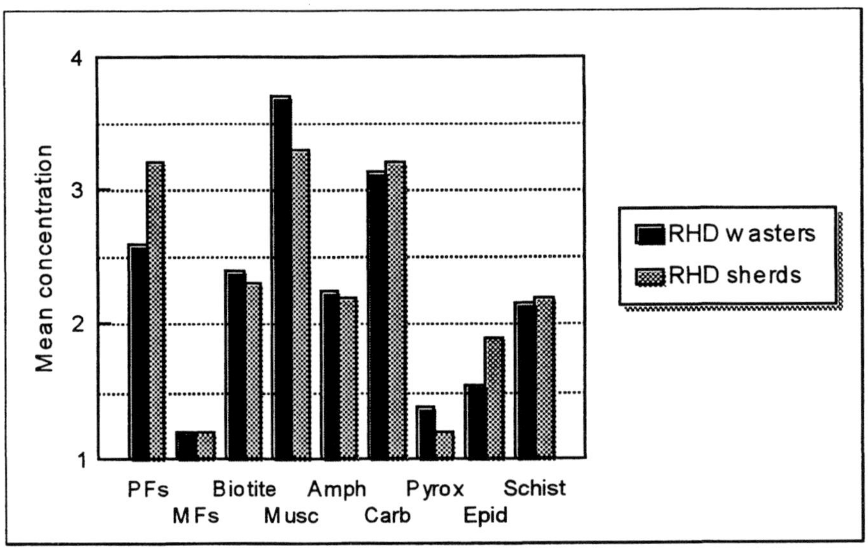

FIGURE 7.3

COMPARISON OF MINERAL CONCENTRATIONS - REHMAN DHERI

PFs = plagioclase feldspar MFs = microcline feldspar Musc = muscovite
Amph = amphibole Carb = carbonate Pyrox = pyroxene
Epid = epidote

The sample of waster sherds collected from the site of Rehman Dheri, which are assumed to have been manufactured at the site, contain a mineral suite identical to that of the sherds excavated from the site. Figure 7.3 shows the mean mineral concentrations observed in the waster sherds compared with the mean mineral concentrations observed in the Rehman Dheri excavated sherds. It is seen that all minerals observed in the excavated sherds are present in the waster sherds in approximately the same proportions.

Overall, the mineral suite present in the Rehman Dheri sherds, both wasters and excavated samples, strongly suggests that they are of local origin, probably of material derived from adjacent formations. The minerals listed in Table 7-I are likely to have been derived from formations to the west, northwest, and north of the site, as these are the directions of natural fluvial action. As discussed in Chapter 5, formations surrounding the Gomal plain which are the most likely source of clay inclusions found there include the Sulaiman range and the Waziristan highlands to the west, the Bhittanni and Marwat ranges and the Khisor range to the north. The formations to the west consist of sedimentary rocks of marine origin, including limestones, and intrusive rocks consisting of mafic and ultramafic rocks and metamorphosed sediments. The amphibolite facies of the Nuristan Complex include biotite, biotite schist, quartzitic gneisses, and pyroxene-amphibole gneisses. The Bain diamictite in the Bhittanni and Marwat ranges includes extensive metamorphic rock fragments which include gneiss, mica schist, and hornblendes. All of these formations are part of the drainage basin producing detrital materials found in the Gomal plain.

Because of the broad geological similarities present over thousands of square kilometres of the Northwest Frontier Province, it could also be said, on the basis of the mineral inclusions seen, that the Rehman Dheri pottery could have been sourced hundreds or even thousands of kilometres away. However, because of the vast quantity of sherds and the presence of kilns with wasters at the site, it would be expected that the pottery was of local manufacture. Moreover, the fact that the wasters consisted of identical mineral inclusions strongly suggests the same. As discussed under 'Intersite analysis' below, some more local differentiation can be made on the basis of the relative concentrations of the minerals, even though the suites have been shown to be the same.

A comparison of the mineral inclusions observed in the samples as a chronological function was also made. Details are shown in Figure 7.4. It presents mean concentrations of mineral inclusions, using the author's scale as previously defined, in three data series, one for each of the major chronological periods at the site. Periods I and II have been defined as two divisions of the Early Harappan, and Period III the Mature Harappan. Calibrated radiocarbon dates for each are as follows (Durrani 1988: 139): Period I, 3300 - 2850 BC; Period II, 2850 - 2500 BC; and Period III, 2500 - 1900 BC. The figure demonstrates that there is very little evidence, at least on a mineralogical basis, to indicate that clay sources utilised by the craftspeople of Rehman Dehri changed significantly over the time period 3300 to 1900 BC. Each of the minerals observed has remained relatively constant from period to period.

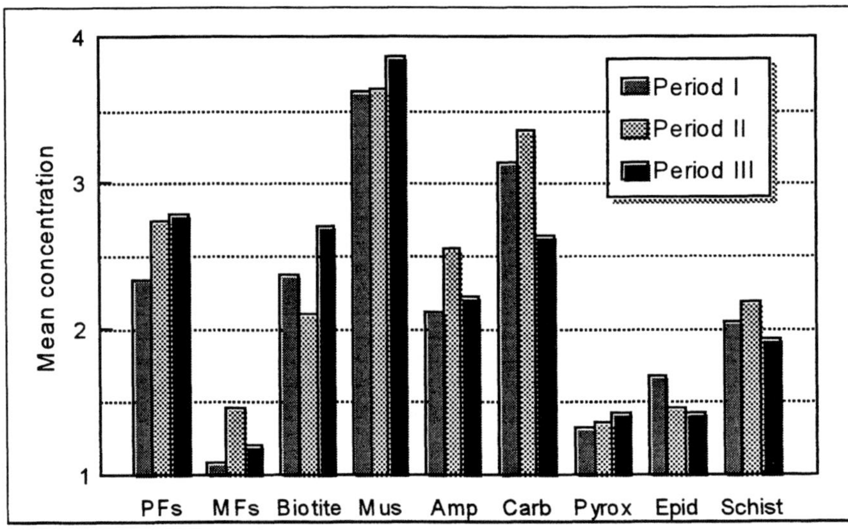

FIGURE 7.4

COMPARISON OF RHD MINERAL INCLUSIONS FOR PERIODS I - III

◆ Excludes quartz, which is discussed under 'Fabric' ◆

PFs = plagioclase feldspar MFs = microcline feldspar Mus = muscovite
Amp = amphibole Carb = carbonate Pyrox = pyroxene
Epid = epidote

Fabrics

Charts summarising concentrations of size ranges in two categories were first prepared. Both the quartz size range data (as obtained from the area count), and the all-inclusions size data (from the point count analysis) were used. These were prepared to examine differences in concentrations of the smaller, silt-size range data with the larger, sand-size inclusions. In general, added temper is of a larger size range, and examining it as a separate category can be revealing. The two data sets used, from the area and point counting, are closely related and tend to show similar textural information. The main differences are that one set considers only quartz size data and the other includes all inclusions in the measured size range. Also, in the case of the quartz data, the smallest size range is <50 microns, compared with <62.5 microns (silt-size inclusions of the standard Udden-Wentworth scale) which was used for the point-count data. Therefore, although it will be informative to compare the two, they are not directly comparable on the same basis. Both sets were prepared in order to provide two independent bases for fabric categorisation and analysis.

For the quartz size data derived from the area count, it was first divided into two ranges: 30-50 microns, and >50 microns. Charts were prepared for each range, to show the number of occurrences falling into each of the smaller and larger concentration ranges. These are presented in Figure 7.5. These data approximate categories of silt- and sand-sizes, respectively.

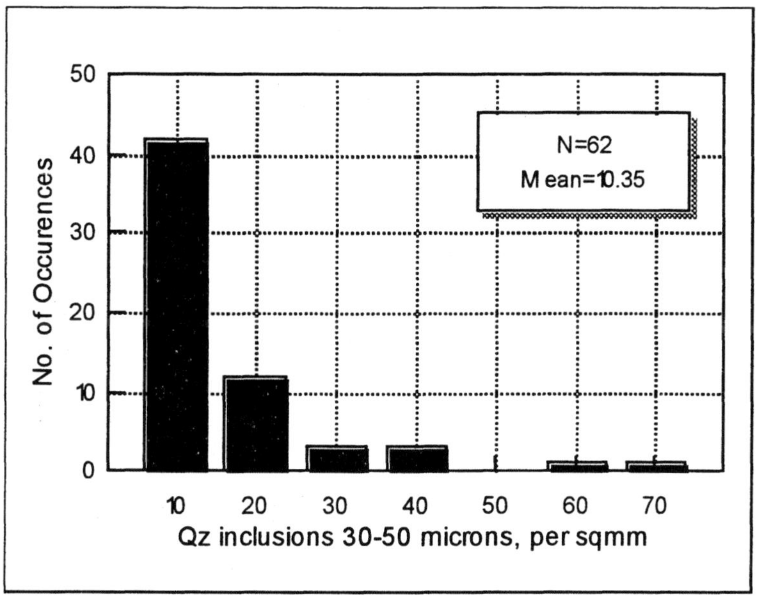

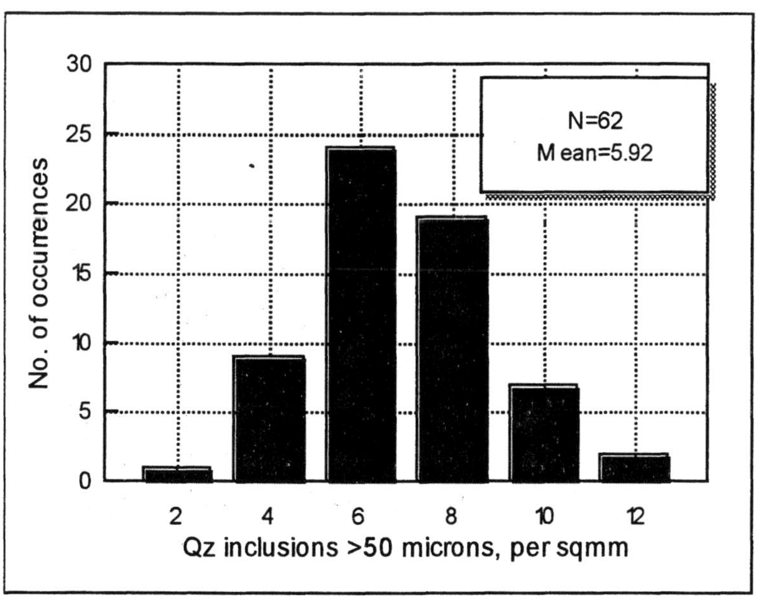

FIGURE 7.5

DISTRIBUTION OF QUARTZ INCLUSIONS BASED ON AREA COUNT DATA

Rehman Dheri

Vertical scales represent number of sherds containing concentrations in given size intervals

The upper chart of Figure 7.5 shows that the mean concentration of quartz inclusions of size <50 microns is 10.35 per square millimetre. Relative to the other sites sampled, this is a low concentration. Comparison with the other sites is discussed later in Section 7.2. There appears to be no set of comparative quartz temper concentration data from the area, apart from the other sites of this research, to provide an indication of what this means. Other thin-section studies of Harappan ceramics did not include detailed fabric analyses, nor point- or area-counts. Krishnan and Hegde (1987: 43-7), in their study of Mature Harappan ceramics from Gujarat, reported only mean grain size. Wright (1984: 226-31), in her research into ceramics of the Indo-Iranian borderlands, limited reporting of grain size distribution to the average and the largest grain size. Some particle-size distribution studies of Pakistan body temper sands were conducted by Rye and Evans (1974: 128-9), but the NWFP sands analysed were from potters at Zakhel Bala (near Peshawar), Kharmathu (near Kohat) and Shahbaz Ahmed Khel (near Bannu), and did not include any from the Gomal plain. Also, they consisted of analysing only raw sands which had been collected by local potters and prepared to be used as temper, and not as part of an analysis of the fabrics themselves.

Very low concentrations of fine silt-size particles suggest a source with a low concentration or that the clay has been meticulously levigated prior to the addition of temper. This is, to a degree, speculation, because the threshold level of the potter cannot be known. One cannot be sure at what point in selecting the upper, finer layers of the levigated clay, the potter decides that too much silt has been included. No published studies of this topic were revealed in the literature survey. However, a level of 10 inclusions per square millimetre would appear to be a reasonably detectable threshold. At levels higher than this, coarseness in the clay could be detected by feel. It is therefore suggested, at this point in the discussion, that the Early Harappan potters of Rehman Dheri meticulously levigated the local clays before further processing them.

Data obtained from point counting each thin section were used to prepare similar charts illustrating relative concentrations of silt- and sand-size inclusions. These showed all inclusions, not only quartz as in the above discussion. The other major difference between the two sets is that the size ranges are not directly comparable; this second set uses the two categories of silt (<62.5 microns) and sand (>62.5 microns) instead of the above 30-50 and >50 micron categories respectively. Also, instead of absolute concentrations per unit area, as was presented for the quartz data, number of occurences in each of a defined range of paste percentages were used. These charts are shown at Figure 7.6. Examining this second set provides a degree of corroboration with the quartz size concentration data, even though the two sets are not directly comparable because of the aforementioned differences.

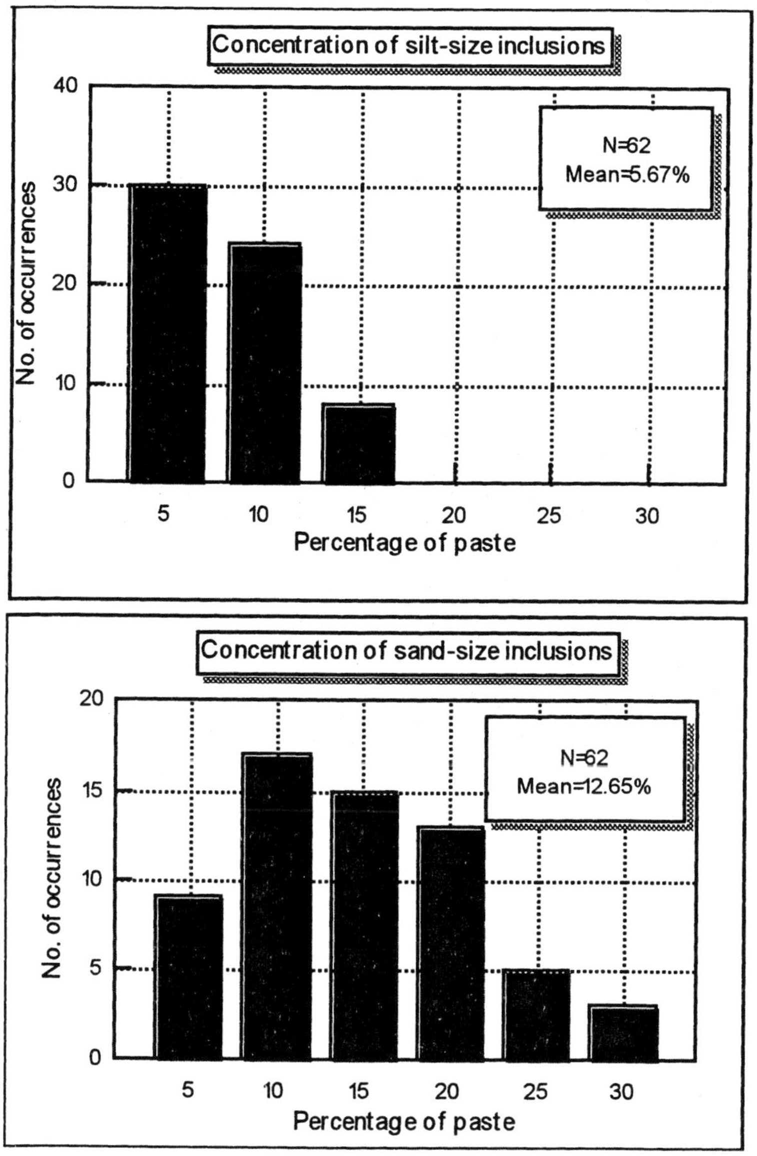

FIGURE 7.6

DISTRIBUTION OF INCLUSIONS BASED ON POINT COUNT DATA

Rehman Dheri sherds

Vertical scale represents number of sherds containing concentrations in given percentage intervals

The upper chart of Figure 7.6 shows that silt-size inclusions form a mean of 5.67% of the paste in Rehman Dheri sherds, and that the distribution has a similar form as the quartz concentrations in the 30-50 micron range. As mentioned above, as there is no comparative data available from the region, the significance of this is not known. Measured as a percentage of paste content, such a mean can be considered a reasonable indication that the clay has been levigated. This is supported by the presence of high concentrations of fine mica particles in most of the thin sections. It has been observed that levigation enriches the suspension of fine detrital micas (Krishnan and Hegde 1987: 42). It would be expected therefore that the two would occur together, i.e. a very low occurrence of fine silt-size inclusions would tend to be associated with high concentrations of fine mica particles.

If it can be assumed that the clays were levigated, then these fine particles that are present in the pastes would represent the remaining natural inclusions. The larger fractions, those shown in the lower charts of Figures 7.5 and 7.6, would then represent either natural inclusions or added temper. If they are natural inclusions, they would be expected to be subangular to subrounded. If they are added temper, their angularity could be the same as the natural inclusions, or very angular (if crushed or primary source quartz was being used as temper). Most of the quartz inclusions (and most of the feldspars) observed in the thin sections were subangular to subround. Figures 7.7 and 7.13 illustrate typical distributions. Many of the sections exhibited a distinct two-part angularity distribution. Examples are shown in Figures 7.8 and 7.14. When this two-part angularity distribution was present, one of them was in every case a very angular one (a scale 4 classification). The high angularity seen in these cases is consistent with a freshly broken inclusion; one that has not been subject to the erosional influences of alluvial or aeolian motion. It suggests that the potters were using crushed quartz (or a primary quartz sand deposit) as temper, perhaps as well as river or aeolian sands. There were no cases of a lone or single angularity when the very angular (4) inclusions were present. They were always in association with similar-sized subangular quartz sands. That is, where the larger quartz fractions were seen, they consisted of either entirely subangular (3) to subrounded (2) inclusions; or these plus angular (4). Moreover, where the larger inclusions were in the less angular categories, their size distributions were usually approximately positively skewed (see also

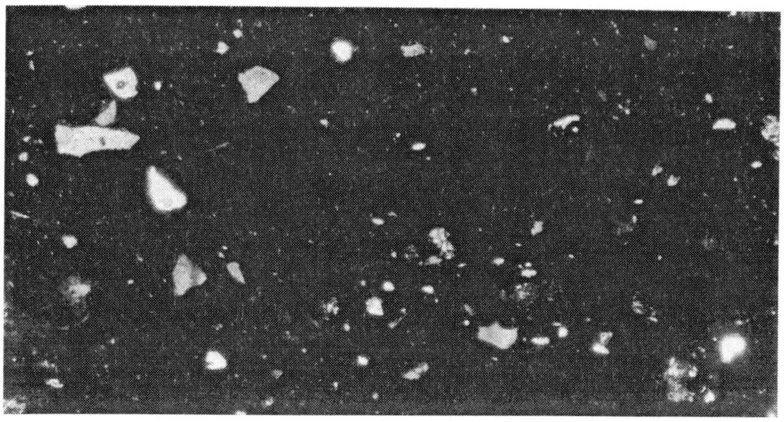

FIGURE 7.7

TYPICAL QUARTZ ANGULARITY DISTRIBUTION

RHD 53. Crossed polars, 40X, Field size 1.0 x 2.0 mm.

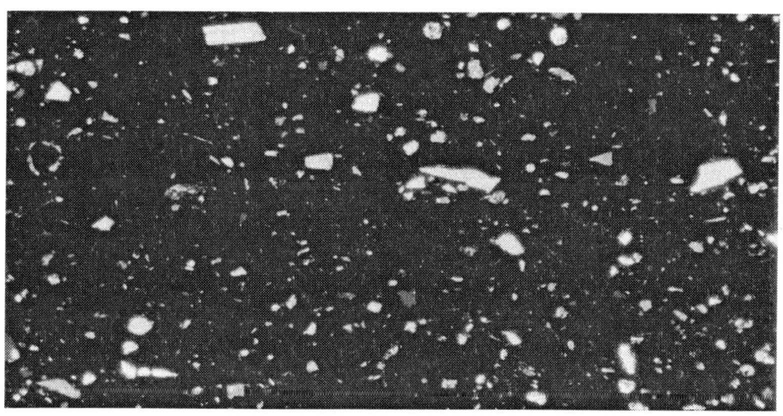

FIGURE 7.8

BIMODAL QUARTZ ANGULARITY DISTRIBUTION

RHD 1. Crossed polars, 40X, field size 1.0 x 2.0 mm.

Figure 7.11). Examples are seen in the quartz size distribution charts contained in Appendix A, for example RHD 35, 64, 69, and 70. In Chapter 5, research was cited which has shown that river sand distributions tend to show positive skewness. This has also been observed in river sands collected and prepared for use as tempers by Pakistani potters (Rye and Evans 1974: 128-9). The distributions observed are therefore consistent with use of river sands as temper.

In terms of quartz temper addition, the foregoing factors suggest three possible scenarios. One, that clays are used with naturally-occurring quartz inclusions left in them. Two, that clays were levigated to a fine silt level and then river sands added as temper. The third scenario is that the clays were finely levigated and then river sands plus newly-crushed or primary quartz temper were added. The presence of the very angular quartz fraction was not seen to be a function of level, wall thickness or morphological type. It is possible that the potters were utilising all three methods, according to characteristics desired in the final paste. As mentioned in Chapter 4, in order to provide the best strength in a finished pot, the most angular temper available is the most desired. Thus, if adequate strength could be achieved using river sand tempers, they would have been used because of their ready availability and lower labour input as compared with preparing temper. In the cases where a stronger fabric was needed than could be provided by river sands alone, quartz sands were crushed and added to provide additional strength. These occasions evidently were at a minimum, because the very angular quartz was observed in only 15 of the Rehman Dheri sherds, or about 21%. It was used only in association with the less angular, probably river, sands. It was present primarily in the larger size ranges, i.e. greater than 50 microns.

With respect to the micaceousness and birefringence observed in each of the thin section fabrics, this was used primarily to examine them with a view to possible categorising into fabric types; and providing further indications of firing temperatures. Observed distribution of the two variables is shown in Figure 7.9. The majority of the fabrics were classified as low to medium micaceousness and birefringence. A close correlation between the two is seen, therefore it is possible that the level of concentration of very fine detrital micas contributes to birefringence in these fabrics. No correlation was seen between either of these variables with respect to morphological type, thickness, rim diameter, or mineral concentrations other than the micas.

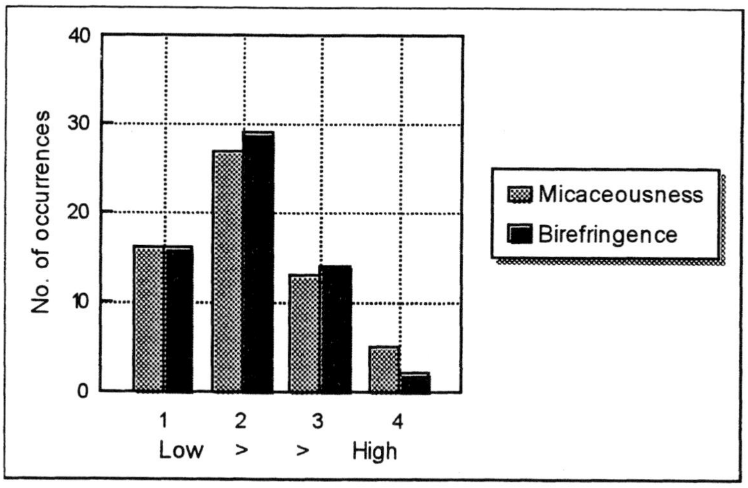

FIGURE 7.9

MICACEOUSNESS AND BIREFRINGENCE OF REHMAN DHERI FABRICS

With respect to further clues of firing temperatures, Figure 7.9 shows that, in the majority of the thin sections (77%), the matrix exhibited low (1-2) birefringence. This may suggest that most sherds were fired to a temperature in excess of 850 degrees C. An anisotropic matrix is nonvitrified and retains its optical properties (chiefly birefringence), suggesting firing temperatures below the vitrification point (below ca. 850 degrees C; Brisbane 1981: 234, citing Peacock 1971). The sixteen samples (23%) that exhibited moderate to high birefringence may have experienced firing temperatures of less than 850 degrees C. It may be suggested from these data that most were fired at or above this temperature, which correlates with thc firing temperature ranges inferred from changes observed in the carbonate inclusions discussed above.

Fabric colours seen in thin section ranged from light brown, medium brown, dark brown through to dark grey. The dark greys were observed only in the waster selection, and may have been darker than the rest because of their reducing firing conditions. Such wasters were selected because they were the most obvious as wasters and one could be certain of their status as wasters. Number of occurrences in each of these colour categories is shown in Figure 7.10. It can be seen that colour observed in the majority of the samples was medium to dark brown. These were also recorded with a view to possible assistance with categorisation into fabric types, and for documentation purposes.

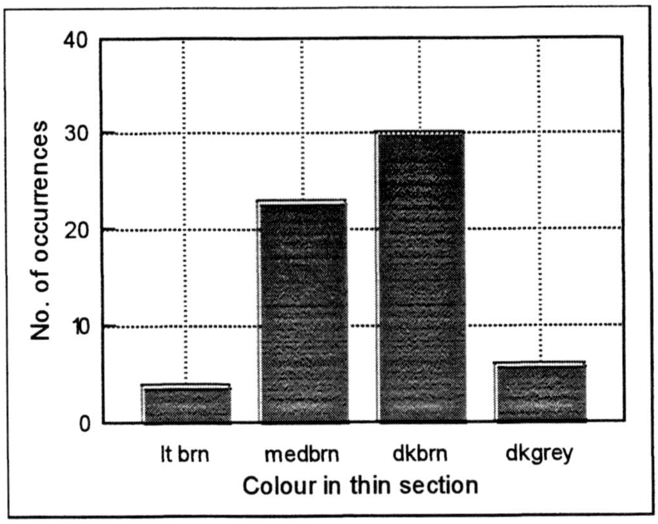

FIGURE 7.10

DISTRIBUTION OF FABRIC COLOURS OBSERVED IN REHMAN DHERI THIN SECTIONS

On the basis of differences observed in the quartz size distributions of the samples, an attempt was made to classify the fabrics into groups. Mineral suites or concentrations do not differ enough to allow the mineralogy to be used as a classifier; and if, as it appears, potters were using local clay deposits, variations in the fabrics would likely have been a result of their treatment by the potter. It must be emphasized that classifying archaeological ceramic fabrics within a local industry can be very misleading as an analytical tool. In doing so, one is making the implicit assumption (or should be) that the potters *intended* to create such groupings, apart from accidental inclusions. It certainly can be much more meaningful when comparing fabrics from two distant sites. Once fabrics are so classified, it can often lead to inferences as to the potters' behaviour in making them. Therefore, in terms of seeking the 'mind behind the artefact' as discussed in Part One, one must be careful to remember that fabric grouping labels assigned by a researcher are only just that, i.e. labels. They are used simply to characterise the material. This material-oriented approach has been criticised in the Harappan literature as well (Jansen 1992: 218). Unless the fabric grouping can be shown to have been purposefully created by the potter in order to achieve certain characteristics, e.g. the addition or non-addition of a particular distinctive temper type, these should be only cautiously relied upon for interpretation of the craftspersons' behaviour. In the present thesis, the major differences observed in the fabrics were with respect to the nature of the quartz inclusions, and, to a lesser degree, colour and birefringence of the fabric. Thus, although fabric groups are being created here, no representation is made that they were the intent of the Rehman Dheri potter. The groups are formed in order to facilitate material comparisons and determine if there exist any relationships between different fabric groupings and pot morphologies, for instance.

The initial attempt at fabric classification was to plot the fabric composition data on ternary diagrams; each point representing three independent percentages of the paste: matrix, sand or silt. All 62 Rehman Dheri samples were plotted, and the result examined for fabric groupings. Apart from confirming the information presented in Figure 7.6 with respect to concentrations of occurrences in silt- and sand-size inclusions independently, this exercise produced no evidence that the fabrics group into predictable combinations of the three paste components. Other researchers have used this tool successfully to reveal fabric classifications (most notably Stoltman 1991), but it is only effective where the groups actually exist. The ternary plot showed only a very slight, tentative, grouping around the 6% silt/12% sand range, which is seen in Figure 7.6 as representing the approximate means of the respective groups. No other discernable groupings could be detected visually. With respect to potter behavioural implications, this lack of groupings on the ternary plot suggests that Rehman Dheri potters were not purposefully creating fabric groupings based on set different combinations of sand and silt-size particles. This led to examination of the silt- and sand-size categories independently for within-group variations which would provide useful classification criteria.

With the exception of the waster sherds, all of the Rehman Dheri thin sections on which counts had been conducted were first reviewed as to concentrations of silt-size inclusions and concentrations of quartz inclusions 30-50 microns. The number of quartz inclusions 30-50 microns per square millimetre cannot be directly correlated with the percentage concentration of silt-size inclusions in the paste unless the count per unit area is converted to a count per unit volume via a process which involves an estimation of the size of an equivalent sphere, which would not have been particularly useful for the present research. It was noted in the analysis discussed earlier however that a reasonable correlation existed between the two measures of silt-size range particles. That is, their respective distribution shapes are observed to be similar and that, in general, a low concentration of quartz inclusions 30-50 microns per square millimetre co-existed with a low percentage concentration of silt-size inclusions in the paste. Comparing Figure 7.5 with Figure 7.6, it is seen that 54 sherds (87% of the sample) had a concentration of quartz between 0 and 20 inclusions per square millimetre; and that the same number had inclusions which constituted less than 10% of the paste. It was felt then, on a purely physical basis, that either of the indicators of very small particle concentration could have been used for an initial grouping of fabrics. Both could be readily related to the procedural thinking of the potter. Grouping based on the concentrations of quartz inclusions 30-50 microns was chosen because, as seen in Figure 7.5, the section counts divided quite definitively between those with concentrations <10 per square millimetre and those with concentrations >10 per square millimetre. Based on this observation, and the possibility that this split suggested levigation as part of the clay preparation process, these two levels were chosen to form the initial basis of the fabric classifications.

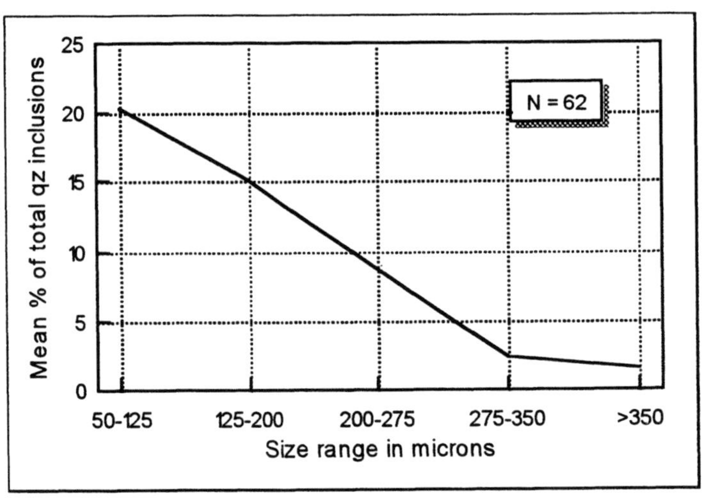

FIGURE 7.11

MEAN DISTRIBUTION OF REHMAN DHERI QUARTZ INCLUSIONS >50 MICRONS IN SIZE

See also footnote on next page

Next, each of these groupings was examined for variations in the nature of the distribution or characteristics of those inclusions in the larger, sand-size, range, and the distribution of quartz inclusions >50 microns. Figure 7.5 (lower) shows the distribution of concentrations of quartz inclusions >50 microns per square millimetre, and Figure 7.11 shows the mean distribution of quartz inclusions >50 microns[1]. The latter figure was created by taking the mean of all of the thin section data in each of the size range categories above 50 microns. It is shown to illustrate the general shape of the size distribution which represents all quartz inclusions >50 microns. This and the concentration data (Figure 7.5) show the distribution of all size ranges >50 microns as a collective group, but they are not very informative with respect to coarseness of temper, or the nature of the larger particles that may have been used as temper. Therefore, within each of the two groups made up on the basis of the inclusions 30-50 microns described previously, data were further sub-divided based upon the differences observed in the distribution of inclusions >50 microns.

All of the size distribution charts were visually examined and compared This revealed that the quartz inclusions >50 microns, whose mean distribution is shown in Figure 7.11, fell into three quite distinct subgroups within each of the two silt-size categories already made. The subgroups are based on observed consistencies in the quartz size distribution curves, and generally indicate three levels of coarseness within this larger size group. These may or may not represent the nature of sand tempers, but it enabled a total of six fabric categories to be formed, according to combinations of quartz size inclusions. These six categories are termed X1, X2, X3; and Y1, Y2, Y3. The first letter designates the concentration of quartz inclusions < 50 microns in size, which is the most likely to indicate whether or not the clay has been levigated. For this designation, X denotes a concentration of greater than 10 quartz inclusions 30-50 microns in size per square millimetre (may not have been levigated), and Y denotes those clays containing less than 10 per square millimetre (which are more likely to have been levigated). Alternatively, both groups could both have been levigated: the first levigated purposefully to achieve that range of concentration of fine quartz sand inclusions; or it may simply represent a selection at a physically lower point in the settled clays. The latter is plausible in light of the observed less consistent nature of the concentrations greater than 10 per sq mm. They are broadly distributed between 20 and 70 per sq mm. The number following each X or Y denotes the coarseness of that component of quartz inclusions >50 microns in size, which is the more likely to indicate the nature of temper sands. These are shown graphically in Figures 7.12 to 7.17. As with Figure 7.11, these are presented as line graphs rather than

[1] Figure 7.11 should properly be a histogram. It is presented as a line graph in order to give a visual impression to match the fabric type presentations in Figures 7.12 through 7.17.

histograms to convey a visual impression of the mean concentration of quartz inclusions in each fabric.

In the absence of differences in the amounts of other minerals in the clay mixtures, the nature of the quartz inclusions is the only variable showing significant differences within the site collection. Therefore, it might quite safely be assumed that the potters were manipulating the clay mixtures with respect to the sand inclusions. The variations seen and the fabric groupings observed are unlikely to have been present in the naturally occurring clays. With respect to other variables recorded, grouping on the basis of these could not consider variations in the sand content. For example, differences observed in the birefringence of the matrices could generally be assigned a value from 1 to 4, but there appeared to be no correlation of this variable with textural variables. Also, it is more difficult to manipulate firing temperatures (of which matrix birefringence is an indicator) than clay paste preparation, and therefore intent of the potter would be less of a factor.

Fabric types as defined are shown in Figures 7.12 to 7.17. Examination of the charts accompanying each of the figures describing the fabrics generally shows a different coarseness grading within each of the main groups, from low to high. In the X-group, where the clays may not have been levigated, the quartz inclusions >50 microns range from almost non-existent to a concentration in the lower end of the size range. That is, the larger fraction has a distribution which is positively skewed but has its median close to the 50-125 micron end of the range. On the other hand, the Y-group, where the clays may have been levigated and sand temper added, exhibits a grading within the coarse fraction which varies within a much coarser range. It is also skewed positively, but has its median closer to the 125-200 micron range. All of these distributions are generally consistent with those found in river sands. The distribution charts accompanying fabrics Y2 and Y3 appear to be similar, but an important distinguishing difference is in the concentration of the quartz inclusions >275 microns, as seen in the accompanying photomicrographs.

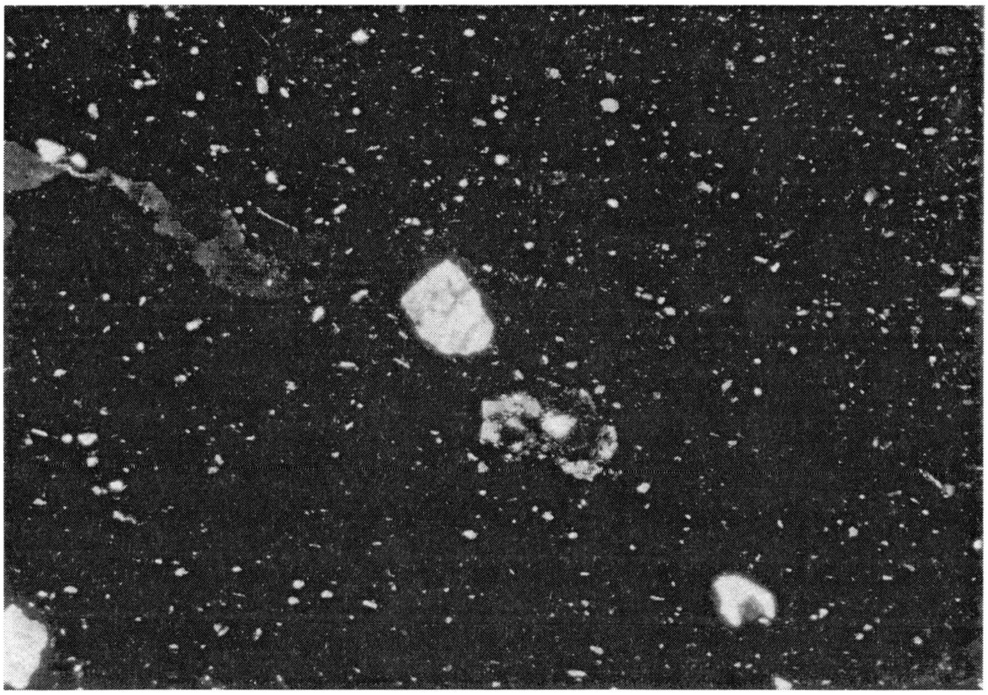

RHD44. Crossed polars. Magnification 40X. Field width 2.50 mm.

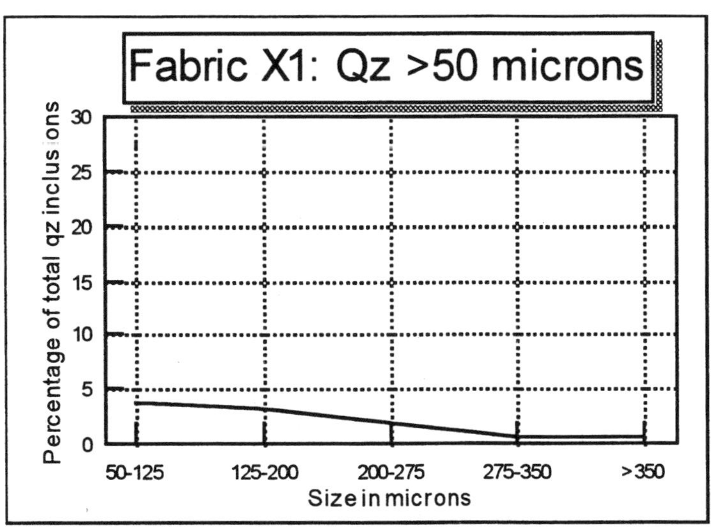

Mean concentration of Qz inclusions 30-50 microns = 32.8 per mm^2

FIGURE 7.12

REHMAN DHERI FABRIC TYPE XI

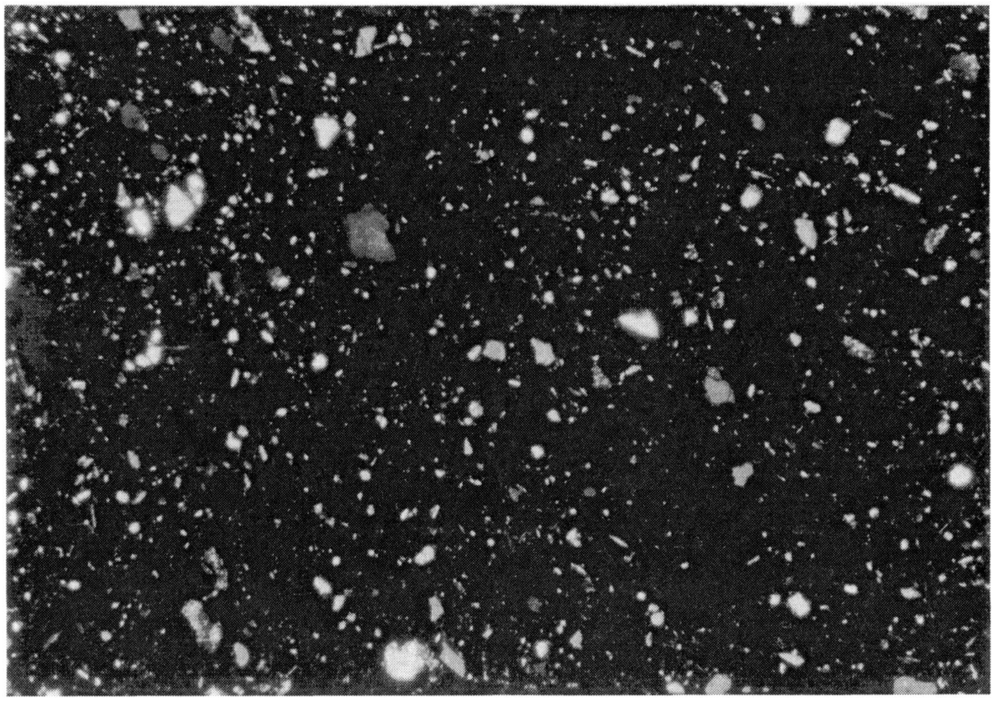

RHD 50. Crossed polars. 40X magnification. Field width 2.50 mm.

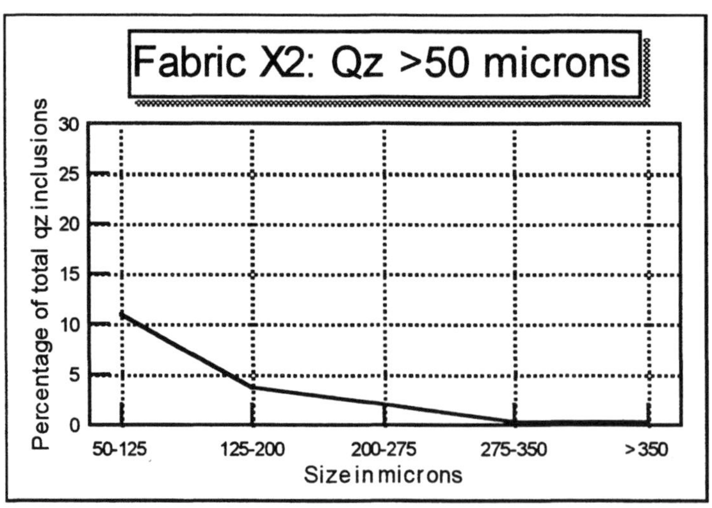

Mean concentration of Qz inclusions 30-50 microns = 34.7 per mm^2

FIGURE 7.13

REHMAN DHERI FABRIC TYPE X2

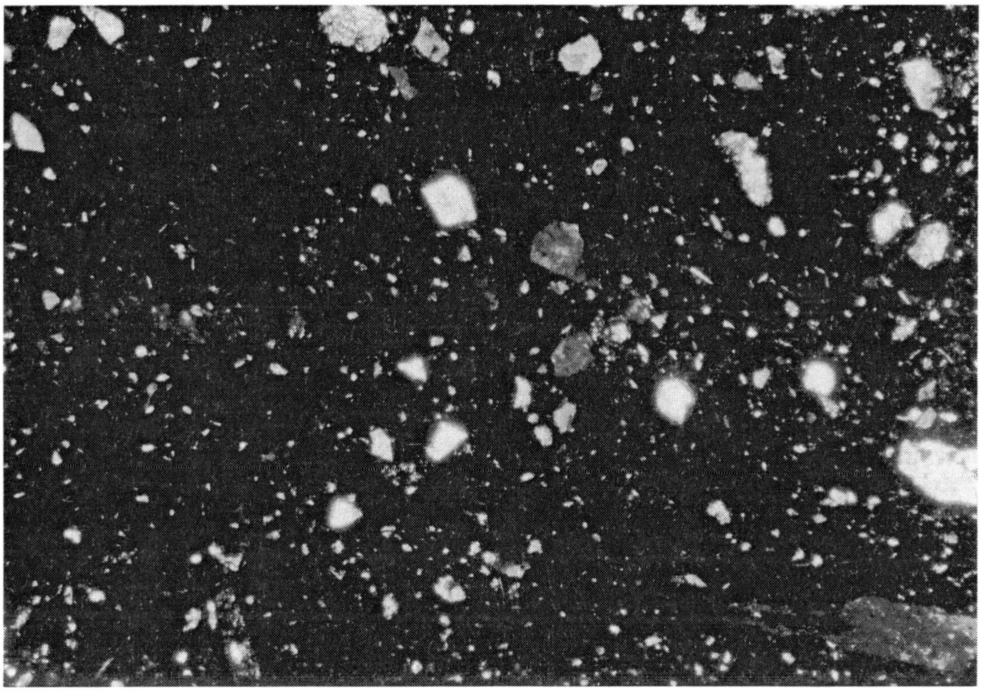

RHD 38. Crossed polars. 40X magnification. Field width 2.50 mm.

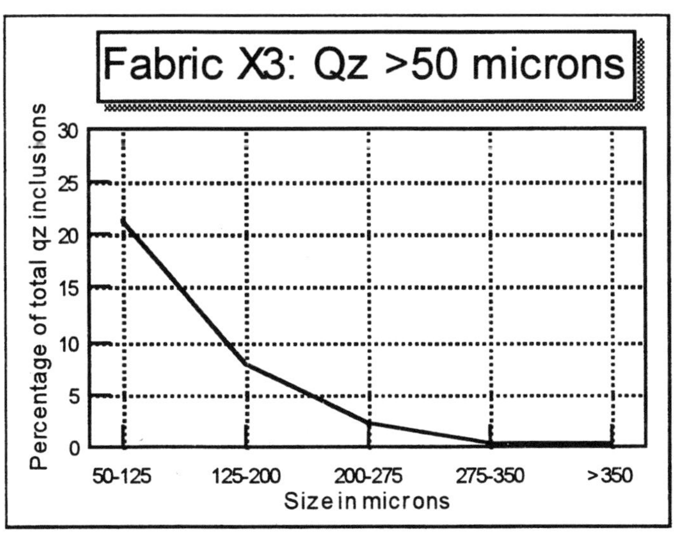

Mean concentration of Qz inclusions 30-50 microns = 13.3 per mm^2

FIGURE 7.14

REHMAN DHERI FABRIC TYPE X3

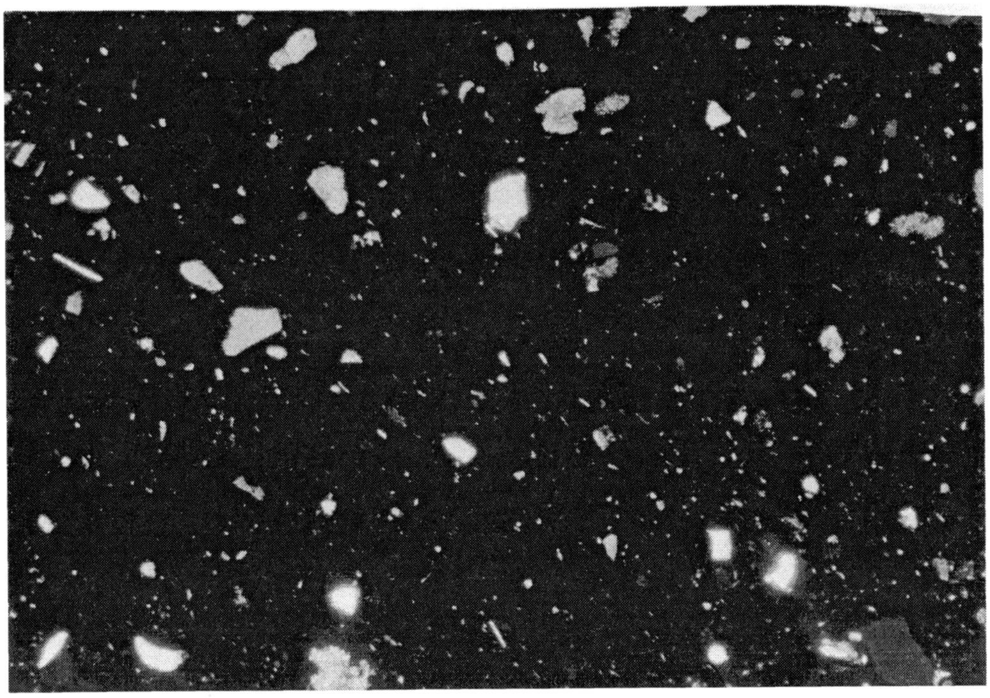

RHD 43. Crossed polars. 40X magnification. Field width 2.50 mm.

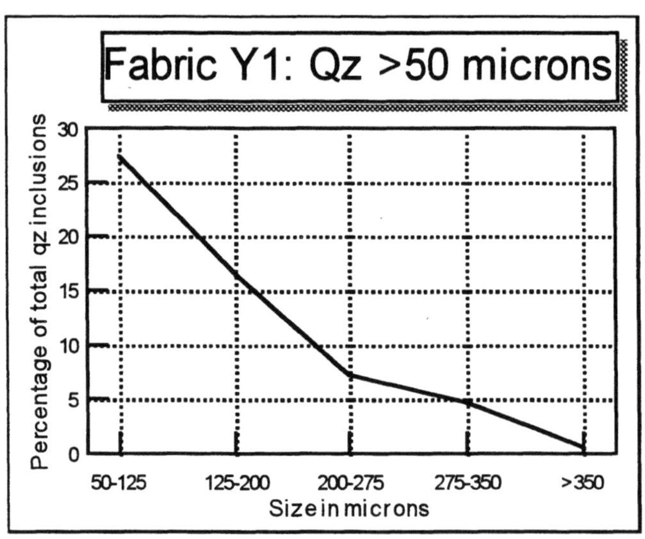

Mean concentration of Qz inclusions 30-50 microns = 5.5 per mm^2

FIGURE 7.15

REHMAN DHERI FABRIC TYPE Y1

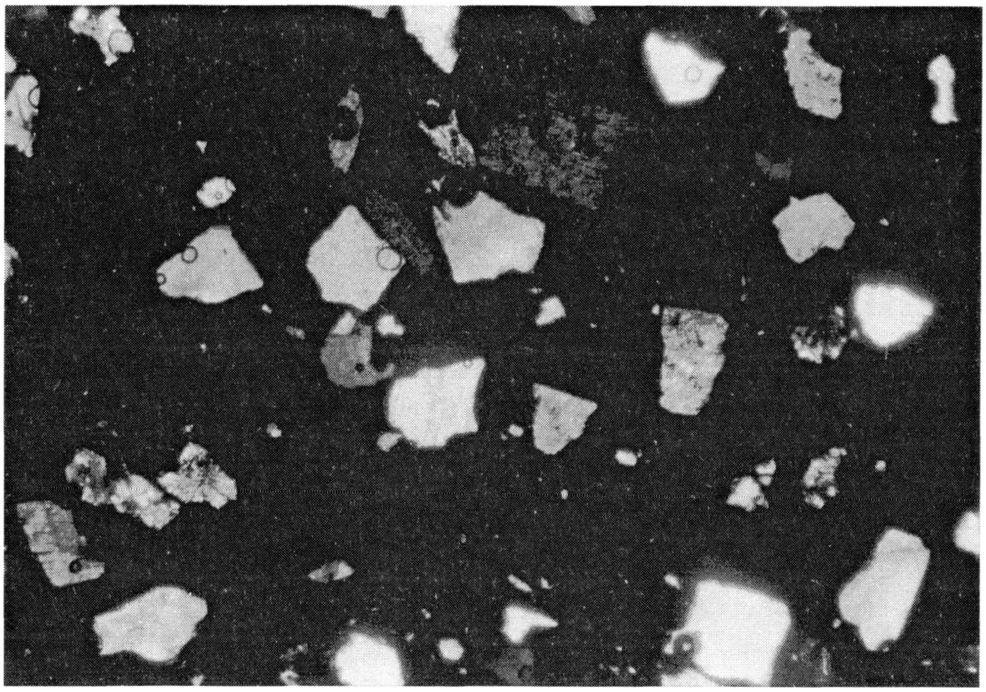

RHD 19. Crossed polars. 40X magnification. Field width 2.50 mm.

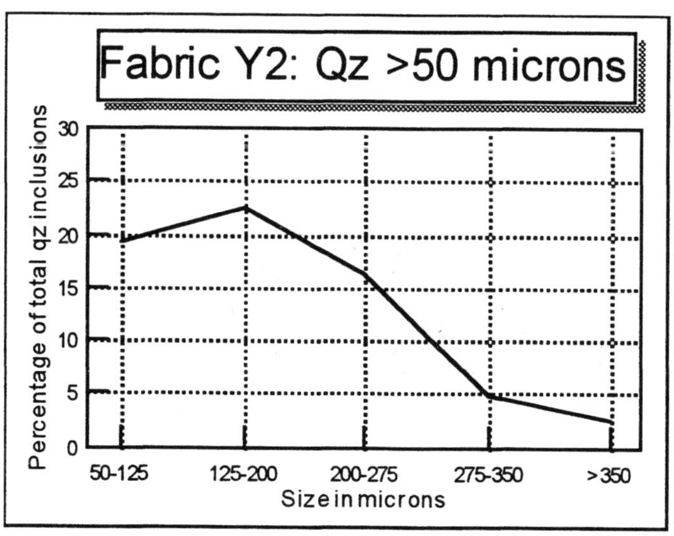

Mean concentration of Qz inclusions 30-50 microns = 3.4 per mm^2

FIGURE 7.16

REHMAN DHERI FABRIC TYPE Y2

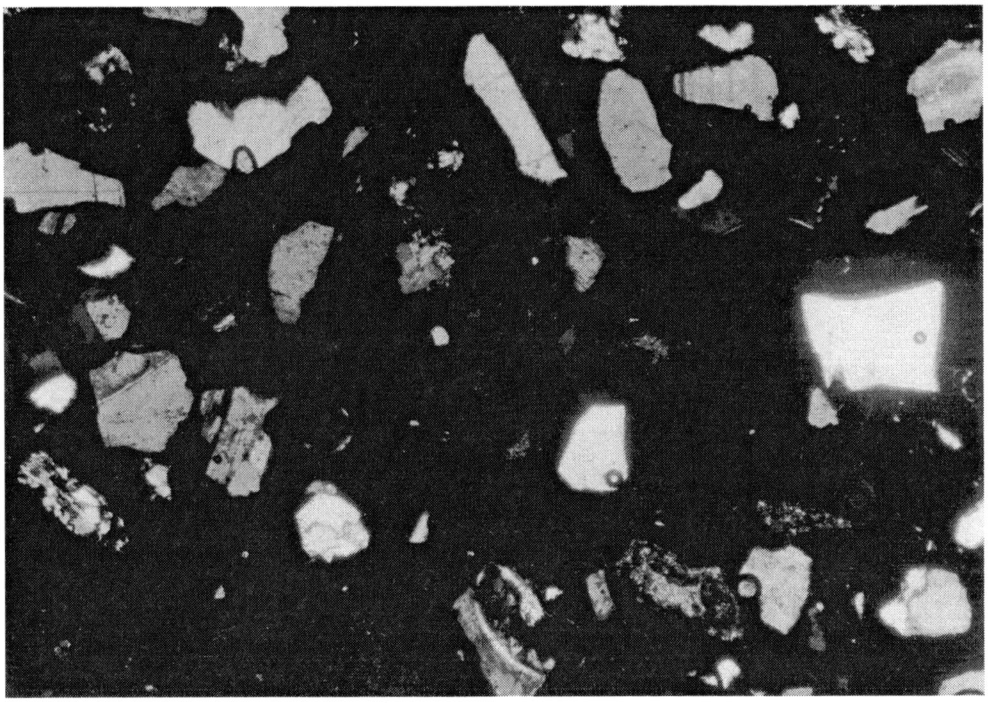

RHD 21. Crossed polars. 40X magnification. Field width 2.50 mm.

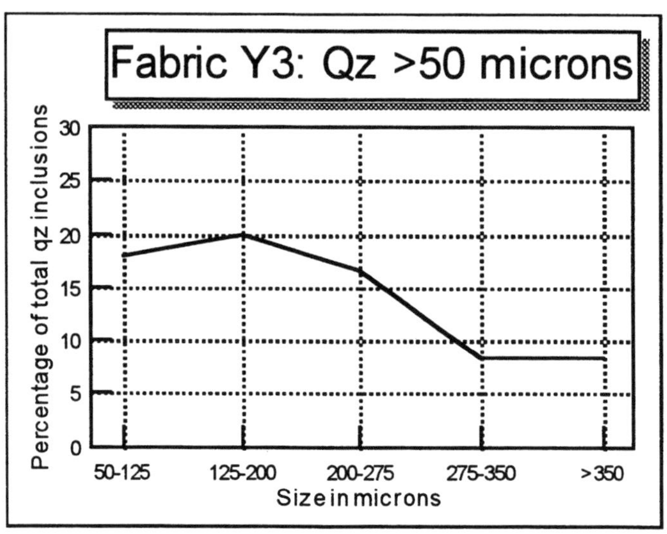

Mean concentration of Qz inclusions 30-50 microns = 2.0 per mm^2

FIGURE 7.17

REHMAN DHERI FABRIC TYPE Y3

The most common fabric types were Y1 and Y2, which together constituted 62% of the total. Figure 7.18 shows the distribution of fabric types at Rehman Dheri and Table 7-II shows fabric types occurring by morphological type. Type X1 is seen to be relatively rare. In those instances where type X1 fabric was used, two occurrences of use on BI bowls and one CII polychrome were recorded. Type X2 was also relatively rare, and only seen used for AII jars, BIII bowls and one BV dish on stand. Fabric type X3, the most coarse of the possibly unlevigated clays, was somewhat more common and was in the majority of cases used for BI bowls, which represented 5 of the 9 occurrences of that fabric noted. No instances of fabric type X3 were encountered in BIII, BIV or BV morphological types. The Y1 fabric occurred in all morphological types except BII bowls. Fabric type Y2 occurred in all morphological types with the exception of BIII bowls. Finally, fabric type Y3 was in a clear minority within the Y fabrics, with only 4 occurrences. This type had the most coarse of the sand tempers, with a significant portion of the larger inclusions >275 microns. Because the thin section sampling programme was not designed for it, the range of variation of fabric within a single vessel was not established.

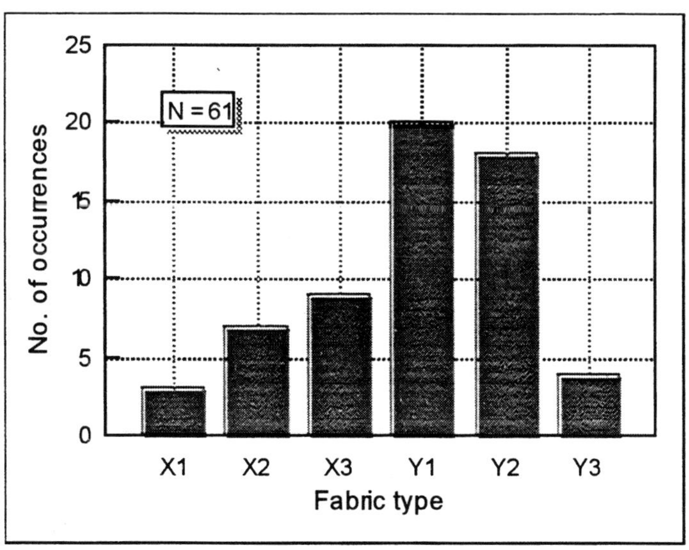

FIGURE 7.18

DISTRIBUTION OF REHMAN DHERI FABRIC TYPES

Some suggestions can be made with respect to technology and potter behaviour by considering the distribution of these fabric groups. Firstly, there does not appear to be any significant correlation between the defined fabric groups and morphological types. It is not apparent from these data that different fabric groups were forming different recipes for manufacture of certain types of jars or bowls. Each morphological type sampled was represented by three or more different fabric types, with the exception of BII bowls. However, only three BII bowls were analysed, therefore this morphological group may not be representative of the range of fabric types. There is the possibility that the methodology used in grouping the fabrics did not reveal deliberate attempts by the potter to vary the recipes. But, as discussed above, the fabrics were grouped on the main observed differences in composition.

Secondly, there was no correlation observed between those sections where a separate distribution was seen in the angularity of the quartz inclusions, as discussed earlier in the chapter, and either fabric type or morphological type. Those which contained, in addition to subangular to subrounded quartz grains, angular (4 on the author's angularity scale) grains, occurred at least once within each of the fabric groups. With respect to morphological groups, they occurred once in AII jars; four times in BI bowls; once in each of BII bowls, BIII bowls and BIV bowls; and three times in waster sherds. The angular inclusions, suggestive of crushed sand or primary deposit quartz temper, appear randomly distributed throughout fabric and morphological types.

Thirdly, there was no correlation observed between the defined fabric types and the measured thickness of the sherds. It could be expected that, in general, a more coarse fabric would be associated with a thicker-walled vessel. A thicker-walled, larger vessel requires greater strength in its fabric. Correlation with respect to rim diameter was not attempted, as the Kot Dijian ceramics (especially the AI and AII jars) have widely varying rim diameters associated with given body diameters because of the different mouth and rim shapes.

TABLE 7-II

REHMAN DHERI FABRIC TYPES OCCURRING BY MORPHOLOGICAL TYPE

Fabric Type	A1 Jar	AII Jar	BI Bowl	BII Bowl	BIII Bowl	BIV Bowl	BV Dish	CII Polych
X1	0	0	2	0	0	0	0	1
X2	0	3	0	0	3	0	1	0
X3	1	1	5	2	0	0	0	0
Y1	2	4	4	0	1	3	3	3
Y2	2	3	2	1	0	3	1	1
Y3	1	0	0	0	1	1	1	0

Variations in fabric types can be attributed to other sources. Because they are identified by differences in sand inclusions, it is possible that the variations are caused by use of different sand temper sources. This is unlikely, because over a broad areal expanse such as the Gomal plain it would be expected that sand found in one comparatively small area, such as the site of Rehman Dheri, would be of quite uniform angularity and size distribution, having been subjected to similar erosional influences and source environments respectively.

Another possible source of fabric variation is that each potter or groups of potters may have had their own distinctive fabric recipe. This phenomenon has been observed in ethnographic studies. Day (1995) reported that potters from Crete often remained with one paste recipe only, regardless of pot type. When they changed locations, mineralogical changes in their product could be observed, but paste recipes remained the same, possibly as a symbol of group cohesion. His research traced these observations back to 2000 BC. In ethnographic studies of potters from the Northwest Frontier Province of Pakistan, many family or kinship groups of potters have their own clay mixture recipes (Rye and Evans 1976). These cases could also represent simply that groups of potters learned one particular recipe and continued working with it throughout their careers, and passed the same one on to their trainees; and not necessarily a symbol of group cohesion.

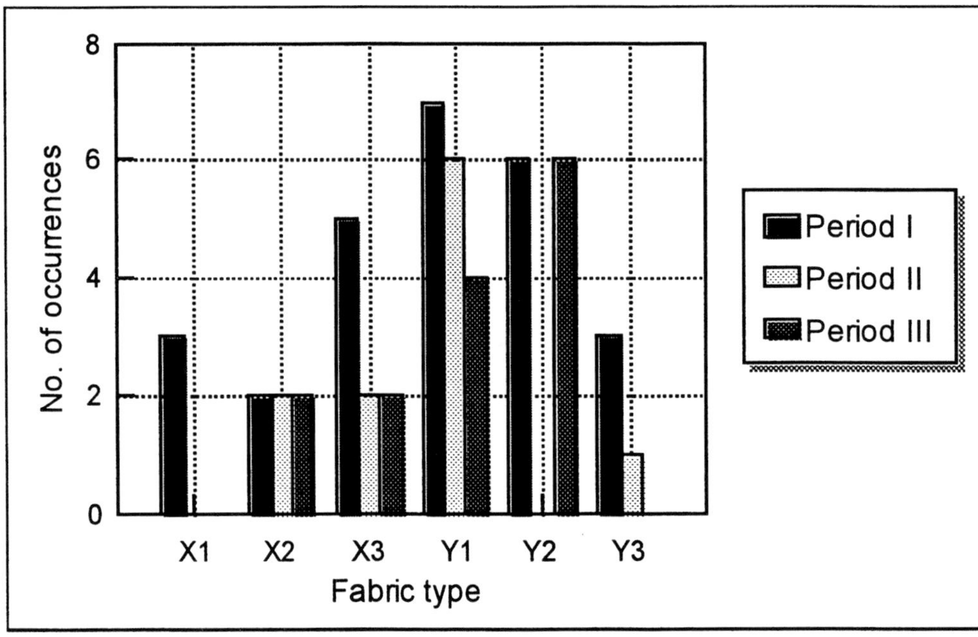

FIGURE 7.19

FREQUENCY OF REHMAN DHERI FABRIC TYPES: PERIODS I-III

Figure 7.19 shows the variation in fabric type distribution for Periods I to III at Rehman Dheri. The main observation to be made from this relationship is that Period I had generally more of the finer fabrics, and by Period III fabrics generally were towards the coarser end of the spectrum. Period II appeared to lie somewhere in between the two; however, the smaller number of samples taken from Period II does not provide a very sound basis for this observation. The gradual change from finer to coarser fabrics may not necessarily imply that attention to technological detail was lessening, although this has been observed as a gradual change which occurred between the Early and Mature Harappan periods at Harappa (Jenkins 1992: 327). Jenkins noted at Harappa that there was an increase in the quantity of plain wares and a decrease in the quantity of slipped and painted wares towards the Mature Harappan period. In the present study, Period I contained the broadest number of different fabric groups, being the only period where all six occurred. It is possible that as the quantity of slipped and painted wares declined towards the later period, the requirement for the finer fabrics declined with it. Alternatively, in the earlier phases of occupation at the site there were perhaps more different potter groups producing ceramics. If each had an individual fabric recipe (see previous paragraph) and the number of separate groups declined as the social organisation became more complex toward Period III, the number of separately definable fabric types may have declined. A shift in social organisation may be correlated with an increasing mass production of ceramics. This phenomena has been suggested to explain changes observed from the Early to the Mature period at Harappa (Wright 1991).

Further suggestions of fabric technologies are provided by the Rehman Dheri sample. There are some observations made in the thin sections which shed light on forming methods. A distinctive alignment of the inclusions was seen in most of the sherds, particularly with respect to those particles with a high aspect ratio such as the micas and some of the coarser quartz inclusions. Rearrangement of particles in a plastic clay occurs under stress deformation. Compression stress results in clay particles aligned normal to the major compressive stress, while shear stress results in clay particles aligned along the direction of shear (Kirby and Blunden 1991, cited in Courty and Roux 1995: 26). Abundance of rigid silt-size mica grains induces the same mechanical behaviour as coarse kaolinite with formation of a preferential parallel arrangement under hydric or mechanical stress (Tessier, pers. comm., cited in Courty and Roux 1995: 26). The majority of thin-sections where this feature was observed exhibited preferential mica particle alignment with the length of the section. Most of the sections were cut parallel to a plane bisecting the vessel perpendicularly, and at right angles to the wall surfaces. The observed particle orientation therefore strongly suggests a drawing or pulling of the plastic fabric in a dirction vertical to the vessel's base; or, alternatively, the vessels may have been thrown (Rye 1977 and 1981, Carr 1990; cited in Courty and Roux 1995: 21). No indications such as those observed by Courty and Roux (1995) to indicate the possibility of combined coil and smoothing techniques in some Harappan wares were seen in any of the thin sections examined. In order to effectively sample a vessel to examine forming technique via thin-sectioning, however, it is necessary to select three or four sherds from the same vessel wall. One can then be more certain that a location where a coil joining occurs is sampled.

Specific surface treatments are suggested by other thin section features. Compression or concentration of mica inclusions near the outer surface of the vessels was observed in some of the thin sections. This may be an indication of burnishing, which probably took place after drying but before firing. Burnishing compacts and reorients the fine clay particles, and is done on leather-hard or dry clays (Rice 1987: 138). The parallel facets observed on the surface of some of the sherds before sawing into thin sections support these indications. Moreover, cross-sections of micaceous slips were observed, shown in Figure 7.20. It could not be conclusively told from the images whether the slip was applied before or after the burnishing. Either technique is effective, but most slips are burnished or polished with a stone after application, to compact and orient the particles and impart a lustre (Rice 1987: 150). Correlation with chronological periods at the site was not observed.

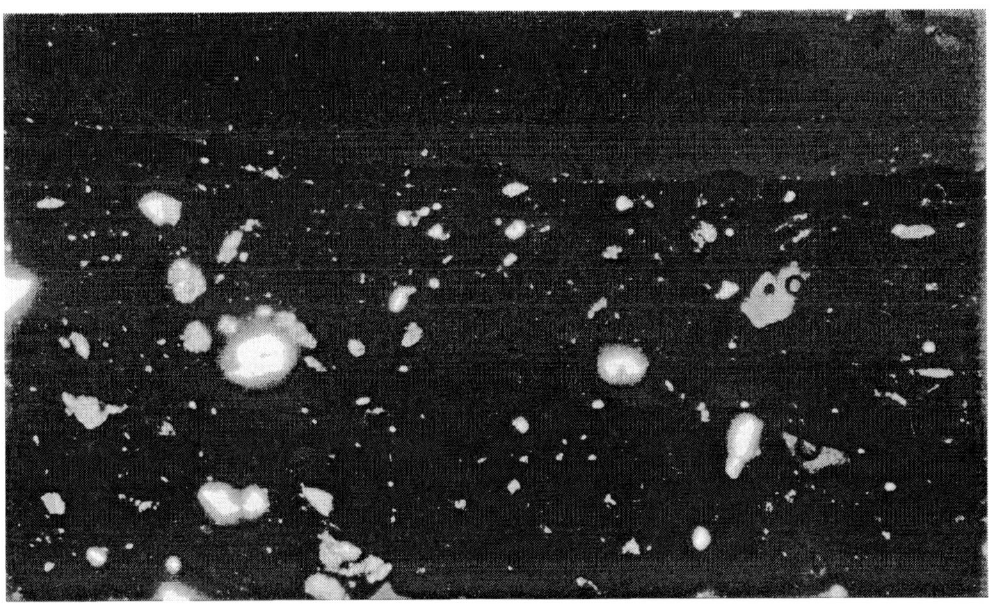

FIGURE 7.20

THIN SECTION SHOWING SLIP LAYER

RHD 53. Crossed polars, 40X, Field width 2.5 mm.

7.1.2. Bannu sites

Data obtained from the two sites located in the Bannu basin, Lewan and Tarakai Qila, are discussed together in this section, because they are situated in close proximity to each other (about 13 km apart). At distances such as these, the presence of intersite interaction may almost be assumed; without resorting to artefactual or data analysis. Indeed, it would be more of a task to demonstrate the *absence* of intersite interaction. Lewan produced ceramics, while no evidence for pottery manufacture was found at Tarakai Qila by the excavators (see discussion Chapter 5). The mineral suites of the two sites are essentially the same; however, distinct differences were noted in the size distribution characteristics of their respective inclusions. Because the similarities are greater than the dissimilarities, data from the two sites are discussed here in the same section. Importantly, however, the respective data are shown and discussed independently. Differences noted in technology between sites situated in such close proximity can be revealing in terms of site interactions.

Raw data from Lewan and Tarakai Qila are presented at Appendix B, respectively. For a detailed explanation of the information presented, the reader is referred to Section 6.4 and the introductory page to Appendix A. One individual sheet is presented for each thin section analysed. There are seven cases where the sample was too small or of too poor quality to complete a useful analysis with respect to textural data; these pages present only provenance, metric and mineralogical information.

Mineralogy

Concentrations of the minerals observed in the thin sections from Lewan and Tarakai Qila are summarised in Table 7-III. The table uses the author's scale described in Section 6.3, where

1= absent	2= trace
3= rare	4= common

Data shown in table 7-III were obtained by taking the means of the each of the individual mineral concentrations recorded on this scale. Standard error of the mean is as described by Shennan (1988: 302-3), using each estimate from the above scale as the individual mean of each thin section. This provides a reasonable estimate which is particularly useful for making intersite comparisons where mineral suites are undifferentiated as to type but not as to concentrations, as is discussed in Section 7.2. The table averages all sherds from each site, both excavated and waster sherds. Data for the quartz grains has not been included on the table, because they were in high concentrations in all of the sherds; and, as with Rehman Dheri, they are the subject of a separate investigation, described under 'Fabric' below.

TABLE 7-III

MINERAL CONCENTRATION DATA - BANNU SITES

◆ Excludes quartz ◆

Mineral	Conc/ Lewan[1]	S.E.Mean/Lewan[2]	Conc/ TRQ	S.E.Mean/TRQ
Plagioclase feldspar	1.85	0.07	1.71	0.10
Microcline feldspar	1.08	0.05	1.04	0.04
Biotite	1.85	0.14	2.00	0.12
Muscovite	3.62	0.11	3.89	0.06
Amphibole	1.35	0.10	1.32	0.09
Carbonate	3.15	0.18	3.32	0.10
Pyroxene	1.54	0.10	1.50	0.10
Epidote	1.46	0.10	1.39	0.09
Schist	1.58	0.11	1.50	0.11

1 Concentration

2 Standard error of the mean

Physical characteristics of the quartz inclusions from both of the Bannu sites are generally the same as those described from Rehman Dheri in Section 7.1.1, so their details are not be repeated in this section. Their characteristics are, as with the Rehman Dheri material, typical of sedimentary, igneous and metamorphic sources.

Minerals recorded in Table 7-III show that a close correlation exists between concentrations of each of the minerals of the two sites, indicating a high probability that they are from the same source geology. Moreover, the characteristics of each mineral from each site are closely similar. The descriptions of each mineral type are very similar to those made for the Rehman Dheri samples (but not their concentrations, which differed in some respects, and are discussed under 'Intersite analysis'), and so will not be repeated here.

One exception to this is weathering of biotite and muscovite inclusions. Both biotite and muscovite grains exhibited a much lesser amount of weathering than that observed on the Rehman Dheri material. Weathering of the biotite inclusions, using the author's scale as desribed in Chapter 6, averaged 2.10 and 2.13, respectively, for the Lewan and the Tarakai Qila sherds. This means that they are in the majority fresh to only slightly weathered. This is compared with 2.98 observed in the Rehman Dheri material. Weathering of the muscovite inclusions averaged 2.15 and 2.00 for Lewan and Tarakai Qila, respectively, about the same as the observations on biotite. This compares with 2.74 observed in the Rehman Dheri samples. While intersite comparisons are discussed in Section 7.2, such differences between the Bannu basin and the Gomal plain may be expected because of the greater age of the sedimentary deposits in the Gomal compared with the Bannu basin (M. Javed Khan, pers. comm.).

With respect to the nature of the carbonate inclusions observed in the sample from the Bannu basin, they were seen to be very similar to those seen in the Rehman Dheri sample. Carbonates of the oolitic type were present in 90% of the sherds from Lewan , 96% of those from Tarakai Qila, and in 83% of the wasters. In association with these there was a much higher incidence of the type '3', or the type which was unidentified from Rehman Dheri. These were present in 76% of the Lewan sherds, 79% of the Tarakai Qila sherds, and in 50% of the wasters. (The lower incidence seen in the wasters may be attributed to the higher firing temperatures they probably were subjected to - see discussion below on implied firing temperatures). There was one observation of a bioclast in the Lewan sherds, two in the Tarakai Qila collection and one in the Lewan waster selection. These were not identified but were of similar skeletal structure as those seen in the Rehman Dheri sherds.

Most of those oolitic type of carbonates that were present exhibited moderate decomposition, and were subjectively rated as the same seen in the Rehman Dheri collection. This would imply that the range of firing temperatures averaged around 850 degrees C. The state of decomposition was transitory, that is, most exhibited some level of decomposition, but none except three of the wasters was in an advanced state. These three wasters showed spalling and blistering on their surfaces, as did those from Rehman Dheri, indicating firing temperatures in excess of 850 degrees C. Further data are given under 'Fabrics' below, in discussion of the temperatures as implied by birefringence of the matrix.

Not included in the concentration table, because their specific concentrations were not recorded, were volcanic rock fragments. These were observed in both the Lewan and Tarakai Qila thin sections as well as the Lewan wasters. They were generally in the fine to very fine sand size range on the Udden-Wentworth scale (62.5 to 250 microns). Their individual constituents could not be readily identified, but plagioclase laths could usually be observed set in an altered groundmass, very similar to an example illustrated in Adams, et al. (1984: figure 24). They were observed in 15 of the Lewan sherds (71%), 17 of the Tarakai Qila sherds (61%), and 4 of the Lewan wasters (67%). In these, concentrations ranged from one particle to three particles in each section. These concentrations, as well as the particles' individual characteristics, are seen to be consistent from group to group, and were present only in sherds from the Bannu basin.

Other inclusions observed and placed in the 'Other' category include one incident of garnet. This was seen in thin section number LWN 202, an excavated sherd from Lewan. Garnets can occur in a variety of rock types, from medium-grade metamorphics (schists and gneisses) and igneous rocks to metamorphosed limestones (Rutley 1988: 355-6), all of which are present in the Bannu drainage basin. Possible grog was observed in 3 of the Lewan sections (14%), none of the wasters from Lewan, and in one of the Tarakai Qila sections. The grog-like inclusions appeared to be of clay, but did not contain inclusions within their individual grains, nor voids. Therefore they may be some other argillaceous inclusions such as natural clay concentrations, and not necessarily grog.

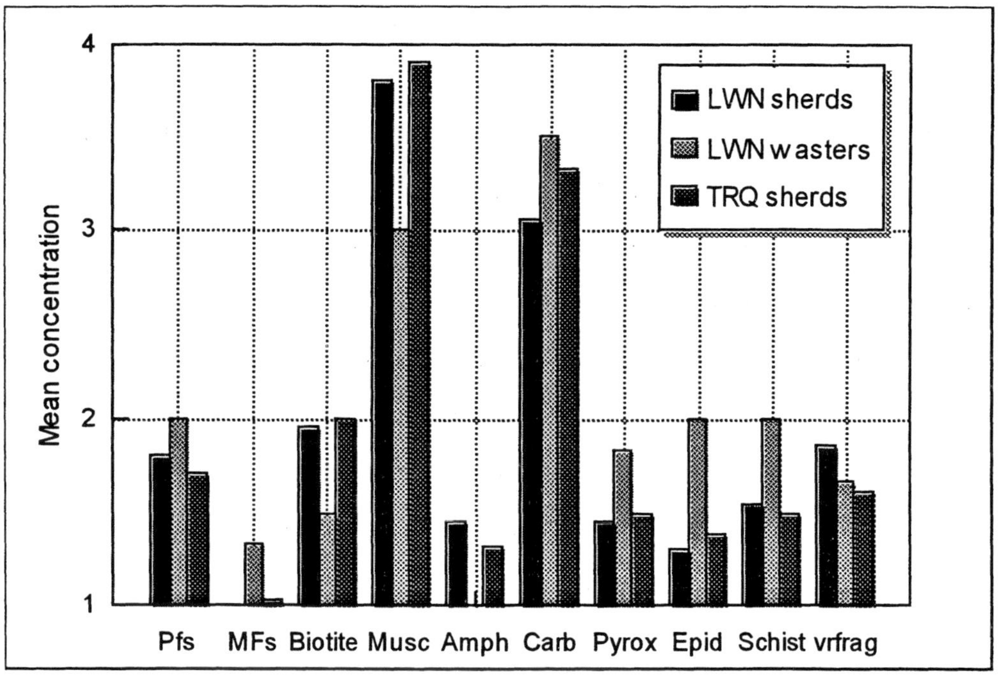

FIGURE 7.21

COMPARATIVE MEAN CONCENTRATIONS OF MINERALS OBSERVED IN LEWAN AND TARAKAI QILA SHERDS, WITH LEWAN WASTERS

PFs= plagioclase feldspar   MFs= microcline feldspar   Musc= muscovite
Amph= amphibole   Carb= carbonate   Pyrox= pyroxene
Epid= epidote   vrfrag= volcanic rock fragment

The wasters collected from the surface of Lewan, which are assumed to have been manufactured at the site, show a mineral suite very similar to that of both the Lewan and the Tarakai Qila excavated material. With two exceptions, all of the minerals are present in approximately the same concentrations. One of these, amphibole, was not observed in the sample of Lewan wasters. Because the sample of wasters is small (six) and the standard error of the mean of the amphibole concentrations in the excavated sherds is 0.09 to 0.10, it is possible that a waster sherd containing one was not selected. Amphiboles are relatively rare in the Bannu material. The other exception is microcline feldspars. As with the amphiboles, these inclusions are rare in the material from the Bannu sites, and it is probable that the waster sherd collection missed an example with a microcline feldspar inclusion.

The high degree of similarity in the mineral concentrations and their characteristics, and their respective similarity with the wasters, suggests that the Lewan and Tarakai Qila sherds are sourced from the same geology. Moreover, that geology is consistent with adjacent geological formations feeding detrital material into the Bannu basin. As described in Section 5.1.3, the Bhittanni and Marwat ranges, the Waziristan highlands, the Sulaiman range and the Parachinar area all form part of the drainage basin. These formations provide (not an all-inclusive listing) common sedimentary rocks and fragments including quartzites and limestones; igneous rocks including basalt and pumice; and metamorphics including gneiss, mica schist, and hornblende. Volcanic rock fragments uniquely present in the Bannu basin samples may have been derived from clasts present in the Bain diamictite (Khan et. al. 1985: 61). Overall, there is no evidence in the mineralogy that would suggest the sherds are derived from ceramics produced anywhere other than the Bannu basin.

Fabrics

The same procedure as was used for the Rehman Dheri inclusion size data was employed on the Bannu sites data. Because the fabrics in that case were classified primarily on the basis of quartz size data, this was used again for Lewan and Tarakai Qila so that a comparison between the two ceramic production areas was facilitated. Figure 7.22 shows relative size distributions of quartz inclusions 30-50 microns from the two sites.

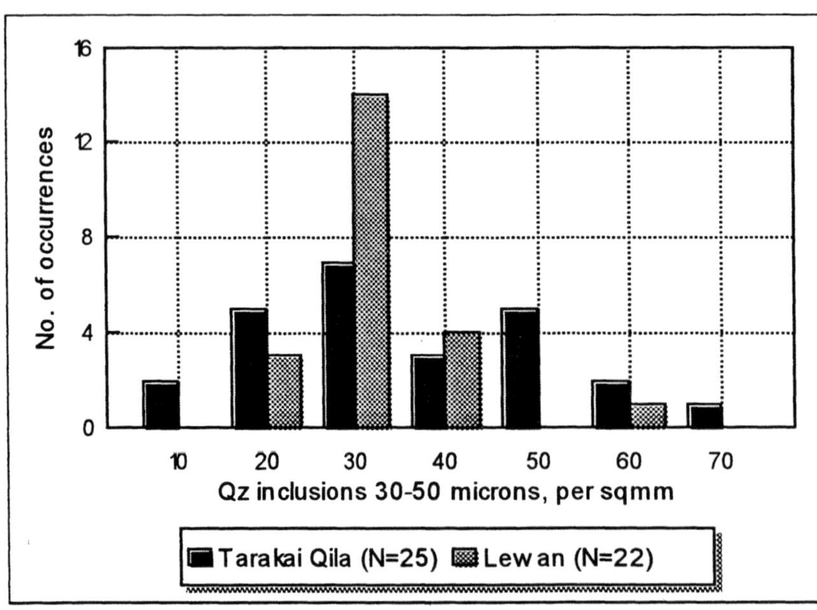

FIGURE 7.22

DISTRIBUTION OF QZ INCLUSIONS 30-50 MICRONS: BANNU SITES

Figure 7.22 shows that the Lewan sherd quartz inclusions of the smaller size range concentrate around the 30 per sq mm value; and exhibit a platykurtic dispersion which is to some extent positively skewed. There is a possibility that another mode exists further to the right, around the 60 per sq mm concentration level. If one does exist, a larger sample would be required to show it, as it appears that it would be a very small segment of the population sampled. The Tarakai Qila data of the same chart shows that the small quartz inclusions there also have a mean at around the 30 per sq mm level. These data also appear to be approximately positively skewed, but are much more leptokurtic in distribution than the Lewan data. The possibility of a second mode is also seen in the Tarakai Qila data. A second peak is seen at the 50 per sq mm level which may indicate a second mode; or just some missing data around the 40 per sq mm level. As the sample size is reasonable to define the distribution, it is more likely that there exist two distributions. Thus, it is seen that the respective distributions of the two sites are very similar in some respects and quite dissimilar in others.

These similarities and differences can be examined in terms of technology and potter behaviour. The factor of two independent distributions present in the material suggests that the clay may have been levigated to achieve two different fabrics. The differences noted in the dispersions of the distributions from the two sites may show that both clays were being levigated, but the potters making the Tarakai Qila vessels were perhaps doing a 'better job' of it; with more sherds in the lower concentration ranges. At Lewan, the most common occurrence was also at the 30 per sq mm range, but with much fewer occurring below it. This may suggest either no levigation, or levigation to a point which includes a larger concentration of the finer particles. In their study of Pakistan river and loess sands, Rye and Evans (1974: 128-9) found that the particle size distribution of sands from the Bannu region (Shahbaz Ahmad Khel village) had by far the highest concentrations of particles in the silt-size range of eight tested from various regions of the country. In a sieve test they were found to contain 20% silt-size particles, compared with 0 to 2% silt-size particles for the others. Thus it is possible that the sands found near the sites are already relatively high in the smaller particle size range.

Although no evidence of a local pottery industry was reported at the site of Tarakai Qila, there may have been one. As seen above, the close similarities in the mineral suites of the two sites indicates that their respective clay sources are in proximity; probably the Bannu basin. The differences noted in the respective distributions of the smaller sized quartz inclusions discussed above are significant in that it indicates a different industry, or at least a different mixture recipe, perhaps individual potter groups each with their own recipe. Alternatively, pottery manufactured at Lewan, where there is a known ceramic industry, may have supplied the needs of Tarakai Qila. If the latter is the case, pottery transported to that site was of a lesser quality, at least in respect to preparation by levigation. It was noted in Section 5.3.4 that Tarakai Qila was much less developed technologically than Lewan, and was very much an agricultural community founded on pastoral and arable elements. Further discussion can be provided following examination of the quartz size component >50 microns, where temper, if used, would be classified.

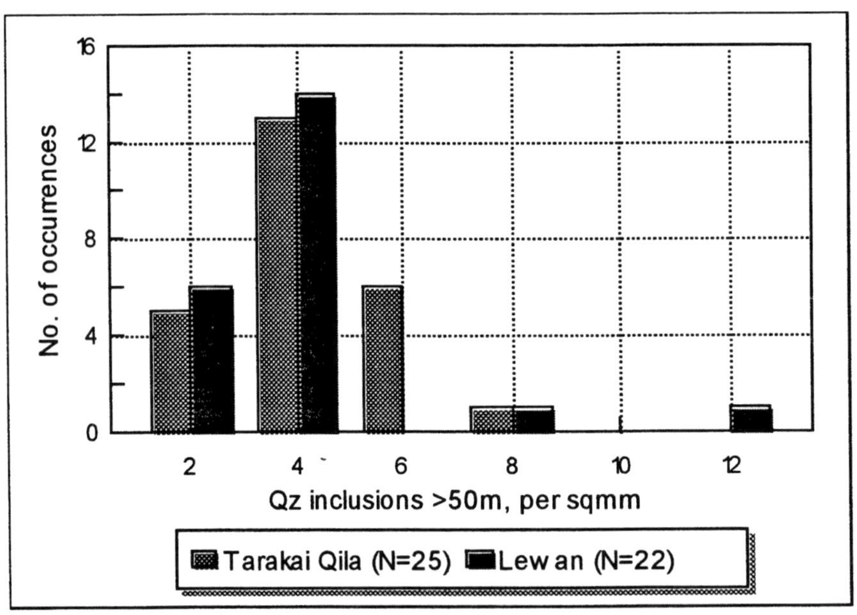

FIGURE 7.23

CONCENTRATIONS OF QZ INCLUSIONS >50 MICRONS: BANNU SITES

Figure 7.23 shows the distribution of quartz inclusions >50 microns for both sites. As with the smaller inclusion component described above, both exhibit approximately the same mode (4 per sq mm, in this case). Both are roughly positively skewed, and unlike the smaller set of inclusions, show similar dispersions. The Tarakai Qila data are strictly unimodal, while those from Lewan indicate some small possibility of a second distribution. It cannot, however, be confirmed by these data.

Only one of the sherds sampled from the Bannu sites, TRQ 222, showed evidence of the different angularity distributions in the fraction >50 microns seen in many of the Rehman Dheri sherds (see discussion for Rehman Dheri). At Rehman Dheri, these suggested limited use of crushed quartz temper. None was observed in the Lewan sample.

The fabrics from Lewan and Tarakai Qila showed close similarities with, and fit into, the categories defined above for the Rehman Dheri fabrics. Under this scheme, the Lewan fabrics all fell into the X category, i.e. they all had a concentration of quartz inclusions 30-50 microns of greater than 10 per sq mm. Of these, 17 (77.3%) were X1, having a distribution of quartz inclusions >50 microns as shown in Figure 7.12; 4 (18.2%) were X2, having a distribution as shown in Figure 7.13; and one (4.5%) was X3, having a distribution as shown in Figure 7.14. The Tarakai Qila fabrics grouped similarly, with 20 (80.0%) X1, 3 (12.0%) X2, but with 2 (8.0%) not fitting any of the six defined categories. In the latter two, the larger fraction fitted the X1 distribution, but had a concentration of the smaller fraction of less than 10 per sq mm.

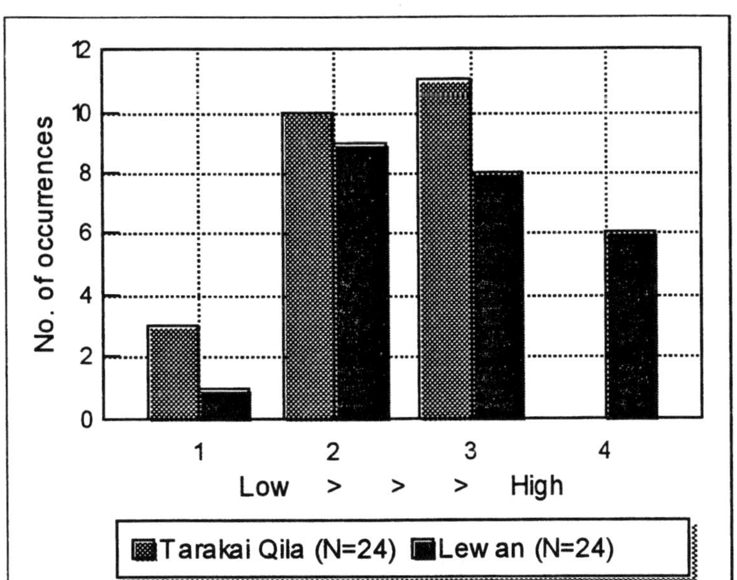

FIGURE 7.24

BIREFRINGENCE OF FABRICS FROM BANNU SITES

As was seen with the Rehman Dheri collection, no correlation of fabric types was observed with vessel morphology, or wall thickness. The variation in fabric groups may have been a function of potter groups, the basis of which was discussed in the Rehman Dheri section.

Relative birefringence seen in thin sections from the Bannu sites is shown in Figure 7.24. Both sites have a mean birefringence in the 2 to 3 range, with the Lewan material showing a broader distribution and generally higher values. 54.2% of Tarakai Qila thin sections exhibit low birefringence (1 to 2), while the remainder show intermediate (3) levels. Based upon the research discussed in the Rehman Dheri section, this would suggest that the lower segment were subjected to firing temperatures above 850 degrees C, and the remainder probably less than the vitrification point. The Lewan sherds showed a broader distribution, significantly skewed towards the higher birefringence levels. This suggests that firing temperatures were less consistent and generally lower than 850 degrees C at Lewan; although this trend was not suggested from observation of the carbonate decomposition levels in the same sections.

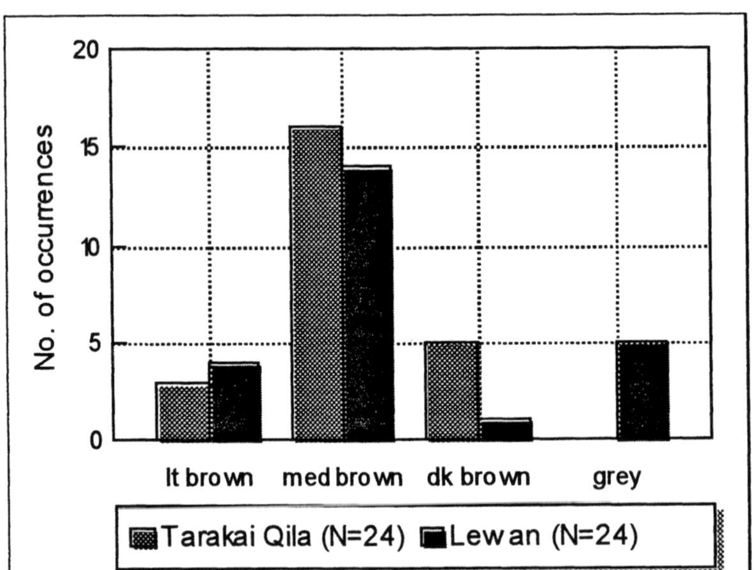

FIGURE 7.25

DISTRIBUTION OF COLOURS IN BANNU SITES FABRICS

Figure 7.25 presents a summary of the colours observed in the fabrics from Tarakai Qila and Lewan in thin section. Most of those from Tarakai Qila (67%) were medium brown, as were 58% from Lewan. The greys observed in the Lewan sections were mostly seen in the waster collection, where a high proportion were fired in a reducing atmosphere.

Finally, thin section observed from the Bannu sites revealed much the same clues as to other ceramic technologies. The same indications of burnishing and slipping as described for those sections were observed in much of the material from both Tarakai Qila and Lewan. Additionally, based on observations of particle alignment in the thin sections, forming techniques appeared to be very similar to those seen from Rehman Dheri.

7.1.3. Harappa

Raw data for the sherds and bangles sampled from Harappa are presented at Appendix C. As described for the sites above, an explanation of the information presented on each sheet is given in Section 6.4 and the introductory page to Appendix A. There is one sherd and four bangle thin sections which were too small or of too poor quality for accurate point or area counts; only basic and mineralogical information is recorded for these. In the case of the bangles, they were particularly vulnerable to difficulties caused by size because of the very small cross sections of these artefacts.

Mineralogy

Concentrations of the minerals observed in the thin sections from Harappa are summarised in Table 7-IV. The concentration scale as used for the previous sites, and discussed in Section 6.3, is used.

TABLE 7-IV

HARAPPA MEAN MINERAL CONCENTRATION DATA

◆ Excludes quartz ◆ see text for explanation of concentration scale ◆

Mineral	Conc /sherds	S.E.Mean[1] /sherds	Conc /bangles	S.E.Mean[1] /bangles
Plagioclase Feldspar	2.13	0.11	1.59	0.12
Microcline Feldspar	1.00	0.00	1.00	0.00
Biotite	2.78	0.13	2.06	0.13
Muscovite	4.00	0.00	4.00	0.00
Amphibole	1.35	0.10	1.06	0.06
Carbonate	1.33	0.11	1.18	0.10
Pyroxene	1.04	0.04	1.12	0.08
Epidote	1.43	0.11	1.24	0.11
Schist	2.70	0.12	1.47	0.12

[1] Standard error of the mean

Quartz concentration data have not been included in the table, as they are treated separately under the section 'Fabrics' below. As with the other sites, quartz was the most abundant mineral present in all of the sections. Both monocrystalline and polycrystalline grain types were observed; in an estimated ratio of 75:25 respectively. Observations with respect to the physical nature of the grains (type of sutures, extinction etc.) were essentially the same as those reported for the Rehman Dheri thin sections, so will not be repeated here.

It is noted that in all of the mineral categories observed, with the exception of pyroxene, concentrations are consistently higher in the Harappa sherd thin sections than the Harappa bangle thin sections. A contributing factor to this is that the average area of the sherd thin sections is much larger than that of the bangle thin sections, generally about four times the size of the bangle thin section areas. The methodology used to estimate the concentrations of minerals in each was subjective (with the exception of quartz) and did not directly take into account measured area. A concentration estimate was made based upon the absolute number of inclusions of a given mineral rather than the number per unit area. The technique was used because the very low concentrations of many of the minerals would not have allowed distinguishing different areas by differences in mean concentration levels without prohibitively high point counts. Often, in the pottery thin sections there were only one or two inclusions of a given mineral such as amphibole. When a thin section about one-quarter to one-eighth the size is examined, such as was the case with the bangles, there is a high probability that the given mineral could be absent in that particular section. The method is valid for series of thin sections which are approximately the same size as was the case with pottery thin sections, but less so for comparison with a group of sections which was orders of magnitude smaller such as the bangles. Because of this, the mean concentration of a more rare mineral may appear higher for a group of larger area sections than smaller ones. This is true for all thin sections examined in the present research, but, as the number of pottery thin sections is large and they are within a fairly standard size range, the effect is lower. This is borne in mind throughout the analysis.

On the other hand, the differences seen in the mean mineral concentrations between the pottery and the bangle groups from Harappa could signify different clay sources for the two. Referring again to Table 7-IV, plagioclase feldspar, biotite, and schist are seen to be completely independent from one another within ± two standard errors of the mean (95% confidence interval). This may be due to more extensive levigation of the clays used for the bangles. For the remainder of the minerals, concentrations shown by the two groups overlap at the same confidence interval. These differences may be significant, but, using the present methodology, as discussed in the last paragraph, the effect of the size of sample cannot be eliminated as a factor, therefore a less subjective method of measuring the mineral concentrations would be needed to analyse those differences. At this point, therefore, the possibility that the bangles have originated from a different clay source than the pottery sherds cannot be ruled out. Some support for the suggestion that excavated bangles were manufactured from local clays is obtained from the observation that bangles of local Ravi River clays made by the experimental potter at the Harappa Project (HRP 148) contained the same mineral suite and concentrations, within the $\pm 2\sigma$ (plus or minus two standard errors of the mean) confidence interval, as the excavated bangle sample.

Feldspar inclusions were mostly plagioclase, and usually of the same angularity and sphericity as the quartz. They were mostly fresh to moderately weathered. Concentrations were slightly higher in the sherd fabrics than in the bangles (but see above). No microcline feldspars were observed in the Harappa fabrics. Concentrations of muscovite were high in all of the samples, and biotites were generally between 'trace' and 'rare'. Minerals of the amphibole group were rare occurrences in the Harappa material, and when present were usually brown hornblendes such as were seen in the Rehman Dheri sherds.

Carbonates were present in much lower concentrations than seen in the samples from the previously analysed sites, west of the Indus River. Despite the soil characteristics observed around the site of Harappa (see discussion Chapter 5) which show concentrations of pedogenic carbonates, instances of carbonates observed in the thin sections were few. It is possible that carbonates were present, but in very minute inclusions which are not visible in the thin sections. Alternatively, complete decomposition may have occurred from high firing temperatures; however, another indicator of firing temperatures, the fabric birefringence, does not corroborate this (see discussion under 'Fabric'). The soils of the area may be rich in carbonates, but the clays as prepared for making ceramics appear not to have been. River sands used as temper by present-day potters at Multan (near Harappa) have been found not to contain carbonates (Rye and Evans 1974: 128). Also, present-day potters near Harappa, as well as a local potter carrying out experimental ceramics at the Harappa Project (Kenoyer 1994: 348), remove calcium carbonate nodules during clay preparation. Early Harappan potters at the site may have also done this. When observed, the grain type was similar to the oolitic shape seen in the Rehman Dheri and Bannu materials. They exhibited mild decomposition only in some cases, and none in others. Alone, this factor suggests that firing temperatures were probably not often in excess of 850 degrees C (but see discussion in Section 7.1.1.). Only one occurrence of bioclast was observed, and none of the third type of carbonate described for the Rehman Dheri and Bannu samples was seen. This latter type was fully described in Section 7.1.1 (Rehman Dheri) under 'Mineralogy'.

Schist occured relatively often in the Harappa sherds, but was not as common in the bangle thin sections. Apart from the thin section size differences discussed above, this may be attributable to the finer fabric textures seen in the bangle selection. The schist grains were generally in the larger, sand-size categories which may have been removed via levigation to achieve the fine inclusion sizes needed for small artefacts such as bangles. Schists were generally muscovite with polycrystalline quartz or plagioclase feldspars; or epidote schists.

Fabric

A summary of the concentration of quartz inclusions 30-50 microns for both bangles and sherds from Harappa is presented in Figure 7.26. The concentration in the bangles shows a broadly dispersed distribution between 20 and 60 per sq mm. Although the sample is not large (N = 15), two maxima are seen, the first at 20 per sq mm, and the second at 40. The sherd distribution data (N = 22) show a concentration around 30 per sq mm, with a positive skewness. A smaller, possibly separate, distribution is seen at 50 per sq mm. River sands used as tempers by present-day potters at Multan, near Harappa, have been shown to contain only 2% particles of silt-size or less (<62.5 microns) (Rye and Evans 1974: 125); much coarser than those observed in the above discussion on the Bannu data. The characteristics and distributions seen in both the bangle and the sherd concentrations are consistent with their being levigated. Especially with respect to the bangle data, there are possibly two separate fabric groups based upon the concentrations of quartz inclusions in the 30-50 micron size range. This may be attributable to two separate types of bangles.

Although some were classified as terracotta by the excavators (specifically samples HRP 130-133, 139, 141-145, and 147) the groups did not divide according to this factor.

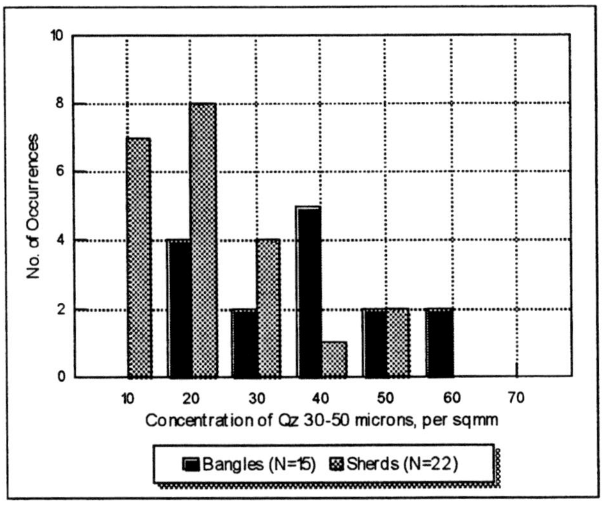

FIGURE 7.26

DISTRIBUTION OF QZ INCLUSIONS 30-50 MICRONS FROM HARAPPA

The distribution of concentrations of quartz inclusions >50 microns for the Harappa bangles and sherds is shown in Figure 7.27. The bangles show a mode around 4 per sq mm, with either a very positive skewness or a second, smaller, mode in the 10 to 12 per sq mm range. As the second group consisted of only two samples, it cannot be conclusively determined whether or not two separate distributions exist. The data are consistent with the addition of sand tempers to achieve a particular concentration level of around 4 per sq mm, with possibly another group of bangles designed to have a higher concentration level of 10 to 12 per sq mm. A larger sample would have to be studied to more accurately indicate the existence of a second mode.

The sherd data in Figure 7.27 may also contain two separate distributions, one with a mode of 4 per sq mm and another of 8 per sq mm. Separating the two and describing each of their individual distribution characteristics would entail sampling a larger number of sherds. Both are positively skewed. These data are consistent with those which would be observed in a group of sherds containing two separate fabrics, one with added river sand tempers designed to achieve a concentration level of approximately 4 quartz inclusions per sq mm, and another designed to achieve a concentration level of about 8 quartz inclusions per sq mm. This split would be consistent with a potter having a 'basic' fabric and simply doubling the amount of river sand temper added to achieve the second fabric mix. Within this fabric size group, only one sherd (5%) was observed to contain the very angular quartz described for some of the Rehman Dheri sherds earlier. Five (23%) contained grog temper or clay pellets.

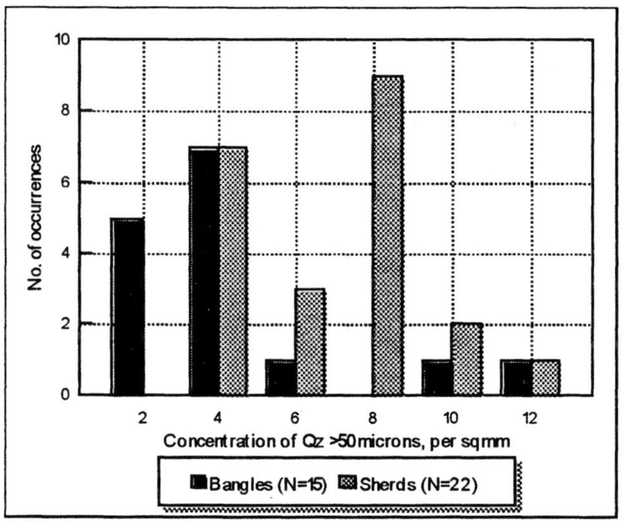

FIGURE 7.27

DISTRIBUTION OF QZ INCLUSIONS >50 MICRONS FROM HARAPPA

The sherd data appear to fall into four identifiable fabric groups based on the foregoing observations. These groupings are defined by, firstly, the two groups in the distribution of quartz inclusions 30-50 microns, which is consistent with two different levels of clay levigation. These two concentrations are at 20 per sq mm and 40 per sq mm. Secondly, two distributions seen in the quartz inclusions >50 microns chart (Figure 7.27) can define two groups consistent with the addition of coarser quartz tempers. These two concentrations are at 4 and 8 per sq mm. It is important to note here that these groupings may not be complete nor said to be representative of the entire Early Harappan ceramic industry at the site of Harappa. Nor is any representation implied that these were fabric groupings that the potters set out to achieve. They are only convenient categorisations of the fabrics of the sherds examined which fit a plausible scenario. A sample of the size and representativeness as that taken at Rehman Dheri would be more appropriate on which to draw such inferences. The foregoing four fabric groupings are, however, more relatable to potter behaviour than those groupings formed for the Rehman Dheri fabrics. The main part of that comparative discussion is included with Section 7.2 below, but it is more reasonable to expect potters to create fabric groupings based upon concentrations of coarse temper than upon selection of a particular size distribution in the temper.

The Harappa sherd fabrics do not fit into the categories which were defined for the Rehman Dheri material. The size distribution of quartz inclusions in the Harappa sample does not closely match those upon which the original Rehman Dheri groups were based. An obvious reason is that those original groups were defined based upon grouping of similarities seen in the Rehman Dheri size distributions. The forcing of the Harappan fabrics into a such a classification system would not be a useful exercise. However, the fact that the Harappan sherds form their own distinctive fabric groups and do not group naturally into the same classifications as those produced at Rehman Dheri provides some indication of the nature of intersite technological information exchange during the Early Harappan period. That discussion is presented below in Section 7.2.

With respect to the bangles, the data are consistent with a craftsperson attempting to achieve a state of levigation which would give a concentration of fine quartz or sand inclusions represented by approximately 20 particles per sq mm, with the occasional need for a higher concentration level of about 50. The level of larger quartz inclusions, those >50 microns, to be achieved appear to be about 4 per sq mm, with the occasional need for a higher concentration of 10 - 12 per sq mm. A comparison of these data with the quartz size and concentration data observed in the bangle made by the experimental potter at the Harappa Project showed remarkable consistency with some of the excavated bangle fragments. Comparing the concentration and distribution of quartz inclusions in the experimental bangle thin section, HRP 148 (Appendix C), with those of excavated examples HRP 140 and 144 for example, shows that the experimental potter has achieved a close copy of one of the ancient bangle fabric recipes. The potter at the Harappa Project prepared the clay used for the experimental bangles by levigating it several times (Kenoyer 1994: 348). Thus we have a graphical representation of the characteristics of quartz inclusions of a fine fabric which has been prepared by levigation; and some further support for the suggestion that Early Harappan clays were meticulously levigated.

It would be expected that a relationship between bangle thickness and fabric coarseness would exist. However, when plotted as scatter diagrams, no clear correlation was evident between bangle cross-sectional area and coarseness of fabric measured either as concentration of quartz inclusions 30-50 microns or >50 microns. It is noted, importantly, that the sample of bangles was too small (only 15 were subjected to the textural analysis) to make useful statistical inference regarding the numbers and characteristics of different fabric types. A proper sampling would consider a representative selection of the various bangle types found at the site. Nevertheless, the present sample provides much insight and permits comparisons to be made with the sherd fabrics. Also, some technological information can be derived.

Birefringence observed in the Harappan fabrics is summarised in Figure 7.28. It shows that the majority (89%) of the bangle fabrics exhibited a medium to high birefringence, as well as the majority (86%) of the sherd fabrics. Two of the bangle fabrics, HRP 142 and 147, were so highly birefringent in thin section that the fabric colour was barely discernable. Based upon the discussion presented earlier with respect to the Rehman Dheri fabrics, this would suggest that most of the bangles and sherds examined from Harappa were fired to temperatures lower than 850 degrees C. However, this cannot be conclusive without some other indicators to corroborate it. Birefringence alone cannot be relied upon to provide a definitive measure of firing temperature. As indicated above, the absence of significant amounts of carbonate in the fabrics did not allow their decomposition to be used as an indicator, as was possible with the Rehman Dheri fabrics. Most of the bangles sampled were classified as terracotta by the excavators. Many of the bangles from the Mature Harappan Period are of stoneware (Schneider 1987) which is typically fired at temperatures higher than 1100 degrees C (Rice 1987: 5), but none of the present samples was of this type.

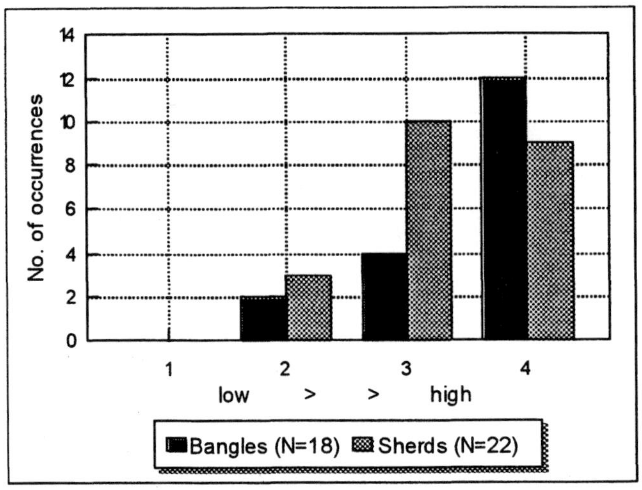

FIGURE 7.28

BIREFRINGENCE OBSERVED IN HARAPPAN FABRICS

Colours observed in the fabrics (in thin section) from Harappa are presented in Figure 7.29 for both bangles and sherds. It is seen that the bangles exhibit a broad range of colours, with the most common being medium brown. The sherds were all medium to dark brown in thin section, with the mode (most common) being medium brown.

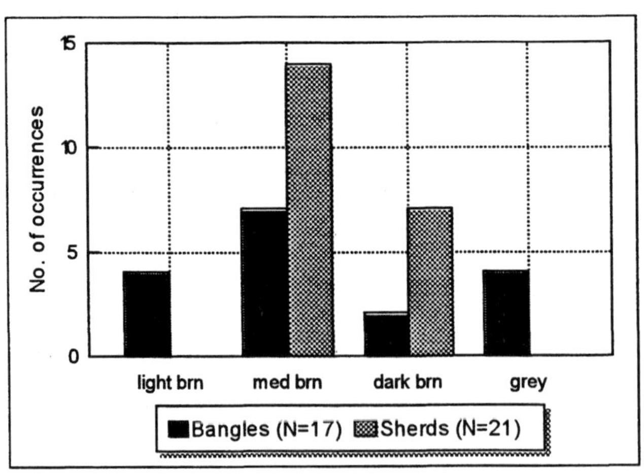

FIGURE 7.29

DISTRIBUTION OF COLOURS OBSERVED IN HARAPPA SECTIONS

Close similarities were seen between technologies evidenced in the Harappan sherd thin sections and those seen in the Rehman Dheri thin sections. Evidence with respect to particle alignment and orientation, burnishing and slipping were all seen in the Harappan sherd sections. The reader is referred to the discussion contained in Section 7.1.1. for details of these.

Some technological aspects of Harappa bangle manufacture were seen in the bangle thin sections. Many of the bangle samples were examined in two orientations: one cut through the cross-section, and one in a plane parallel to the bangle's circumference. The cross-section in some instances showed an amorphic void approximately in its centre, with a slightly discernable alignment of inclusions (primarily the muscovites) in a concentric pattern around it. In the longitudinal section, a preferred orientation of the same inclusions was evident parallel to the circumference of the bangle. Referring again to the discussion in Section 7.1.1, these two observations suggest that these bangles were made by rolling the clay, while plastic, between two surfaces. As the clay is formed into a thin tube-like form, it became longer, causing the inclusions to align with the direction of forming. These observations were seen only in bangles circular in cross-section. In many others, including the rectangular cross-section and multi-piece bangles, a longitudinal alignment of particles was seen but not the circumferential alignment. In cross-section, these exhibited alignment but it was angular, from one side to the other. Experimental studies carried out at the Harappa Project have suggested that bangle manufacture consisted of starting with a hollow clay column on a wheel and removing a series of blanks with a cord (Kenoyer 1994: 349-51). The angular alignment of inclusions would be consistent with this methodology. It is suggested that the Early Harappan craftspeople employed at least two different methods of forming bangles.

The multi-section bangles appear to have been manufactured by making a series of these thinner bangles, and then pressing them together side by side. This was evidenced by the joints observed in the cross-sections of these bangles where the flat sides of each individual bangle met. None of the circular cross-section bangle fragments thin sectioned contained an end-to-end join, however, so it cannot be told from the thin sections how the lengths, once rolled out, were joined to form the circular shape. Another common observation in the bangle thin sections was a slight variation in colour (a darkening) extending about 0.5 to 1.0 mm into the inside surface of the bangle. This may have been caused by body oils from the wearer being absorbed into the ceramic fabric.

7.2. INTERSITE ANALYSIS

With the analyses of the individual sites and areas now complete, their respective similarities and dissimilarities are now examined. All of the variables noted and recorded for each of the thin sections from each of the sites are considered. Firstly, the mineralogies of each of the collections are considered in order to determine the probability of their being sourced from different populations. This provides suggestions as to the likelihood of pottery being moved from site to site. As was discussed in Section 4.2, the next level of interaction is exchange of technological information, or innovations. Examination and comparison of technological attributes as inferred by the individual site analyses are used to determine the degree of this exchange amongst the sites sampled.

7.2.1. Mineralogy

It was observed that all of the thin sections analysed in this research contained basically the same mineral suite. Some significant differences were seen, however, in their respective concentrations from site to site or from area to area.

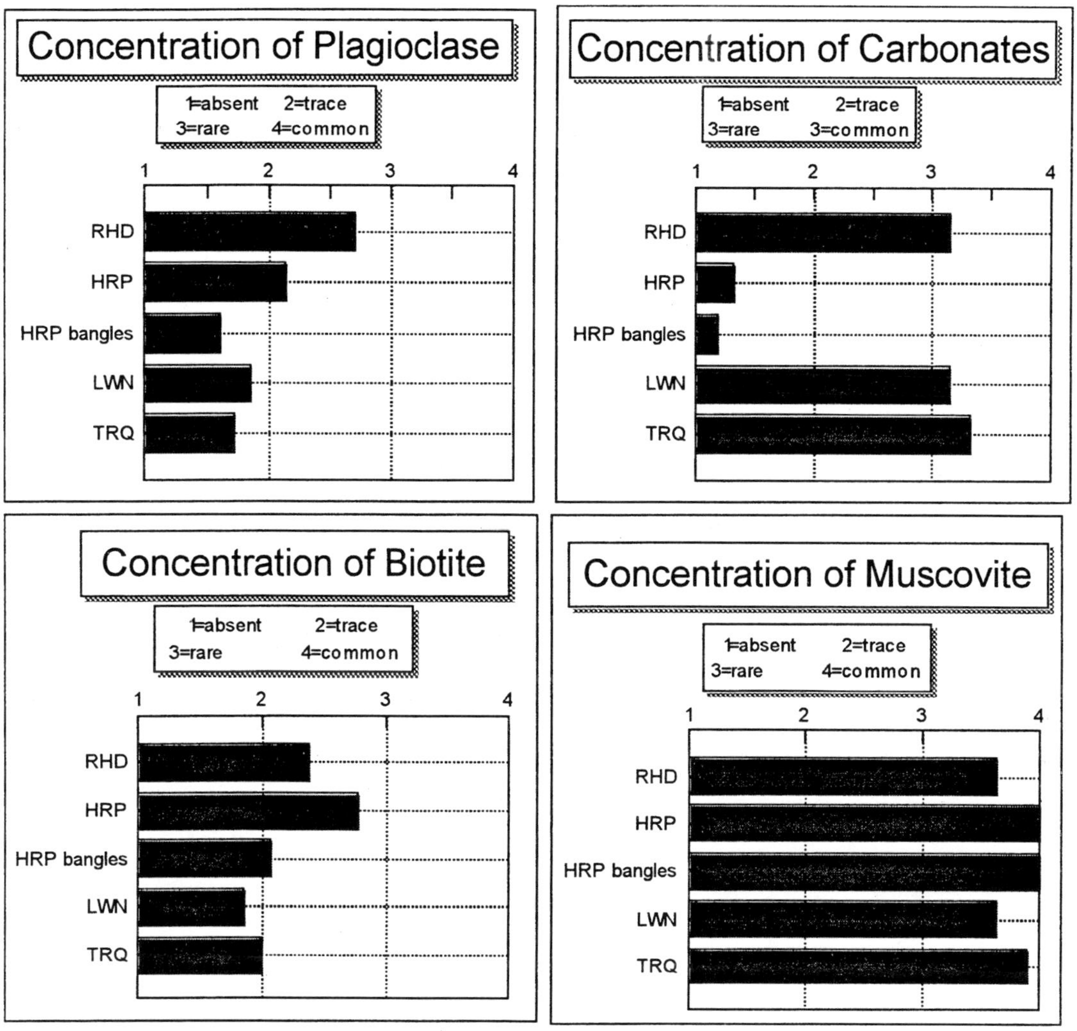

FIGURE 7.30

COMPARATIVE MEAN CONCENTRATIONS OF PLAGIOCLASE, CARBONATE AND MICAS FROM FIVE GROUPS

RHD= Rehman Dheri HRP= Harappa
LWN= Lewan TRQ= Tarakai Qila

Figures 7.30 and 7.31 present comparative concentrations of eight of the minerals observed in the five different site and type groupings, treating Harappan bangles as a separate group. Not included in either of the figures are quartz and microcline feldspar concentration data. Quartz was common in all of the fabrics and has been treated independently via the fabric analysis. Microcline feldspars were extremely rare in all of the sites and not useful as a provenancing tool.

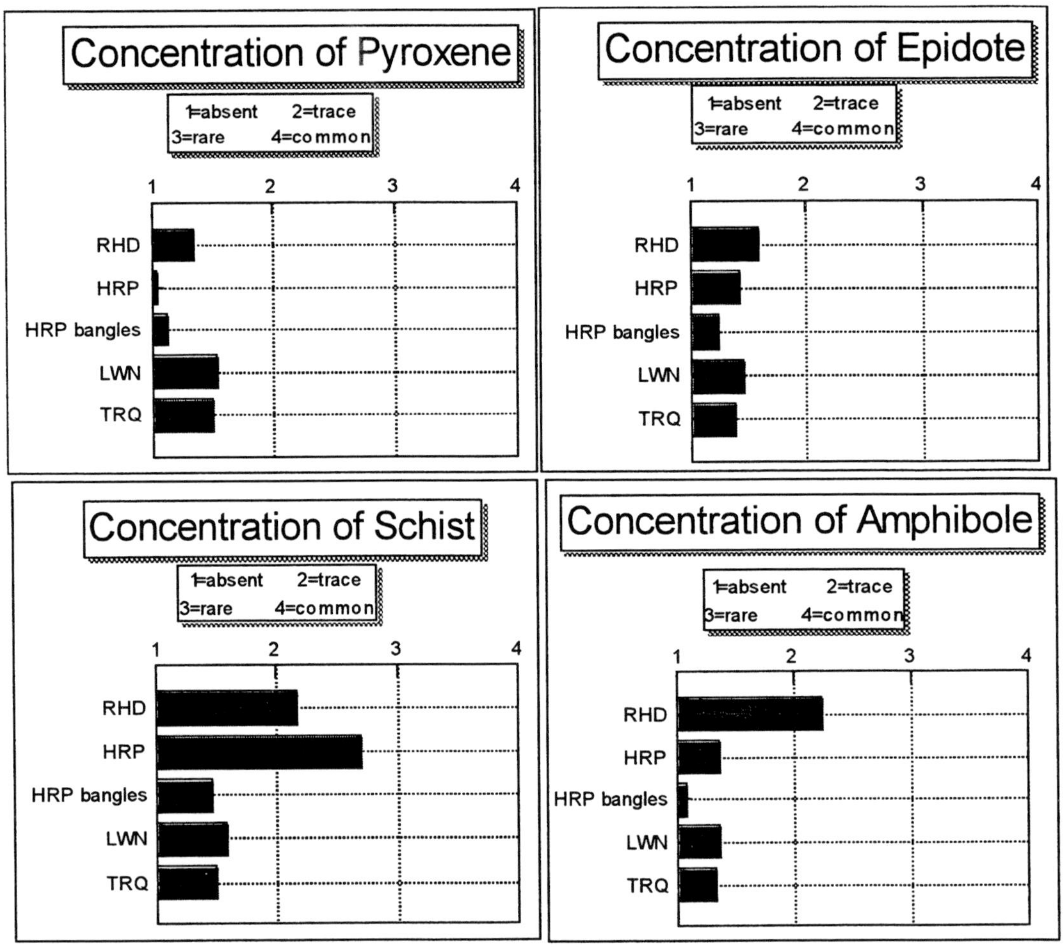

FIGURE 7.31

COMPARATIVE MEAN CONCENTRATIONS OF PYROXENE, EPIDOTE, SCHIST AND AMPHIBOLE FOR FIVE GROUPS

RHD= Rehman Dheri HRP= Harappa
LWN= Lewan TRQ= Tarakai Qila

Examination of these charts shows that certain of the minerals occur in sufficient differences of concentrations so as to demonstrate with reasonable confidence that the groups are from different sources. Others are present in similar concentrations from site to site. In order to examine and compare these in a valid sense, some sort of confidence interval needs to be considered for each concentration level.

A note to be made here first is that the concentrations presented for the Harappan bangle sections cannot, strictly speaking, be compared on the same basis as the pottery thin sections. As discussed in Section 7.1.3 above, most of the thin sections made from the Harappan bangles were significantly smaller in cross-sectional area than the others.

Because of the methodology employed, their mean concentrations, may appear relatively less. The same effect is reduced considerably in the sherd thin sections because in general there is not a consistent difference in size from group to group.

The standard error of the mean defines the confidence interval. Values for the standard error of the mean for the concentrations of each mineral for each of the groups were given in association with the individual site analyses. Table 7-I contains standard errors of the mean for concentrations of the minerals found at Rehman Dheri, 7-III for Lewan and Tarakai Qila (the Bannu Basin sites) and 7-IV for Harappa. For convenience while referring to Figures 7.30 and 7.31, the standard errors of the mean for each concentration at each site are reproduced in Table 7-V. As explained in Chapter 6, the concentration value for each mineral obtained from each thin section was treated as the mean of that concentration for that thin section. Although this is an approximation of the mean, the use of the standard error of the mean is still valid (C. Orton, pers. comm.). The actual mean could be determined by counting all particles in the section, and determining the percentage of the total that each individual mineral represented. Alternatively, one could accurately measure the area of the thin section and determine the concentration of the mineral per unit area. Because these concentrations are so low (less than 1% in many cases), the number of points which would need to be counted would approach 2000 or 3000, as shown in Section 6.4. It can be shown however, that the method used would provide a reasonably good approximation of the actual mean concentrations. Using these approximations of the mean, there is a large enough sample taken from each site to use the standard error of the mean for each. This statistic is essentially the standard deviation of the distribution of the individual means taken from each group. Therefore in using two standard deviations ($\pm 2\sigma$) as the confidence interval, we can be certain that, in a very large experiment, the same concentrations would result in about 95 cases out of 100. This confidence interval has been chosen for the discussions that follow.

TABLE 7-V

STANDARD ERROR OF THE MEAN FOR MINERAL CONCENTRATIONS GIVEN IN FIGURES 7.30 AND 7.31

Mineral	**RHD**	**LWN**	**TRQ**	**HRP**	**HRP (bangles)**
Plagioclase feldspar	0.14	0.07	0.10	0.11	0.12
Microcline feldspar	0.05	0.05	0.04	0.00	0.00
Biotite	0.11	0.14	0.12	0.13	0.13
Muscovite	0.08	0.11	0.06	0.00	0.00
Amphibole	0.13	0.10	0.09	0.10	0.06
Carbonate	0.10	0.18	0.10	0.11	0.10
Pyroxene	0.06	0.10	0.10	0.04	0.08
Epidote	0.06	0.10	0.09	0.11	0.11
Schist	0.11	0.11	0.11	0.12	0.12

If Figures 7.30 and 7.31 are examined while giving consideration to the respective standard errors of the mean, it is seen that some concentrations are consistently different from group to group. (The Harappa bangle group will not be included in the following discussion because of the bias introduced by their smaller thin section size.) The Lewan and Tarakai Qila groups, the Bannu sites, are clearly not distinguishable from each other based on their respective mineral concentrations. In all cases, their concentration values overlap one another at the 2σ level. This shows that in 95% of cases, we could not be confident that the two come from different populations. These similarities were discussed in Section 7.1.2.

The mineral concentrations can then be divided into three source groups (initially for discussion purposes only): Rehman Dheri, Bannu, and Harappa. Significant differences at the 2σ confidence level are seen amongst these three groups in some minerals, while in others, differences are evident between two of the areas but not all three of them. For example, lesser differences are seen between Rehman Dheri and the Bannu sites in the concentrations of carbonates than between either of them and Harappa. Also, some are not significant at the 2σ level, but are at the 1σ level (repeatable about 67 times out of 100, or 67% confidence interval). The 1σ level is too small a confidence interval for the present purposes because of the aforementioned method for estimating the mean. By restricting the discussion to the 2σ confidence interval, the end result of the interpretation is more valid.

Firstly, each of the individual charts in Figures 7.30 and 7.31 are discussed. Concentrations of plagioclase feldspars are seen to distinguish Rehman Dheri material from both the Bannu sites and Harappa at the 2σ level. Harappa material is only distinguishable from the Bannu sites material at the 1σ level.

Concentrations of carbonates offer the most significant discriminator of source between Rheman Dheri and Harappa, or between the Bannu sites and Harappa. Carbonates were often very common in the sherds from sites west of the Indus, and usually only in traces in Harappa sherds. However, there is not a significant difference at even the 1σ level between Rehman Dheri and the Bannu sites with respect to this mineral. The individual site discussions of the Rehman Dheri and the Bannu sherds noted the similarities in limestone concentrations in the geological source formations surrounding both areas.

With respect to the micas, biotite concentrations did not offer any differences at the 2σ level between Rehman Dheri and Harappa, but were significant between the Bannu sites and Harappa. Muscovite was common in most of the sherds from all areas, and could not be relied upon to provide any discrimination amongst groups.

Pyroxenes were mostly rare in the Harappan material, and at the 2σ level could be used to distinguish it from Rehman Dheri or Bannu. However, differences were not significant at this level between Rehman Dheri and the Bannu sites for concentrations of pyroxene. Epidote did not prove to be useful difference indicator of difference amongst any of the sites. Most of the epidotes observed were clinozoisite (especially those from Rehman Dheri and the Bannu sites), which occurs in low-grade regionally metamorphosed rocks forming from micaceous minerals, so similarities could be expected amongst the three groups.

Concentrations of schist showed significant differences amongst all three site groups at the 2σ confidence level. The pottery from Rehman Dheri was different from the Harappan, which in turn was different from the Bannu sites. The Bannu site concentrations were significantly different from the Rehman Dheri concentrations. This, at first consideration, may indicate that it might be useful to differentiate the respective site areas. However, it should be used with some caution. As previously discussed, the category of schist included much of what may be termed as gneiss. Some inclusions which were mostly quartz but contained a small amount of mica or epidote were often termed schist.

Amphibole concentration differences were significant at the 2σ level to differentiate between Rehman Dheri sherds and either Harappa or Bannu. The high concentrations evident in the Rehman Dheri pottery were not seen in the Bannu or Harappa sherds. Differences in amphibole concentrations between Harappa and Bannu material were not significant at the 2σ level.

Finally, one concentration which is not included on the concentration charts discussed above is that of the volcanic rock fragments which were observed in the majority of the Bannu sherds. These were present in 71% of the Lewan sherds and 61% of the Tarakai Qila sherds; and were not seen in any from the other sites. Thus their unique presence can be an important factor in discriminating them from the Rehman Dheri and the Harappa material.

Consideration of the foregoing similarities and dissimilarities shows that the differences in concentrations at the 2σ confidence level can be used to suggest that the groups are from different populations. Apart from the volcanic rock fragments seen in the Bannu sherds there were no minerals or rocks present in any of the groups which would immediately label them as 'foreign' or unique. Therefore, if any one of the sherds in the sample from a given site was actually an intrusive sherd from one of the other sites, it would not be detected as such. The concentration values of its minerals would be lost in the calculation of the mean for that site. Differences can only be told on the basis of mean concentration levels of each mineral. Therefore, in order to determine whether or not a particular set of sherds is intrusive, it is required that a statistically valid sample of them be thin sectioned and analysed.

7.2.2. Technological aspects and innovations

Examination of similarities and dissimilarities seen in aspects of technology amongst the sites is now discussed. As with the mineral concentrations data, there were many of each. The technological discussion is divided into sections in the same order as the ceramic processing: clay preparation, vessel forming, and firing.

Data from all of the sites are consistent with clay levigation. It appeared to have occurred with a high degree of skill at Rehman Dheri. Sherd data from the Bannu sites were less suggestive of either levigation or the absence of it. Those clays may or may not have been levigated, as they recorded a higher level of fine (30-50 micron) sand remaining in the mixture. If they were levigated, it was done to a coarser level. At Harappa, size distribution data in the clays were also consistent with their being levigated to higher levels of remaining fine quartz sand inclusions than Rehman Dheri. At Rehman Dheri, data suggest the use of river sands primarily as temper, with occasional crushed quartz and grog. The Bannu sites appeared also to have employed river sand tempers, but showed a very low incidence of crushed quartz or grog. The Harappa data suggested that potters there were also using sand tempers primarily; with rare crushed quartz and about the same incidence of grog usage as that seen at Rehman Dheri.

Six different fabric types were observed at Rehman Dheri; based on suggested levigation levels and the nature and concentrations of quartz sand tempers seen in the fabrics. At Tarakai Qila and Lewan, all of the fabrics except two fitted into three of these classifications. The Harappa fabrics formed an entirely different set of groups from those of the Rehman Dheri or the Bannu sherds. None of the sites showed a correlation between fabric groups and morphological vessel type; nor between the occurrence of crushed quartz temper/grog with vessel type. There was also no correlation observed between fabric type and vessel wall thickness.

With respect to forming techniques, all of the sherds from all of the sites showed similar evidence of forming methods. All generally exhibited the same characteristic alignment of inclusions indicative of drawing the fabric in a direction vertical to the vessel's base, or wheel-thrown. Occasional evidence of burnishing was seen in sherds from all of the sites.

Although not conclusive, indications of firing temperature ranges varied from site to site. Evidence from the Rehman Dheri sherds was suggestive of firing temperatures consistently very near or above 850 degrees C. Although firing temperature data from the Bannu sites were less consistent, the sherds from Tarakai Qila appeared to achieve this figure in about half of the samples. Firing temperatures in the Lewan sherds appeared less consistent, but generally less than 850 degrees. At Harappa, the individual samples were smaller, but based on birefringence as an indicator of firing temperatures (few carbonates were present), the data suggest that most of the bangles and sherds sampled may have been fired to less than 850 degrees C. As mentioned above, however, this is not considered to be a reliable indication on its own.

Overall, the foregoing intersite comparison of similarities and dissimilarities suggests that, although there was no evidence to imply intersite transportation of ceramics, there was some communication with respect to ceramic technologies. Dealing firstly with the interactions between Rehman Dheri and the Bannu sites that the data suggest, it can be inferred that the ceramic industry at Rehman Dheri was more advanced and sophisticated than at the latter two sites. A possibly finer degree of, and standardisation in, clay levigation; a broader range of separate fabric types; wider use of different temper types; and more consistent and higher firing temperatures would support this. The same processes were likely being used at Lewan and Tarakai Qila but to a less sophisticated degree. It is suggested that at these Bannu sites the clays were also being levigated, but to a less meticulous level; similar fabric types were being produced, but a smaller range of them; similar tempers were being added, but crushed quartz and grog were used only rarely; and firing temperatures generally approached 850 degrees, but less consistently achieved it. The lower consistency in firing temperatures inferred from the data from the

Bannu sites may suggest that the potters at those sites were using open bonfire techniques and not kilns. No evidence of kilns was recorded by the researchers there.

If we take into account the greater size of the site of Rehman Dheri, the extensive evidence of early urbanisation and city features and other evidence of advanced systems as compared to the Bannu sites (see Section 5.3), the above can fit into some of the information exchange models discussed in Section 4.2. In particular, it will be recalled that Hudson (1969) provided a model which states that goods and ideas move downward through a central-place system with a rank-size hierarchical structure. Hantman and Plog (1982: 241) state that in innovation diffusion of this type, the innovation is likely to occur first in the settlement that has the greatest amount of intraregional and interregional interaction, usually the largest settlement or political structure. In this model, the innovations would have occurred at Rehman Dheri (the largest settlement or political structure) and diffused to the sites of Lewan and Tarakai Qila. The direction of innovation flow fits with the differences in levels of sophistication of the respective ceramic industries seen between the regions.

Comparing possible exchange of technological information and innovations between Rehman Dheri and Harappa, the same relationship is not suggested by the data. Their respective site types and natures were much more similar in terms of evidence of early urbanisation and city-like features; and sophistication seen in their respective ceramic industries. They were both large settlements and political structures. Harappa also appeared to be using levigation as a clay preparation technique (as well as removing carbonate inclusions); their potters were regularly using a broader range of temper types such as grog; and their fabrics formed standardised groups but in a different manner to those from Rehman Dheri (they had their own fabric groups), suggesting parallel development of the respective ceramic industries. With respect to firing temperatures, the data are less conclusive but they may not have achieved the same levels of sophistication as either Rehman Dheri or the Bannu sites, at least in the Early Harappan period sampled. Thus, it would seem that there was much less technological information and innovations being diffused between Rehman Dheri and Harappa at the time. According to the models just discussed, as much information diffusion would not be expected between these two sites, at least based on their similarity in size of political centre. Moreover, although the models did not consider geographical distances separating the centres, Harappa is more than 240 kilometres in a straight line away from Rehman Dheri, while the Bannu sites are about 110 kilometres from Rehman Dheri, through a natural pass (the Pezu Gap). Thus more intense information exhange and diffusion of innovations would be expected between Rehman Dheri and Bannu than between Rehman Dheri and Harappa. Equivalent levels of complexity and sophistication in the ceramic industries may well exist in each of these two larger centres, but would not necessarily be expected to be diffused between them. On the other hand, Lewan and Tarakai Qila, being smaller and less urbanised sites, would be expected to emulate the industry of the larger centre nearest to it.

Innovations which may have diffused from Rehman Dheri to the Bannu sites, if they did take place, have been altered in some way. That is, the innovations represent some of the similarities observed between the regions, but they are not identical between regions. It is difficult to make a judgment on just what types of information were being exchanged from site to site. Some technical knowledge such as methodology would be the most likely information exchanged. The receiver site of the innovation either practised the new technology but didn't quite achieve the same result as the site initiating the innovation, or else modified it to suit their particular (undefined) requirements. Either way, techniques could be expected to differ from site to site even when innovations have diffused. For instance, in the case of clay levigation, it was inferred that the Rehman Dheri potters accomplished a high degree of sophistication in the technique. The data from the Bannu sites, on the other hand, suggested there may have been less attention paid to achieving the same concentrations of fine quartz inclusions as Rehman Dheri. This may have been true, but it also may have been the case that the Bannu potters adapted the levigation technique to suit their own particular requirements and didn't wish to achieve exactly the same concentration of fine quartz inclusions as their counterparts at Rehman Dheri. The Bannu potters also produced the same range of morphological types, but used a smaller number of different fabric types. None of the coarser-textured fabrics were observed in the Bannu samples. If innovations were diffusing unchanged from Rehman Dheri to Bannu as the model described in Section 2.2 suggested (which is highly unlikely), it would be expected that the same range of fabrics would have been encountered at the latter sites.

A further explanation for standardisation in some areas and the lack of it in others may li in political control. Although his theorising is based upon observations made from th Mature Harappan period only, Kenoyer (1992) divided Harappan crafts into 4 categories:

1. Local raw material and simple technologies
2. Non-local raw material and simple technologies
3. Local raw material and complex technologies (includes 'elaborated' ceramics)
4. Non-local raw material and complex technologies (beads, seals, metallurgy, etc.)

Categories 3 and 4 would involve standardisation, while the other categories would shov higher degrees of regional variation. Regional variation in categories 3 and 4, however could be caused by political control of craft production assuming the form of control o technological know-how, according to patterns to be contextually investigated in th archaeological record (Bhan et al. 1994: 143). More highly specialised crafts in cities an towns reflect centralised political control of production and distribution (*ibid.:* 142) Adapting this argument to the Early Harappan investigation just completed, one coul make a point that the sites of Rehman Dheri and Harappa were both large centres wit independent political control and therefore influenced each other's ceramic production an technology to only a small degree. On the other hand, the two Bannu sites investigate were smaller centres which may have been subjected to some political control of craf production from their nearest large centre, which was Rehman Dheri. This politica control would have been in the form of craft production, and probably not of technology because in an economic sense control of production is generally more effective tha control of technological information.

Some of the observations made in the present research provide supporting evidence of th discontinuity in craft specialisation between the Early Harappan and the Mature Harappa periods identified by Possehl (1990: 269-70) and discussed in Section 5.2. Possehl's articl did not elaborate on which particular aspects of craft specialisation were seen to b discontinuous. The present observations from Rehman Dheri with respect to the Matur Harappan are limited, but it was seen that the number of different fabric types observe decreased from period to period for the data studied. Although this may be attributable t the fewer number of samples selected from the Mature Harappan, a reduction in sophistication has been observed in Harappa ceramic technology from the Early to the Mature Harappan as well (see discussion Section 7.1.1).

Thus, in terms of the objective of the thesis, it is seen that socioeconomic interaction amongst the Early Harappan sites examined may have been restricted to the exchange of technical information and innovations, at least to the extent of what can be told from petrographic and textural analysis of ceramic sherds. Smaller, lighter craft items may have been traded, however. Bangles were shown to have been traded in the Mature Harappan period between Harappa and Mohenjo-Daro (Blackman and Vidale 1990). Other smaller artefacts such as shell, beads, or figurines may also have been exchanged, but there have been no provenancing studies yet published on these items for the Early Harappan Period.

CHAPTER EIGHT
SUMMARY OF PART TWO, AND DIRECTIONS FOR FURTHER RESEARCH

8.1 SUMMARY

At the outset, one objective of Part Two was to demonstrate that data obtained from the field portable petrographic analysis system developed in Part One could be used effectively to examine an archaeological problem. Another was to examine the nature and extent of intersite social and economic interactions of the Early Harappan Period in Northwestern Pakistan using the data. Both objectives were accomplished to a large degree. The first was achieved by using the data to achieve the second. The results showed no evidence that pottery was being traded amongst Rehman Dheri, Harappa, Tarakai Qila and Lewan during the Early Harappan period. They suggested that Rehman Dheri and Harappa independently developed fairly sophisticated ceramic industries during the period; and that Tarakai Qila and Lewan had less developed ceramic industries which were probably influenced by innovations from Rehman Dheri.

The Early Harappan Period was chosen as the application of the field analysis system because extensive stylistic and morphological similarities seen in Early Harappan pottery over a broad geographical area had prompted researchers to suggest that it was being traded or exchanged throughout the region. Large craft production centres such as Rehman Dheri have been suggested as the sources of the pottery. However, there had been no provenance studies conducted to examine sources of the stylistically similar pottery to corroborate the hypothesis. This thesis therefore set out to conduct a petrographic field study of ceramic sherds from four sites where excavated Early Harappan pottery had demonstrated stylistic and morphological similarities.

Chapter Four first reviewed theory and models of socioeconomic interactions in prehistory, to gain an appreciation and understanding of the forces which shaped intersite and inter-regional behaviour. Factors influencing the distribution of styles, and information and innovation diffusion, were examined in the context of the models proposed by prominent theorists. It was seen that the existence of a common style over a broad geographical extent can be caused and influenced by a variety of factors. To recapitulate, Hantman and Plog (1982) theorise that a specific style can exist simply as an indicator of socioeconomic status or of belonging to a particular group. Innovations, stylistic or technological, occurring in socially and spatially differentiated societies are likely to occur first in the settlement that has the greatest amount of intraregional and interregional interaction, usually the largest settlement or political centre. They then move downward through the central hierarchy, followed by centres next in rank, and finally diffuse among households or individuals.

The use of chemical and petrographic data to study patterns of stylistic distribution was reviewed. These were seen to be more accurate than the traditional 'criterion of abundance' concept for determining provenance of excavated pottery. Respective advantages and disadvantages of chemical and petrographic methods were discussed. Neither is inherently superior; they are complementary methods. The discussion then concentrated on describing the use of petrographic analysis. It was seen how mineral inclusions identified in archaeological ceramic thin sections can often be matched to source lithologies to determine provenance, provided sufficient variations exist in the geology of the different regions being compared. Where these differences are not adequate to differentiate sources, other analyses may be used to complement the mineralogical data. These include textural analysis, study of relative mineral concentrations, observation of characteristics such as morphology or weathering of the mineral inclusions, and examination of technological characteristics. In textural analysis, data can sometimes be used to match tempers (especially quartz sands) to local sources. A shortcoming of reliance on textural data to source ceramics was seen to result from the necessary assumptions as to the potters' treatments of the natural clays before forming and firing the vessels. For example, most researchers using this method have tended to assume that coarser, more angular inclusions always represented added temper.

As well as using petrographic data for provenancing, technological attributes of pottery can be examined via thin sectioning. This is an important part of examining intra- and inter-site interactions. If petrography can show that stylistically and morphologically similar pottery was not being traded or exchanged, then it is important to determine the nature and extent of communication which existed with respect to innovations and aspects of a technological nature. Parallels in the manufacturing process from site to site, for example, might suggest that technological information was being exchanged amongst the potters; and that the common style signalled more than group identity. Technological attributes which can be examined via thin sectioning include amounts and types of temper used, including preferred quartz sand temper sizes, observation of slips and surface finishes, inference of firing temperature and forming techniques.

Chapter Five presented the background to the problem being studied. The geography, climate and geology of the Indus Valley and its adjacent areas were described. In addition to describing the present climate of the region, research into the nature of the palaeoclimate was reviewed. Those studies suggested that approximately 10,000 years ago, the climate of the area changed from arid to semi-arid, and since that time has remained much the same as that prevailing today.

The geology of the Indian subcontinent falls into two main parts: Peninsular India, with rocks mainly of Precambrian to Cretaceous age, and some Cenozoic strata; and the extra-peninsular region, which consists of a continuous series of highly fossiliferous marine sedimentary rocks ranging in age from early Palaeozoic to Eocene, and a zone composed almost entirely of Tertiary sedimentary river deposits. The depositional histories of the site areas were reviewed and it was seen that the Gomal Plain (where Rehman Dheri is located) and the Bannu Basin (where Tarakai Qila and Lewan are located) derive an influx of detrital material and sediments from largely similar adjacent geological formations to the west and northwest of the two areas. Detrital and sedimentary material to be found in Harappan clays, on the other hand, are thought to be more recent material brought by the Ravi River from mountains 200 km to the northeast.

The Harappan Civilisation and its antecedents were then described, beginning with the Early Food-Producing Era as defined by the aceramic Mehrgarh Phase at the beginning of the seventh millennium. Research has suggested that settlements of this phase were the earliest sedentary settlements of the Indus Valley Tradition. Domesticated animals and crops gradually replaced the importance of hunting and gathering on local economies during this period, and these factors offer a strong case for an indigenous process of domestication within the Indian subcontinent. The earliest pottery was recorded from around 5500 BC. Early interaction with other social groups was evidenced by the limited use of imported craft materials.

Continuities and discontinuities in developments leading up to the Mature Harappan Period were examined from the viewpoints of two of the more prominent researchers in Harappan archaeology, Mughal and Possehl. A discussion of both was central to placing the present research into its perspective. Mughal,. who introduced the term Early Harappan into the literature, saw in the evidence a smooth transition from sedentary village farming life to a complex urban civilisation. Possehl's interpretation is also reflected in his terminology; he uses the term pre-Urban to describe the period leading up to the transition to the Urban Phase. Possehl interprets the data to suggest a much more abrupt change with some continuities and some discontinuities. Continuities seen by both include items related to subsistence and expressions having to do with style. Discontinuities seen by Possehl include those features that make it distinctively urban in character such as social stratification and craft specialisation.

The sites from which samples were selected were described, with a concentration on Rehman Dheri. Rehman Dheri has provided evidence suggesting some of the earliest developments towards urbanism as seen at Mohenjo-Daro and Harappa, and appeared to have been a major ceramic and craft production centre during the Early Harappan. This evidence was reviewed, and characteristics of the three major periods at the site described. The early levels at Harappa have also shown evidence of early urbanisation, and the site appeared to have been a major ceramic and craft production centre, on a scale similar to that of Rehman Dheri. The two sites situated in the Bannu Basin, Lewan and Tarakai Qila, appeared to have had a less sophisticated craft and ceramic industry than the other two sites. Ceramics from all four sites showed remarkable standardisation amongst them. Morphological types could be placed into the same classifications and typologies.

Following a description of the typologies first defined by the original excavator at Rehman Dheri, ceramic morphologies were grouped from all four sites. The seven most common types were then chosen for sampling from each site. This included two types of jars: AI and AII; four types of bowls: BI, BII, BIII and BIV; and the BV dish-on-stand. Additionally, CII type (polychromes, of no particular morphology) were sampled from Rehman Dheri.

In Chapter Six, the methodology was described and discussed. The largest sample was taken from Rehman Dheri because of its status as a larger centre and evidence for an extensive ceramic and craft industry. It was well-excavated and provided a complete sequence from the earliest part of the Early Harappan Period to levels contemporary with the Mature Harappan Period (Period III). Sampling was designed to provide a complete representation of each of the seven morphological types from each of the three chronological periods at the site; with a concentration on the Early Harappan levels. All were randomly selected from provenanced material. Smaller samples representing the same types were chosen from the remaining three sites. From Harappa, a selection of bangles was also sampled for thin-sectioning.

One hundred and sixty-five thin sections were prepared from the samples at Peshawar, using the field kit developed in Part One. They were examined and analysed using the field microscope developed as part of the kit. Mineral inclusions in the sections were identified and recorded using standard petrographic techniques. In addition, other characteristics of the minerals such as concentration, morphological data, and weathering condition were recorded. Concentration data relevant to all of the minerals except quartz were recorded on a scale of 1 to 4, representing 'absent' to 'common' respectively. An area count to define the size distribution of the quartz inclusions and a point count to determine the characteristics of the matrix with respect to all inclusions (classed as sand, silt, or matrix) were conducted on each of the sections. This entailed counting 250 to 550 points for each sherd in the area count, and, in addition, approximately 400 per slide for the point count. These counting levels were designed to provide a 95% confidence interval which was acceptable at the 2σ level. All data were then entered into a Lotus 123R5 database using a Zenith Z-Star 433VL personal computer.

The database provided automation, manipulation and statistical analysis of the recorded data. It enabled charting of the frequency distribution of the size ranges of the quartz inclusions with a confidence interval, concentration levels of each size distribution, and percentage composition charts of sand, silt and matrix in each ceramic sample. These charts proved to be powerful tools for grouping and classifying fabrics based on concentration characteristics of quartz sand inclusions. In addition, mean concentrations of each of the minerals recorded were plotted for site-to-site comparisons. Standard errors of the mean were calculated for these concentration charts to provide a measure of confidence in differentiating sources based on the data.

Chapter Seven presented the analysis and discussion of the data. Firstly, data were analysed individually by site to interpret them in light of individual site ceramic industries. Criteria included relative concentration of minerals, characteristics of the matrix and quartz sand size distributions, and observation of technological characteristics. These individual analyses were then compared site-to-site to examine the possibility of movement of ceramic products from site to site, and to assess the degree of compositional and technological similarities and dissimilarities which might reveal the nature of intersite interaction amongst the four sites.

Commencing with Rehman Dheri, characteristics of the mineral inclusions were described, including those observed in the waster sherds from the site. It was observed that the mean concentrations of minerals found in the excavated sherds closely matched those of the waster collection sampled, suggesting that the excavated sample was of local origin. Moreover, the mineral concentrations were consistent with a clay source containing inclusions derived from adjacent geological formations. Conditions of the minerals observed suggested firing temperatures consistently in the region of 850 degrees C. No significant differences in mineral concentrations were noted between the earliest and the latest levels. Quartz size concentration data were divided into two ranges, 30-50 microns and >50 microns. The number of occurrences falling into each range showed a high frequency of sherds with very low concentrations of the smaller quartz inclusions, suggesting that the clays may have been meticulously levigated. Characteristics of the frequency of the larger size range were consistent with use of river sands and occasionally crushed quartz or quartz from a primary source as temper. On the basis of these size distribution characteristics, the Rehman Dheri fabrics fell into six separate fabric types, according to the concentration of inclusions 30 - 50 microns in size, and the nature of the distribution of those quartz inclusions >50 microns in size. The resulting fabric groupings showed no correlation with morphological vessel types, or with sherd thickness. Occurrences of fabric types appeared to change between Periods I and III, lessening in number present and tending towards more of the coarser fabrics. Finally, some indications of forming techniques and finishes employed at Rehman Dheri were provided by the thin section examinations.

Minerals observed in the thin sections from Lewan and Tarakai Qila were the same as those seen in the Rehman Dheri samples. Individual minerals varied in concentration but exhibited generally the same characteristics. The presence of volcanic rock fragments in the majority of the samples from the Bannu Basin, which were not present in any of the samples from the other two sites, provided a distinguishing aspect of the pottery from that area. There was a close correlation observed in the mineral concentrations between the two sites of the basin, and those seen in the wasters collected from the surface at Lewan. Based on characteristics of the carbonate inclusions, firing temperatures were estimated to be around 850 degrees C, as in the Rehman Dheri sherds, but appeared to be less consistent than those from the latter site. Birefringence of the matrices, however, did not corroborate this; this indicator suggested that firing temperatures were generally lower than 850 degrees.

Analysis of the size distribution characteristics of the quartz inclusions in the samples from the Bannu Basin revealed all but two fit into the fabric groups defined on the basis of the Rehman Dheri sherds. The large majority from both the Bannu sites matched a single Rehman Dheri fabric type. The remainder, apart from the two mentioned above, fell into two other Rehman Dheri fabric types.

Mineral suites observed in the Harappa thin sections were again generally the same as those observed in the sites west of the Indus River. The most significant difference seen was the much lower concentrations of carbonates, which may be attributable to removal of these minerals in clay preparation. Mineral concentrations in the bangles were generally smaller than those seen in the sherd samples. This was attributed partly to the choice of methodology in recording the mineral concentrations, but the use of different clay sources for the bangles could not be ruled out. Study of the fabrics through quartz size concentration data suggested that there exist two basic bangle fabric groups and four identifiable pottery fabric groups. The ceramic fabric groups did not match those defined for the Rehman Dheri pottery samples.

The very low incidence of carbonate inclusions noted in the Harappa samples precluded use of their characteristics as an indicator of firing temperature ranges. Birefringence observed in the fabrics suggested that most were fired to less than 850 degrees C, but this was not seen as a particularly reliable indicator. Examination of the inclusion alignment pattern in the bangle samples suggested two different forming techniques may have been used in manufacturing the bangles.

Mineral concentrations from each of the four sites were examined and compared. It was seen that differences observed in certain of the mineral concentrations were significant at the 95% confidence level such that they could be reliably used to differentiate sources of material. The exception was in differences between concentrations seen within the two Bannu Basin sites, which were not distinguishable on this basis. Using the standard error of the mean as the confidence interval, significant differences in some of the mineral concentrations existed to form three source groups: Rehman Dheri, Bannu and Harappa. This result suggested that material from each of the areas was sourced from within that area. Apart from the unique occurrence of volcanic rock fragments in the Bannu Basin sherds, there were no minerals or rocks observed that could be used to identify an intrusive sherd in any of the groups. A larger intrusive group may be able to be identified, however, on the basis of relative concentrations of constituent minerals.

With respect to technological aspects, data were consistent with clay levigation at all of the sites. At Rehman Dheri, it may have occurred with a higher degree of skill and standardisation than that inferred from data from the Bannu sites. Harappa appeared to achieve a different level of levigation, but with consistency comparable with that seen at Rehman Dheri. River sands were suggested to be the primary temper employed, with additional use of crushed quartz or grog more common at Rehman Dheri and Harappa. Six different fabric types were observed at Rehman Dheri with the Bannu data fitting primarily into three of these. The Harappa data suggested four fabric types; at a similar level of sophistication but different from the Rehman Dheri groups.

Overall, intersite comparison of similarities and dissimilarities suggested that, although no evidence of intersite ceramics transport was seen, there was an exchange of technology and innovations. It is suggested that the ceramic industry at Rehman Dheri was more advanced and sophisticated than at the Bannu sites. This was shown by a possibly finer degree of, and standardisation in, clay levigation; a broader range of fabric types; wider use of different tempers; and more consistent and higher firing temperatures. Based upon the relative sizes of the respective sites, this interpretation is consistent with information exchange theories earlier reviewed.

Between Rehman Dheri and Harappa, the same relationship was not suggested by the data. Both were large, structured urban settlements. A similar degree of sophistication in ceramic technology was found at both sites, but because they differed in absolute form, parallel development of the respective technologies is inferred, rather than extensive exchange of information and innovations. The low incidence of carbonate inclusions in the Harappa material may suggest a more advanced technology; in that these inclusions were removed during clay preparation.

8.2. DIRECTIONS FOR FURTHER RESEARCH

The foregoing programme of research has accomplished the objectives set out initially. It has provided some basic understanding of the nature of intersite socioeconomic communication in one area of the Indus Valley during the Early Harappan Period. In doing so, further questions have arisen.

Firstly, as the research has revealed no evidence to suggest that any exchange of ceramic products was taking place amongst the three regions examined, some corroboration of this is needed from larger samples and with other craft products. The inference was made based upon mean concentrations of minerals observed in thin sections of pottery from each site. Although the original objective was to utilise only data obtained by way of the portable petrographic analysis kit, further delineation could be derived from chemical analysis of a similar sample set. Identical mineral suites occurring at each of the regions, especially between Rehman Dheri and the Bannu Basin sites, did not allow differentiation to be based upon mineralogical differences; only on differences in concentrations of certain minerals. Chemical characterisation of the clays or certain of the minerals may reveal more distinctive differences. Konta (1985, cited in Chamley 1989: 65) observed that Indus River silts contain concentrations of both kaolinite and mica-illite clays. Chemical characterisation may therefore be able to reveal differences in their respective sources. Furthermore, chemical characterisation of the plagioclase feldspar inclusions observed in the thin sections via Scanning Electron Microscope or Electron Microprobe may reveal compositional data that might assist in their sourcing.

Chemical characterisation may also assist in more accurately defining the amount of calcium carbonate removed by the Early Harappan potters at Harappa. The present research revealed very low levels of this mineral in the Harappan ceramics, despite the area soils being rich in it. The lack of carbonates which can be identified by petrographic analysis in the material could be corroborated via a chemical analysis, as the inclusions may have been too small to detect easily via petrography. Removal of significant amounts of carbonates could signify a more complex ceramic industry at the site.

As the sampling from the production site of Rehman Dheri was restricted to one excavation trench in order to keep the sample size at a workable level, it cannot be assumed that this trench is representative of the site as a whole. It is suggested that a programme of sampling that encompasses at least three of the excavated areas at the site be implemented. The same methodology should be used to examine these and compare them with the present results.

A programme of research, either petrographical or chemical, but preferably both, conducted on bangles excavated from the same four sites may reveal much more extensive information with respect to intersite product exchange. It has been shown that bangles were traded between Mohenjo-Daro and Harappa during the Mature Harappan (Blackman and Vidale 1992). Bangles are found in large quantities at the early and transitional levels at all four of the sites examined. Bangles as personal adornments may be more significant indicators of group affiliation than pottery. Although the present study is inconclusive in this respect, differences were suspected in mineral concentrations between pottery sherds and bangle fragments from Early Harappan contexts at Harappa. Similar studies could be conducted on ceramic figurines which were common in form and style amongst the same sites. Figurines may have functioned as icons in cults widespread in the Harappan system and thus may have moved from site to site. Demonstration of the exchange of figurines however, would likely have different implications with respect to social behaviour than pottery.

A wider sampling of similar pottery would assist in defining the extent of the interactions that have been suggested by the present research. In particular, pottery excavated from Sarai Khola, near Taxila, is stylistically, morphologically and chronologically similar to that examined here. It had been part of the original intent to include samples of this pottery in the present research, but official permission was not obtained in time. A similar examination of a sample of this pottery may reveal whether or not Sarai Khola was a 'satellite' region with respect to technological and innovation diffusion from a larger site, or if it appeared to develop in a parallel sense as has been suggested at Harappa and Rehman Dheri.

Finally, some further studies suggested by the present results could assess the important differences between the two schools of thought on the continuity/discontinuity argument for the transition from the Early to the Mature Harappan. As discussed in Section 5.2, the discontinuities relate to social stratification and craft specialisation. A larger sampling from the Mature Harappan levels at Rehman Dheri would better define the changes observed in the number of fabric types from the early to the later levels. A similar programme is under way at Harappa. Since Harappa is also a major ceramic production centre, a programme undertaken to assess technological changes from the earliest to the latest levels there would also be most informative when compared with Rehman Dheri technological changes throughout the same period. Additionally, the Harappan Civilisation is frequently interpreted as having been a major trading civilisation during the Mature Period and pottery, claimed to be Harappan, has been recognised as far afield as Oman and Shortugai on the edge of Central Asia. These arguments have been typically based upon similarities in style and morphology. With respect to pottery, there have been no provenance studies published showing the Mature Harappan Period to have been characterised by extensive trade in pottery. Since the present study has revealed no evidence of ceramic exchange during the Early Harappan, it would be valuable to examine whether or not long distance trade or exchange in ceramic products was taking place during the Mature Harappan Period.

REFERENCES CITED

Adams, A.E., W.S. Mackenzie, and C. Guildford 1984. *Atlas of Sedimentary Rocks under the Microscope.* New York: John Wiley & Sons.

Adams, W.A. and E.A. Adams 1991. *Archaeological Typology and Practical Reality.* Cambridge.

Adan-Bayewitz, D., and I. Perlman 1985. Local pottery provenience studies: a role for clay analysis. *Archaeometry* 28(2): 203-17.

Ahmad, K.S. 1969. *A Geography of Pakistan.* Karachi: Oxford University Press.

Alauddin, M., H.U. Khan, S. Nasreen, H. Masood, N. Jan, and M.A. Khattak 1993. Characteristics of glass making sands of Khisore-Marwat ranges of D.I. Khan Division, NWFP, Pakistan. *Pakistan Journal of Scientific and Industrial Research* 36 (5): 187-92.

Allchin, B. 1981. Stone industries of Lewan, Tarakai Qila and Lak Largai in the Bannu Basin. In Härtel, H. (ed.), *South Asian Archaeology 1979* pp 233-9. Berlin: Dietrich Reimer Verlag.

Allchin, B. and R. Allchin 1982. *The Rise of Civilisation in India and Pakistan.* Cambridge World Archaeology.

Allchin, B. and R. Allchin 1992. Lewan - A stone tool factory of the fourth to third millennium B.C. In G.L. Possehl (ed.), *Harappan Civilization: A Recent Perspective,* 2nd Revised Edition, pp 521-53. New Delhi: Oxford and IBH Publishing.

Allchin, B. and J.R. Knox 1981. Preliminary report on the excavation at Tarakai Qila (1978-79). In Härtel, H. (ed.), *South Asian Archaeology 1979* pp 245-50. Berlin: Dietrich Reimer Verlag.

Allchin, B. and H. Rendell 1986. The site of Lewan. In Allchin, F.R,. B. Allchin, F.A. Durrani, and M.F. Khan (eds.), *Lewan and the Bannu Basin* pp 35-87. Cambridge: BAR International Series 310.

Allchin, F.R. and R. Knox 1986. The Lewan pottery. In Allchin, F.R., B. Allchin, F.A. Durrani, and M.F. Khan (eds.), *Lewan and the Bannu Basin* pp 103-12. Cambridge: BAR International Series 310.

Amundson, R. and E. Pendall 1991. Pedology and late Quaternary environments surrounding Harappa: a review and synthesis. In Meadow, R.H. (ed.), *Harappa Excavations 1986-1990,* Monographs in World Archaeology No. 3, pp 13-27. Madison: Prehistory Press.

Arnold, D.E. 1980. Localized exchange; an ethnoarchaeological perspective. In Fry, R.E. (ed.), *Models and Methods in Regional Exchange* pp 147-50. SAA Papers 1. Washington DC: Society for American Archaeology.

Beg, A.R. 1975. *Wildlife Habitats of Pakistan.* Bulletin No. 5, Sciences Research Division, Pakistan Forest Institute, Peshawar.

Betts, I.M. 1982. Roman brick and tile: A study in fabric variability. In Freestone, I., C. Johns and T. Potter (eds.) *Current Research in Ceramics: Thin-section Studies.* British Museum Occasional Paper No. 32: 123-34.

Bhan, K.K., M. Vidale and J.M. Kenoyer 1994. Harappan technology: theoretical and methodological issues. *Man and Environment* 19(1-2): 141-57.

Bignot, G. 1985. *Elements of Micropaleontology: Microfossils--Their Geological and Palaeobiological Applications.* London: Graham & Trotman.

Binford, L.R. 1965. Archaeological systematics and the study of cultural process. *American Antiquity* 31: 203-10.

Bishop, R.L. 1980. Aspects of ceramic compositional modelling. In Fry, R.E. (ed.), *Models and Methods in Regional Exchange.* Society for American Archaeology Papers No. 1 pp 47-66.

Bishop, R.L., G. Harbottle and E.V. Sayre 1982. Chemical and mathematical procedures employed in the Maya fine paste ceramics project. In Sabloff, J.A. (ed.) *Analysis of Fine Paste Ceramics.* Cambridge, Mass.: Memoirs of the Peabody Museum of Archaeology and Ethnology 15(2): 272-82.

Bishop, R.L., R.L. Rands and G.R. Holley 1982. Ceramic compositional analysis in archaeological perspective. In Schiffer, M.B. (ed.), *Advances in Archaeological Method and Theory* Vol 5: 275-330. New York: Academic Press.

Blackman, M.J., G.J. Stein, and P.B. Vandiver 1993. The standardisation hypothesis and ceramic mass production: technological, compositional and metric indexes of craft specialization at Tell Leilan, Syria. *American Antiquity* 58 (1): 60-80.

Blackman, M.J., and M. Vidale 1992. The production and distribution of stoneware bangles at Mohenjo-Daro and Harappa as monitored by chemical characterization studies. In Jarrige, C. (ed.), *South Asian Archaeology 1989* pp 37-44. Monographs in World Archaeology, Madison: Prehistory Press.

Blatt, H., G Middleton and R. Murray 1980. *Origin of Sedimentary Rocks.* New Jersey: Prentice-Hall.

Bousfield, B. 1992. *Surface Preparation and Microscopy of Materials.* Chichester: Wiley.

Braun D.P. 1982. Radiographic analysis of temper in ceramic vessels: goals and initial methods. *Journal of Field Archaeology* 9: 183-92.

Brisbane, M.A. 1981. Incipient markets for early Anglo-Saxon ceramics: variations in levels and modes of production. In Howard, H. and E. Morris (eds.), *Production and*

Distribution: A Ceramic Viewpoint pp 229-42. BAR International Series 120. Oxford: British Archaeological Reports.

Buko, A. 1984. Problems and research prospects in the determination of the provenance of pottery. *World Archaeology* 15 (3): 348-65.

Buller, A.J. and J. McManus 1972. Simple metric sedimentary statistics used to recognize different environments. *Sedimentology* 18: 1-21.

Caldwell, J.R. 1965. Interaction spheres in prehistory. In Caldwell, J.R. and R.L. Hall (eds.), *Hopewellian Studies*. Springfield: Illinois State Museum Scientific Papers Vol 12, pp133-43.

Carr, C. 1990. Advances in ceramic radiography and analysis: applications and potentials. *Journal of Archaeological Science* 17: 13-4.

Chamley, H. 1989. *Clay Sedimentology*. Berlin: Springer-Verlag.

Chandler, G.M. 1994. New applications of archaeological microscopy in the field: ceramic petrography and microwear analysis. *Papers from the Institute of Archaeology* 5: 39-48. Institute of Archaeology, University College London.

Chaolong, X. 1990. The Kot Dijians and the Harappans: their similarity: another possible interpretation. In Taddei, M. (ed.), *South Asian Archaeology 1987*. Rome.

Chayes, F. 1956. *Petrographic Modal Analysis*. New York: John Wiley.

Chayes, F. 1965. Reliability of point counting results. *American Journal of Science* 263: 719-21.

Childe, V.G. 1950. The urban revolution. *The Town Planning Review* 21 (4): 3-17.

Clark, J. and W. Parry 1990. Craft specialization and cultural complexity. *Research in Economic Anthropology* 12: 289-346.

Clarke, D.L. 1973. Archaeology: the loss of innocence. *Antiquity* 47: 6-18.

Cordell, L., and F. Plog 1979. Escaping the confines of normative thought: a reevaluation of Pueblan prehistory. *American Antiquity* 44: 405-29.

Costin, C.L. 1991. Craft specialisation: issues in defining, documenting, and explaining the origin of production. In Schiffer M.B. (ed.), *Archaeological Method and Theory* Vol. 3: 1-56. Tucson: University of Arizona Press.

Courty, M.A. and V. Roux 1995. Identification of wheel throwing on the basis of ceramic surface features and microfabrics. *Journal of Archaeological Science* 22: 17-50.

Crawford, A.R. 1979. Gondwanaland and the Pakistan Region. In Farah, A. and K. Dejong (eds.), *Geodynamics of Pakistan* Quetta: Geological Survey of Pakistan.

Crook, K.A.W. 1968. Weathering and roundness of quartz sand grains. *Sedimentology* 11: 171-82.

Dakshini, K.M.M. 1985. Indian subcontinent. In Goodin, J.R. and D.K. Northington (eds.), *Plant Resources of Arid and Sem-Arid Lands.* Orlando: Academic Press.

Dales, G.F. 1962. Harappan outposts on the Makran coast. *Antiquity* 36: 86-92.

Dales, G.F. 1965. A Suggested Chronology for Afghanistan, Baluchistan, and the Indus Valley. In R.W. Ehrich (ed.), *Chronologies in Old World Archaeology,* pp 255-87. University of Chicago Press.

Dales, G.F. 1979. The Balakot Project: Summary of Four Years of Excavations in Pakistan. In Taddei, M. (ed.) *South Asian Archaeology 1977* pp. 241-74. Naples: Instituto Universitario Orientale.

Dani, A.H. 1971. Excavations in the Gomal Valley. *Ancient Pakistan* No. 5. Peshawar.

Day, P.M. 1995. Spoilt for choice: paste recipes as social markers in Crete. Paper presented at the Archaeomaterials Workshop, University of Sheffield, 29th April 1995.

de Cardi, B. 1965. Excavations and Reconnaissance in Kalat, West Pakistan. *Pakistan Archaeology* 2: 86-182.

de Cardi, B. 1983. *Archaeological Surveys in Baluchistan, 1948 and 1957.* Institute of Archaeology Occasional Papers No. 8, London.

Dickinson, W.R. and R. Shutler, Jr. 1971. Temper sands in prehistoric pottery of the Pacific islands. *Archaeology and Physical Anthropology in Oceania* 6(3): 191-203.

Donahue, J., D.R. Watters, and S. Millspaugh 1990. Thin section petrography of northern Lesser Antilles Ceramics. *Geoarchaeology: An International Journal* 5(3): 229-54.

Durrani, F.A. 1988. Excavations in the Gomal Valley: Rehman Dheri Excavation Report No. 1. *Ancient Pakistan* VI. Department of Archaeology, University of Peshawar.

Durrani, F.A., I. Ali, and G. Erdosy 1991. Futher excavation at Rehman Dheri. *Ancient Pakistan* VII: 61-151.

Durrani, F.A., I. Ali, and G. Erdosy 1994. The beads of Rehman Dheri. *Ancient Pakistan* 10: 15-81.

Durrani, F.A. and R.P.Wright 1992. Excavation at Rehman Dheri: the pottery typology and technology. In Possehl, G.L. (ed.), *South Asian Archaeology Studies* pp145-162. New Delhi: Oxford and IBH Publishing.

Echallier, J.C. 1991. Common and pseudo-Ionian ware from Le Pegue (Drome, France): An analytical and archaeological problem. In Middleton, A. and I. Freestone (eds.), *Recent Developments in Ceramic Petrology* pp 17-37. British Museum Occasional Paper No. 81.

Ehrenreich, R.M. 1995. Archaeometry into archaeology. *Journal of Archaeological Method and Theory* 2(1): 1-6.

Fairservis, W.A. 1956. *Excavations in the Quetta Valley, West Pakistan.* Anthropological Papers of the American Museum of Natural History, Vol XLV, Part 2, pp 169-402. New York.

Fairservis, W.A. 1959. *Archaeological Surveys in the Zhob and Loralai Districts, West Pakistan.* Anthropological Papers of the American Museum of Natural History, Vol. XLVII, Part 2. New York.

Fairservis, W.A. 1961. The Harappan Civilization: New Evidence and More Theory. *Novitates* 2055: 1-35. New York: American Museum of Natural History.

Feathers, J.K. 1989. Effects of temper on strength of ceramics: response to Bronitsky and Hamer. *American Antiquity* 54(3): 579-88.

Flam, L. 1993. Fluvial geomorphology of the lower Indus basin (Sindh, Pakistan) and the Indus civilization. In Shroder, J.F. Jr. (ed.), *Himalaya to the Sea.* London and New York: Routledge.

Flannery, K.V. 1967. The Olmec and the Valley of Oaxaca: a model for interregional interaction in Formative times. In Benson, E.P. (ed.), *Dumbarton Oaks Conference on the Olmec*, pp 79-110. Washington: Dumbarton Oaks Research Library and Collection.

Flavin, K., I. Ali, and G. Erdosy 1994. The terracotta human figurines of Rehman Dheri. *Ancient Pakistan* 10: 82-133.

Folk, R.L. and W.C. Ward 1957. Brazos River bar: a study in the significance of grain size parameters. *Journal of Sedimentary Petrology* 27(1): 3-26.

Freestone, I.C. 1991. Extending Ceramic Petrology. In Middleton, A. and I. Freestone (eds.), *Recent Developments in Ceramic Petrology* pp 399-410. British Museum Occasional Paper No. 81.

Friedman, G.M. 1961. Distinction between dune, beach, and river sands from their textural characteristics. *Journal of Sedimentary Petrology* 31(4): 514-29.

Friedman, G.M. 1967. Dynamic processes and statistical parameters compared for size frequency distribution of beach and river sands. *Journal of Sedimentary Petrology* 37(2): 327-54.

Friedman, G.M. 1979. Differences in size distributions of populations of particles among sands of various origins. *Sedimentology* 26: 3-32.

Gerrard, C. and A. Gutierrez 1991. The thin-section analysis and macroscopic characterization of some medieval and post-medieval pottery from northern Spain. In Middleton, A. and I. Freestone (eds.), *Recent Developments in Ceramic Petrology* pp 113-35. British Museum Occasional Paper No. 81.

Goudie, A. 1973. The environmental background to early man in the dry zone of North-Western India: The geomorphic evidence for climatic change. In Hammond, N. (ed.), *South Asian Archaeology,* pp 29-37. London: Duckworth.

Gribble, C.D. and Hall, A.J. 1985. *A Practical Guide to Optical Mineralogy.* London: Allen & Unwin.

Hagerstrand, T. 1966. Aspects of the spatial structure of social communication and the diffusion of information. *Papers of the Regional Science Association* 16. (Cracow Congress 1965)

Hagerstrand, T. 1967. *Innovation Diffusion as a Spatial Process.* Translated by A.R. Pred. Chicago: University of Chicago Press.

Hantman, J.L. and S. Plog 1982. The relationship of stylistic similarity to patterns of material exchange. In Ericson, J. E. and T. J. Earle (eds.), *Contexts for Prehistoric Exchange,* pp 237-63. New York: Academic Press.

Harvey, M.D. and L Flam 1993. Prehistoric soil and water detention structures (gabarbands) at Phang, Sindh Kohistan, Pakistan: an adaptation to environmental change? *Geoarchaeology: An International Journal* 8 (2): 109-26.

Hashimi, H.H. and R.R. Nair 1980. Holocene climatic inferences from the sediments of the western Indian continental shelf. *Proceedings of the Indian Academy of Sciences (Earth Planet Science)* 89: 229-315.

Hashimi, H.H. and R.R. Nair 1986. Climatic aridity over India 11,000 years ago: evidence from feldspar distribution in shelf sediments. *Palaeogeograpy, Palaeoclimatology, and Palaeoecology* 53 (2/4): 309-19.

Helmig, D., E. Jackworth and A Hauptmann 1989. Archaeometallurgical fieldwork and the use of a portable X-ray spectrometer. *Archaeometry* 31 (2): 181-91.

Hemphill, W.R. and A.H. Kidwai 1973. Stratigraphy of the Bannu and Dera Ismail Khan areas, Pakistan. *U. S. Geological Survey / Geological Survey of Pakistan Professional Paper* 716-B. Washington: United States Government Printing Office.

Horie, C.V. 1987. *Materials for Conservation.* London: Butterworths.

Hudson, J.C. 1969. Diffusion in a central place system. *Geographical Analysis* 1: 45-58.

Hunt, P.N. and D.R. Griffiths 1989. Optical petrology in the field. *World Archaeology* 21(1): 163-72.

Hunting Survey Corporation, Limited 1960. *Reconnaissance Geology of Part of West Pakistan.* Oshawa: Maracle Press Limited.

Husain, J. 1992. Potter's craft at Shaikhan Dheri: an ethnoarchaeological reconstruction. *Pakistan Archaeology* 27: 171-95.

Hussain, A. 1975. Silica sand deposits of Khisore and Marwat ranges, D.I. Khan, NWFP, Pakistan. *Geological Survey of Pakistan Information Release* No. 91. Quetta.

Inman, D.L. 1952. Measures for describing the size distribution of sediments. *Journal of Sedimentary Petrology* 22(3): 125-45.

Isaacson, J.S. and T.F. Aleto 1989. Petrographic analysis of ceramic thin sections from La Puna Island, Ecuador. *Archaeomaterials* 3: 61-7.

Jacobsen, T.W. 1984. Seasonal pastoralism in southern Greece: a consideration of the ecology of Neolithic Urfirnis pottery. In Rice, P.M. (ed.) *Pots and Potters: Current Approaches in Ceramic Archaeology.* London: Institute of Archaeology Monograph.

Jansen, M. 1992. Non-contemporaneity of the contemporaneous in the Indus culture. In Possehl, G.L. (ed.), *South Asian Archaeology Studies* pp209-22. New Delhi: Oxford and IBH Publishing.

Jansen, M.R.N. 1993. Mohenjo-Daro: type site of the earliest urbanization process in south Asia. In Spodek, H. and D.M. Srinivasan (eds.), *Urban Form and Meaning in South Asia: The Shaping of Cities from Prehistoric to Precolonial Times* pp 35-51. National Gallery of Art, Washington. Connecticut: Herlin Press.

Jarrige, J.-F. 1979. Economy and society in the early Chalcolithic/Bronze Age of Baluchistan: new perspectives from recent excavations at Mehrgarh. In Härtel, H. (ed.), *South Asian Archaeology* 1979: 93-114. Berlin: Dietrich Reimer Verlag.

Jarrige, J.-F. 1981. Chronology of the earlier periods of the greater Indus as seen from Mehrgarh, Pakistan. In Allchin, B. (ed.), *South Asian Archaeology 1981,* pp 21-8. Cambridge: University Press.

Jarrige, J.-F. 1986. Excavations at Mehrgarh-Nausharo. *Pakistan Archaeology* 10-22: 63-131.

Jarrige, J.-F. 1989. Excavations at Nausharo 1987-88. *Pakistan Archaeology* 24: 21-67.

Jenkins, P.C. 1994. Continuity and change in the ceramic sequence at Harappa. In Parpola, A. and P. Koskikallio (eds.), *South Asian Archaeology 1993* (I): 315-28. Helsinki: Suomalainen Tiedeakatemia.

Jim, C.Y. 1985. Impregnation of moist and dry unconsolidated clay samples using Spurr resin for microstructural studies. *Journal of Sedimentary Petrology* 85(1): 597-99.

Johnson, J.M. 1992. The ceramics at Ban Don Ta Phet: A preliminary report with special reference to the classification of ceramic fabrics. In Glover, I.C. (ed.), *Southeast Asian Archaeology 1990,* pp 179-86. University of Hull: Centre for South-East Asian Studies.

Jones, L.D. 1986. Lowland Maya pottery: the place of petrographic analysis. *British Archaeological Reports International Series* 288. Oxford.

Jones, L.D. 1991. Tempering trends in lowland Maya pottery. In Middleton, A. and I. Freestone (eds.), *Recent Developments in Ceramic Petrology,* pp 165-83. British Museum Occasional Paper No. 81.

Kapur, S., N. Sakarya and E.A. FitzPatrick 1992. Mineralogy and micromorphology of Chalcolithic and Early Bronze Age Ikiztepe ceramics. *Geoarchaeology* 7(4): 327-37.

Kamilli, D.C., and A. Steinberg 1985. New approaches to mineral analysis of ancient ceramics. In Rapp, G. and J. Gifford (eds.), *Archaeological Geology.* Yale University Press.

Kenoyer, J.M. 1991. Urban process in the Indus tradition: a preliminary model from Harappa. In Meadow, R.H. (ed.), *Harappa Excavations 1986-1990* Monographs in World Archaeology No. 3, pp 29-60. Madison: Prehistory Press.

Kenoyer, J.M. 1992. Harappan craft specialisation and the question of urban segregation and stratification. *The Eastern Anthropologist* 45(1-2): 39-54.

Kenoyer, J.M. 1994. Experimental studies of Indus valley technology at Harappa. In Parpola, A. and P. Koskikallio (eds.), *South Asian Archaeology 1993* (I): 345-62. Helsinki: Suomalainen Tiedeakatemia.

Kerr, P.F. 1977. *Optical Mineralogy.* Fourth edition. San Francisco: McGraw Hill.

Khan, M.F. 1986. Archaeological sites in the Bannu basin. In Allchin, F.R., B. Allchin, F.A. Durrani and M.F. Khan (eds.), *Lewan and the Bannu Basin* pp 183-195. BAR International Series 310. Cambridge.

Khan, M.J. 1987. Sedimentation and tectonics of the Siwalik group of the Trans-Indus Salt range, northwestern Pakistan. *Geological Bulletin of the University of Peshawar* 20: 129-41.

Khan, M.J., N.D. Opdyke, and J.F. Shroder, Jr. 1985. Bain diamictite: lithology, age and origin. *Geological Bulletin of the University of Peshawar* 18: 53-64.

Kilikoglou, V., Y. Maniatis, and A.P. Grimanis 1988. The effect of purification and firing of clays on trace element provenance studies. *Archaeometry* 30(1): 37-46.

Kingery, W.D. 1987. Microstructure analysis as part of a holistic interpretation of ceramic art and achaeological artefacts. *Archaeomaterials* 1: 91-9.

Knox, J.R. 1986. Catalogue of pottery from Lewan. In Allchin, F.R., B. Allchin, F.A. Durrani, and M.F. Khan (eds.), *Lewan and the Bannu Basin* pp 209-53. Oxford: BAR International Series 310.

Kohl, P. 1987. The ancient economy, transferable technologies and the Bronze Age world-system: a view from the northeastern frontier of the ancient Near East. In Rowlands, M., M. Larsen and K. Kristiansen (eds.), *Centre and Periphery in the Ancient World,* pp 13-24. Cambridge University Press.

Konta, J. 1985. Mineralogy and chemical maturity of suspended matter in major rivers samples under the SCOPE/UNEP project. *Mitt. Geol. Paläontol. Inst. Univ. Hamburg* 58: 563-9.

Krishnan, K. and K.T.M. Hegde 1987. Chemical and petrographic studies in pottery of Harappa culture in Gujarat. *Journal of the Maharaja Sayajirao University of Boroda* 35-36(1): 27-56.

Lahiri, N. 1992. *The Archaeology of Indian Trade Routes up to c. 200 B.C.* New Delhi: Oxford University Press.

Lechevallier, M. and G. Quivron 1981. The neolithic in Baluchistan: new evidences from Mehrgarh. In Härtel, H. (ed.), *South Asian Archaeology 1979* pp 71-92. Berlin: Dietrich Reimer Verlag.

Lombard, J.P. 1987. Provenance of sand temper in Hohokam ceramics, Arizona. *Geoarchaeology* 2: 91-119.

Mackay, E.J. 1937. *Further Excavations at Mohenjo-Daro.* Delhi: Government of India.

MacKenzie, W.S. and Adams, A.E. 1994. *A Colour Atlas of Rocks and Minerals in Thin Section.* New York: John Wiley & Sons.

MacKenzie, W.S. and Guilford, C. 1980. *Atlas of Rock-forming Minerals in Thin Section.* New York: John Wiley & Sons.

Magetti, M. 1982. Phase analysis and its significance for technology and origin. In Olin, J.S. and A.D. Franklin (eds.), *Archaeological Ceramics.* Washington, D.C.: Smithsonian Institution.

Mallett, F.R. 1887. *A Manual of the Geology of India.* Part IV Mineralogy. Geological Survey of India.

Marshall, J. 1924. First light on a long forgotten civilisation. *Illustrated London News* September 20: 528-32.

Marshall, J. 1931. *Mohenjo-daro and the Indus Civilisation.* London: A. Probsthain.

Mason, R.B. 1995. Criteria for the petrographic characterization of stonepaste ceramics. *Archaeometry* 37 (2): 307-21.

Masson, C. 1844. *A Narrative of Various Journeys in Baluchistan, Afghanistan and the Punjab.* Vol I. London: Richard Bently.

Matthew, A.J. 1991. Spots before the eyes: New comparison charts for visual percentage estimation in archaeological material. In Middleton, A. and I. Freestone (eds.), *Recent Developments in Ceramic Petrology,* pp 211-63. British Museum Occasional Paper No.81.

Meadow, R.H. 1984. Notes on the faunal remains from Mehrgarh with a focus on cattle (Bos). In Allchin, B. (ed.), *South Asian Archaeology 1981* pp 34-40. Cambridge: University Press.

Meadow, R.H. 1989. Continuity and change in the agriculture of the greater Indus Valley: the palaeoethnobotanical and zooarchaeological evidence. In Kenoyer, J.M. (ed.), *Old Problems and New Perspectives in the Archaeology of South Asia,* pp 61-74. Madison: University of Wisconsin.

Meadow, R.H. (ed.) 1991. *Harappa Excavations 1986-1990.* Monographs in World Archaeology No. 3. Madison: Prehistory Press.

Meissner, C.R. and H. Rahman 1973. Distribution, thickness, and lithology of Palaeocene rocks in Pakistan. *U.S. Geological Survey / Geological Survey of Pakistan Professional Paper* 716-E. Washington: United States Government Printing Office.

Meissner, C.R., M. Hussain, M.A. Rashid, and U.B. Sethi 1975. Geology of the Parachinar quadrangle, Pakistan. *U.S. Geological Survey / Geological Survey of Pakistan Professional Paper* 716-F. Washington: United States Government Printing Office.

Mian, M.A. and M.N. Syal 1986. Geomorphology of Pakistan. In M. Ahmad, M. Akram, M.S. Baig, M.Y. Javed, and Riaz-ul-Amin (eds.), *Proceedings of XII International Forum on Soil Taxonomy and Agrotechnology Transfer,* pp 26-42. Lahore and Washington, D.C.: Soil Survey of Pakistan and Soil Management Support Service.

Middlemost, E.A.K. 1985. *Magmas and Magmatic Rocks.* London: Longman Group.

Middleton, A.P., I.C. Freestone and M.N. Leese 1985. Textural analysis of ceramic thin sections: Evaluation of grain sampling procedures. *Archaeometry* 27(1): 64-74.

Miller, G.L. 1980. Classification and economic scaling of 19th century ceramics. *Historical Archaeology* 14: 1-41.

Millot, G. 1970. *Geology of Clays.* Paris: Masson et Cie.

Misra, V.N. 1984. Climate, a Factor in the Rise and Fall of the Indus Civilization - Evidence from Rajasthan and Beyond. In Lal, B.B. and Gupta, S.P (eds) *Frontiers of the Indus Civilization.* New Delhi: Books & Books.

Morris, T.O. 1938. The Bain Boulder-bed: a glacial episode in the Siwalik series of the Marwat Kundi range and Sheikh Budin, Northwest Frontier Province, India. *Quarterly Journal of the Geological Society of London* 94: 385-421.

Mughal, M.R. 1970. *The Early Harappan Period in the Greater Indus Valley and Northern Baluchistan (ca. 3000 - 2400 B.C.).* Unpublished Ph.D. Dissertation, University of Pennsylvania. University Microfilms International.

Mughal, M.R. 1990. Further evidence of the Early Harappan culture in the Greater Indus Valley: 1971-1990. *South Asian Studies* 6: 175-99.

Mughal, M.R. and M.A. Haleem 1972. Excavations at Sarai Khola: Chapter Two: The Pottery. *Pakistan Archaeology* 34-110.

Neff, H. 1993. Theory, sampling, and analytical techniques in the archaeological study of prehistoric ceramics. *American Antiquity* 58 (1): 23-44.

Neff, H., R.L. Bishop and E.V. Sayre 1988. A simulation approach to the problem of tempering in compositional studies of archaeological ceramics. *Journal of Archaeological Science* 15: 159-72.

Neff, H., R.L. Bishop and E.V. Sayre 1989. More observations on the problem of tempering in compositional studies of archaeological ceramics. *Journal of Archaeological Science* 16: 57-69.

Orton, C., P. Tyers, and A. Vince 1993. *Pottery in Archaeology.* Cambridge: Cambridge University Press.

Parpola, A. 1994. Deciphering the Indus script: a summary report. In Parpola, A. and P. Koskikallio (eds.), *South Asian Archaeology 1993* pp 571-86. Helsinki: Suomalainen Tiedeakatemia.

Peacock, D.P.S. 1971. Petrography of certain coarse pottery. In Cunliffe, B.E. (ed.) *Excavations at Fishbourne 1961-1969, Volume II: The Finds.* London: Reports of the Research Committee of the Society of Antiquaries of London No. XXVII pp 255-59.

Pedersen, P. 1970. Innovation diffusion within and between national urban systems. *Geographical Analysis* 2: 203-54.

Pendall, E. and R.G. Amundson 1990a. Soil/landform relationships surrounding the Harappa archaeological site, Pakistan. *Geoarchaeology: An International Journal* 5(4): 301-22.

Pendall, E. and R.G. Amundson 1990b. The stable isotope chemistry of pedogenic carbonates in an alluvial soil from the Punjab, Pakistan. *Soil Science* April 1990.

Pervil'ev, Y.S. 1982. The metamorphism in the Afghan-Pakistan segment. In Lepezin, G., and N.L. Dobretsov (eds.), *Metamorphic Complexes of Asia* pp 260-3. Oxford: Pergamon Press.

Pettijohn, F.J. 1975. *Sedimentary Rocks.* Third edition. New York: Harper and Row.

Pettijohn, F.J., P.E. Potter and R. Siever 1973. *Sand and Sandstone.* Berlin: Springer.

Piggott, S. 1950. *Prehistoric India to 1000 B.C.* Middlesex: Penguin Books.

Plog, S. 1980. *Stylistic Variation in Prehistoric Ceramics.* Cambridge University Press.

Porat, N., J. Yellin, L Heller-Kallai and L. Halicz 1991. Correlation between petrography, NAA and ICP analyses: application to Early Bronze Age Egyptian pottery from Canaan. *Geoarchaeology* 6(2): 133-49.

Possehl, G.L. 1990. Revolution in the urban revolution: The emergence of Indus urbanization. *Annual Review of Anthropology* 19: 261-82.

Possehl, G.L. 1991. A short history of archaeological discovery at Harappa. In Meadow, R.H. (ed.), *Harappa Excavations 1986-1990,* Monographs in World Archaeology No.3 pp 5-11. Madison: Prehistory Press.

Press, F. and R. Siever 1986. *Earth.* New York: Freeman and Company.

Pye, K. and J. Mazzullo 1994. Effects of tropical weathering on quartz grain shape: an example from Northeastern Australia. *Journal of Sedimentary Research* A64(3): 500-7.

Raikes, R.L. 1968. Archaeological explorations in Southern Jhalawan and Las Bela. *Originii II,* pp 103-71. University of Rome.

Raikes, R.L., and R.H. Dyson 1961. The prehistoric climate of Baluchistan and the Indus valley. *American Anthropologist* 63: 265-81.

Rapp, G. 1985. The provenance of artefactual raw materials. In Rapp, G. and J.A. Gifford (eds.), *Archaeological Geology* pp 353-75. New Haven: Yale University Press.

Rendell, H. 1981. A preliminary investigation of the sedimentary history of the Bannu Basin in the late Holocene. In Härtel, H. (ed.), *South Asian Archaeology 1979* pp 219-25. Berlin: Dietrich Reimer Verlag.

Rendell, H. 1986. Geology and geomorphology of the Bannu basin. In Allchin, F.R., B. Allchin, F.A. Durrani, and M.F. Khan (eds.), *Lewan and the Bannu Basin,* pp 1-12. Cambridge: BAR International Series 310.

Rice, P.M. 1987. *Pottery Analysis: A Sourcebook.* Chicago: University of Chicago Press.

Rigby, V., A.P. Middleton, and I.C. Freestone 1989. The primary workshop: technical examination of La Tene bichrome painted pottery from Champagne. *World Archaeology* 15(3): 348-65.

Roff, W.J. and J.R. Scott 1971. *Fibres, Films, Plastics and Rubbers.* London: Butterworths.

Rubertone, P. 1978. *Social organization in an Islamic town: a behavioral explanation of ceramic variability.* Unpublished PhD dissertation, Department of Anthropology, State University of New York at Binghamton.

Rutley, F. 1988. *Rutley's Elements of Mineralogy.* Revised by C.D. Gribble. London: Unwin Hyman.

Rye, O.S. 1977. Pottery manufacturing techniques: X-ray studies. *Archaeometry* 19(2): 205-11.

Rye, O.S. (ed.) 1981. *Pottery Technology: Principles and Reconstruction.* Washington, D.C.: Taraxacum Press.

Rye, O.S. and C. Evans 1976. *Traditional Pottery Techniques of Pakistan.* Smithsonian Contributions to Anthropology 21. Washington: Smithsonian Institute Press. Reprinted 1990 by Lok Virsa, Islamabad.

Sakarya, N., S Kapur, V.S. Cavusgil and E.A. FitzPatrick 1990. Basaltic ceramics. In *Third International Working Group Meeting on Soil Submicroscopy.* Bangor.

Samzun, A. and P. Sellier 1985. First anthropological and cultural evidence for the funerary practices of the chalcolithic population of Mehrgarh, Pakistan. In Schotsmans, J. and M. Taddei (eds.), *South Asian Archaeology 1983* pp 91-120. Naples: Instituto Universitario Orientale.

Schneider, G. 1987. Chemical analysis of stoneware bangles and related material from Mohenjo-Daro. In Jansen, M. and G. Urban (eds.), *Interim Reports* pp 73-7. Aachen.

Scholz, F. 1983. Baluchistan: a brief introduction to the geography of Pakistan's mountainous province. *Newsletter of Baluchistan Studies* 1: 13-8.

Schubert, P. 1986. Petrographic modal analysis: A necessary complement to chemical analysis of ceramic coarse ware. *Archaeometry* 28 (2): 163-78.

Sehgal, J.H. and G. Stoops 1972. Pedogenic calcite accumulation in arid and semi-arid regions of the Indo-Gangetic alluvial plains of erstwhile Punjab (India)--their morphology and origin. *Geoderma* 8: 59-72.

Selley, R.C. 1988. *Applied Sedimentology.* London: Academic Press.

Shaffer, J.G. 1992. Indus Valley, Baluchistan and the Helmand. In Ehrich, R.W. (ed.), *Chronologies in Old World Archaeology,* Third Edition, Volume I. Chicago and London: The University of Chicago Press.

Shaffer, J.G. and B.K. Thapar 1992. Pre-Indus and early Indus cultures of Pakistan and India. In Dani, A.H. and V.M. Masson (eds.), *History of Civlizations of Central Asia* Volume I: 247-81. Paris: UNESCO Publishing.

Shennan, S. 1988. *Quantifying Archaeology.* Edinburgh: University Press.

Singh, G. 1971. The Indus Valley culture (seen in the context of post-glacial climate and ecological studies in North-West India). *Archaeology and Physical Anthropology in Oceania* 6(2): 177-89.

Singh, G., Joshi R.D., Chopra, S.K., and Singh, A.B. 1974. Late Quaternary history of vegetation and climate of the Rajasthan Desert, India. *Philosophical Transactions of the Royal Society of London* 267 (889): 467-501.

Smithson, F. 1939. Statistical methods in sedimentary petrology, Part II: Grain-size measurements and their graphic study. *Geology Magazine* 76: 348-60.

Spate, O.H.K., A.T.A. Learmonth, and B.H. Farmer 1972. *India, Pakistan and Ceylon: The Regions.* Part IV: The Face of the Land. London: Methuen & Co Ltd.

Stein, Sir Aurel 1931. An archaeological tour of Gedrosia. *Memoirs of the Archaeological Survey of India* 43. New Delhi: Government of India Press.

Stoltman, J.B. 1989. A quantitative approach to the petrographic analysis of ceramic thin sections. *American Antiquity* 54(1): 147-60.

Stoltman, J.B. 1991. Ceramic petrography as a technique for documenting cultural interaction: An example from the Upper Mississippi Valley. *American Antiquity* 56 (1): 103-20.

Stoltman, J.B., J.H. Burton, and J. Haas 1992. Chemical and Petrographic Characterisations of Ceramic Pastes: Two Perspectives on a Single Data Set. In H. Neff (ed.), *Chemical Characterization of Ceramic Pastes in Archaeology.* Monographs in World Archaeology No. 7, pp 85-92. Madison: Prehistory Press.

Streeten, A.D.F. 1982. Textural analysis: an approach to the characterisation of sand-tempered ceramics. In Freestone, I., T. Potter and C. Johns (eds.), *Current Research in Ceramics: Thin-section Studies,* pp 123-34. British Museum Occasional Paper No. 32, London.

Tankersley, K. and J. Meinhart 1982. Physical and structural properties of ceramic materials utilized by a Fort Ancient group. *Midcontinental Journal of Archaeology* 7: 225-43.

Thomas, K.D. 1983. Tarakai Qila: site, economy and environment. In Proudfoot, B. (ed.), *Site, Environment and Economy* pp 127-44. Oxford: BAR International Series 173.

Thomas, K.D. 1986. Environment and subsistence in the Bannu basin. In Allchin, F.R., B. Allchin, F.A. Durrani, and M.F. Khan (eds.), *Lewan and the Bannu Basin,* pp 13-33. Cambridge: BAR International Series 310.

Tite, M.S. 1991. Archaeological science: past achievements and future prospects. *Archaeometry* 33 (2): 139-51.

Topping, P.G., and A.B. Mackenzie 1988. A test of the use of neutron activation analysis for clay source characterization. *Archaeometry* 30(1): 92-101.

U.S. Geological Survey and Geological Survey of Pakistan 1964. *Geological Map of Pakistan.* 1:2,000,000. Government of Pakistan.

Van der Leeuw, S. 1977. Towards a study of the economics of pottery making. *Ex Horreo.* Cingvla IV. Albert Egges van Giffen Institute of Prehistory, University of Amsterdam.

Van der Plas, L. 1962. Preliminary note on the granulometric analysis of sedimentary rocks. *Sedimentology* 1: 145-57.

Van der Plas, L. and A.C. Tobi 1965. A chart for judging the reliability of point counting results. *American Journal of Science* 263: 87-90.

Vats, M.S. 1940. *Excavations at Harappa.* Delhi: Government of India Press.

Verstappen, H.T. 1970. Aeolian geomorphology of the Thar Desert and palaeoclimates. *Zeitschrift fur Geomorphologie Supplementand* 10: 104-20.

Waagen, W. 1889. Salt Range fossils; geological results. *Palaeont. Indica.* 13 (4) Pt 1.

Waagen, W. 1891. Salt Range fossils; geological results. *Palaeont. Indica.* 13 (4) Pt 2.

Wadia, D.N. 1966. *Geology of India.* London: MacMillan and Company Ltd

Watt, I. 1993. Light microscopes for use in unorthodox situations. *Microscopy and Analysis* 38: 27-9.

Wheeler, R.E.M. 1947. Harappa 1946: The defenses and cemetery R-37. *Ancient India* 3: 58-130.

Wheeler, Sir Mortimer 1953. *The Cambridge History of India: The Indus Civilization.* Cambridge University Press.

Whitbread, I. K. 1986. The characterisation of argillaceous inclusions in ceramic thin sections. *Archaeometry* 28(1): 79-88.

Willey, G.R. and P. Phillips 1958. *Method and Theory in American Archaeology.* Chicago: University of Chicago Press.

Willey, G.R. and J.R. Sabloff 1980. *A History of American Archaeology*. Second Edition. New York: W.H. Freeman and Company.

Willis, B.J. and A.K. Behrensmeyer 1994. Architecture of Miocene overbank deposits in northern Pakistan. *Journal of Sedimentary Research* B64(1): 60-7.

Wilmsen, E.N. 1973. *Lindenmeir: A Pleistocene hunting society*. New York: Harper and Row.

Wobst, H.M. 1977. Stylistic behavior and information exchange. *Michigan Anthropological Papers* 61: 317-42.

Wright, R.P. 1984. *Technology, Style, and Craft Specialization: Spheres of Interaction and Exchange in the Indo-Iranian Borderlands, Third Millennium B.C.* Unpublished Ph.D. Thesis, Harvard University. University Microfilms International.

Wright, R.P. 1991. Patterns of technology and the organization of production at Harappa. In Meadow, R.H. (ed.), *Harappa Excavations 1986-1990: A Multidisciplinary Approach to Third Millennium Urbanism.* Monographs in World Archaeology 3: 71-88. Madison, Wisconsin.

Wynne, A.B. 1878. On the geology of the Salt Range in the Punjab. *India Geological Survey Memoirs* 14.

Yardley, B.W.D., W.S. Mackenzie, and C. Guilford 1990. *Atlas of Metamorphic Rocks and their Textures.* New York: John Wiley & Sons.

Yanguas, J.E. and S.T. Paxton 1986. A new technique for preparation of petrographic thin sections using ultra-violet curing adhesive. *Journal of Sedimentary Petrology* 56(1): 539-40.

KEY TO ABBREVIATIONS USED IN APPENDICES A to C

(in order of appearance, by row)

For a detailed description, refer to Sections 6.3 and 6.4

morph type	= parent vessel morphological type (described in Section 5.4)
rim dia	= for rim sherds, rim diameter in cm
thick	= average wall thickness, or, for bangles, cross-section dimensions
sqmm	= area of thin section used for area count, in mm^2
Qz size	= refers to size range of quartz inclusions, in microns, next 6 columns
numpts	= total number of points counted for area count
%	= percentage of total counts occurring in each size range
sigma	= one standard deviation, σ
2psigma	= $\% + 2\sigma$
2msigma	= $\% - 2\sigma$
Conc	= concentration, number of each quartz size range, per mm^2
Micac	= micaceousness of matrix in thin section (1 = low, 4 = high)
Biref	= birefringence of matrix in thin section (1 = low, 4 = high)
Colour	= ltbrn (light brown), medbrn (medium brown), dkbrn (dark brown)
Ang	= angularity (1= low, 4 = high)
Sph	= sphericity (1 = low, 4 = high)
Wth	= weathering (1 = low, 4 = high)
Conc	= concentration (other than quartz, 1=absent, 2=trace, 3=rare, 4=common)
CarOpa	= carbonate opacity (refer to section 6.3 for explanation)
PFs	= plagioclase feldspar
MFs	= microcline feldspar
Bio	= biotite
Mus	= muscovite
Amp	= amphibole
Car	= carbonate
m, p	= for quartz, monocrystalline or polycrystalline, respectively
Pyrox	= pyroxene
ind	= indeterminate
arf	= argillaceous rock fragment
vrf	= volcanic rock fragment
conc<50	= concentration of quartz inclusions <50 microns in size, per mm^2
conc>50	= concentration of quartz inclusions >50 microns in size, per mm^2
%<50	= quartz inclusions <50 microns in size as percentage of total quartz
%>50	= quartz inclusions >50 microns in size as percentage of total quartz

APPENDIX A

REHMAN DHERI DATA

See page 259 for key to abbreviations used

RHD 1	period	level	morph type	rim d a	thick	sqmm
B IV/4	1	16	BIII	14.00	5.00	6.25

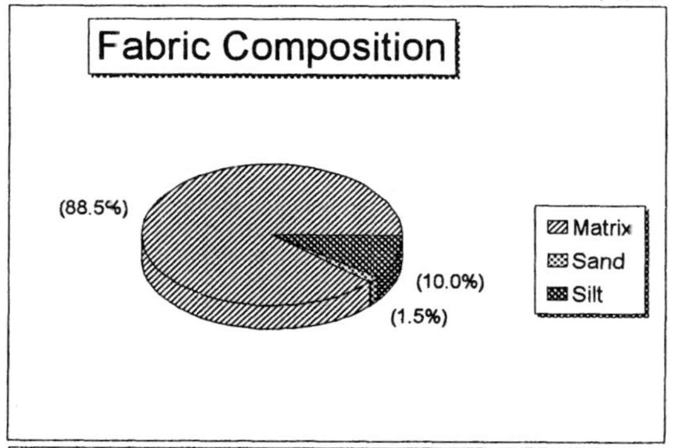

Qz Size	< 50	50-125	125-200	200-275	275-350	> 350	numpts
Number	413	40	5	2	0	0	460
%	89.78	8.70	1.09	0.43	0.00	0.00	
sigma	1.41	1.31	0.48	0.31	0.00	0.00	
2psigma	92.61	11.32	2.05	1.05	0.00	0.00	
2msigma	86.96	6.07	0.12	-0.18	0.00	0.00	
Conc	66.08	6.40	0.80	0.32	0.00	0.00	
Fabric	Matrix	Sand	Silt	Micac	Birefr	Colour	
	354	6	40	2	2	dk brn	400

Minerals	Size	Ang	Sph	Wth	Conc	Type	CarbOpa
PFs	2	3	2		2		
MFs					1		
Bio	3			2	3		
Mus	2			2	4		
Amp					1		
Car	2			2	4	2	2
Quartz		2S 3L	3S 1L			m,p	
Pyrox					2		
Epidote					2		
Schist					2		
Other							

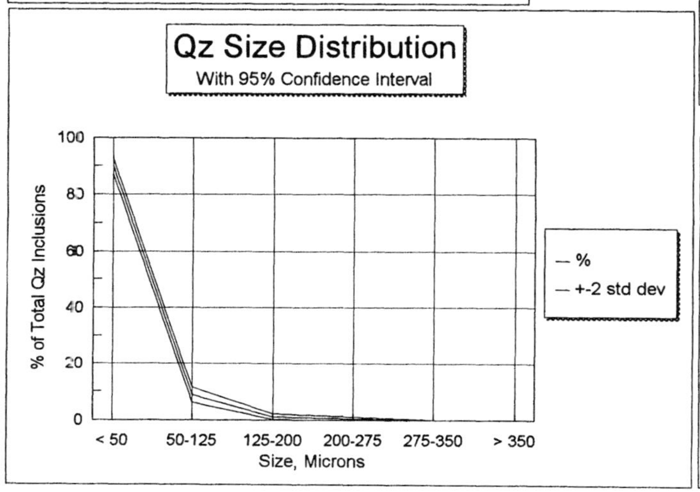

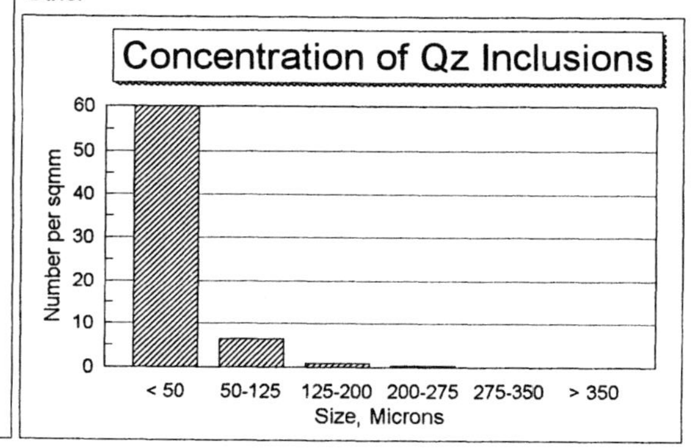

%matrix	88.50
%sand	1.50
%silt	10.00
conc<50	66.08
conc>50	7.52
%<50	89.78
%>50	10.22

APPENDIX A

rhd2	period	level	morphtype	rimdia	thick
BIV/4	1	19	BIII	base	7

Section damaged; not reliable for analysis

RHD 3	period	level	morph type	rim dia	thick	sqmm
B IV/14	2	11	BIII	16	5	12.5

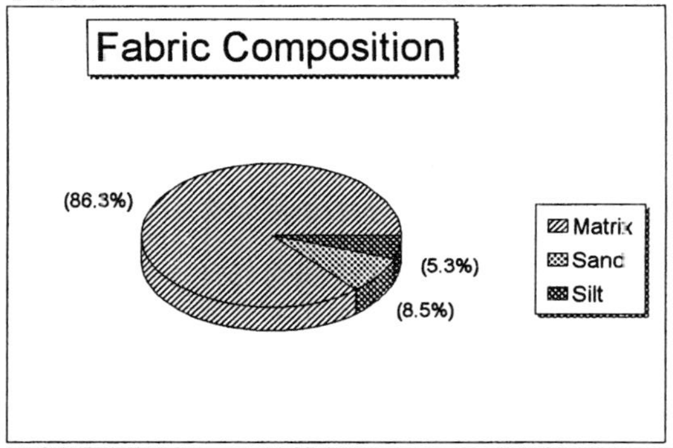

Qz Size	< 50	50-125	125-200	200-275	275-350	> 350	numpts
Number	266	37	13	9	4	2	331
%	80.36	11.18	3.93	2.72	1.21	0.60	
sigma	2.18	1.73	1.07	0.89	0.60	0.43	
2psigma	84.73	14.64	6.06	4.51	2.41	1.46	
2msigma	76.00	7.71	1.79	0.93	0.01	-0.25	
Conc	21.28	2.96	1.04	0.72	0.32	0.16	
Fabric	Matrix	Sand	Silt	Micac	Birefr	Colour	
	345	34	21	3	3	medbrn	400

Minerals	Size	Ang	Sph	Wth	Conc	Type	CarOpa
PFs	4	3	2		4		
MFs					1		
Bio	4			2	2		
Mus	4			2	4		
Amp					3		
Car	4			2	4	2	2
Quartz		3	2			m,p	
Pyrox					1		
Epidote					1		
Schist					4		
Other							

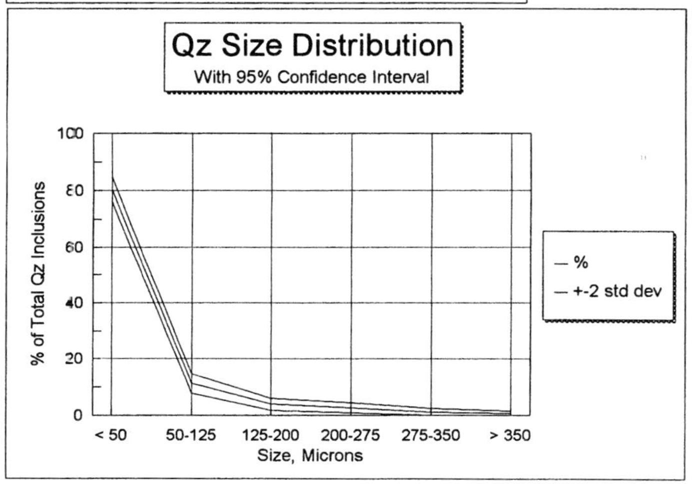

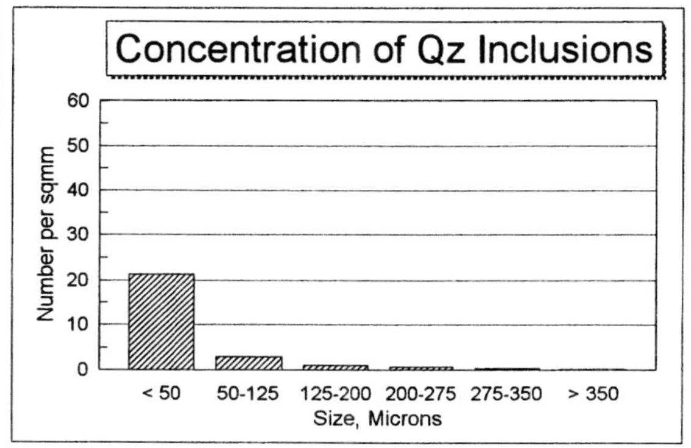

%matrix	86.25
%sand	8.50
%silt	5.25
conc<50	21.28
conc>50	5.20
%<50	80.36
%>50	19.64

RHD 4	period	level	morph type	rim dia	thick	sqmm
B IV/14	1	13	BIII	15	6	18.75

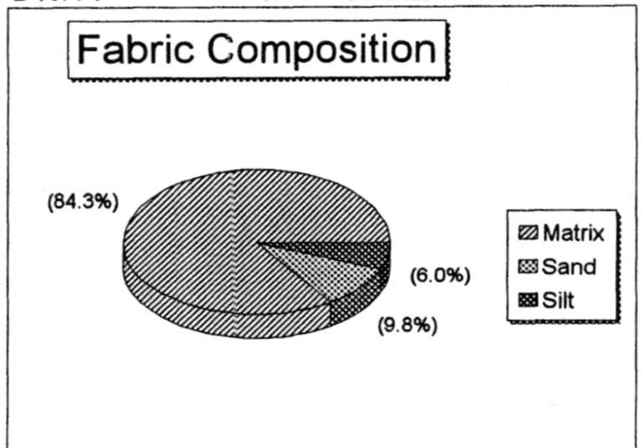

Qz Size	< 50	50-125	125-200	200-275	275-350	> 350	numpts
Number	275	37	30	20	3	0	365
%	75.34	10.14	8.22	5.48	0.82	0.00	
sigma	2.26	1.58	1.44	1.19	0.47	0.00	
2psigma	79.85	13.30	11.09	7.86	1.77	0.00	
2msigma	70.83	6.98	5.34	3.10	-0.12	0.00	
Conc	14.67	1.97	1.60	1.07	0.16	0	
Fabric	Matrix	Sand	Silt	Micac	Biref	Colour	
	337	39	24	3	3	medbrn	400

Minerals	Size	Ang	Sph	Wth	Conc	Type	CarOpa
PFs	2	3	2		2		
MFs		3	2		2		
Bio	2			2	2		
Mus	2			2	4		
Amp					1		
Car	3			2	4	2	2
Quartz		2	2			m,p	
Pyrox					2		
Epidote					1		
Schist					2		
Other							

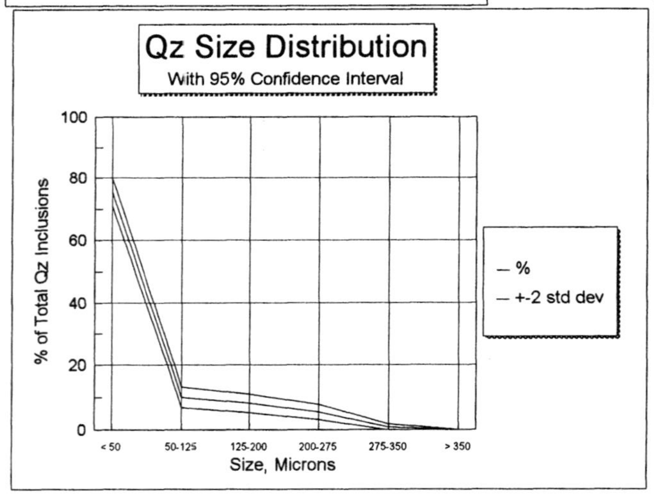

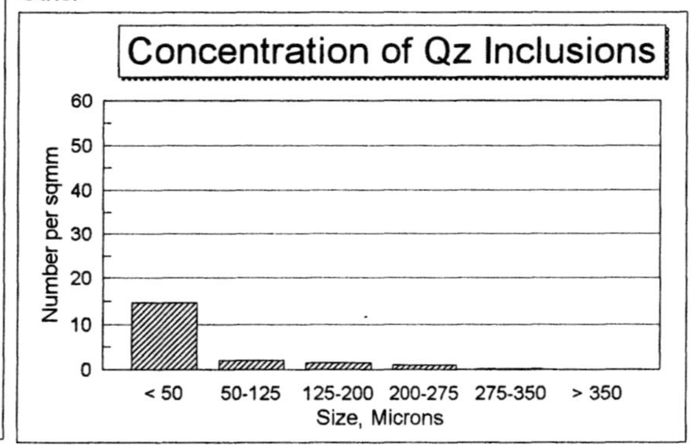

%matrix	84.25
%sand	9.75
%silt	6.00
conc<50	14.67
conc>50	4.80
%<50	75.34
%>50	24.66

RHD 5	period	level	morph type	rim dia	thick	sqmm
B IV/5	3	2	BIII	17	4	12.5

Qz Size	< 50	50-125	125-200	200-275	275-350	> 350	numpts
Number	273	55	18	9	0	0	355
%	76.90	15.49	5.07	2.54	0.00	0.00	
sigma	2.24	1.92	1.16	0.83	0.00	0.00	
2psigma	81.38	19.33	7.40	4.20	0.00	0.00	
2msigma	72.43	11.65	2.74	0.87	0.00	0.00	
Conc	21.84	4.40	1.44	0.72	0	0	

Fabric	Matrix	Sand	Silt	Micac	Birefr	Colour	
	354	25	19	2	2	dk brn	398

Minerals	Size	Ang	Sph	Wth	Conc	Type	CarOpa
PFs		3	1		2		
MFs					1		
Bio	3			2	3		
Mus	3			2	4		
Amp	2				2		
Car	3			3	4	1,2	2
Quartz		2S 4L	2			m,p	
Pyrox					1		
Epidote					1		
Schist					2		
Other							

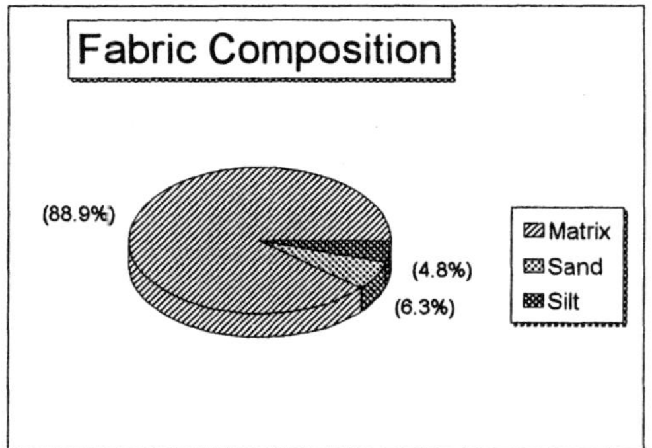

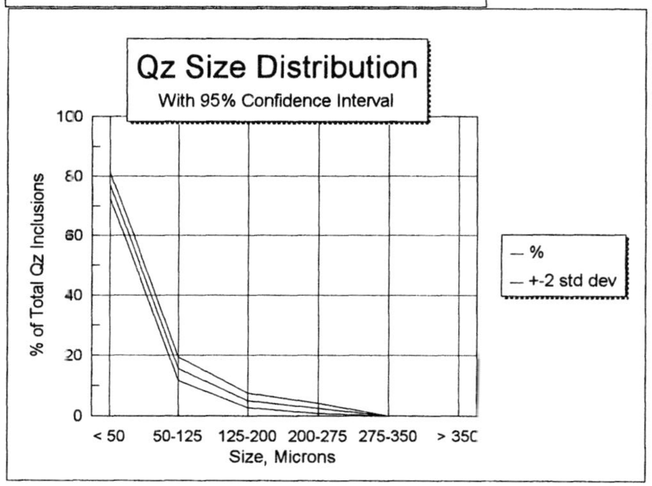

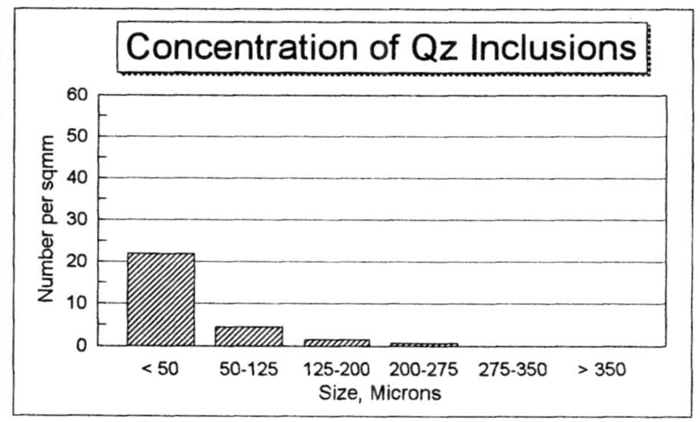

%matrix	88.94
%sand	6.28
%silt	4.77
conc<50	21.84
conc>50	6.56
%<50	76.90
%>50	23.10

APPENDIX A

RHD 6	period	level	morph type	rim dia	thick	sqmm
B IV/24	1	15	BIII		8	25

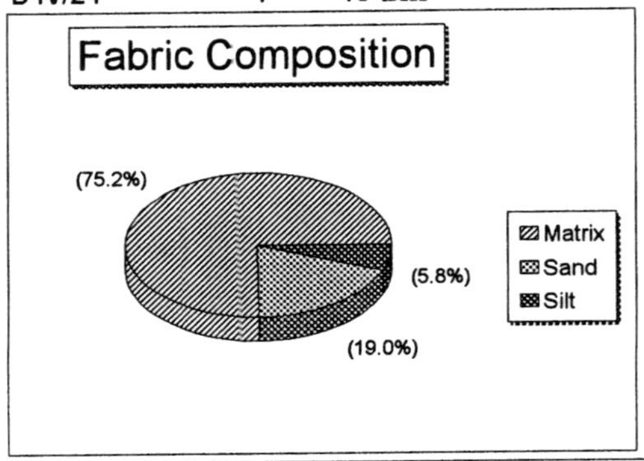

Qz Size	< 50	50-125	125-200	200-275	275-350	> 350	numpts
Number	69	33	19	27	9	18	175
%	39.43	18.86	10.86	15.43	5.14	10.29	
sigma	3.69	2.96	2.35	2.73	1.67	2.30	
2psigma	46.82	24.77	15.56	20.89	8.48	14.88	
2msigma	32.04	12.94	6.15	9.97	1.80	5.69	
Conc	2.76	1.32	0.76	1.08	0.36	0.72	
Fabric	Matrix	Sand	Silt	Micac	Birefr	Colour	
	297	75	23	2	2	dkbrn	395

Minerals	Size	Ang	Sph	Wth	Conc	Type	CarOpa
PFs	4	3	3		3		
MFs					1		
Bio	4			2	2		
Mus	4			2	4		
Amp	3				3		
Car	3				2	2	2
Quartz		4	3			m,p	
Pyrox					2		
Epidote					2		
Schist					3		
Other							

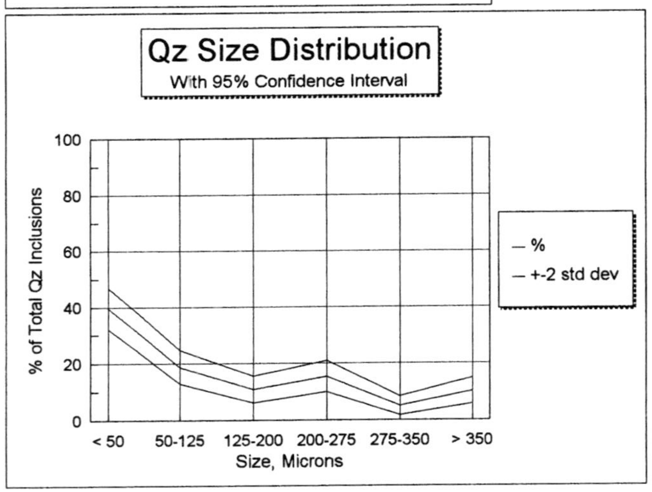

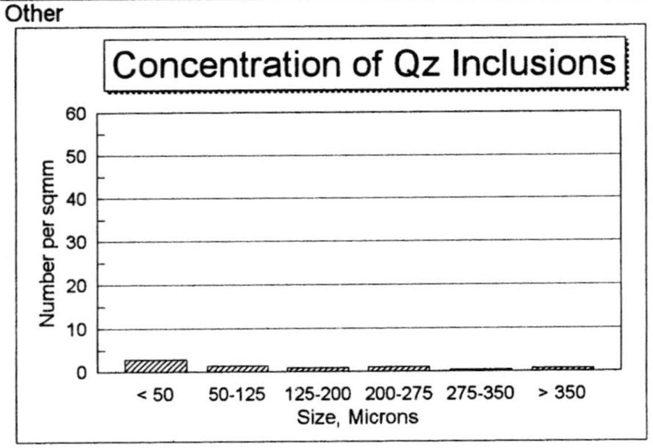

%matrix	75.19
%sand	18.99
%silt	5.82
conc<50	2.76
conc>50	4.24
%<50	39.43
%>50	60.57

RHD 7	period	level	morph type	rim dia	thick	sqmm
B IV/19	1	14	BI	11	6	12.5

Qz Size	< 50	50-125	125-200	200-275	275-350	> 350	numpts
Number	741	21	9	4	4	4	783
%	94.64	2.68	1.15	0.51	0.51	0.51	
sigma	0.81	0.58	0.38	0.25	0.25	0.25	
2psigma	96.25	3.84	1.91	1.02	1.02	1.02	
2msigma	93.03	1.53	0.39	0.00	0.00	0.00	
Conc	59.28	1.68	0.72	0.32	0.32	0.32	
Fabric	Matrix	Sand	Silt	Micac	Biref	Colour	
	317	30	52	4	3	medbrn	399

Minerals	Size	Ang	Sph	Wth	Conc	Type	CarOpa
PFs					2		
MFs					1		
Bio	3			4	2		
Mus	2			2	3		
Amp					2		
Car	3				4	2	2,4
Quartz		4L 3S	2L 3S			m,p	
Pyrox					1		
Epidote					1		
Schist					3		
Other							

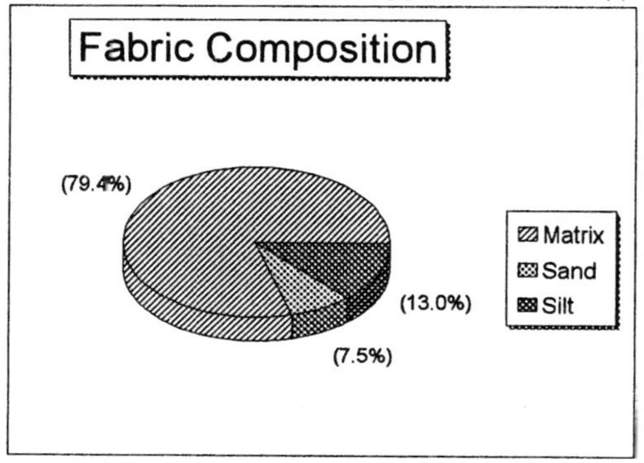

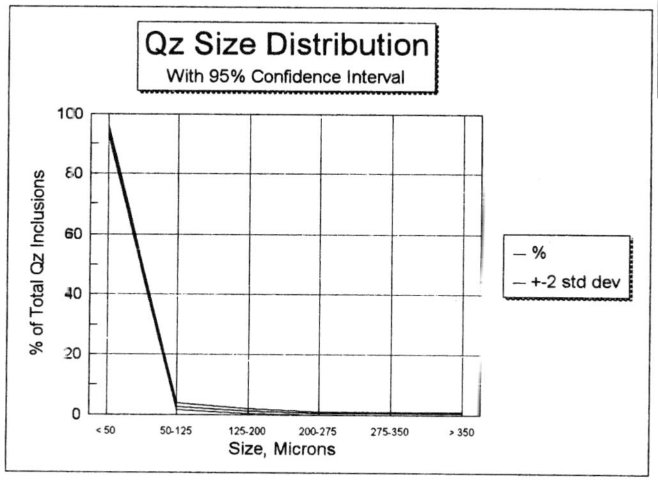

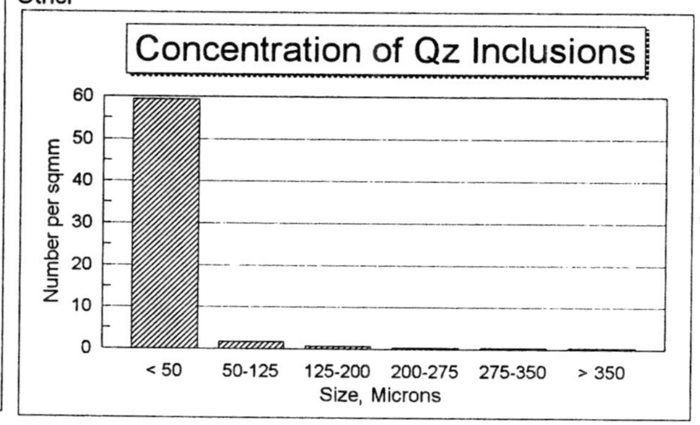

%matrix	79.45
%sand	7.52
%silt	13.03
conc<50	59.28
conc>50	3.36
%<50	94.64
%>50	5.36

RHD 8	period	level	morph type	rim dia	thick	sqmm
B IV/14	2	11	BI	18	6	18.75

Qz Size	< 50	50-125	125-200	200-275	275-350	> 350	numpts
Number	202	59	24	12	4	4	305
%	66.23	19.34	7.87	3.93	1.31	1.31	
sigma	2.71	2.26	1.54	1.11	0.65	0.65	
2psigma	71.65	23.87	10.95	6.16	2.61	2.61	
2msigma	60.81	14.82	4.79	1.71	0.01	0.01	
Conc	10.77	3.15	1.28	0.64	0.21	0.21	
Fabric	Matrix	Sand	Silt	Micac	Biref	Colour	
	295	64	38	2	2	medbrn	397
Minerals	Size	Ang	Sph	Wth	Conc	Type	CarOpa
PFs	3				3		
MFs					2		
Bio	3			4	2		
Mus	3			4	4		
Amp	3				3		
Car					3	1,2	2
Quartz		3	2			m,p	
Pyrox					1		
Epidote					2		
Schist					3		
Other							

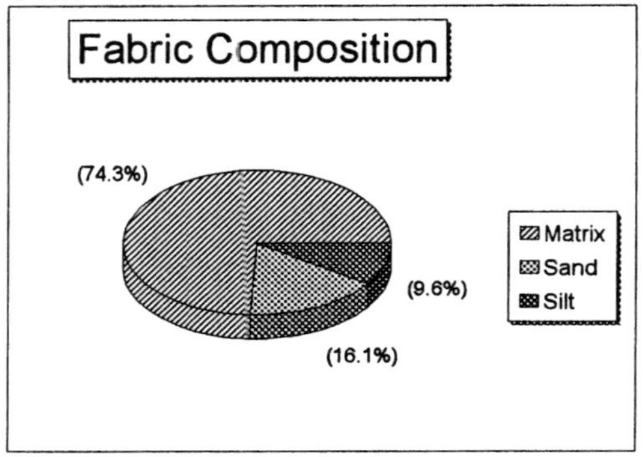

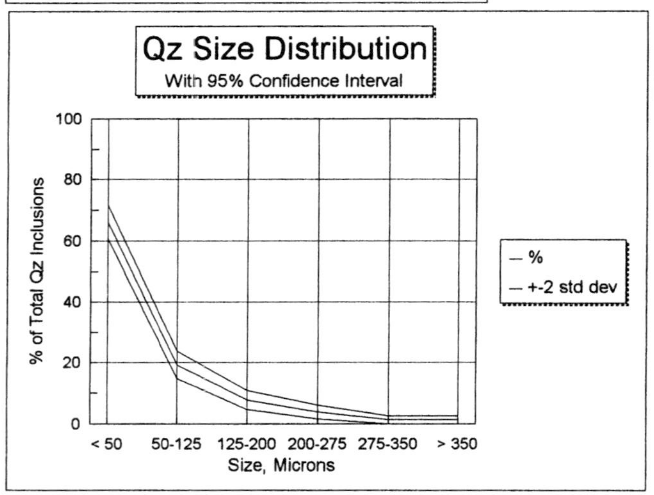

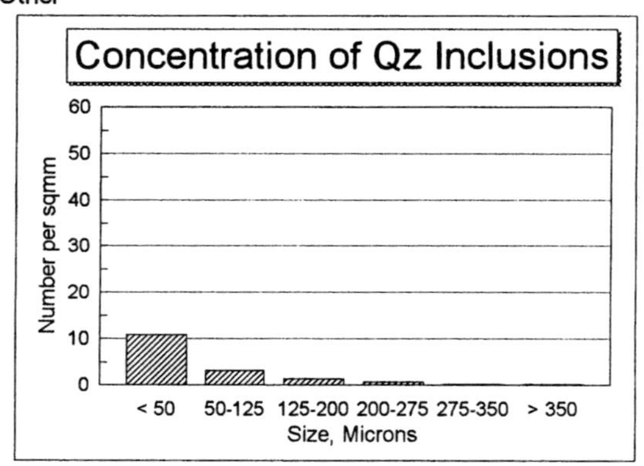

%matrix	74.31
%sand	16.12
%silt	9.57
conc<50	10.77
conc>50	5.49
%<50	66.23
%>50	33.77

RHD 9	period	level	morph type	rim dia	thick	sqmm
B IV/19	1	14	BI	12	7	18.75

Qz Size	< 50	50-125	125-200	200-275	275-350	> 350	numpts
Number	342	24	25	9	5	1	406
%	84.24	5.91	6.16	2.22	1.23	0.25	
sigma	1.81	1.17	1.19	0.73	0.55	0.25	
2psigma	87.85	8.25	8.54	3.68	2.33	0.74	
2msigma	80.62	3.57	3.77	0.76	0.14	-0.25	
Conc	18.24	1.28	1.33	0.48	0.27	0.05	
Fabric	Matrix	Sand	Silt	Micac	Biref	Colour	
	313	36	47	3	3	medbrn	396
Minerals	Size	Ang	Sph	Wth	Conc	Type	CarOpa
PFs	3				3		
MFs					1		
Bio	4			3	4		
Mus	4			3	4		
Amp	2				3		
Car					4	2	2
Quartz		3L 2S	2L 3S			m,p	
Pyrox					2		
Epidote					2		
Schist					3		
Other							

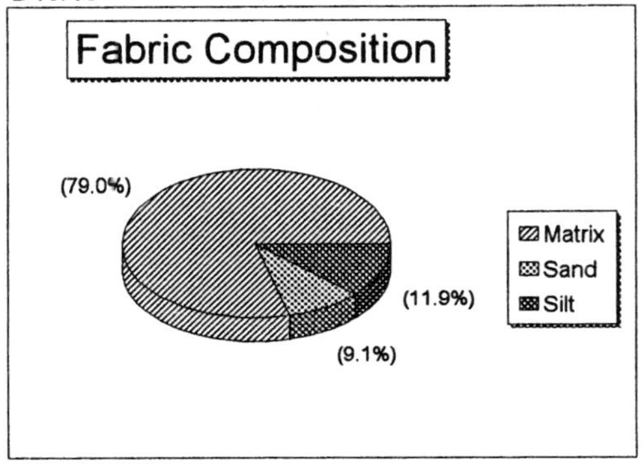

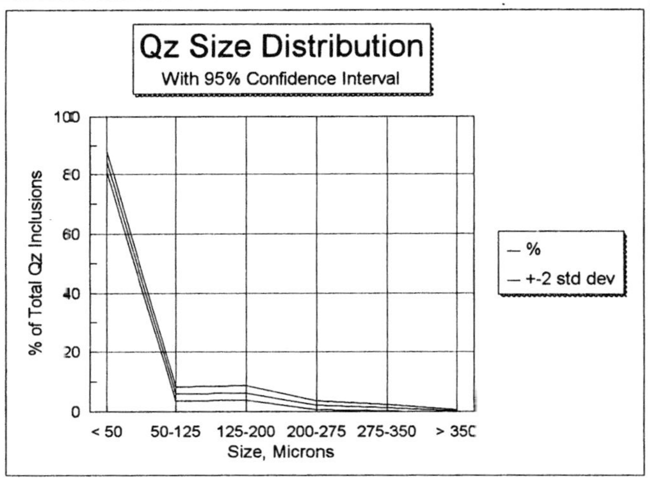

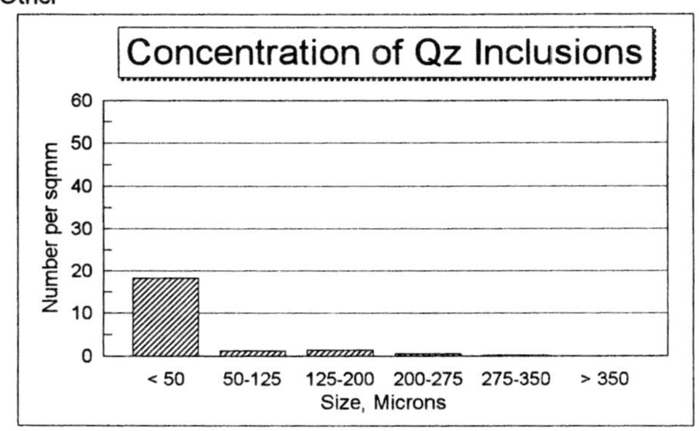

%matrix	79.04
%sand	9.09
%silt	11.87
conc<50	18.24
conc>50	3.41
%<50	84.24
%>50	15.76

APPENDIX A

rhd10	period	level	morphtype	rimdia	thick
BIV/15	3	2	BI		5

Section damaged; not reliable

APPENDIX A

RHD 11	period	level	morph type	rim dia	thick	sqmm
B IV/4	1	18	BI	14	7	25

Qz Size	< 50	50-125	125-200	200-275	275-350	> 350	numpts
Number	122	31	33	12	8	3	209
%	58.37	14.83	15.79	5.74	3.83	1.44	
sigma	3.41	2.46	2.52	1.61	1.33	0.82	
2psigma	65.19	19.75	20.83	8.96	6.48	3.08	
2msigma	51.55	9.92	10.74	2.52	1.17	-0.21	
Conc	4.88	1.24	1.32	0.48	0.32	0.12	
Fabric	Matrix	Sand	Silt	Micac	Biref	Colour	
	328	44	28	2	2	dkbrn	400

Minerals	Size	Ang	Sph	Wth	Conc	Type	CarOpa
PFs	2				2		
MFs					1		
Bio	3			3	2		
Mus	3			3	4		
Amp	2				3		
Car	3				4	1,2	2
Quartz		3L 2S	2			m,p	
Pyrox					1		
Epidote					2		
Schist					3		
Other							

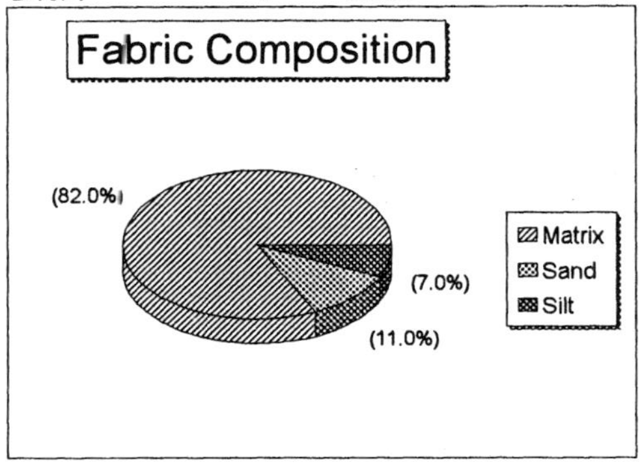

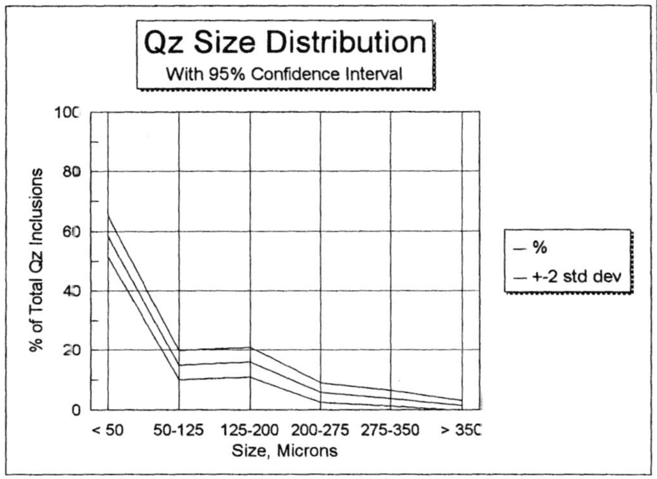

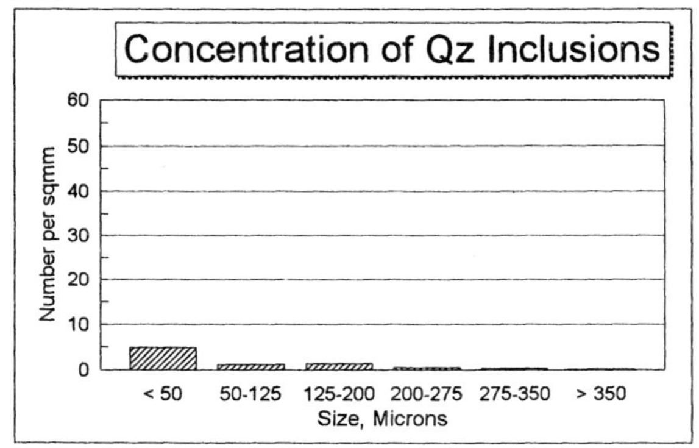

%matrix	82.00
%sand	11.00
%silt	7.00
conc<50	4.88
conc>50	3.48
%<50	58.37
%>50	41.63

APPENDIX A

RHD 12	period	level	morph type	rim dia	thick	sqmm
B IV/23	3	5	BI	11	6	12.5

Qz Size	< 50	50-125	125-200	200-275	275-350	> 350	numpts
Number	190	52	21	7	1	0	271
%	70.11	19.19	7.75	2.58	0.37	0.00	
sigma	2.78	2.39	1.62	0.96	0.37	0.00	
2psigma	75.67	23.97	11.00	4.51	1.11	0.00	
2msigma	64.55	14.40	4.50	0.66	-0.37	0.00	
Conc	15.20	4.16	1.68	0.56	0.08	0	
Fabric	Matrix	Sand	Silt	Micac	Biref	Colour	
	320	36	42	2	2	dkbrn	398

Minerals	Size	Ang	Sph	Wth	Conc	Type	CarOpa
PFs	3				2		
MFs					1		
Bio	2			2	2		
Mus	3			2	4		
Amp	2				2		
Car					3	ind	
Quartz		4L 2S	2			m,p	
Pyrox					1		
Epidote					1		
Schist					2	amph	
Other						arf	

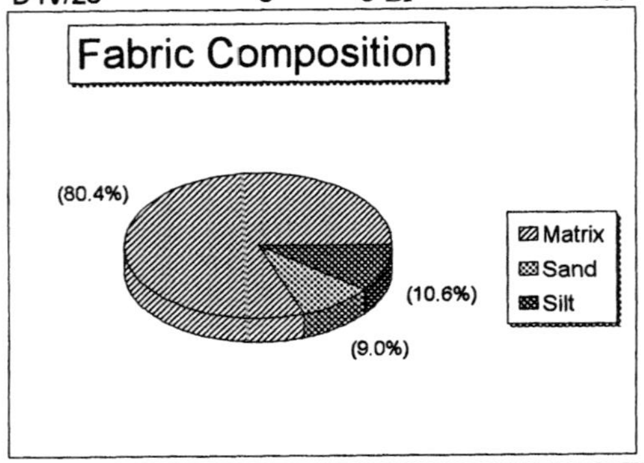

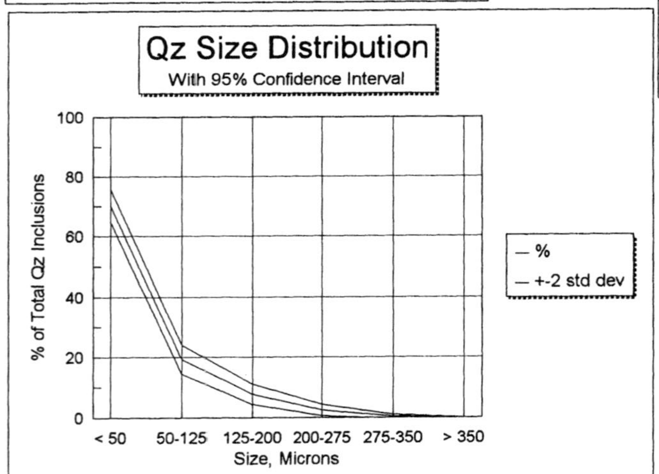

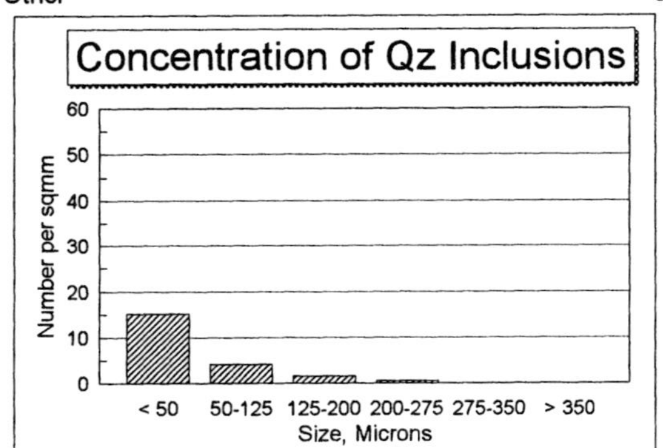

%matrix	80.40
%sand	9.05
%silt	10.55
conc<50	15.20
conc>50	6.48
%<50	70.11
%>50	29.89

RHD 13	period	level	morph type	rim dia	thick	sqmm
B IV/4	1	18	BI	14	5	18.75

Qz Size	< 50	50-125	125-200	200-275	275-350	> 350	numpts
Number	106	66	29	11	0	0	212
%	50.00	31.13	13.68	5.19	0.00	0.00	
sigma	3.43	3.18	2.36	1.52	0.00	0.00	
2psigma	56.87	37.49	18.40	8.24	0.00	0.00	
2msigma	43.13	24.77	8.96	2.14	0.00	0.00	
Conc	5.65	3.52	1.55	0.59	0	0	
Fabric	Matrix	Sand	Silt	Micac	Biref	Colour	
	342	38	18	3	2	medbrn	398

Minerals	Size	Ang	Sph	Wth	Conc	Type	CarOpa
PFs	2				2		
MFs					1		
Bio	2			2	2		
Mus	1			2	4		
Amp	2				2		
Car	3				4	1,2	2
Quartz		3	3			m,p	
Pyrox					1		
Epidote					2		
Schist					3		
Other							

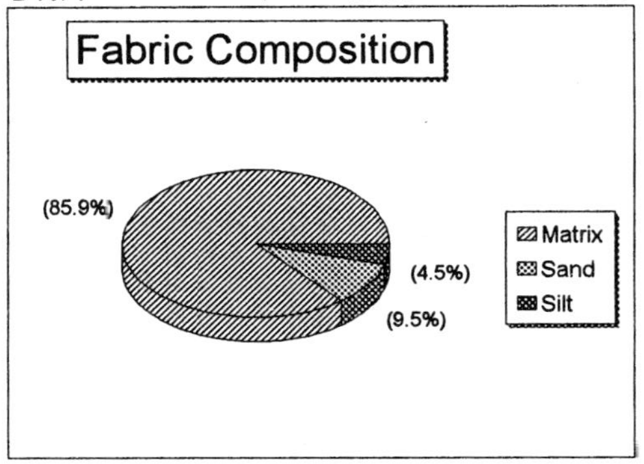

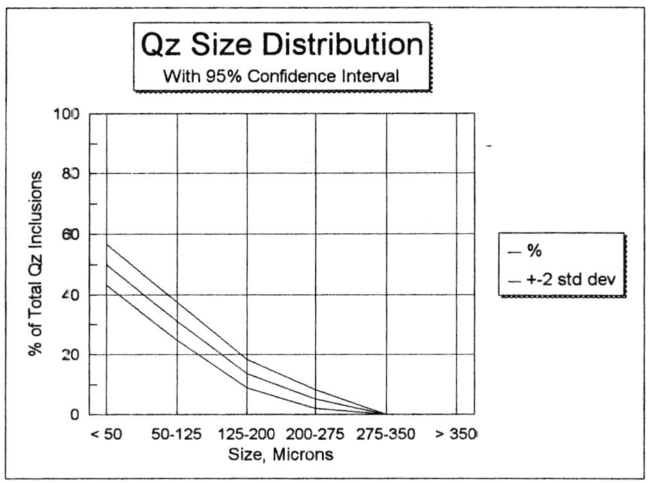

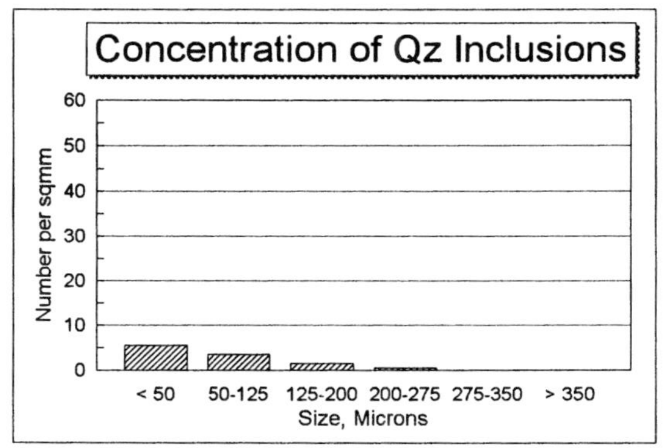

%matrix	85.93
%sand	9.55
%silt	4.52
conc<50	5.65
conc>50	5.65
%<50	50.00
%>50	50.00

RHD 14	period	level	morph type	rim dia	thick	sqmm
B IV/23	3	5	BI	15	6	18.75

Qz Size	< 50	50-125	125-200	200-275	275-350	> 350	numpts
Number	205	77	25	4	0	1	312
%	65.71	24.68	8.01	1.28	0.00	0.32	
sigma	2.69	2.44	1.54	0.64	0.00	0.32	
2psigma	71.08	29.56	11.09	2.56	0.00	0.96	
2msigma	60.33	19.80	4.94	0.01	0.00	-0.32	
Conc	10.93	4.11	1.33	0.21	0.00	0.05	
Fabric	Matrix	Sand	Silt	Micac	Biref	Colour	
	323	28	42	1	1	dkbrn	393

Minerals	Size	Ang	Sph	Wth	Conc	Type	CarOpa
PFs					2		
MFs					1		
Bio					1		
Mus	2			2	4		
Amp					1		
Car	3				3	2,3	2
Quartz		3	2			m,p	
Pyrox					1		
Epidote					2		
Schist					2		
Other						arf,clay	

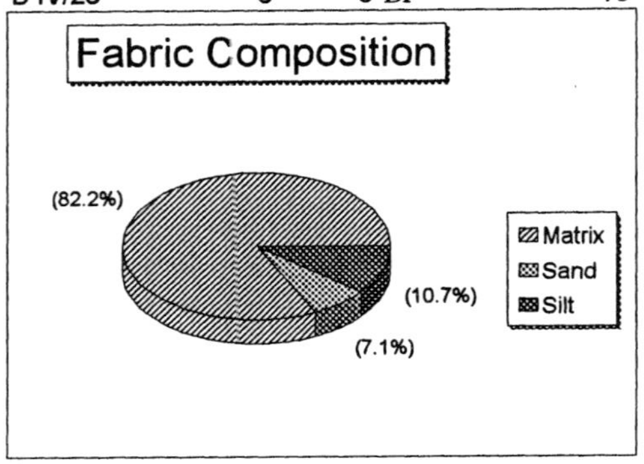

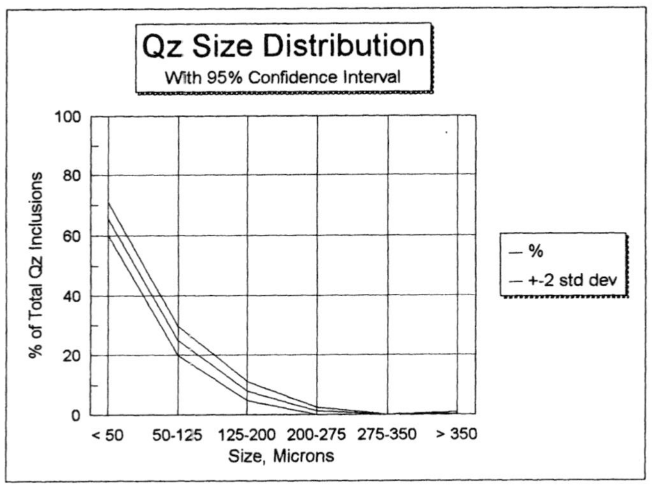

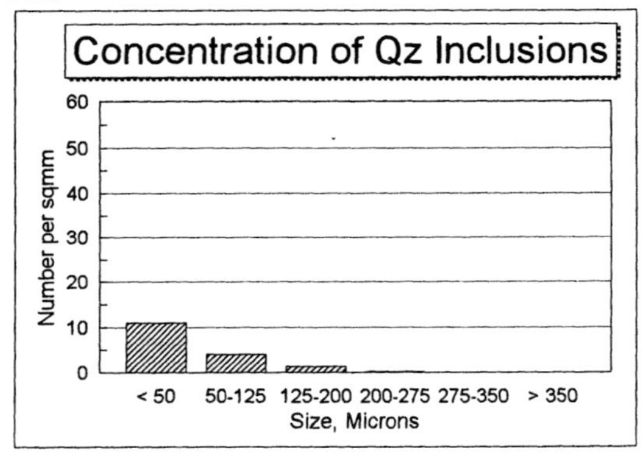

%matrix	82.19
%sand	7.12
%silt	10.69
conc<50	10.93
conc>50	5.71
%<50	65.71
%>50	34.29

RHD 15	period	level	morph type	rim dia	thick	sqmm
B IV/4	1	18	BI	10	6	18.75

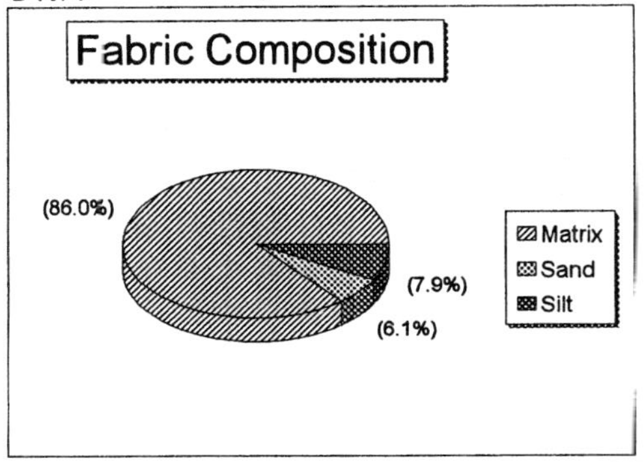

Qz Size	< 50	50-125	125-200	200-275	275-350	> 350	numpts
Number	251	77	22	6	0	0	356
%	70.51	21.63	6.18	1.69	0.00	0.00	
sigma	2.42	2.18	1.28	0.68	0.00	0.00	
2psigma	75.34	25.99	8.73	3.05	0.00	0.00	
2msigma	65.67	17.27	3.63	0.32	0.00	0.00	
Conc	13.39	4.11	1.17	0.32	0	0	

Fabric	Matrix	Sand	Silt	Micac	Biref	Colour	
	339	24	31	4	3	ltbrn	394

Minerals	Size	Ang	Sph	Wth	Conc	Type	CarOpa
PFs					1		
MFs					1		
Bio	2			2	2		
Mus	1			2	4		
Amp					1		
Car	2				4	2,3	2,3
Quartz		3	2			m,p	
Pyrox					1		
Epidote					2		
Schist					1		
Other						arf	

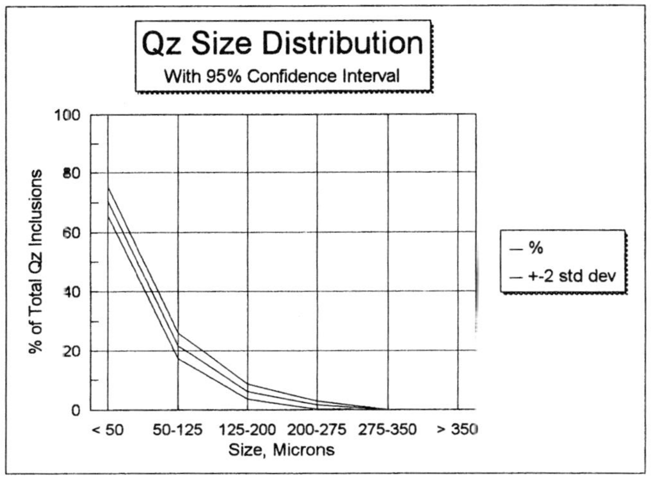

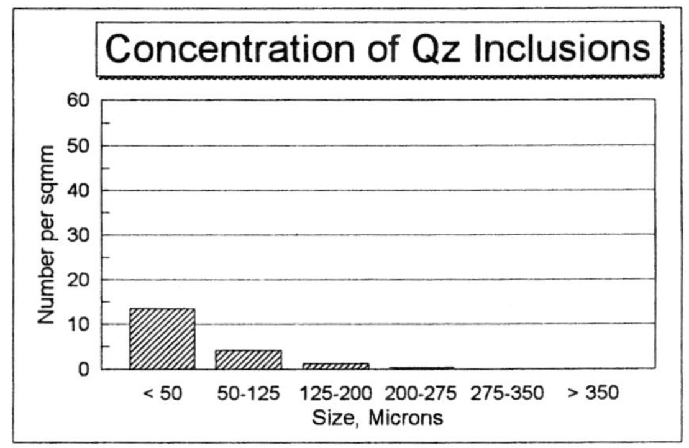

%matrix	86.04
%sand	6.09
%silt	7.87
conc<50	13.39
conc>50	5.60
%<50	70.51
%>50	29.49

APPENDIX A

RHD 16	period	level	morph type	rim dia	thick	sqmm
B IV/14	2	11	BI	16	6	18.75

Qz Size	< 50	50-125	125-200	200-275	275-350	> 350	numpts
Number	96	90	39	15	8	4	252
%	38.10	35.71	15.48	5.95	3.17	1.59	
sigma	3.06	3.02	2.28	1.49	1.10	0.79	
2psigma	44.21	41.75	20.03	8.93	5.38	3.16	
2msigma	31.98	29.68	10.92	2.97	0.97	0.01	
Conc	5.12	4.80	2.08	0.80	0.43	0.21	
Fabric	Matrix	Sand	Silt	Micac	Biref	Colour	
	278	83	30	2	2	dkbrn	391
Minerals	Size	Ang	Sph	Wth	Conc	Type	CarOpa
PFs	3			2	3		
MFs					1		
Bio	3			2	3		
Mus	3			2	2		
Amp	3				3		
Car	3				4	2	2
Quartz		4L 3S	2			m,p	
Pyrox					1		
Epidote					2		
Schist					3	amph	
Other						arf	

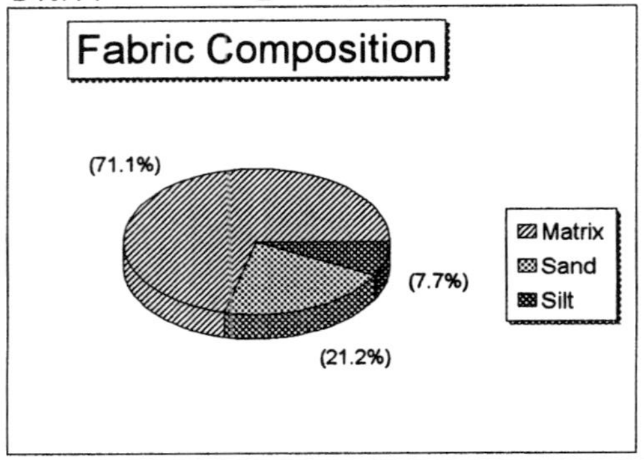

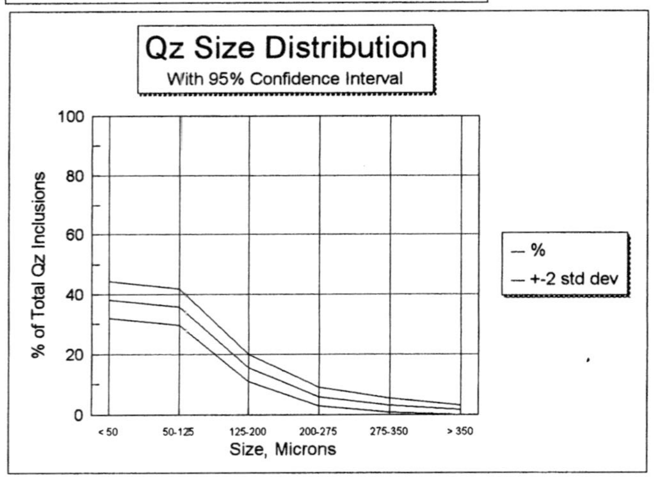

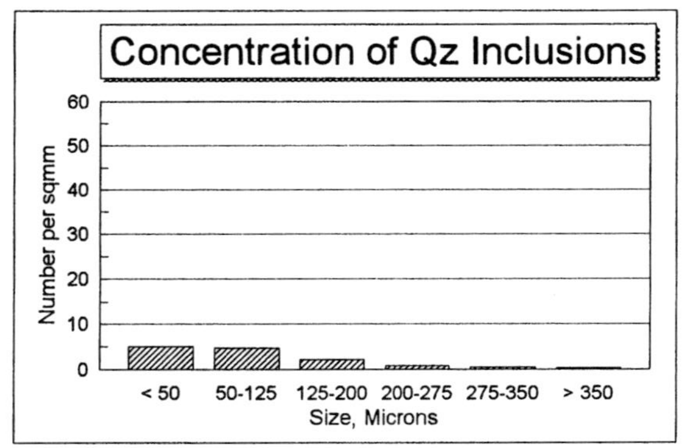

%matrix	71.10
%sand	21.23
%silt	7.67
conc<50	5.12
conc>50	8.32
%<50	38.10
%>50	61.90

RHD 17	period	level	morph type	rim dia	thick	sqmm
B IV/14	2	11	BI	18	6	18.75

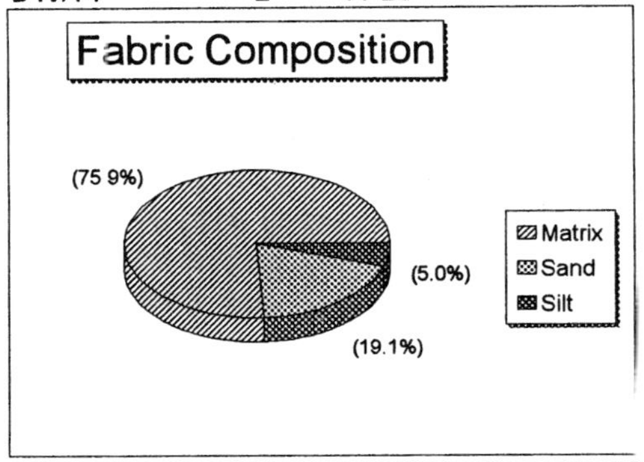

Qz Size	< 50	50-125	125-200	200-275	275-350	> 350	numpts
Number	142	82	53	13	6	0	296
%	47.97	27.70	17.91	4.39	2.03	0.00	
sigma	2.90	2.60	2.23	1.19	0.82	0.00	
2psigma	53.78	32.91	22.36	6.77	3.67	0.00	
2msigma	42.17	22.50	13.45	2.01	0.39	0.00	
Conc	7.57	4.37	2.83	0.69	0.32	0	
Fabric	Matrix	Sand	Silt	Micac	Biref	Colour	
	302	76	20	2	2	dkbrn	398

Minerals	Size	Ang	Sph	Wth	Conc	Type	CarOpa
PFs	3	3			3		
MFs					2		
Bio	3			3	3		
Mus	3			3	4		
Amp	2				4		
Car	3				4	2,3	2
Quartz		3L 2S	2			m,p	
Pyrox					2		
Epidote					2		
Schist					3	biotite	
Other						arf	

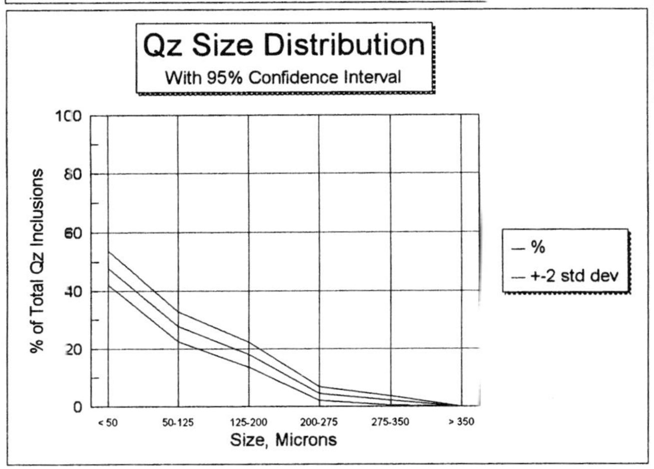

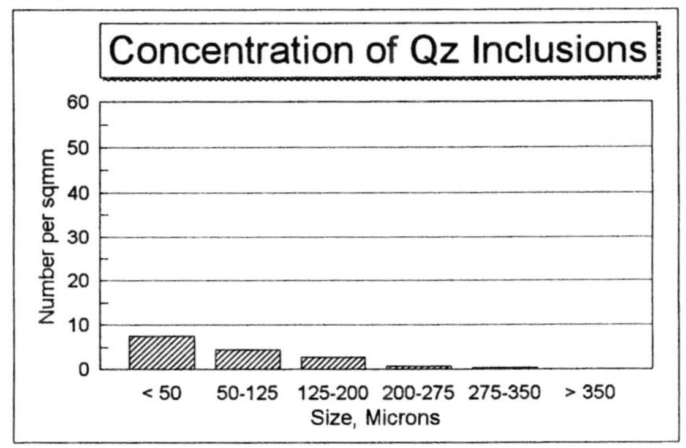

%matrix	75.88
%sand	19.10
%silt	5.03
conc<50	7.57
conc>50	8.21
%<50	47.97
%>50	52.03

APPENDIX A

RHD 18	period	level	morph type	rim dia	thick	sqmm
B IV/24	1	17	BI	14	5	18.75

Qz Size	< 50	50-125	125-200	200-275	275-350	> 350	numpts
Number	243	123	27	5	0	0	398
%	61.06	30.90	6.78	1.26	0.00	0.00	
sigma	2.44	2.32	1.26	0.56	0.00	0.00	
2psigma	65.94	35.54	9.30	2.37	0.00	0.00	
2msigma	56.17	26.27	4.26	0.14	0.00	0.00	
Conc	12.96	6.56	1.44	0.27	0	0	
Fabric	Matrix	Sand	Silt	Micac	Biref	Colour	
	304	50	34	3	3	medbrn	388

Minerals	Size	Ang	Sph	Wth	Conc	Type	CarOpa
PFs	1				2		
MFs					1		
Bio	2			3	2		
Mus	2			3	2		
Amp	2				2		
Car	3				4	1,2,3	2
Quartz		3	3			m,p	
Pyrox					1		
Epidote					2		
Schist					2		
Other						arf	

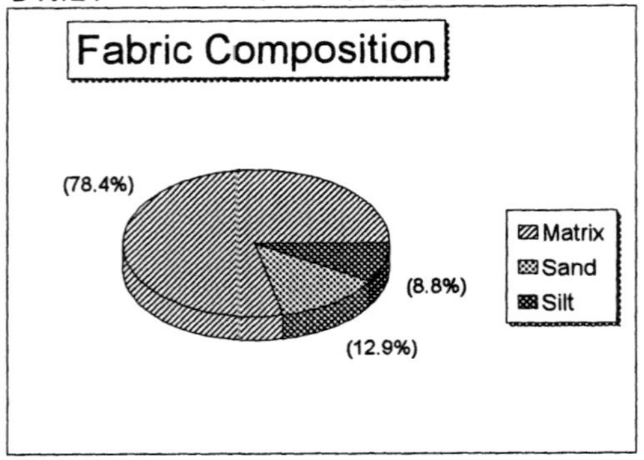

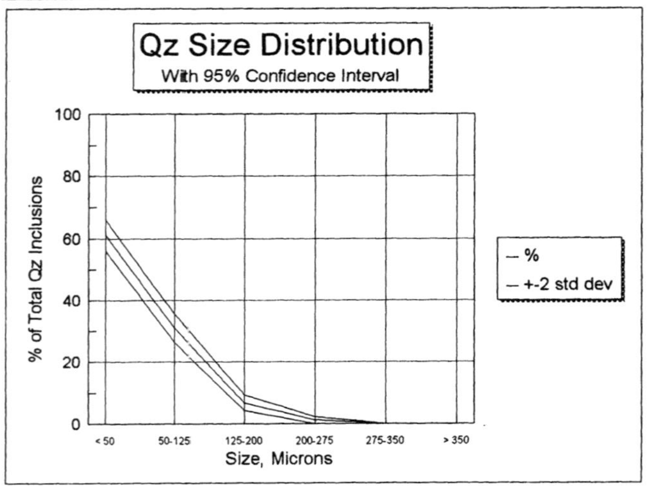

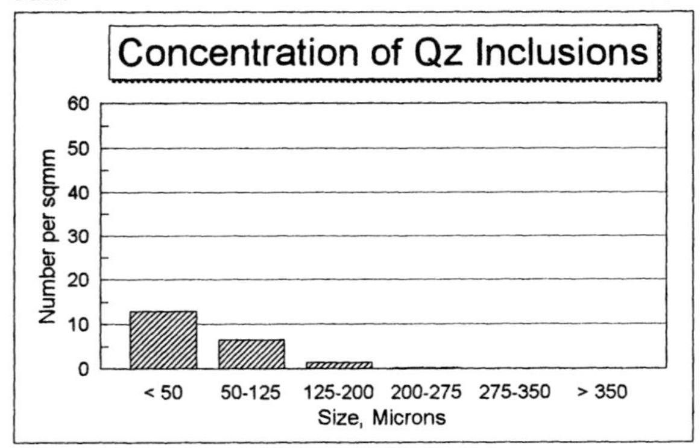

%matrix	78.35
%sand	12.89
%silt	8.76
conc<50	12.96
conc>50	8.27
%<50	61.06
%>50	38.94

RHD 19	period	level	morph type	rim dia	thick	sqmm
B IV/19	3	1	BV	38	8	25

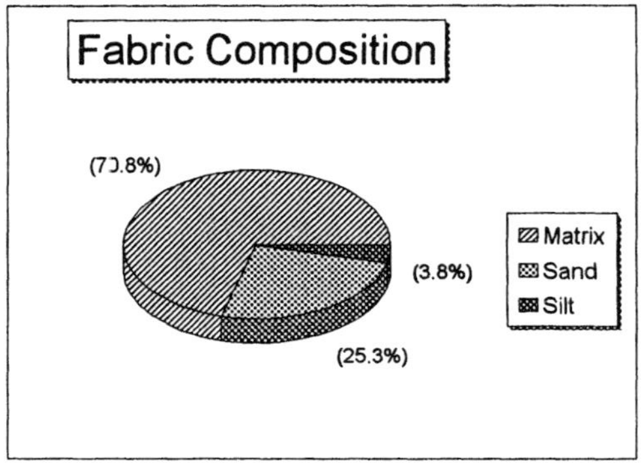

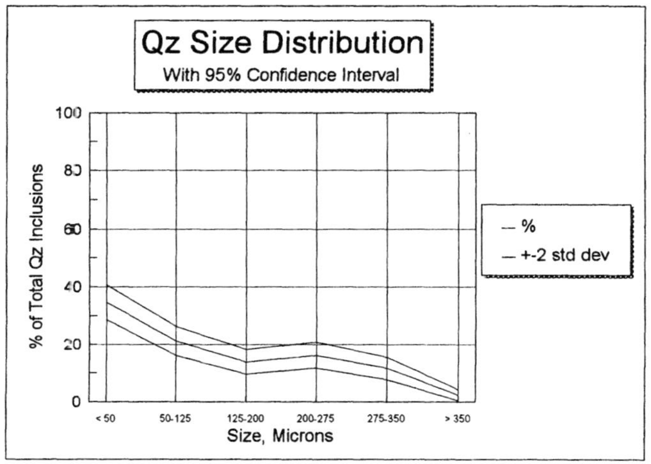

Qz Size	< 50	50-125	125-200	200-275	275-350	> 350	numpts
Number	90	55	36	42	30	6	259
%	34.75	21.24	13.90	16.22	11.58	2.32	
sigma	2.96	2.54	2.15	2.29	1.99	0.93	
2psigma	40.67	26.32	18.20	20.80	15.56	4.19	
2msigma	28.83	16.15	9.60	11.64	7.61	0.45	
Conc	3.60	2.20	1.44	1.68	1.20	0.24	
Fabric	Matrix	Sand	Silt	Micac	Biref	Colour	
	277	99	15	1	1	dkbrn	391
Minerals	Size	Ang	Sph	Wth	Conc	Type	CarOpa
PFs	4	3			4		
MFs					1		
Bio	4			3	3		
Mus	2			3	4		
Amp	3				4		
Car	2				2		
Quartz		3	2			m,p	
Pyrox					2		
Epidote					2		
Schist					3		
Other						arf	

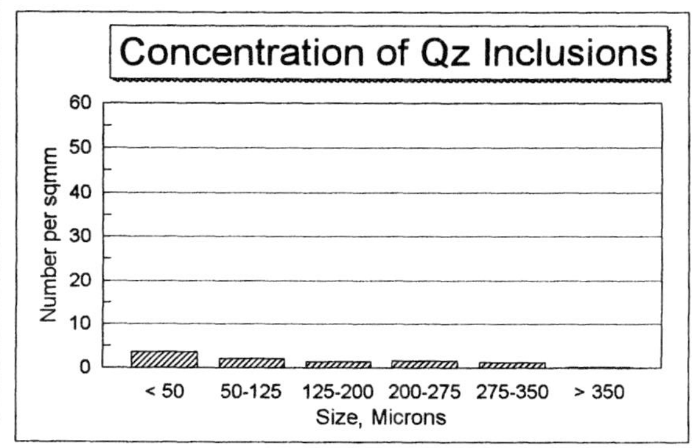

%matrix	70.84
%sand	25.32
%silt	3.84
conc<50	3.60
conc>50	6.76
%<50	34.75
%>50	65.25

RHD 20	period	level	morph type	rim dia	thick	sqmm
B IV/10	3	5	BV	40	9	25

Qz Size	< 50	50-125	125-200	200-275	275-350	> 350	numpts
Number	114	65	50	29	6	6	270
%	42.22	24.07	18.52	10.74	2.22	2.22	
sigma	3.01	2.60	2.36	1.88	0.90	0.90	
2psigma	48.23	29.28	23.25	14.51	4.02	4.02	
2msigma	36.21	18.87	13.79	6.97	0.43	0.43	
Conc	4.56	2.60	2.00	1.16	0.24	0.24	
Fabric	Matrix	Sand	Silt	Micac	Biref	Colour	
	322	53	18	3	3	medbrn	393

Minerals	Size	Ang	Sph	Wth	Conc	Type	CarOpa
PFs	3	3			3		
MFs	3				2		
Bio	4			3	4		
Mus	4			3	4		
Amp	2				2		
Car	2	2	2,3		2	2,3	2
Quartz						m,p	
Pyrox					2		
Epidote					1		
Schist					2		
Other							

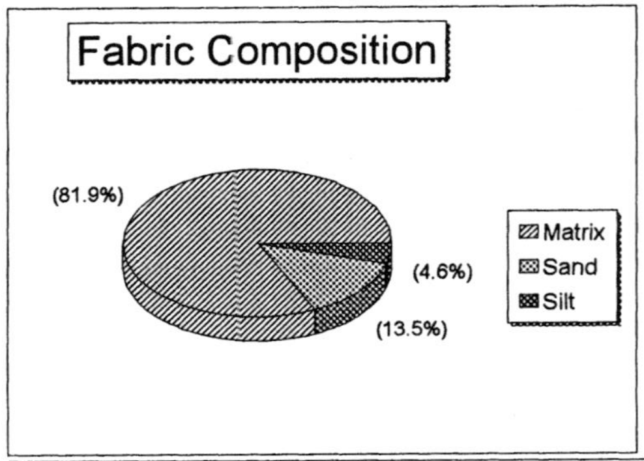

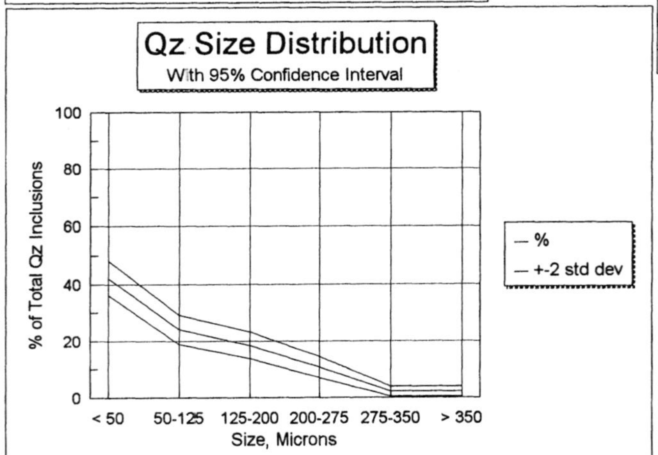

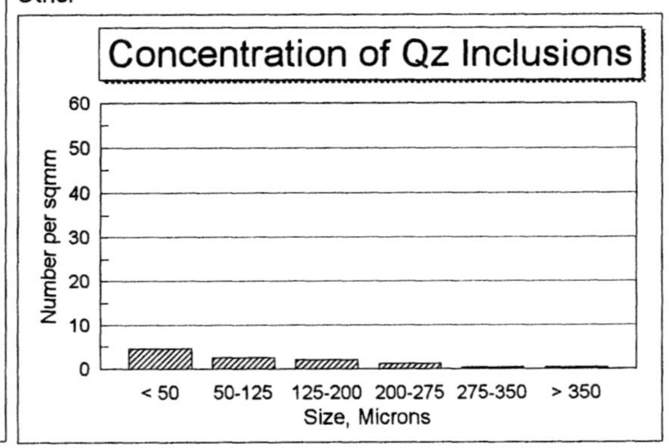

%matrix	81.93
%sand	13.49
%silt	4.58
conc<50	4.56
conc>50	6.24
%<50	42.22
%>50	57.78

RHD 21	period	level	morph type	rim dia	thick	sqmm
B IV/19	1	14	BV	34	10	25

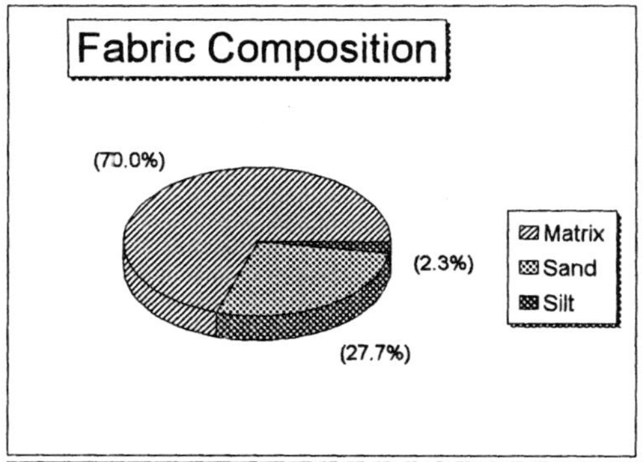

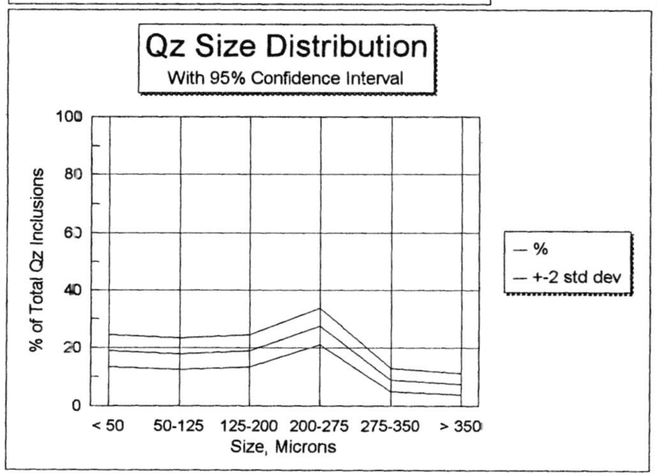

Qz Size	< 50	50-125	125-200	200-275	275-350	> 350	numpts
Number	38	36	38	55	18	15	200
%	19.00	18.00	19.00	27.50	9.00	7.50	
sigma	2.77	2.72	2.77	3.16	2.02	1.86	
2psigma	24.55	23.43	24.55	33.81	13.05	11.22	
2msigma	13.45	12.57	13.45	21.19	4.95	3.78	
Conc	1.52	1.44	1.52	2.20	0.72	0.60	
Fabric	Matrix	Sand	Silt	Micac	Biref	Colour	
	268	106	9	2	3	dkbrn	383
Minerals	Size	Ang	Sph	Wth	Conc	Type	CarOpa
PFs	4	3	3		4		
MFs					1		
Bio	3			3	3		
Mus	3			3	3		
Amp	3				3		
Car	3				3	1,2,3	2
Quartz		3	3			m,p	
Pyrox					2		
Epidote					2		
Schist					4		
Other							

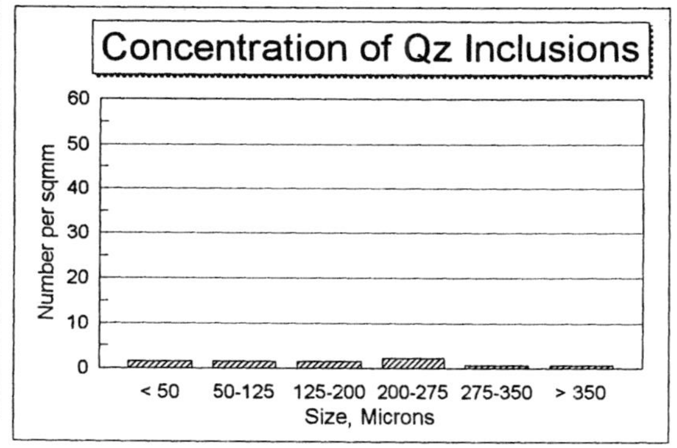

%matrix	69.97
%sand	27.68
%silt	2.35
conc<50	1.52
conc>50	6.48
%<50	19.00
%>50	81.00

APPENDIX A

RHD 22	period	level	morph type	rim dia	thick	sqmm
B IV/14	1	16	BV		9	12.5

Qz Size	< 50	50-125	125-200	200-275	275-350	> 350	numpts
Number	493	82	14	6	0	0	595
%	82.86	13.78	2.35	1.01	0.00	0.00	
sigma	1.55	1.41	0.62	0.41	0.00	0.00	
2psigma	85.95	16.61	3.60	1.83	0.00	0.00	
2msigma	79.77	10.96	1.11	0.19	0.00	0.00	
Conc	39.44	6.56	1.12	0.48	0	0	
Fabric	Matrix	Sand	Silt	Micac	Biref	Colour	
	341	19	37	1	1	medbrn	397

Minerals	Size	Ang	Sph	Wth	Conc	Type	CarOpa
PFs	2				2		
MFs					1		
Bio	2			2	2		
Mus	3			2	4		
Amp					1		
Car	2				2	2	2
Quartz		2	3			m,p	
Pyrox					1		
Epidote					1		
Schist					1		
Other							

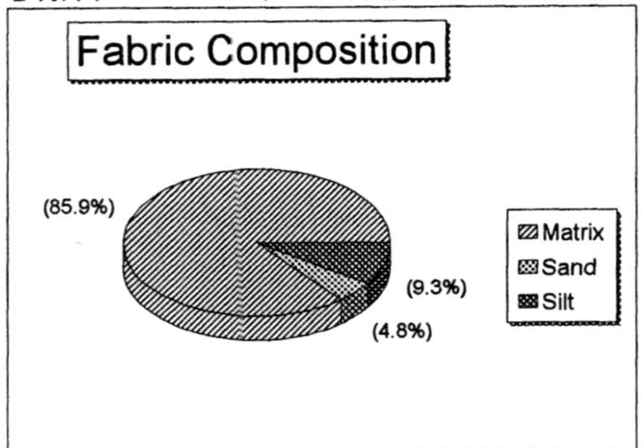

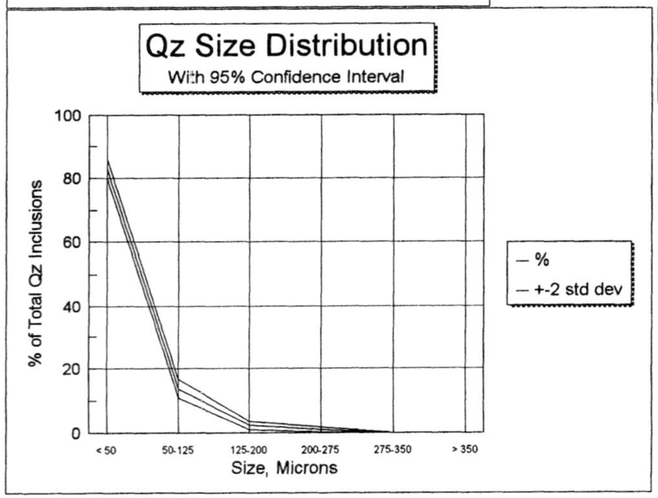

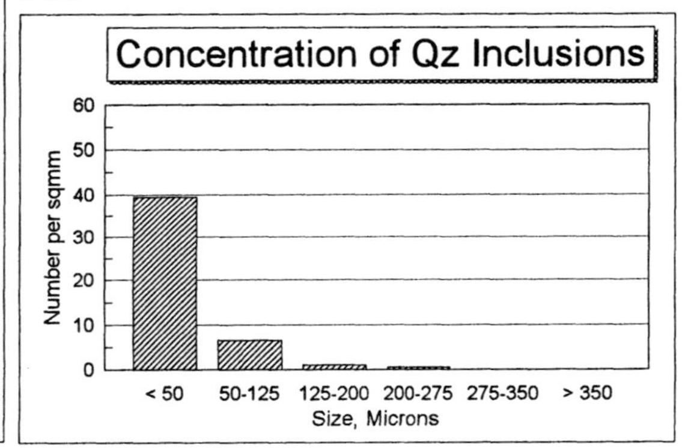

%matrix	85.89
%sand	4.79
%silt	9.32
conc<50	39.44
conc>50	8.16
%<50	82.86
%>50	17.14

RHD 23	period	level	morph type	rim dia	thick	sqmm
B IV/19	2	11	BV	38	7	25

Qz Size	< 50	50-125	125-200	200-275	275-350	> 350	numpts
Number	114	72	64	15	1	1	267
%	42.70	26.97	23.97	5.62	0.37	0.37	
sigma	3.03	2.72	2.61	1.41	0.37	0.37	
2psigma	48.75	32.40	29.20	8.44	1.12	1.12	
2msigma	36.64	21.53	18.74	2.80	-0.37	-0.37	
Conc	4.56	2.88	2.56	0.60	0.04	0.04	
Fabric	Matrix	Sand	Silt	Micac	Biref	Colour	
	332	34	22	3	3	medbrn	388

Minerals	Size	Ang	Sph	Wth	Conc	Type	CarOpa
PFs	2				2		
MFs					1		
Bio	3			3	2		
Mus	2			3	4		
Amp					1		
Car	2				3	2,3	2
Quartz		2	3			m,p	
Pyrox					1		
Epidote					1		
Schist					2		
Other							

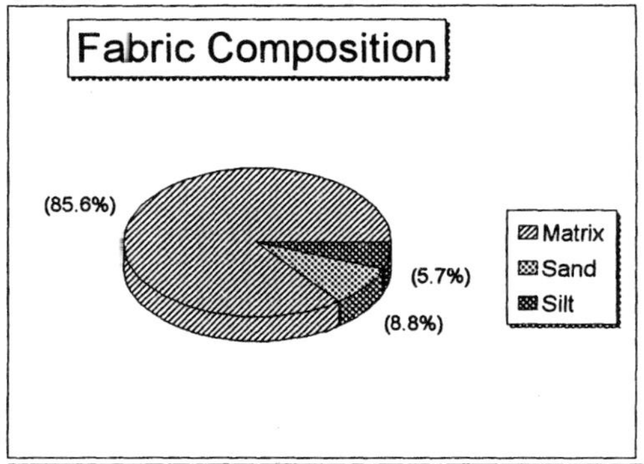

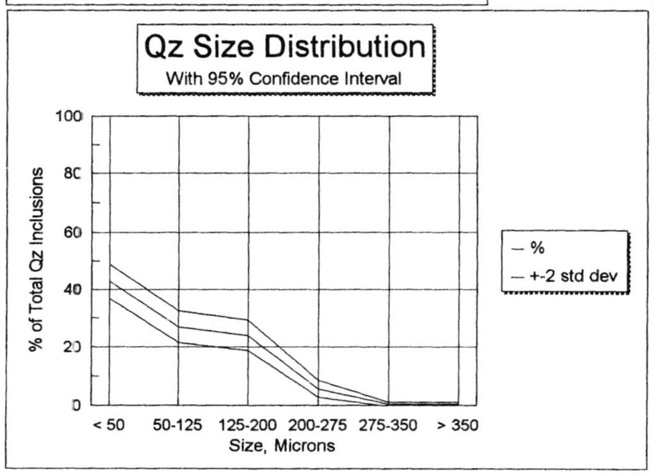

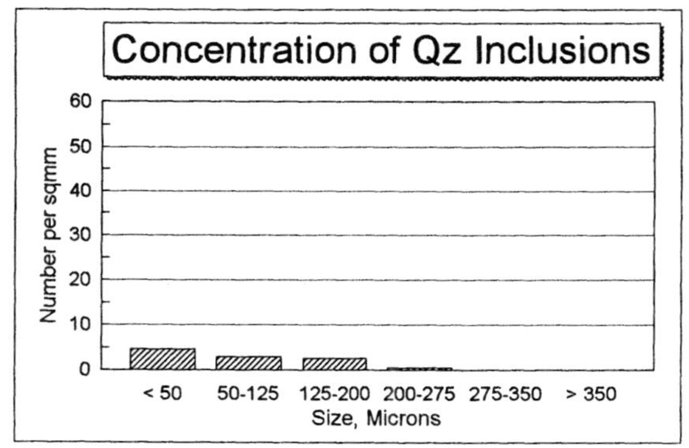

%matrix	85.57
%sand	8.76
%silt	5.67
conc<50	4.56
conc>50	6.12
%<50	42.70
%>50	57.30

APPENDIX A

RHD 24	period	level	morph type	rim dia	thick	sqmm
B IV/14	2	11	BV	31	8	18.75

Fabric Composition

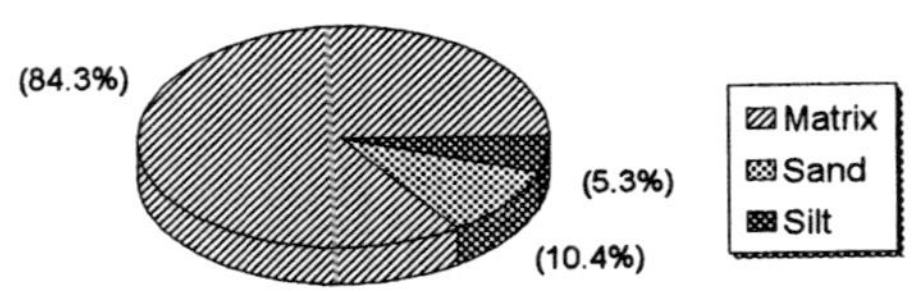

Qz Size	< 50	50-125	125-200	200-275	275-350	> 350	numpts
Number	130	77	50	17	3	1	278
%	46.76	27.70	17.99	6.12	1.08	0.36	
sigma	2.99	2.68	2.30	1.44	0.62	0.36	
2psigma	52.75	33.07	22.59	8.99	2.32	1.08	
2msigma	40.78	22.33	13.38	3.24	-0.16	-0.36	
Conc	6.93	4.11	2.67	0.91	0.16	0.05	
Fabric	Matrix	Sand	Silt	Micac	Biref	Colour	
	333	41	21	2	2	medbrn	395

Minerals	Size	Ang	Sph	Wth	Conc	Type	CarOpa
PFs	2	3	2		2		
MFs	2			4	2		
Bio					1		
Mus	1			3	4		
Amp	1				2		
Car	3				3	2,3	2
Quartz		3	2			m,p	
Pyrox					1		
Epidote					1		
Schist					1		
Other							

Qz Size Distribution

With 95% Confidence Interval

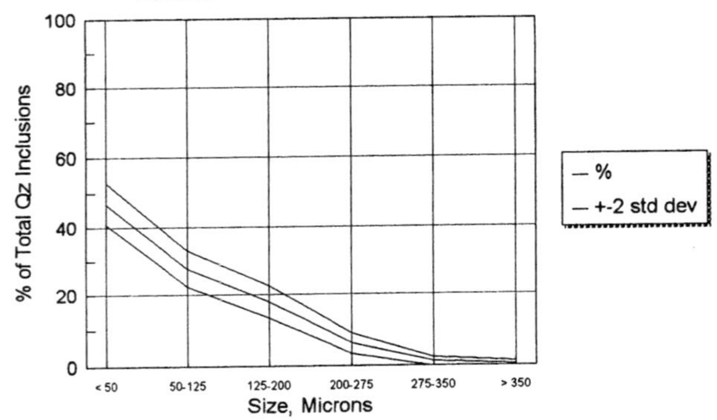

Concentration of Qz Inclusions

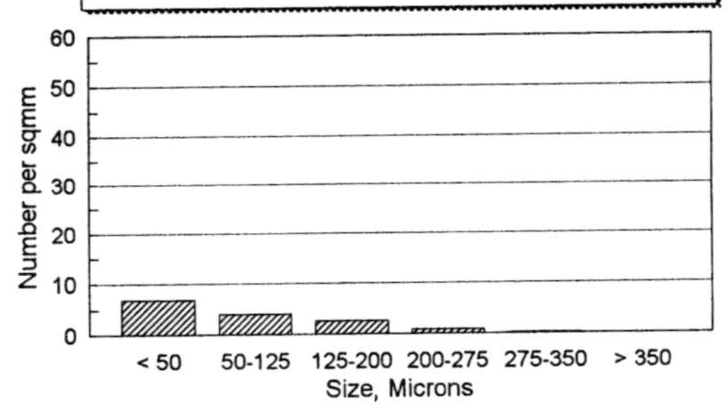

%matrix	84.30
%sand	10.38
%silt	5.32
conc<50	6.93
conc>50	7.89
%<50	46.76
%>50	53.24

RHD 25	period	level	morph type	rim dia	thick	sqmm
B IV/14	2	11	BII		11	18.75

Qz Size	< 50	50-125	125-200	200-275	275-350	> 350	numpts
Number	357	91	35	7	1	1	492
%	72.56	18.50	7.11	1.42	0.20	0.20	
sigma	2.01	1.75	1.16	0.53	0.20	0.20	
2psigma	76.58	22.00	9.43	2.49	0.61	0.61	
2msigma	68.54	15.00	4.80	0.35	-0.20	-0.20	
Conc	19.04	4.85	1.87	0.37	0.05	0.05	
Fabric	Matrix	Sand	Silt	Micac	Biref	Colour	
	313	40	42	4	4	medbrn	395

Minerals	Size	Ang	Sph	Wth	Conc	Type	CarOpa
PFs	1				2		
MFs					2		
Bio	2			3	2		
Mus	2			3	3		
Amp	3				2		
Car	3				4	2	2
Quartz		4	2			m,p	
Pyrox					1		
Epidote					1		
Schist					1		
Other							

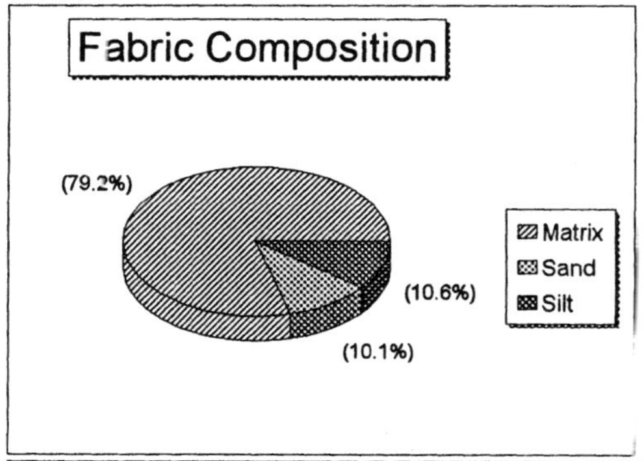

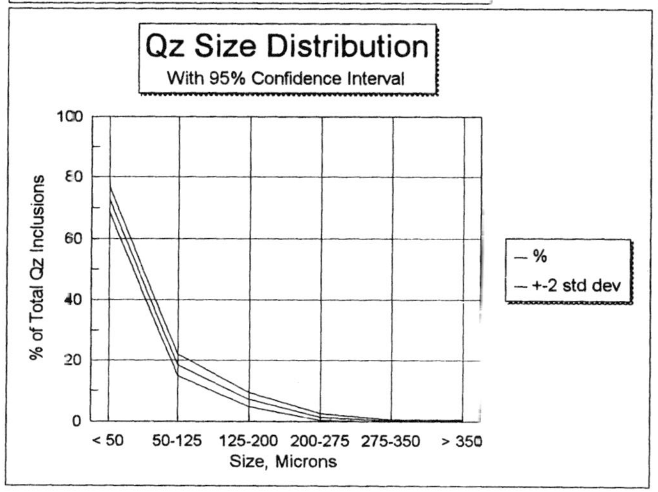

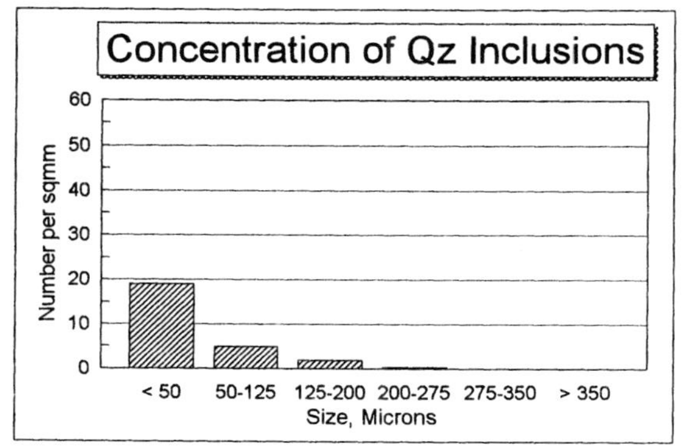

%matrix	79.24
%sand	10.13
%silt	10.63
conc<50	19.04
conc>50	7.20
%<50	72.56
%>50	27.44

RHD 26	period	level	morph type	rim dia	thick	sqmm
B IV/19	1	20	BII	44	11	18.75

Qz Size	< 50	50-125	125-200	200-275	275-350	> 350	numpts
Number	181	49	54	28	6	6	324
%	55.86	15.12	16.67	8.64	1.85	1.85	
sigma	2.76	1.99	2.07	1.56	0.75	0.75	
2psigma	61.38	19.10	20.81	11.76	3.35	3.35	
2msigma	50.35	11.14	12.53	5.52	0.35	0.35	
Conc	9.65	2.61	2.88	1.49	0.32	0.32	
Fabric	Matrix	Sand	Silt	Micac	Biref	Colour	
	290	69	26	2	2	dkbrn	385

Minerals	Size	Ang	Sph	Wth	Conc	Type	CarOpa
PFs	4	3	2		4		
MFs					1		
Bio	3			3	4		
Mus	4			3	4		
Amp	3				4		
Car	2				2	3	
Quartz		4L 3S	2L3S			m,p	
Pyrox					1		
Epidote					2		
Schist					3		

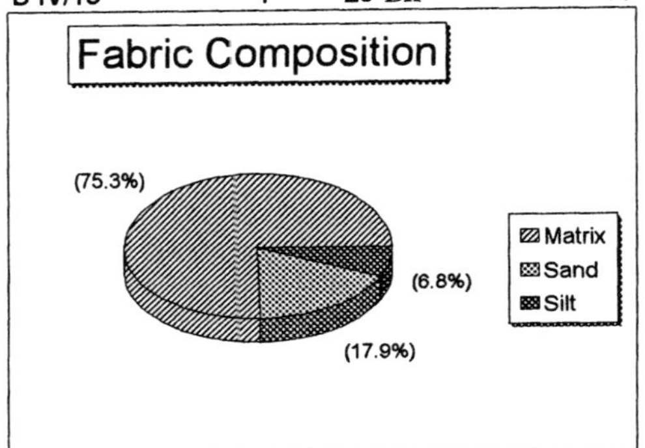

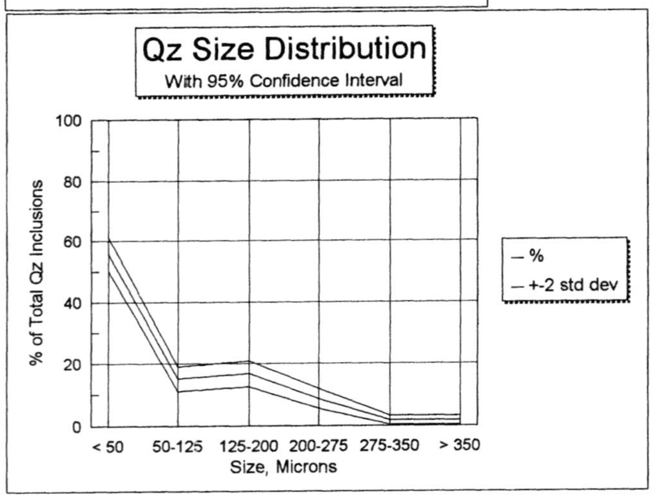

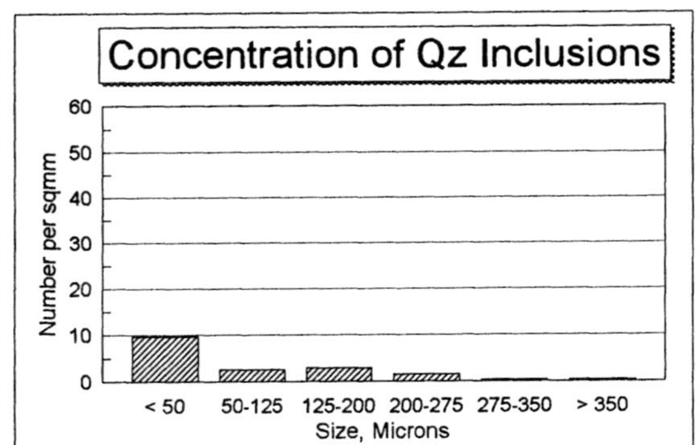

%matrix	75.32
%sand	17.92
%silt	6.75
conc<50	9.65
conc>50	7.63
%<50	55.86
%>50	44.14

APPENDIX A

RHD 27	period	level	morph type	rim dia	thick	sqmm
B IV/19	1	20	BII	34	8	18.75

Qz Size	< 50	50-125	125-200	200-275	275-350	> 350	numpts
Number	287	78	47	24	2	3	441
%	65.08	17.69	10.66	5.44	0.45	0.68	
sigma	2.27	1.82	1.47	1.08	0.32	0.39	
2psigma	69.62	21.32	13.60	7.60	1.09	1.46	
2msigma	60.54	14.05	7.72	3.28	-0.19	-0.10	
Conc	15.31	4.16	2.51	1.28	0.11	0.16	
Fabric	Matrix	Sand	Silt	Micac	Biref	Colour	
	294	57	34	2	2	medbr	385
Minerals	Size	Ang	Sph	Wth	Conc	Type	CarOpa
PFs	1				2		
MFs					1		
Bio	1			3	2		
Mus	2			2	3		
Amp					1		
Car	3				4	2	2
Quartz		3L 2S	3			m,p	
Pyrox					1		
Epidote					1		
Schist					2		

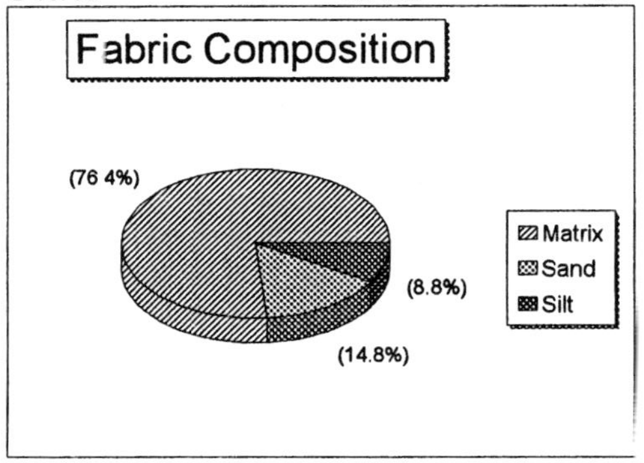

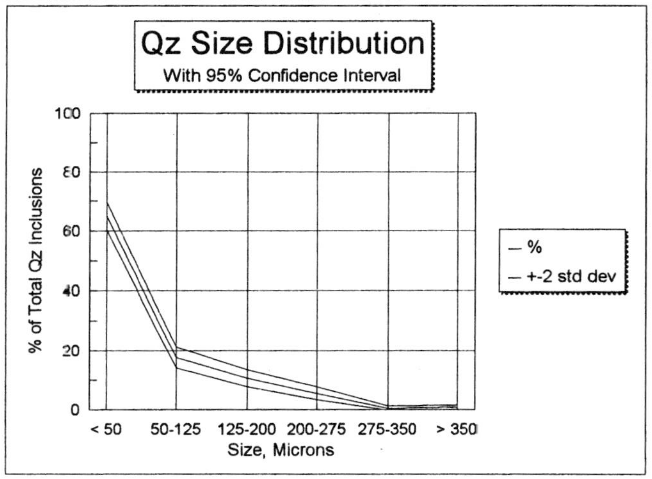

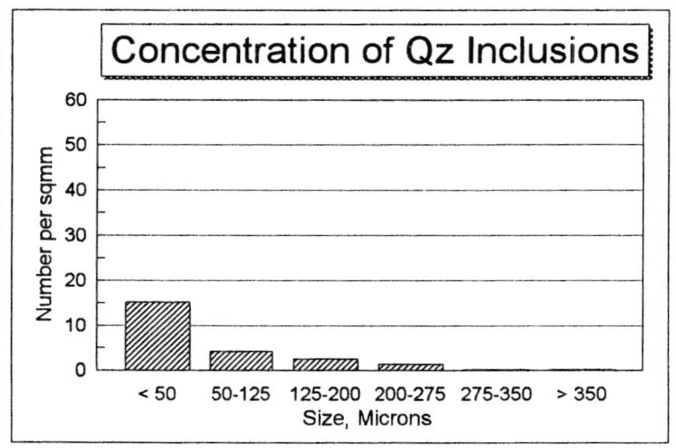

%matrix	76.36
%sand	14.81
%silt	8.83
conc<50	15.31
conc>50	8.21
%<50	65.08
%>50	34.92

APPENDIX A

RHD 28	period	level	morph type	rim dia	thick	sqmm
B IV/19	1	14	BIV	15	6	25

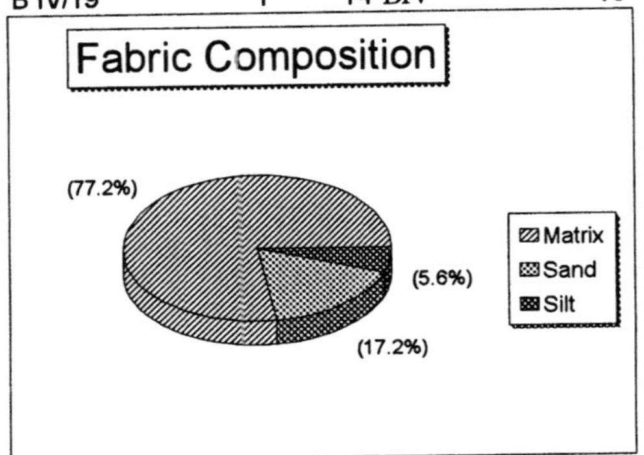

Qz Size	< 50	50-125	125-200	200-275	275-350	> 350	numpts
Number	153	51	50	35	11	3	303
%	50.50	16.83	16.50	11.55	3.63	0.99	
sigma	2.87	2.15	2.13	1.84	1.07	0.57	
2psigma	56.24	21.13	20.77	15.22	5.78	2.13	
2msigma	44.75	12.53	12.24	7.88	1.48	-0.15	
Conc	6.12	2.04	2.00	1.40	0.44	0.12	
Fabric	Matrix	Sand	Silt	Micac	Biref	Colour	
	305	68	22	2	2	medbr	395

Minerals	Size	Ang	Sph	Wth	Conc	Type	CarOpa
PFs	3	3	2		4		
MFs					1		
Bio	3			4	3		
Mus	3			3	4		
Amp	2				2		
Car	3				3	2	2
Quartz		3	2			m,p	
Pyrox					1		
Epidote					1		
Schist					3		

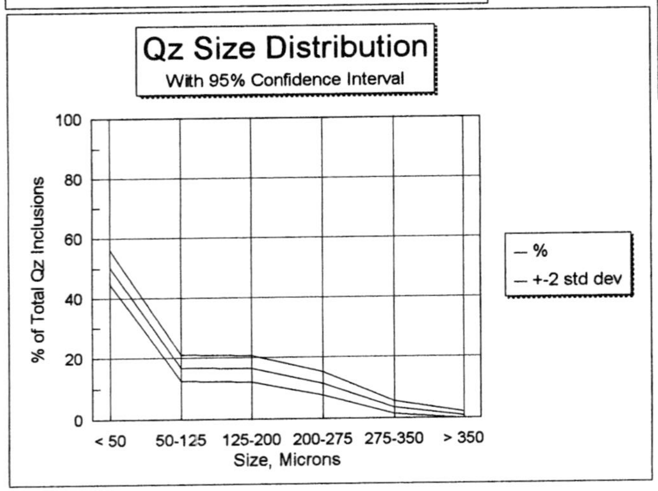

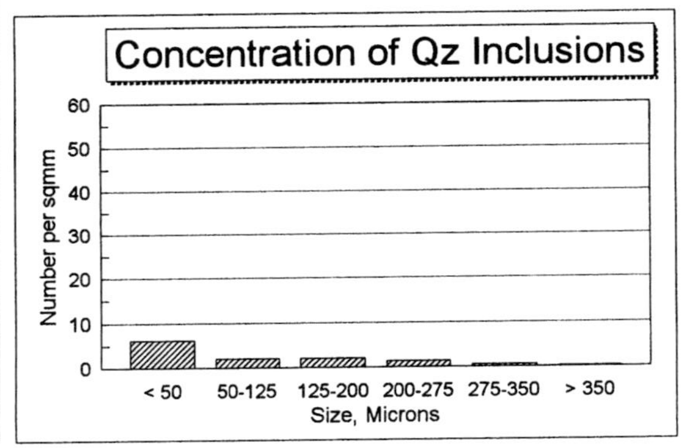

%matrix	77.22
%sand	17.22
%silt	5.57
conc<50	6.12
conc>50	6.00
%<50	50.50
%>50	49.50

APPENDIX A

RHD 29	period	level	morph type	rim dia	thick	sqmm
B IV/19	2	11	BIV		9	25

Qz Size	< 50	50-125	125-200	200-275	275-350	> 350	numpts
Number	70	64	52	14	2	1	203
%	34.48	31.53	25.62	6.90	0.99	0.49	
sigma	3.34	3.26	3.06	1.78	0.69	0.49	
2psigma	41.15	38.05	31.74	10.45	2.37	1.48	
2msigma	27.81	25.01	19.49	3.34	-0.40	-0.49	
Conc	2.80	2.56	2.08	0.56	0.08	0.04	
Fabric	Matrix	Sand	Silt	Micac	Biref	Colour	
	343	39	10	1	1	dkbrn	392

Minerals	Size	Ang	Sph	Wth	Conc	Type	CarOpa
PFs	2			3	2		
MFs					1		
Bio	3			3	2		
Mus	2			3	4		
Amp					1		
Car	2				3	2,3	2
Quartz		3	2			m,p	
Pyrox					2		
Epidote					1		
Schist					1		

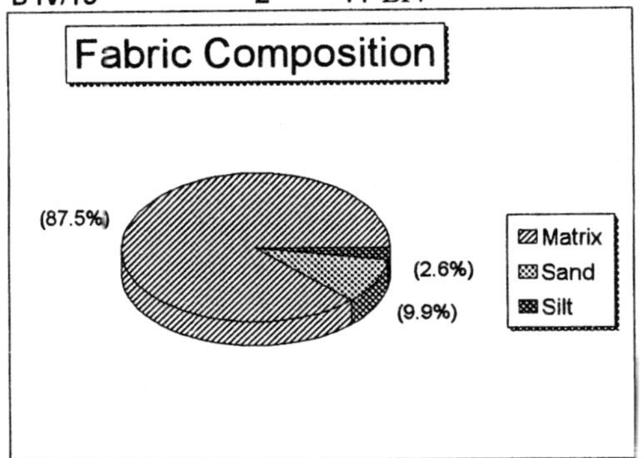

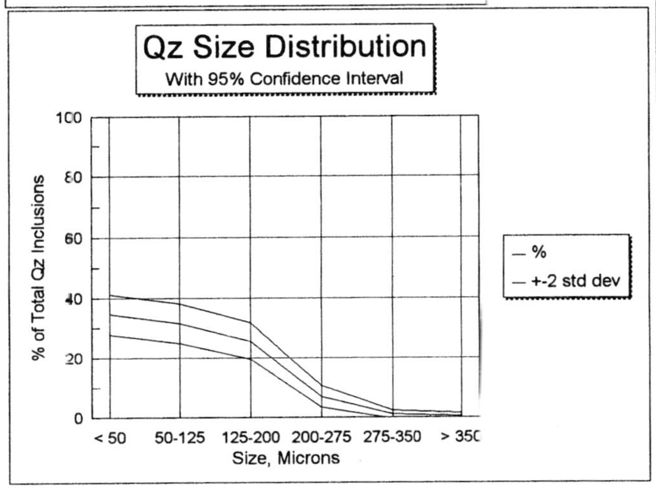

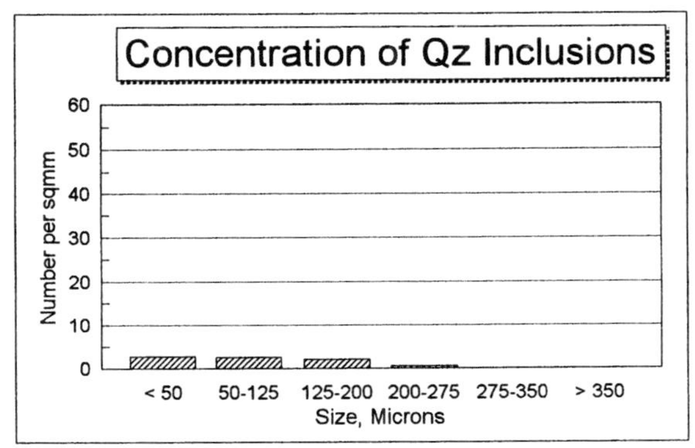

%matrix	87.50
%sand	9.95
%silt	2.55
conc<50	2.80
conc>50	5.32
%<50	34.48
%>50	65.52

APPENDIX A

RHD 30	period	level	morph type	rim dia	thick	sqmm
B IV/4	1	16	BIV	15	4	25

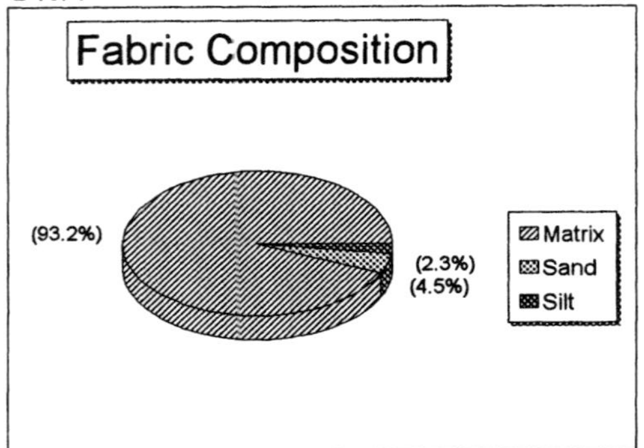

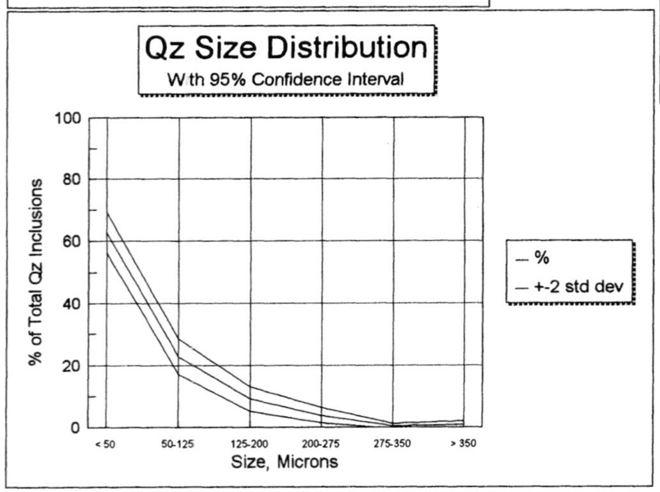

Qz Size	< 50	50-125	125-200	200-275	275-350	> 350	numpts
Number	143	52	21	9	1	2	228
%	62.72	22.81	9.21	3.95	0.44	0.88	
sigma	3.20	2.78	1.92	1.29	0.44	0.62	
2psigma	69.12	28.36	13.04	6.53	1.31	2.11	
2msigma	56.31	17.25	5.38	1.37	-0.44	-0.36	
Conc	5.72	2.08	0.84	0.36	0.04	0.08	
Fabric	Matrix	Sand	Silt	Micac	Biref	Colour	
	371	18	9	1	1	dkbrn	398

Minerals	Size	Ang	Sph	Wth	Conc	Type	CarOpa
PFs					1		
MFs					1		
Bio					2		
Mus	1			3	2		
Amp	1				2		
Car	2				3	ind	
Quartz		3	2			m,p	
Pyrox					1		
Epidote					2		
Schist					1		

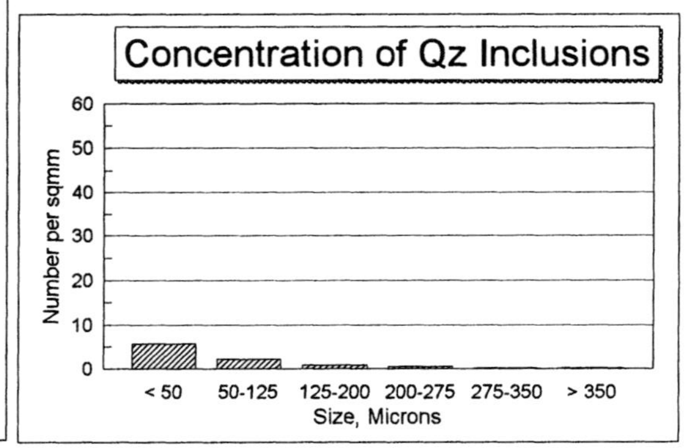

%matrix	93.22
%sand	4.52
%silt	2.26
conc<50	5.72
conc>50	3.40
%<50	62.72
%>50	37.28

RHD 31	period	level	morph type	rim dia	thick	sqmm
B IV/19	3	2	BIV		8	25

Qz Size	< 50	50-125	125-200	200-275	275-350	> 350	numpts
Number	25	24	27	19	5	0	100
%	25.00	24.00	27.00	19.00	5.00	0.00	
sigma	4.33	4.27	4.44	3.92	2.18	0.00	
2psigma	33.66	32.54	35.88	26.85	9.36	0.00	
2msigma	16.34	15.46	18.12	11.15	0.64	0.00	
Conc	1.00	0.96	1.08	0.76	0.20	0	
Fabric	Matrix	Sand	Silt	Micac	Biref	Colour	
	324	51	13	4	3	ltbrn	388

Minerals	Size	Ang	Sph	Wth	Conc	Type	CarOpa
PFs	2				2		
MFs					1		
Bio	3			4	3		
Mus	2			3	4		
Amp	2				3		
Car					1		
Quartz		3	2			m,p	
Pyrox					1		
Epidote					1		
Schist					3		

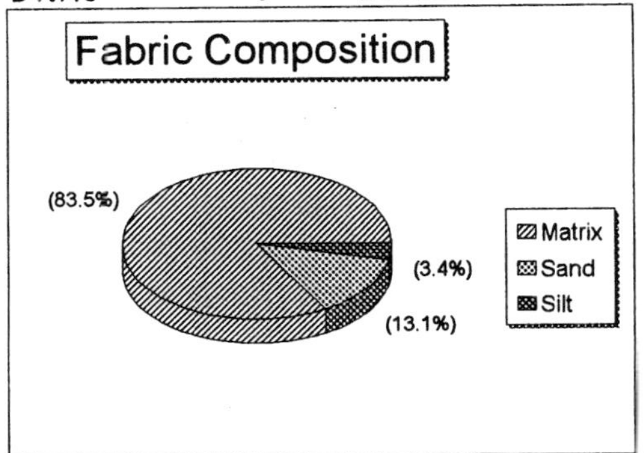

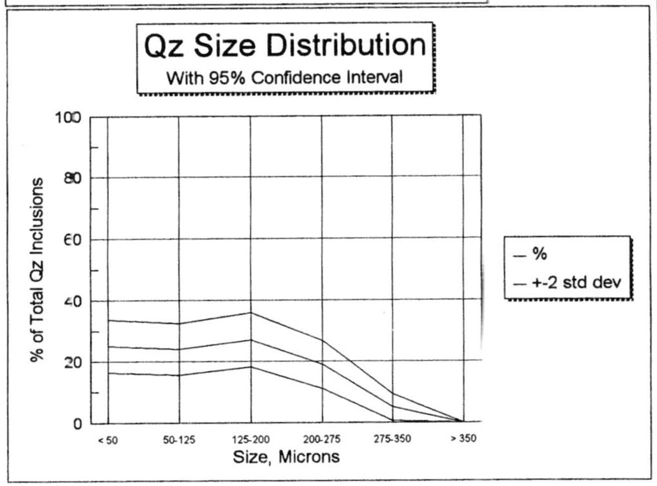

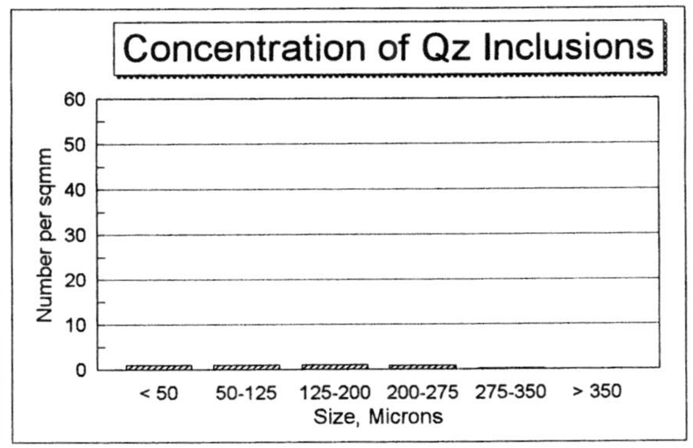

%matrix	83.51
%sand	13.14
%silt	3.35
conc<50	1.00
conc>50	3.00
%<50	25.00
%>50	75.00

APPENDIX A

RHD 32	period	level	morph type	rim dia	thick	sqmm
B IV/5	3	2	BIV		8	25

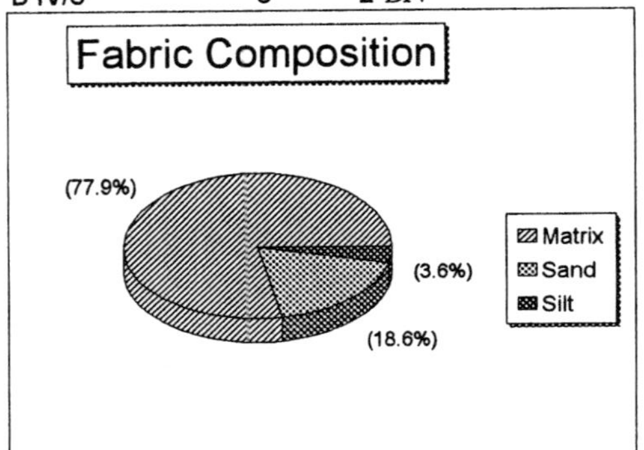

Qz Size	< 50	50-125	125-200	200-275	275-350	> 350	numpts
Number	85	47	72	55	10	6	275
%	30.91	17.09	26.18	20.00	3.64	2.18	
sigma	2.79	2.27	2.65	2.41	1.13	0.88	
2psigma	36.48	21.63	31.48	24.82	5.89	3.94	
2msigma	25.34	12.55	20.88	15.18	1.38	0.42	
Conc	3.40	1.88	2.88	2.20	0.40	0.24	
Fabric	Matrix	Sand	Silt	Micac	Biref	Colour	
	306	73	14	1	1	dkbrn	393

Minerals	Size	Ang	Sph	Wth	Conc	Type	CarOpa
PFs	3	4	2		4		
MFs					1		
Bio	3			3	3		
Mus	3			3	3		
Amp	2				2		
Car	2				2		
Quartz		4	2			m,p	
Pyrox					1		
Epidote					1		
Schist					4		

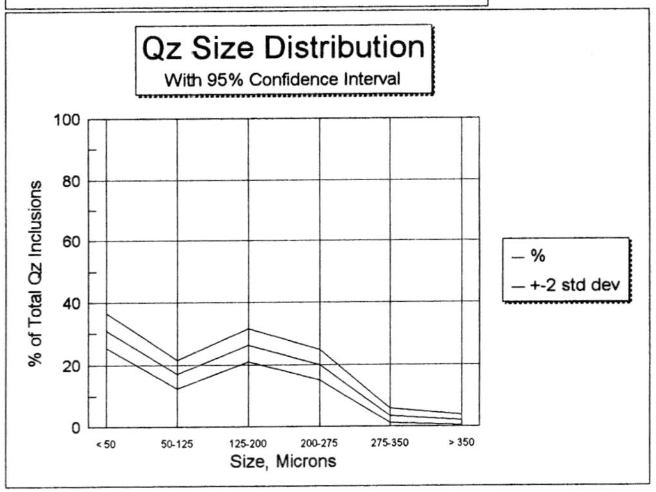

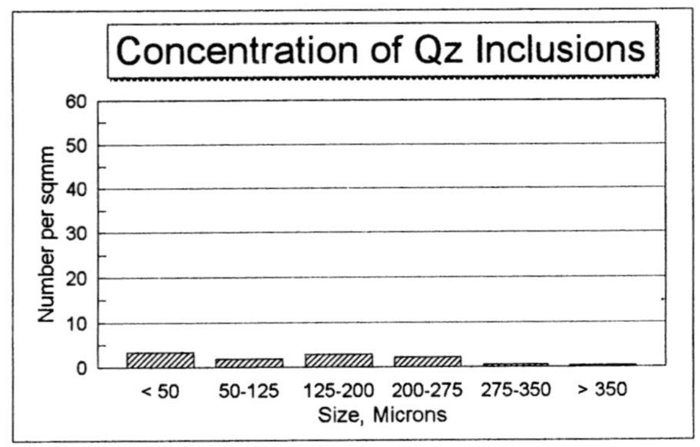

%matrix	77.86
%sand	18.58
%silt	3.56
conc<50	3.40
conc>50	7.60
%<50	30.91
%>50	69.09

RHD 33	period	level	morph type	rim dia	thick	sqmm
B IV/5	1	20	BIV		6	25

Qz Size	< 50	50-125	125-200	200-275	275-350	> 350	numpts
Number	32	22	34	18	15	16	137
%	23.36	16.06	24.82	13.14	10.95	11.68	
sigma	3.61	3.14	3.69	2.89	2.67	2.74	
2psigma	30.59	22.33	32.20	18.91	16.28	17.17	
2msigma	16.13	9.78	17.44	7.37	5.61	6.19	
Conc	1.28	0.88	1.36	0.72	0.60	0.64	
Fabric	Matrix	Sand	Silt	Micac	Biref	Colour	
	288	76	7	1	1	dkbrn	371
Minerals	Size	Ang	Sph	Wth	Conc	Type	CarOpa
PFs	2				3		
MFs					1		
Bio	3			4	3		
Mus	3			3	3		
Amp	2				3		
Car	3				3	2,3	2
Quartz		4L,2S	2			m,p	
Pyrox					1		
Epidote					2		
Schist					2		

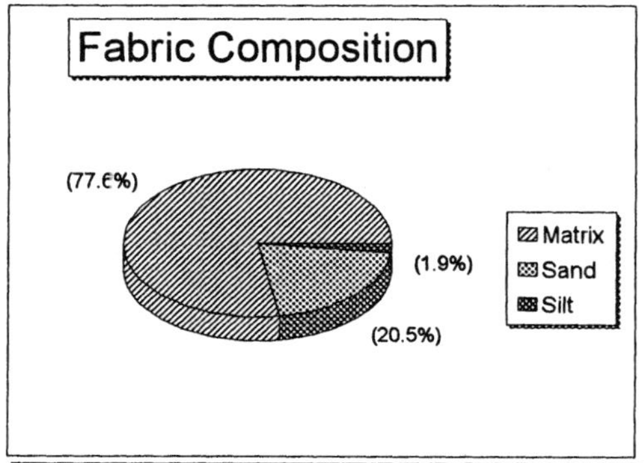

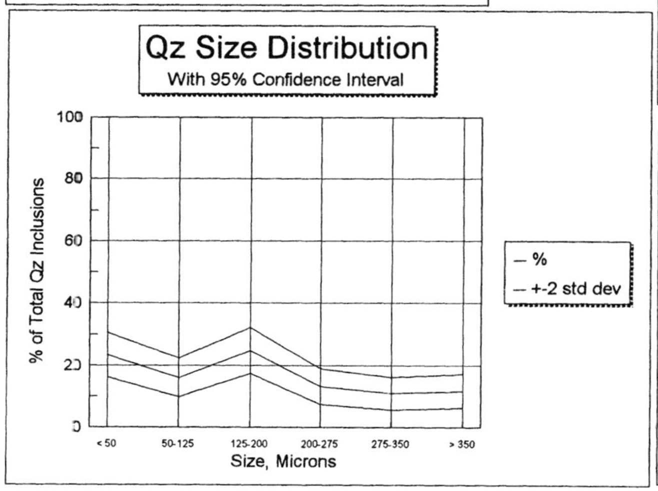

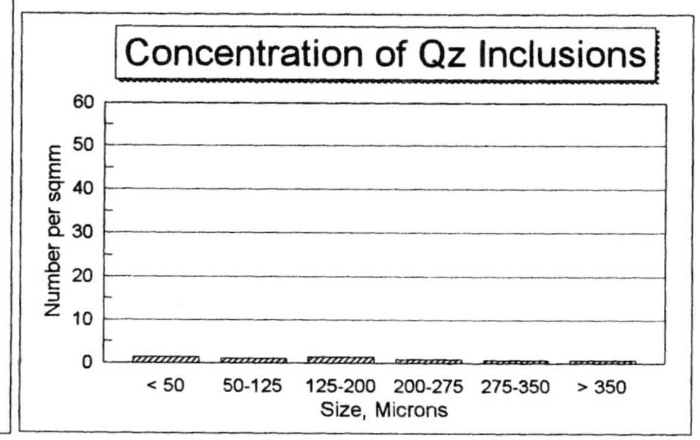

%matrix	77.63
%sand	20.49
%silt	1.89
conc<50	1.28
conc>50	4.20
%<50	23.36
%>50	76.64

APPENDIX A

RHD 34	period	level	morph type	rim dia	thick	sqmm
B IV/9	3	1	BIV	8	9	25

Qz Size	< 50	50-125	125-200	200-275	275-350	> 350	numpts
Number	62	34	46	32	11	4	189
%	32.80	17.99	24.34	16.93	5.82	2.12	
sigma	3.42	2.79	3.12	2.73	1.70	1.05	
2psigma	39.63	23.58	30.58	22.39	9.23	4.21	
2msigma	25.97	12.40	18.10	11.48	2.41	0.02	
Conc	2.48	1.36	1.84	1.28	0.44	0.16	
Fabric	Matrix	Sand	Silt	Micac	Biref	Colour	
	324	48	14	1	1	dkbrn	386

Minerals	Size	Ang	Sph	Wth	Conc	Type	CarOpa
PFs	3	3	2		4		
MFs					1		
Bio	3			4	2		
Mus	3			3	4		
Amp	2				2		
Car	3				3	2,3	
Quartz		3	2			m,p	
Pyrox					2		
Epidote					2		
Schist					1		
Other					arf		

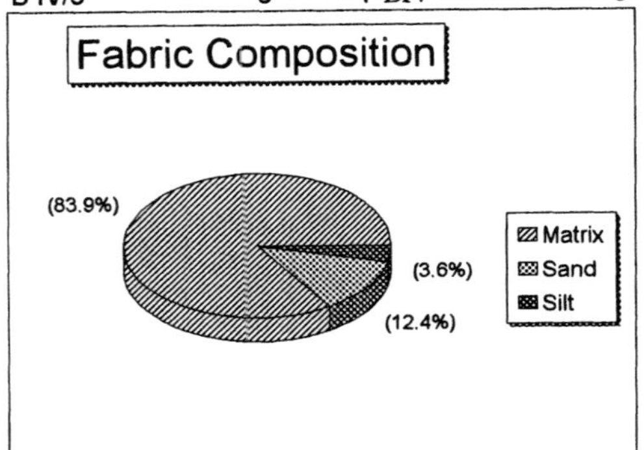

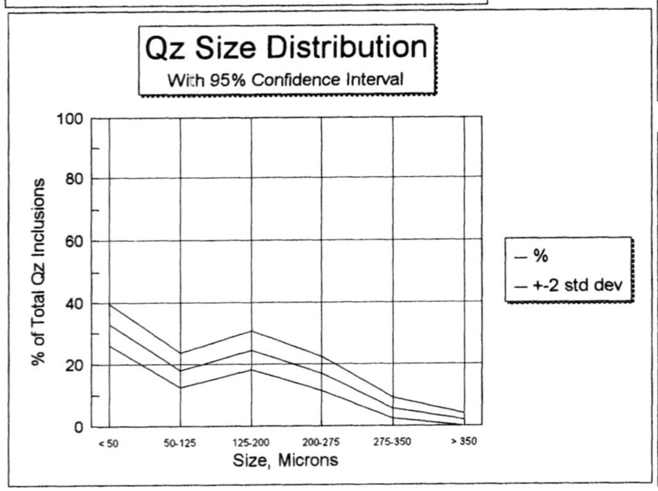

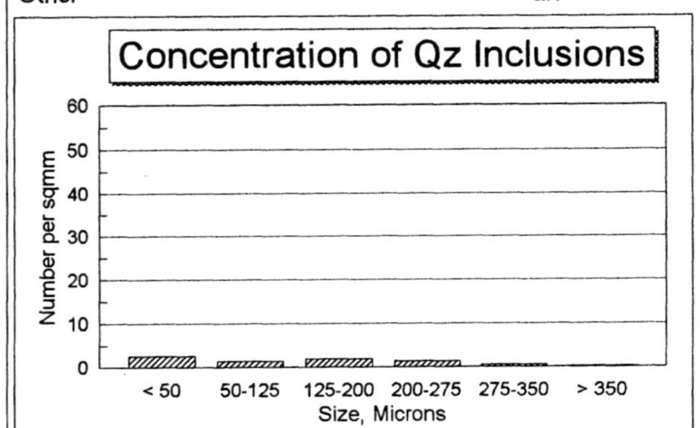

%matrix	83.94
%sand	12.44
%silt	3.63
conc<50	2.48
conc>50	5.08
%<50	32.80
%>50	67.20

RHD 35	period	level	morph type	rim dia	thick	sqmm
B IV/14	2	11	AI	10	4	25

Qz Size	< 50	50-125	125-200	200-275	275-350	> 350	numpts
Number	57	33	44	18	15	7	174
%	32.76	18.97	25.29	10.34	8.62	4.02	
sigma	3.56	2.97	3.30	2.31	2.13	1.49	
2psigma	39.87	24.91	31.88	14.96	12.88	7.00	
2msigma	25.64	13.02	18.70	5.73	4.37	1.04	
Conc	2.28	1.32	1.76	0.72	0.60	0.28	
Fabric	Matrix	Sand	Silt	Micac	Biref	Colour	
	311	73	10	1	1	dkbrn	394

Minerals	Size	Ang	Sph	Wth	Conc	Type	CarOpa
PFs	3	3	2		4		
MFs					2		
Bio	2			3	2		
Mus	2			3	3		
Amp	3				4		
Car	4				3	2	2
Quartz		3	2			m,p	
Pyrox					1		
Epidote					2		
Schist					3		
Other					arf		

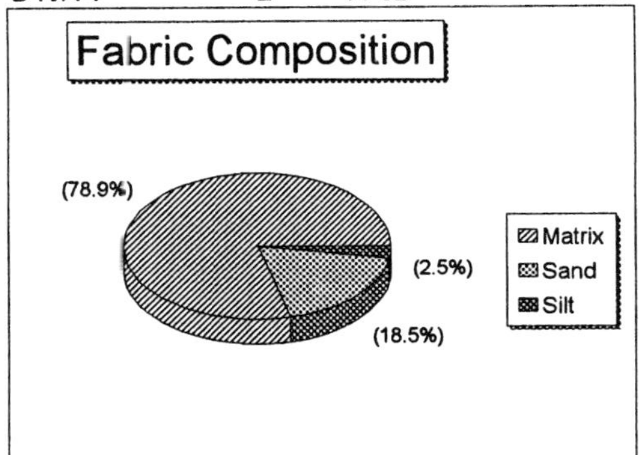

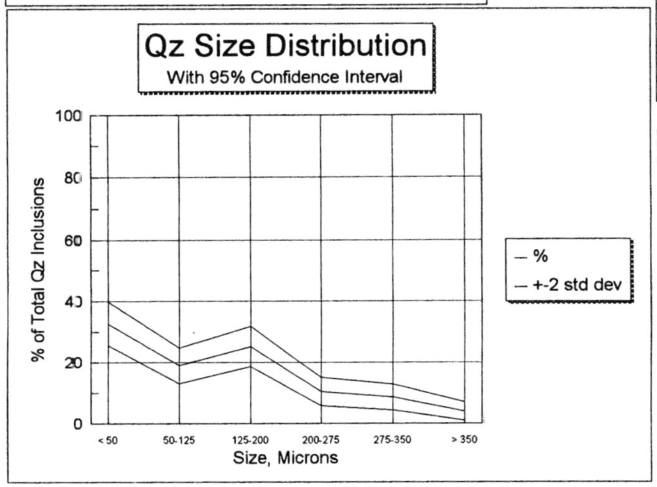

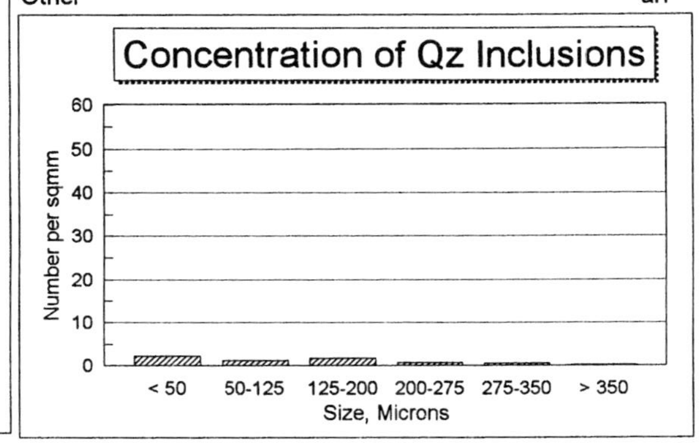

%matrix	78.93
%sand	18.53
%silt	2.54
conc<50	2.28
conc>50	4.68
%<50	32.76
%>50	67.24

RHD 36	period	level	morph type	rim dia	thick	sqmm	Quartz	< 50	50-125	125-200	200-275	275-350	> 350	numpts
B IV/4	1	18	AI	12	8	25	qz							
							%							
Section damaged; not reliable							sigma							
							2psigma							
							2msigma							
							Conc							
							Fabric							
							Minerals							
							PFs							
							MFs							
							Bio							
							Mus							
							Amp							
							Car							

RHD 37	period	level	morph type	rim dia	thick	sqmm
B IV/5	3	2	AI	8	7	25

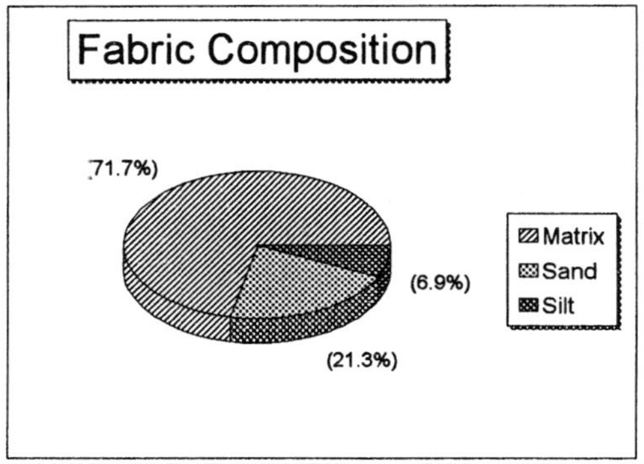

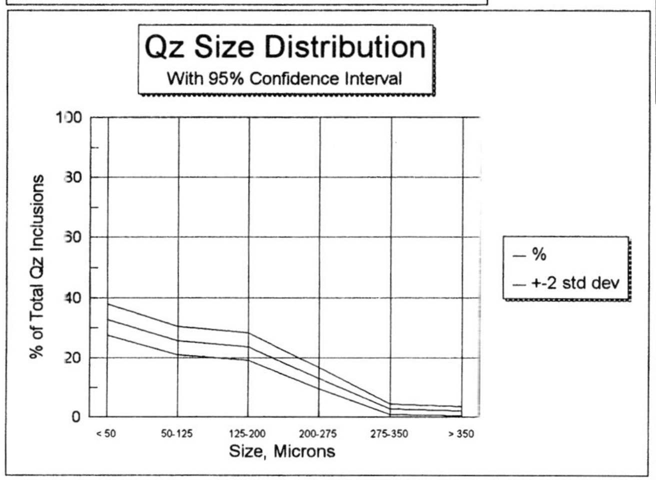

Qz Size	< 50	50-125	125-200	200-275	275-350	> 350	numpts
Number	113	89	82	45	9	7	345
%	32.75	25.80	23.77	13.04	2.61	2.03	
sigma	2.53	2.36	2.29	1.81	0.86	0.76	
2psigma	37.81	30.51	28.35	16.67	4.32	3.55	
2msigma	27.70	21.09	19.18	9.42	0.89	0.51	
Conc	4.52	3.56	3.28	1.80	0.36	0.28	
Fabric	Matrix	Sand	Silt	Micac	Biref	Colour	
	269	80	26	4	4	medbr	375

Minerals	Size	Ang	Sph	Wth	Conc	Type	CarOpa
PFs	3	3	2		4		
MFs					2		
Bio	3			3	3		
Mus	3			3	4		
Amp	3				4		
Car	3				3	2	2
Quartz		3	2			m,p	
Pyrox					2		
Epidote					2		
Schist					3		

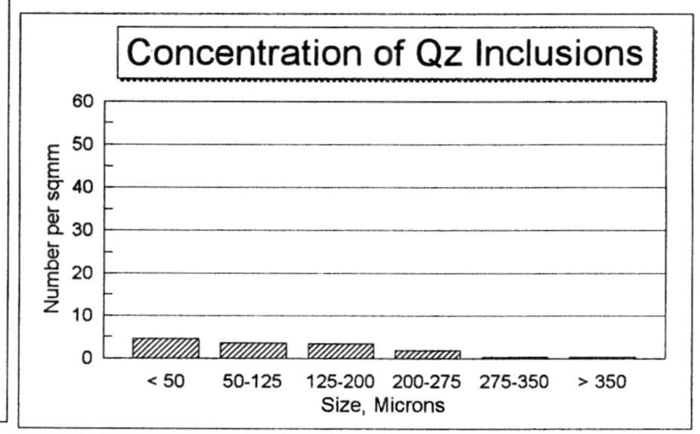

%matrix	71.73
%sand	21.33
%silt	6.93
conc<50	4.52
conc>50	9.28
%<50	32.75
%>50	67.25

APPENDIX A

RHD 38	period	level	morph type	rim dia	thick	sqmm
B IV/19	1	20	AI	11	5	18.75

Qz Size	< 50	50-125	125-200	200-275	275-350	> 350	numpts
Number	207	56	12	2	1	0	278
%	74.46	20.14	4.32	0.72	0.36	0.00	
sigma	2.62	2.41	1.22	0.51	0.36	0.00	
2psigma	79.69	24.95	6.75	1.73	1.08	0.00	
2msigma	69.23	15.33	1.88	-0.29	-0.36	0.00	
Conc	11.04	2.99	0.64	0.11	0.05	0	
Fabric	Matrix	Sand	Silt	Micac	Biref	Colour	
	330	17	43	3	3	ltbrn	390

Minerals	Size	Ang	Sph	Wth	Conc	Type	CarOpa
PFs					1		
MFs					1		
Bio	2			3	2		
Mus	1			3	4		
Amp					1		
Car	2				3	2	2
Quartz		3L 1S	2L 3S			m,p	
Pyrox					1		
Epidote					1		
Schist					1		

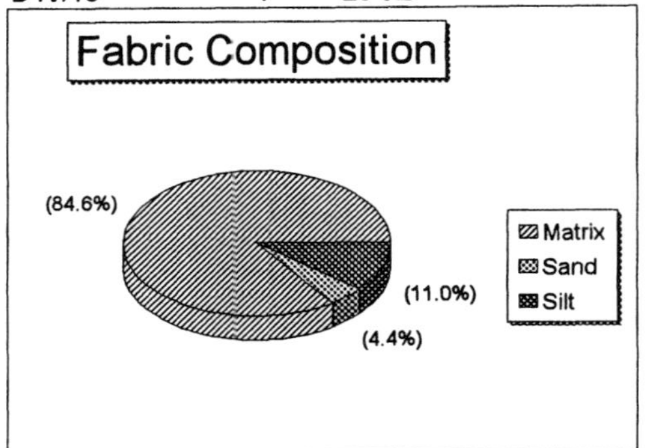

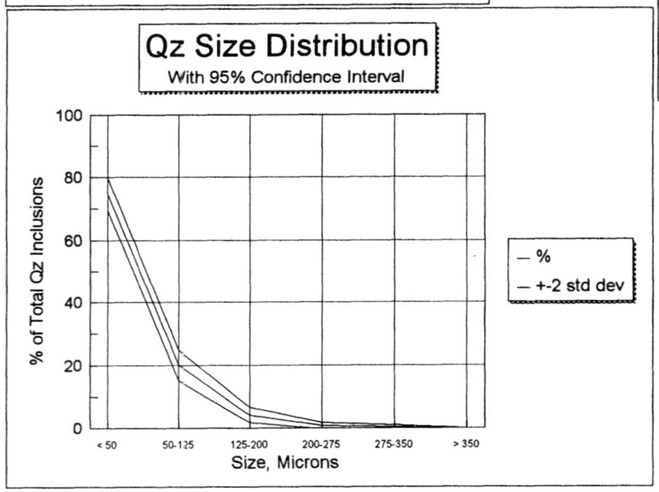

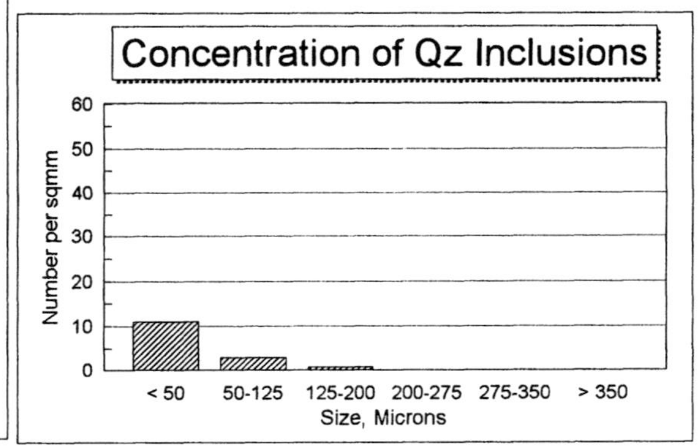

%matrix	84.62
%sand	4.36
%silt	11.03
conc<50	11.04
conc>50	3.79
%<50	74.46
%>50	25.54

RHD 39	period	level	morph type	rim dia	thick	sqmm
B IV/5	3	2	AI	10	7	18.75

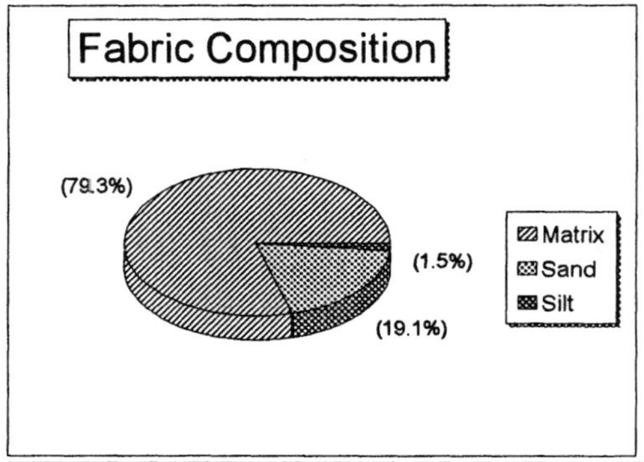

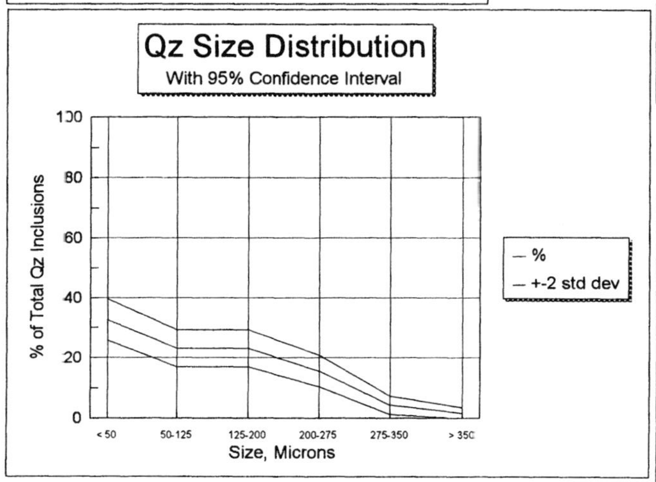

Qz Size	< 50	50-125	125-200	200-275	275-350	> 350	numpts
Number	61	43	43	29	8	3	187
%	32.62	22.99	22.99	15.51	4.28	1.60	
sigma	3.43	3.08	3.08	2.65	1.48	0.92	
2psigma	39.48	29.15	29.15	20.80	7.24	3.44	
2msigma	25.76	16.84	16.84	10.21	1.32	-0.23	
Conc	3.25	2.29	2.29	1.55	0.43	0.16	
Fabric	Matrix	Sand	Silt	Micac	Biref	Colour	
	311	75	6	2	2	dkbrn	392
Minerals	Size	Ang	Sph	Wth	Conc	Type	CarOpa
PFs	3	4	2		4		
MFs					1		
Bio	4			3	4		
Mus	3			3	4		
Amp	2				3		
Car					1		
Quartz		4	2			m,p	
Pyrox					2		
Epidote					1		
Schist					2		

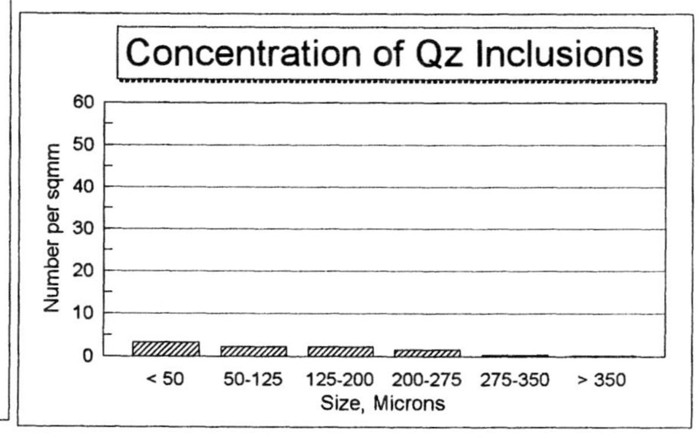

%matrix	79.34
%sand	19.13
%silt	1.53
conc<50	3.25
conc>50	6.72
%<50	32.62
%>50	67.38

APPENDIX A

RHD 40	period	level	morph type	rim dia	thick	sqmm
B IV/24	1	17	AI	11	7	25

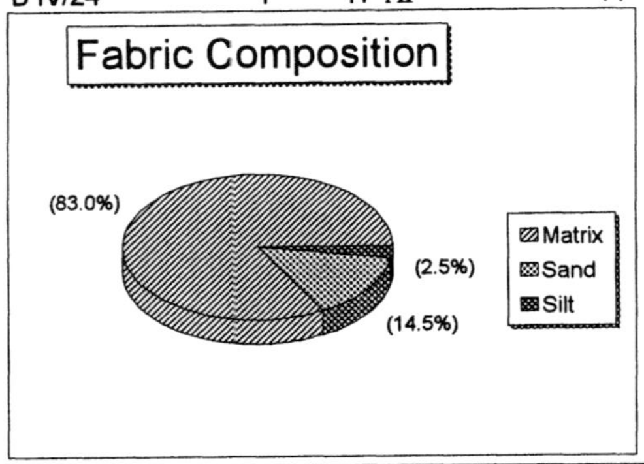

Qz Size	< 50	50-125	125-200	200-275	275-350	> 350	numpts
Number	72	38	43	36	8	4	201
%	35.82	18.91	21.39	17.91	3.98	1.99	
sigma	3.38	2.76	2.89	2.70	1.38	0.99	
2psigma	42.58	24.43	27.18	23.32	6.74	3.96	
2msigma	29.06	13.38	15.61	12.50	1.22	0.02	
Conc	2.88	1.52	1.72	1.44	0.32	0.16	
Fabric	Matrix	Sand	Silt	Micac	Biref	Colour	
	326	57	10	2	2	dkbrn	393

Minerals	Size	Ang	Sph	Wth	Conc	Type	CarOpa
PFs	3	3	2		4		
MFs					1		
Bio	3			4	2		
Mus	3			3	4		
Amp	2				2		
Car	3			4	2	2,3	2
Quartz		3	2			m,p	
Pyrox					2		
Epidote					1		
Schist					3		

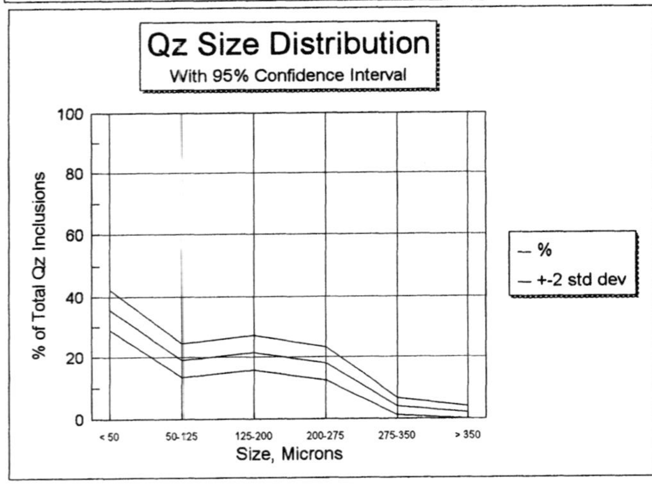

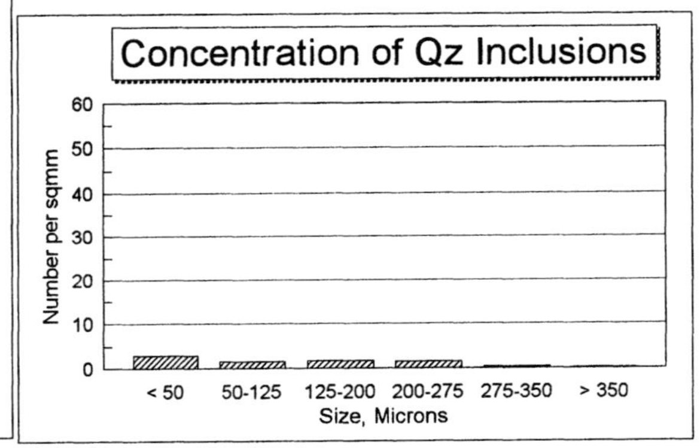

%matrix	82.95
%sand	14.50
%silt	2.54
conc<50	2.88
conc>50	5.16
%<50	35.82
%>50	64.18

RHD 41	period	level	morph type	rim dia	thick	sqmm
B IV/19	1	20	AI	8	4	25

Qz Size	< 50	50-125	125-200	200-275	275-350	> 350	numpts
Number	90	32	19	5	0	0	146
%	61.64	21.92	13.01	3.42	0.00	0.00	
sigma	4.02	3.42	2.78	1.51	0.00	0.00	
2psigma	69.69	28.77	18.58	6.43	0.00	0.00	
2msigma	53.60	15.07	7.44	0.41	0.00	0.00	
Conc	3.60	1.28	0.76	0.20	0.00	0.00	
Fabric	Matrix	Sand	Silt	Micac	Biref	Colour	
	355	11	15	2	2	dkbrn	381

Minerals	Size	Ang	Sph	Wth	Conc	Type	CarOpa
PFs					1		
MFs					1		
Bio					1		
Mus	1			2	4		
Amp					1		
Car	3				4	2	2
Quartz		3	2			m,p	
Pyrox					2		
Epidote					1		
Schist					1		

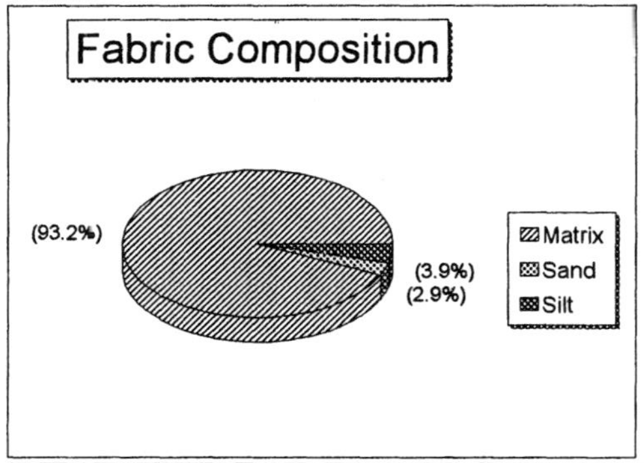

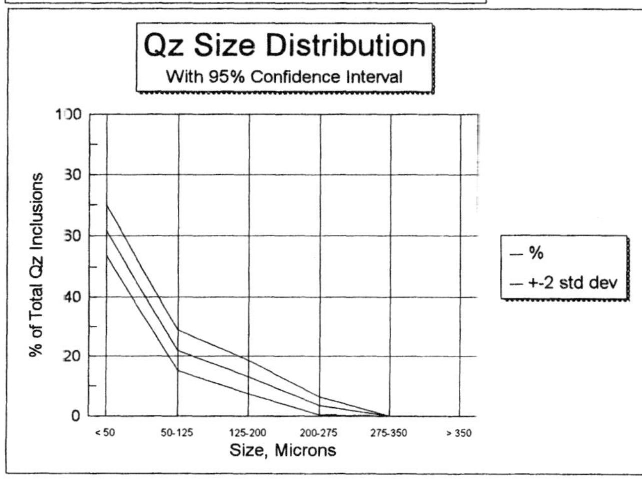

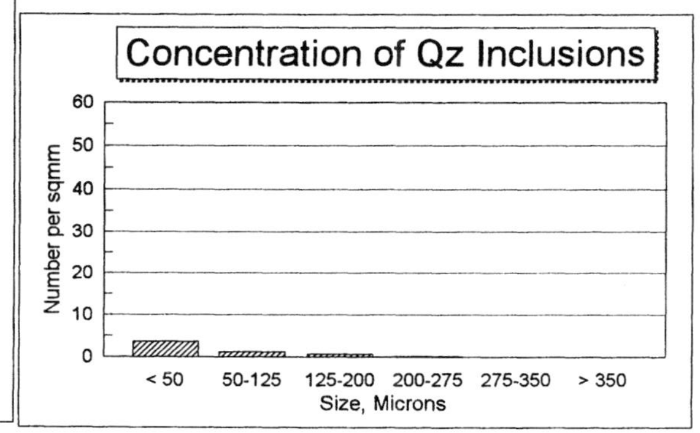

%matrix	93.18
%sand	2.89
%silt	3.94
conc<50	3.60
conc>50	2.24
%<50	61.64
%>50	38.36

RHD 42	period	level	morph type	rim dia	thick	sqmm
B IV/4	1	16	CII	18	6	25

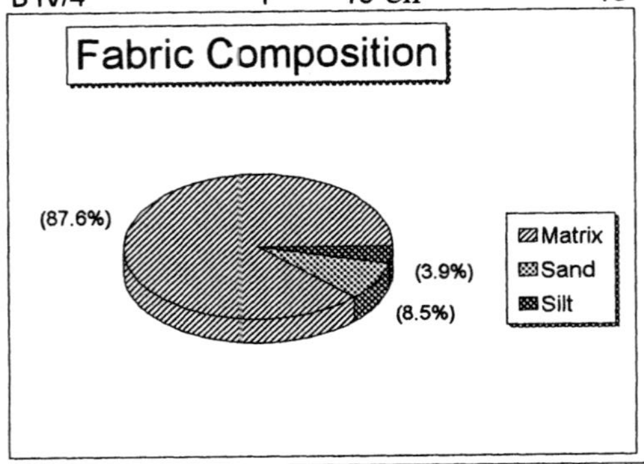

Qz Size	< 50	50-125	125-200	200-275	275-350	> 350	numpts
Number	129	79	43	17	2	2	272
%	47.43	29.04	15.81	6.25	0.74	0.74	
sigma	3.03	2.75	2.21	1.47	0.52	0.52	
2psigma	53.48	34.55	20.23	9.19	1.77	1.77	
2msigma	41.37	23.54	11.38	3.31	-0.30	-0.30	
Conc	5.16	3.16	1.72	0.68	0.08	0.08	
Fabric	Matrix	Sand	Silt	Micac	Biref	Colour	
	338	33	15	3	3	medbr	386

Minerals	Size	Ang	Sph	Wth	Conc	Type	CarOpa
PFs	4	3	2		3		
MFs					1		
Bio				2	2		
Mus	1			2	4		
Amp					1		
Car	4				4	2	2
Quartz		3	2			m,p	
Pyrox					2		
Epidote					1		
Schist					2		

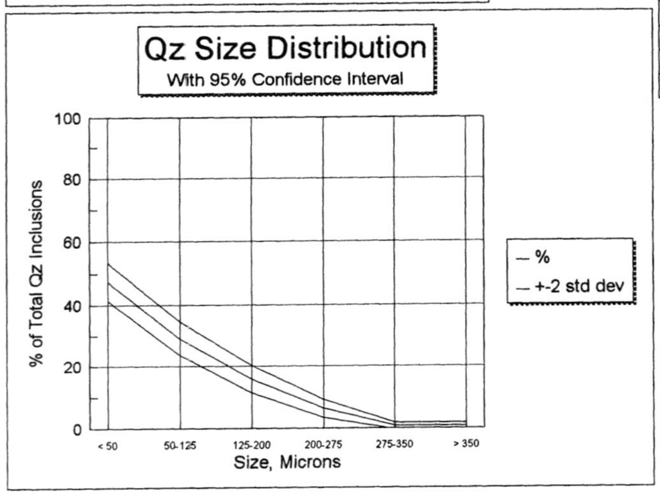

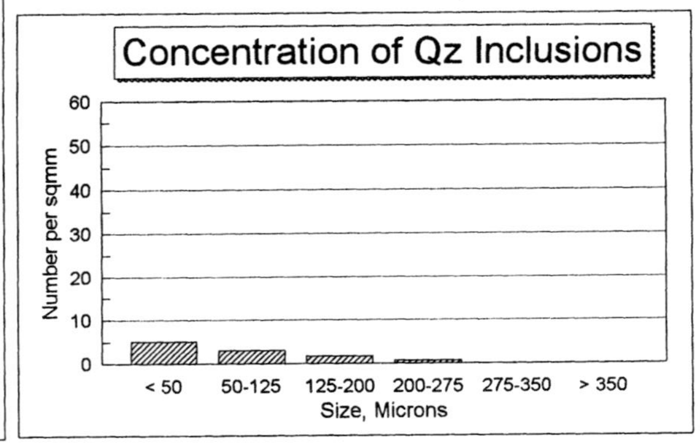

%matrix	87.56
%sand	8.55
%silt	3.89
conc<50	5.16
conc>50	5.72
%<50	47.43
%>50	52.57

RHD 43	period	level	morph type	rim dia	thick	sqmm
B IV/23	3	8	BI/CII	20	6	25

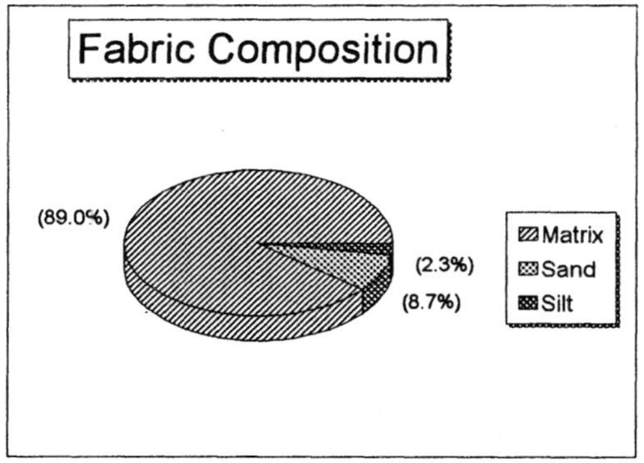

Qz Size	< 50	50-125	125-200	200-275	275-350	> 350	numpts
Number	170	90	36	8	2	2	308
%	55.19	29.22	11.69	2.60	0.65	0.65	
sigma	2.83	2.59	1.83	0.91	0.46	0.46	
2psigma	60.86	34.40	15.35	4.41	1.56	1.56	
2msigma	49.53	24.04	8.03	0.78	-0.27	-0.27	
Conc	6.80	3.60	1.44	0.32	0.08	0.08	
Fabric	Matrix	Sand	Silt	Micac	Biref	Colour	
	349	34	9	2	2	medbr	392
Minerals	Size	Ang	Sph	Wth	Conc	Type	CarOpa
PFs					2		
MFs					1		
Bio	1			2	1		
Mus	2			2	4		
Amp					1		
Car	2				3	1,2	2
Quartz		2,4	2,3			m,p	
Pyrox					1		
Epidote					1		
Schist					1		

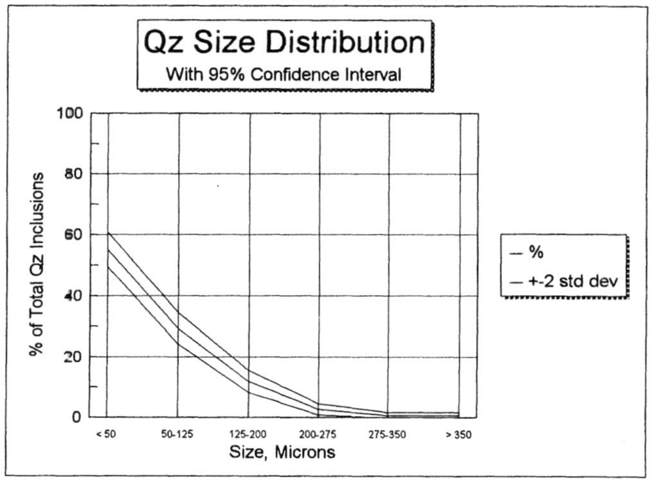

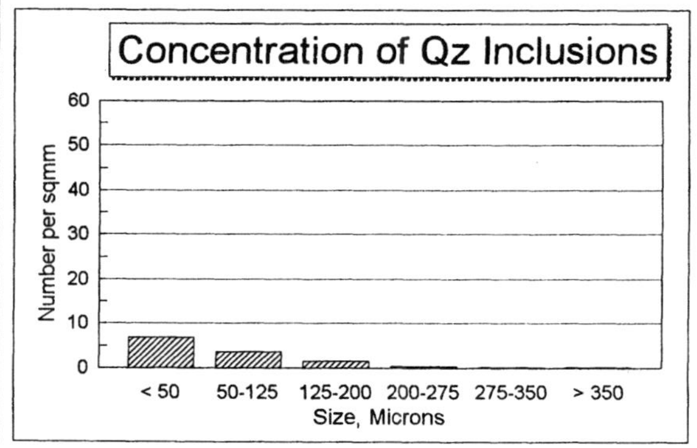

%matrix	89.03
%sand	8.67
%silt	2.30
conc<50	6.80
conc>50	5.52
%<50	55.19
%>50	44.81

RHD 44	period	level	morph type	rim dia	thick	sqmm
B IV/19	1	14	CII	22	4	12.5

Qz Size	< 50	50-125	125-200	200-275	275-350	> 350	numpts
Number	262	7	5	8	1	2	285
%	91.93	2.46	1.75	2.81	0.35	0.70	
sigma	1.61	0.92	0.78	0.98	0.35	0.49	
2psigma	95.16	4.29	3.31	4.76	1.05	1.69	
2msigma	88.70	0.62	0.20	0.85	-0.35	-0.29	
Conc	20.96	0.56	0.40	0.64	0.08	0.16	
Fabric	Matrix	Sand	Silt	Micac	Biref	Colour	
	344	15	27	3	2	dkbrn	386
Minerals	Size	Ang	Sph	Wth	Conc	Type	CarOpa
PFs	3			3	2		
MFs					1		
Bio	3			4	2		
Mus	1			3	4		
Amp	3				3		
Car	3			3	2	2	2
Quartz		4L 3S	2			m,p	
Pyrox					1		
Epidote					2		
Schist					2		

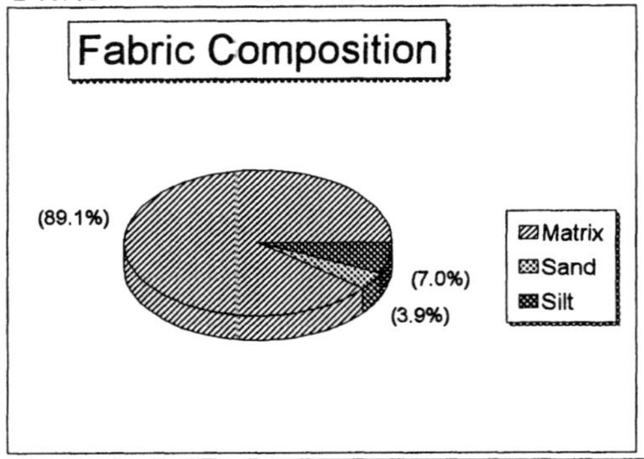

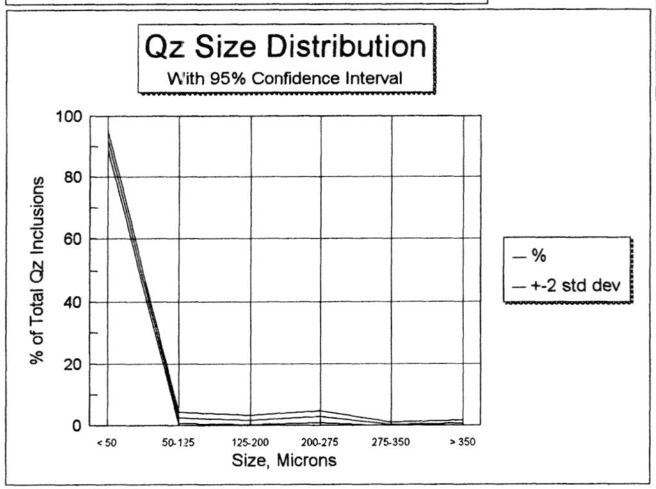

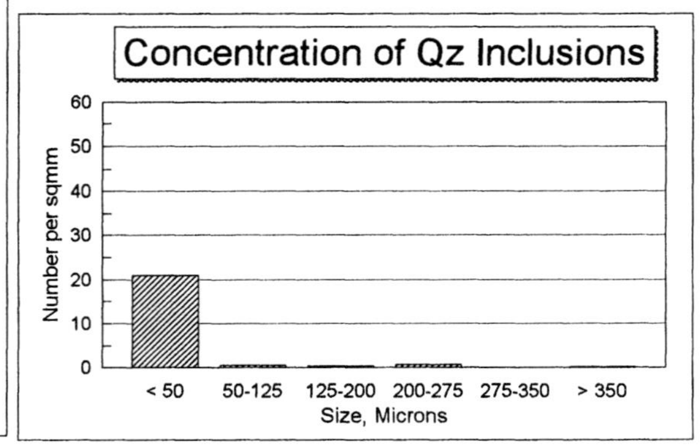

%matrix	89.12
%sand	3.89
%silt	6.99
conc<50	20.96
conc>50	1.84
%<50	91.93
%>50	8.07

RHD 45	period	level	morph type	rim dia	thick	sqmm
B IV/9	1	20	BI/CII	12	2	25

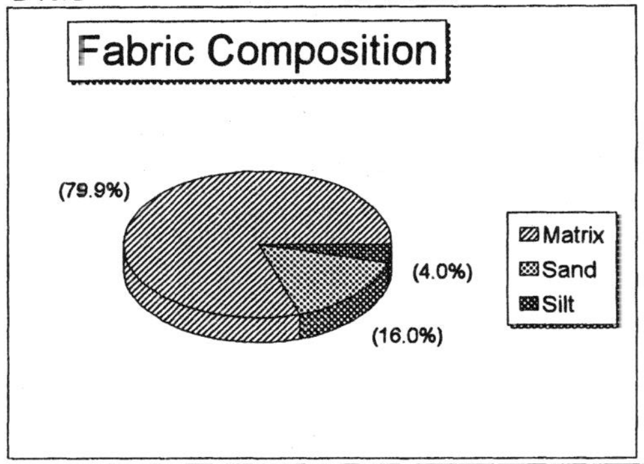

Qz Size	< 50	50-125	125-200	200-275	275-350	> 350	numpts
Number	59	45	55	36	5	8	208
%	28.37	21.63	26.44	17.31	2.40	3.85	
sigma	3.13	2.85	3.06	2.62	1.06	1.33	
2psigma	34.62	27.34	32.56	22.55	4.53	6.51	
2msigma	22.11	15.92	20.33	12.06	0.28	1.18	
Conc	2.36	1.80	2.20	1.44	0.20	0.32	
Fabric	Matrix	Sand	Silt	Micac	Biref	Colour	
	319	64	16	2	2	dkbrn	399
Minerals	Size	Ang	Sph	Wth	Conc	Type	CarOpa
PFs	3	3	2		3		
MFs					1		
Bio	3			4	3		
Mus	2			3	4		
Amp	2				4		
Car	3			3	3	2	2,rare3
Quartz		3	2			m,p	
Pyrox					2		
Epidote					2		
Schist					1		

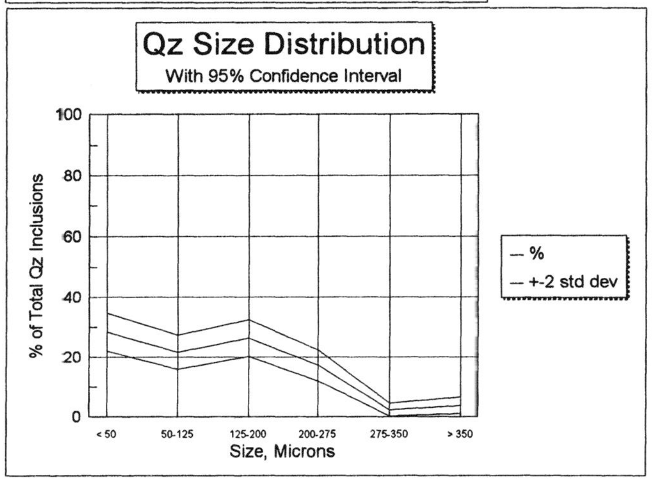

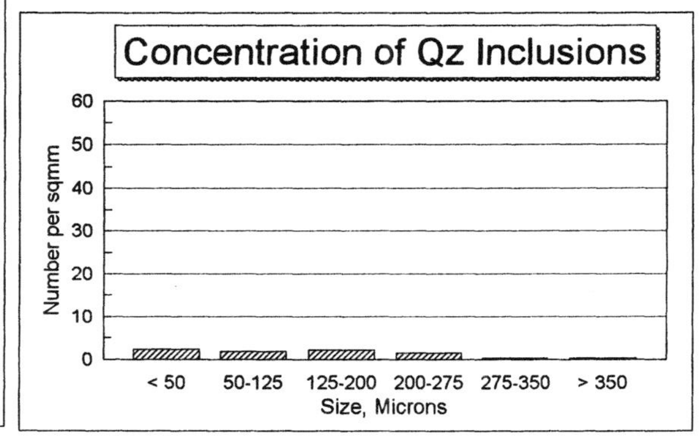

%matrix	79.95
%sand	16.04
%silt	4.01
conc<50	2.36
conc>50	5.96
%<50	28.37
%>50	71.63

RHD 46	period	level	morph type	rim dia	thick	sqmm
B IV/14	3	7	CII	18	6	25

Qz Size	< 50	50-125	125-200	200-275	275-350	> 350	numpts
Number	139	87	67	31	3	1	328
%	42.38	26.52	20.43	9.45	0.91	0.30	
sigma	2.73	2.44	2.23	1.62	0.53	0.30	
2psigma	47.84	31.40	24.88	12.68	1.97	0.91	
2msigma	36.92	21.65	15.97	6.22	-0.14	-0.30	
Conc	5.56	3.48	2.68	1.24	0.12	0.04	
Fabric	Matrix	Sand	Silt	Micac	Biref	Colour	
	341	36	18	2	2	dkbrn	395

Minerals	Size	Ang	Sph	Wth	Conc	Type	CarOpa
PFs	2	3	2		2		
MFs					1		
Bio	2			3	2		
Mus	1			2	3		
Amp					1		
Car	3			2	4	2	2
Quartz		3	2			m,p	
Pyrox					1		
Epidote					1		
Schist					2		
Other						arf	

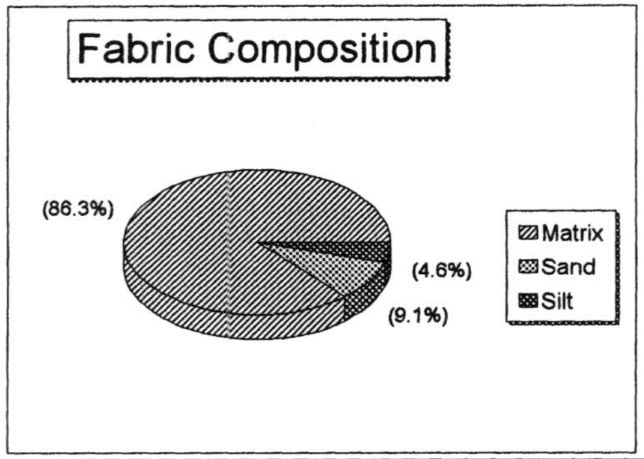

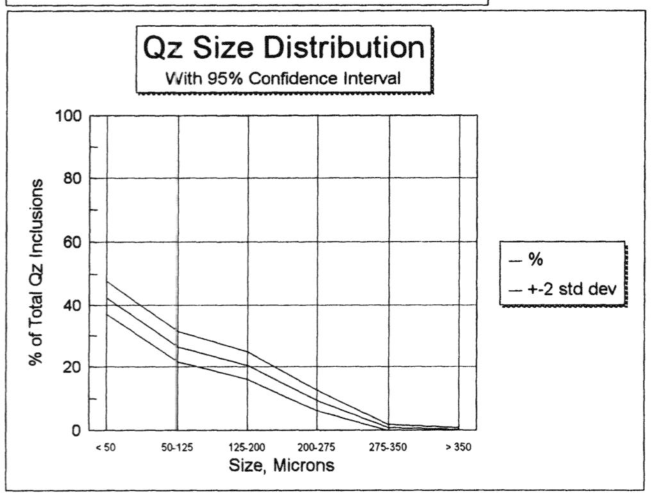

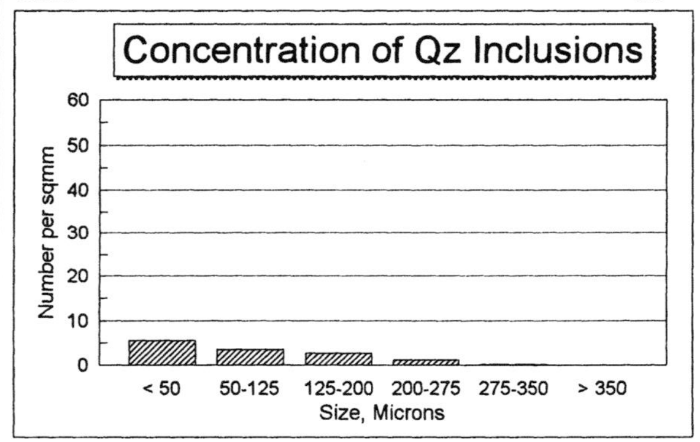

%matrix	86.33
%sand	9.11
%silt	4.56
conc<50	5.56
conc>50	7.56
%<50	42.38
%>50	57.62

APPENDIX A

RHD 47	period	level	morph type	rim dia	thick	sqmm
B IV/5	3	2	AII	24	7	25

Qz Size	< 50	50-125	125-200	200-275	275-350	> 350	numpts
Number	75	45	47	36	6	7	216
%	34.72	20.83	21.76	16.67	2.78	3.24	
sigma	3.24	2.76	2.81	2.54	1.12	1.20	
2psigma	41.20	26.36	27.37	21.74	5.01	5.65	
2msigma	28.24	15.31	16.14	11.60	0.54	0.83	
Conc	3.00	1.80	1.88	1.44	0.24	0.28	
Fabric	Matrix	Sand	Silt	Micac	Biref	Colour	
	311	57	20	2	2	dkbrn	388

Minerals	Size	Ang	Sph	Wth	Conc	Type	CarOpa
PFs	3	3	2		3		
MFs					2		
Bio	4			4	4		
Mus	3			3	4		
Amp					2		
Car	4				4	2	2,rare3
Quartz		3L 2S	2			m,p	
Pyrox					2		
Epidote					2		
Schist					1		

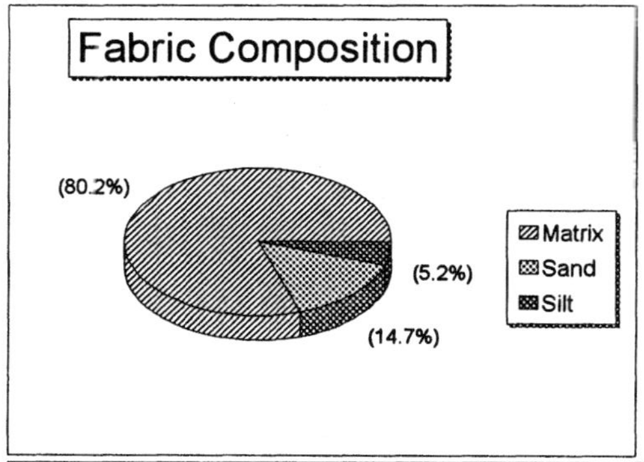

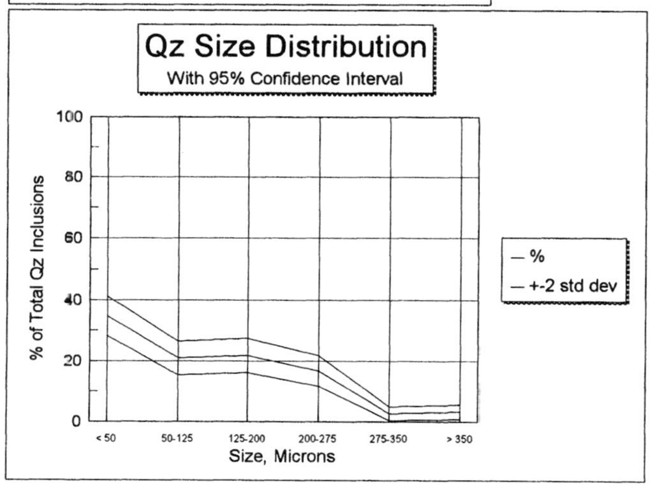

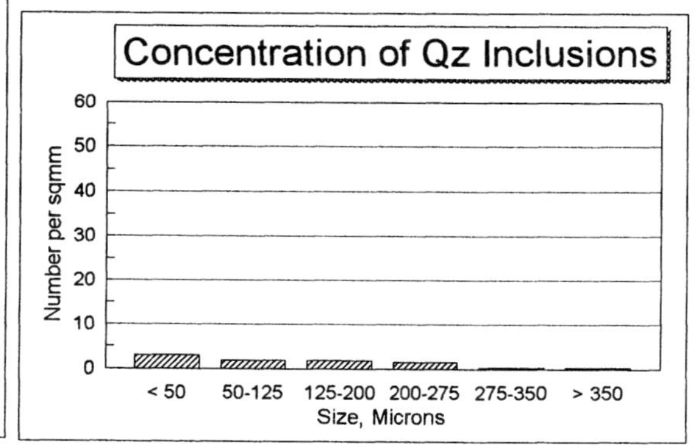

%matrix	80.15
%sand	14.69
%silt	5.15
conc<50	3.00
conc>50	5.64
%<50	34.72
%>50	65.28

RHD 48	period	level	morph type	rim dia	thick	sqmm
B IV/14	1	18	AII	16	7	25

Qz Size	< 50	50-125	125-200	200-275	275-350	> 350	numpts
Number	100	55	63	37	6	1	262
%	38.17	20.99	24.05	14.12	2.29	0.38	
sigma	3.00	2.52	2.64	2.15	0.92	0.38	
2psigma	44.17	26.02	29.33	18.43	4.14	1.14	
2msigma	32.17	15.96	18.77	9.82	0.44	-0.38	
Conc	4.00	2.20	2.52	1.48	0.24	0.04	
Fabric	Matrix	Sand	Silt	Micac	Biref	Colour	
	321	55	15	2	2	dkbrn	391

Minerals	Size	Ang	Sph	Wth	Conc	Type	CarOpa
PFs	3	3	2		4		
MFs					1		
Bio	3			3	2		
Mus	3			3	4		
Amp	3				3		
Car	3				3	2	2
Quartz		3	2			m,p	
Pyrox					2		
Epidote					2		
Schist					3		

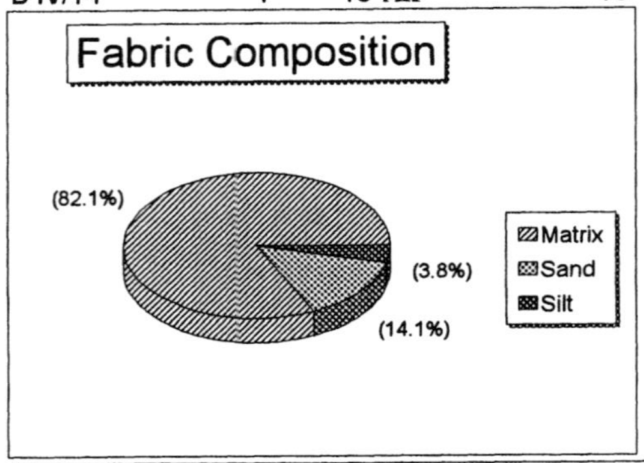

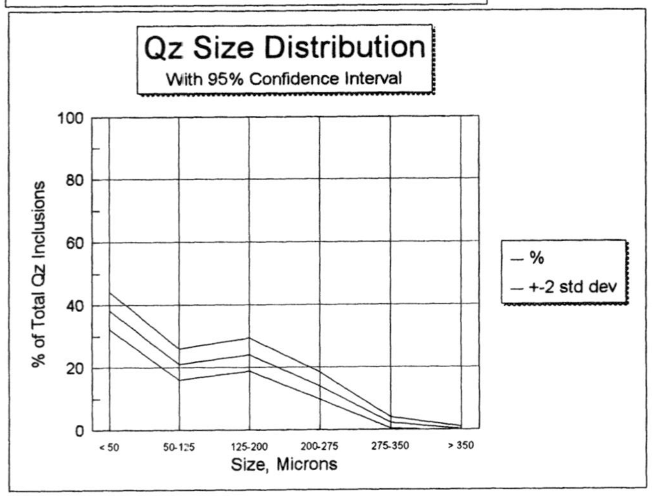

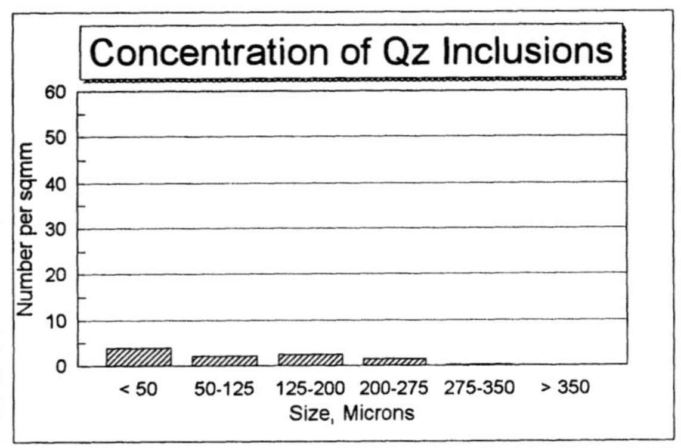

%matrix	82.10
%sand	14.07
%silt	3.84
conc<50	4.00
conc>50	6.48
%<50	38.17
%>50	61.83

RHD 49	period	level	morph type	rim dia	thick	sqmm
B IV/14	2	11	AII	11	3	18.75

Qz Size	< 50	50-125	125-200	200-275	275-350	> 350	numpts
Number	109	55	30	7	4	0	205
%	53.17	26.83	14.63	3.41	1.95	0.00	
sigma	3.49	3.09	2.47	1.27	0.97	0.00	
2psigma	60.14	33.02	19.57	5.95	3.88	0.00	
2msigma	46.20	20.64	9.70	0.88	0.02	0.00	
Conc	5.81	2.93	1.60	0.37	0.21	0	

Fabric	Matrix	Sand	Silt	Micac	Biref	Colour	
	344	31	19	2	2	medbr	394

Minerals	Size	Ang	Sph	Wth	Conc	Type	CarOpa
PFs					1		
MFs					1		
Bio					1		
Mus	3			4	4		
Amp	1				2		
Car					4	2	2
Quartz		2,4				m,p	
Pyrox					1		
Epidote					1		
Schist					1		

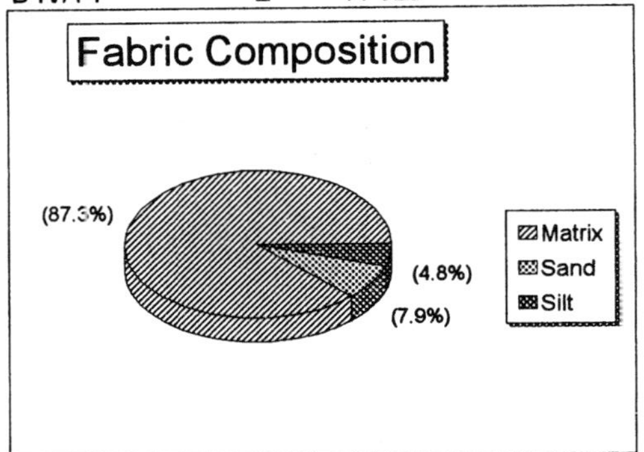

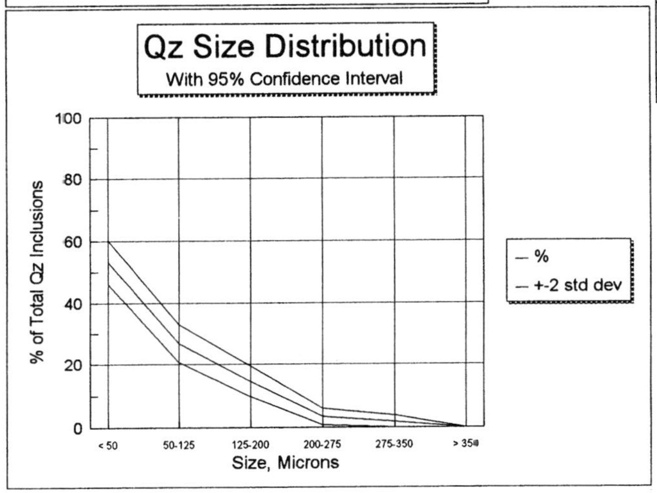

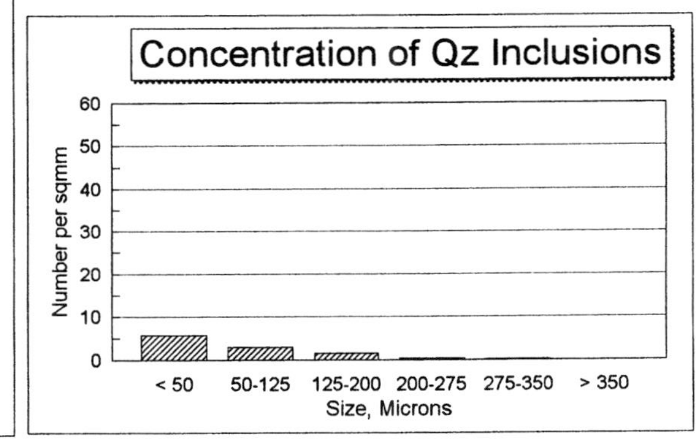

%matrix	87.31
%sand	7.87
%silt	4.82
conc<50	5.81
conc>50	5.12
%<50	53.17
%>50	46.83

RHD 50	period	level	morph type	rim dia	thick	sqmm
B IV/19	1	20	AII	8	5	12.5

Qz Size	< 50	50-125	125-200	200-275	275-350	> 350	numpts
Number	428	42	10	4	0	0	484
%	88.43	8.68	2.07	0.83	0.00	0.00	
sigma	1.45	1.28	0.65	0.41	0.00	0.00	
2psigma	91.34	11.24	3.36	1.65	0.00	0.00	
2msigma	85.52	6.12	0.77	0.00	0.00	0.00	
Conc	34.24	3.36	0.80	0.32	0	0	
Fabric	Matrix	Sand	Silt	Micac	Biref	Colour	
	318	19	54	3	2	medbr	391

Minerals	Size	Ang	Sph	Wth	Conc	Type	CarOpa
PFs					1		
MFs					1		
Bio	3			3	3		
Mus	2			2	3		
Amp					2		
Car					4	2	2
Quartz		3L 2S	2L3S			m,p	
Pyrox					1		
Epidote					2		
Schist					2		

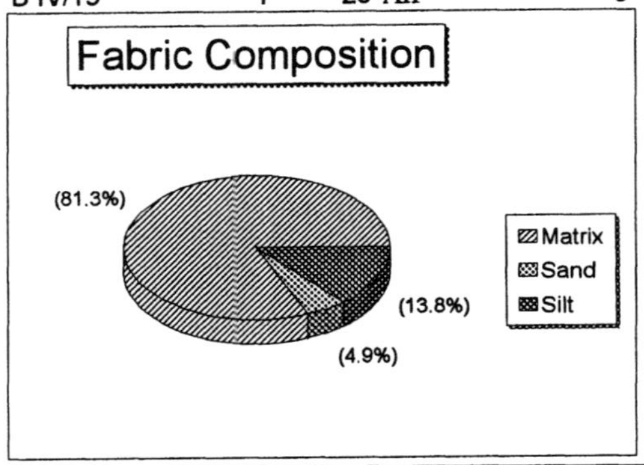

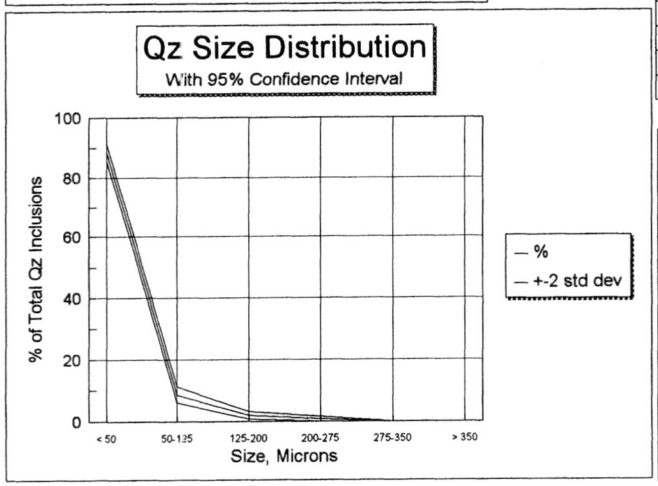

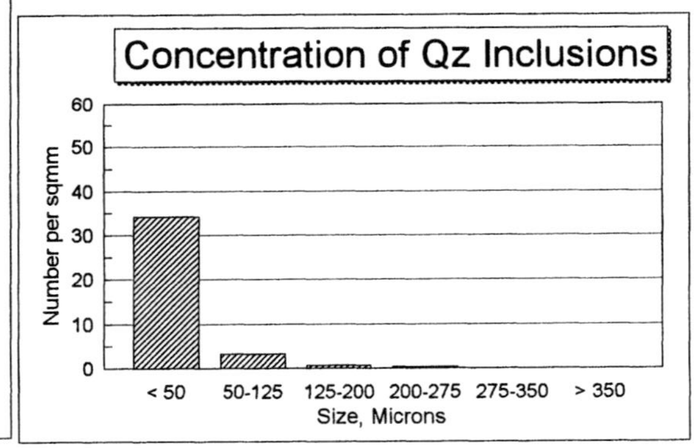

%matrix	81.33
%sand	4.86
%silt	13.81
conc<50	34.24
conc>50	4.48
%<50	88.43
%>50	11.57

RHD 51	period	level	morph type	rim dia	thick	sqmm
B IV/4	1	14	AII	11		18.75

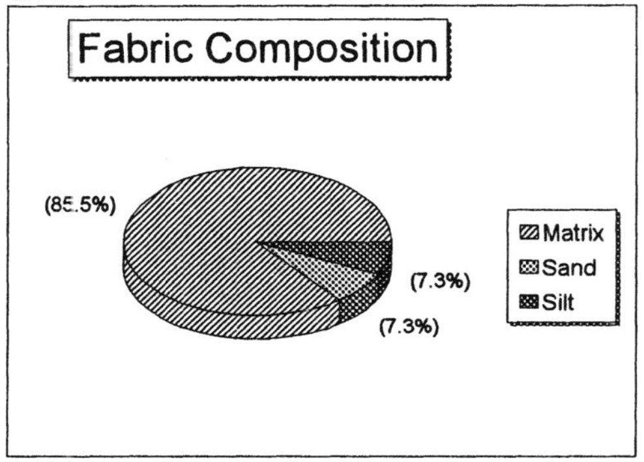

Qz Size	< 50	50-125	125-200	200-275	275-350	> 350	numpts
Number	208	62	41	5	0	0	316
%	65.82	19.62	12.97	1.58	0.00	0.00	
sigma	2.67	2.23	1.89	0.70	0.00	0.00	
2psigma	71.16	24.09	16.76	2.99	0.00	0.00	
2msigma	60.49	15.15	9.19	0.18	0.00	0.00	
Conc	11.09	3.31	2.19	0.27	0	0	
Fabric	Matrix	Sand	Silt	Micac	Biref	Colour	
	342	29	29	2	2	medbr	400

Minerals	Size	Ang	Sph	Wth	Conc	Type	CarOpa
PFs	2	3	2		2		
MFs	2	3	2		2		
Bio					1		
Mus	1			2	4		
Amp					2		
Car					3	2	2
Quartz		3	2			m,p	
Pyrox					1		
Epidote					2		
Schist					1		
Other						arf	

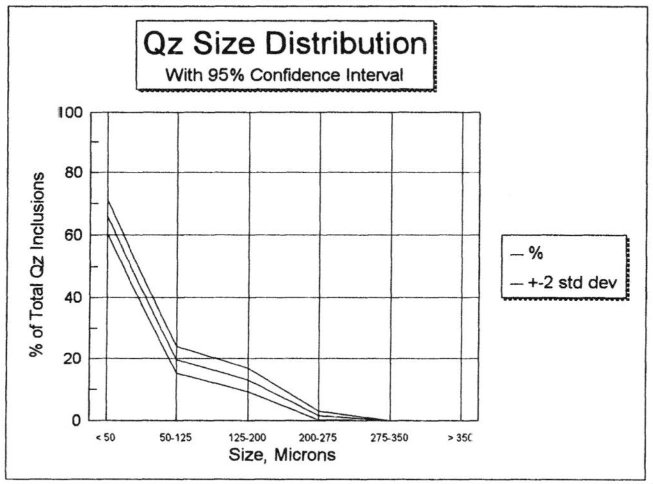

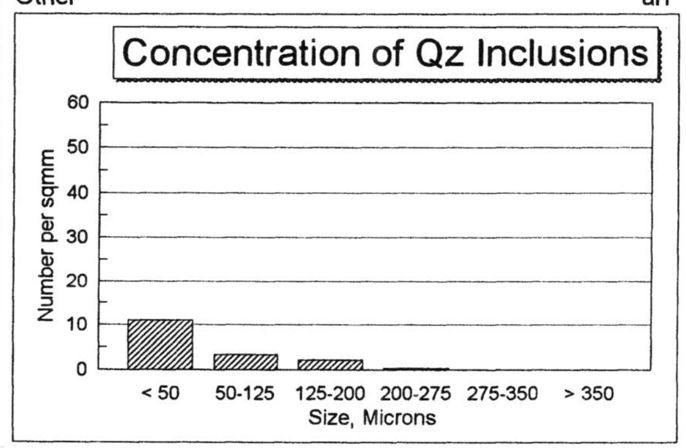

%matrix	85.50
%sand	7.25
%silt	7.25
conc<50	11.09
conc>50	5.76
%<50	65.82
%>50	34.18

RHD 52	period	level	morph type	rim dia	thick	sqmm
B IV/4	1	14	AII	11	5	12.5

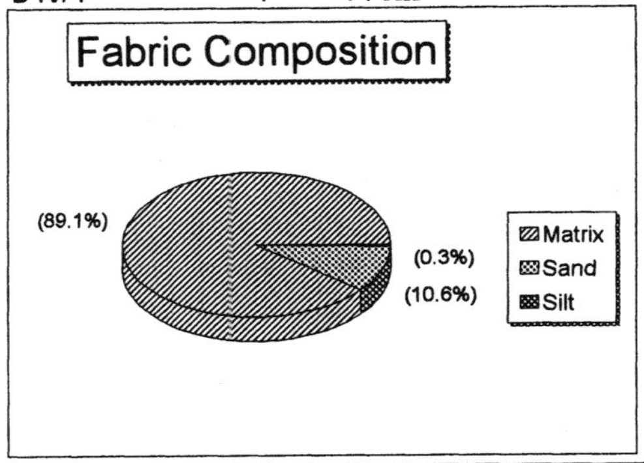

Qz Size	< 50	50-125	125-200	200-275	275-350	> 350	numpts
Number	12	7	12	10	6	3	50
%	24.00	14.00	24.00	20.00	12.00	6.00	
sigma	6.04	4.91	6.04	5.66	4.60	3.36	
2psigma	36.08	23.81	36.08	31.31	21.19	12.72	
2msigma	11.92	4.19	11.92	8.69	2.81	-0.72	
Conc	0.96	0.56	0.96	0.80	0.48	0.24	
Fabric	Matrix	Sand	Silt	Micac	Biref	Colour	
	344	41	1	1	1	dkbrn	386

Minerals	Size	Ang	Sph	Wth	Conc	Type	CarOpa
PFs	4	3	2		4		
MFs					1		
Bio	3			4	3		
Mus	2			3	3		
Amp					3		
Car	3				3	2	2
Quartz		3	2			m,p	
Pyrox					1		
Epidote					2		
Schist					2		

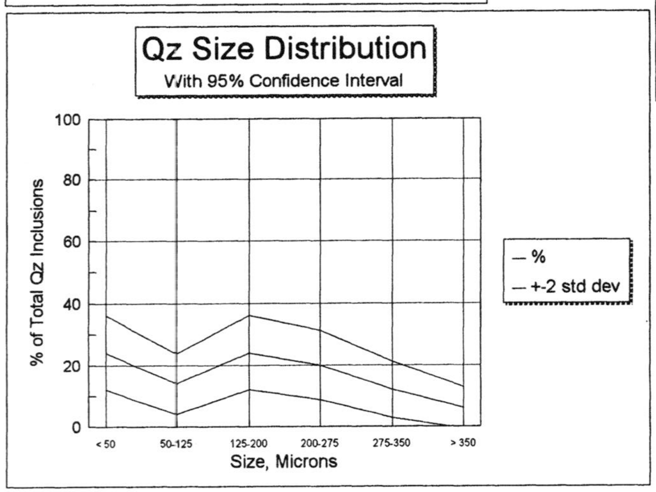

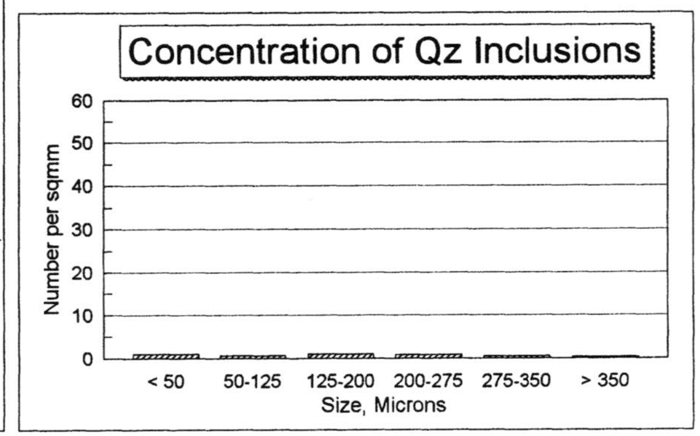

%matrix	89.12
%sand	10.62
%silt	0.26
conc<50	0.96
conc>50	3.04
%<50	24.00
%>50	76.00

RHD 53	period	level	morph type	rim dia	thick	sqmm
B IV/4	1	19	AII	9	4	18.75

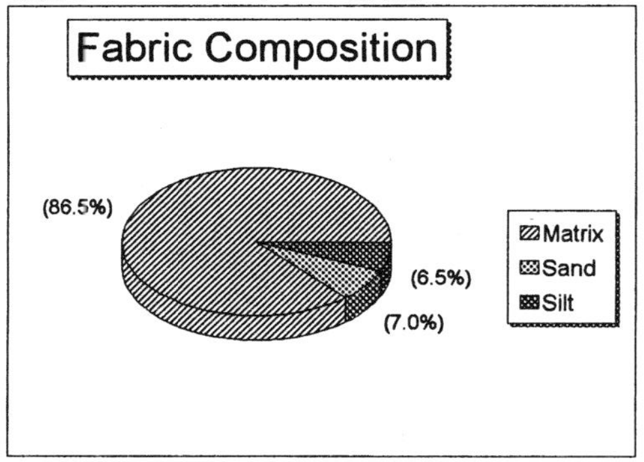

Qz Size	< 50	50-125	125-200	200-275	275-350	> 350	numpts
Number	151	102	20	7	0	0	280
%	53.93	36.43	7.14	2.50	0.00	0.00	
sigma	2.98	2.88	1.54	0.93	0.00	0.00	
2psigma	59.89	42.18	10.22	4.37	0.00	0.00	
2msigma	47.97	30.68	4.06	0.63	0.00	0.00	
Conc	8.05	5.44	1.07	0.37	0	0	
Fabric	Matrix	Sand	Silt	Micac	Biref	Colour	
	333	27	25	2	2	medbr	385

Minerals	Size	Ang	Sph	Wth	Conc	Type	CarOpa
PFs	1	3	3		2		
MFs					1		
Bio	2			3	2		
Mus	2			3	4		
Amp	1				2		
Car	2				4	2	2
Quartz		3	3			m,p	
Pyrox					1		
Epidote					1		
Schist					2		

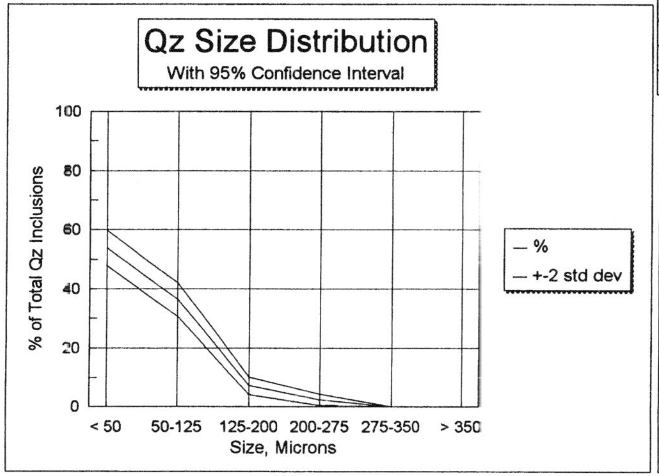

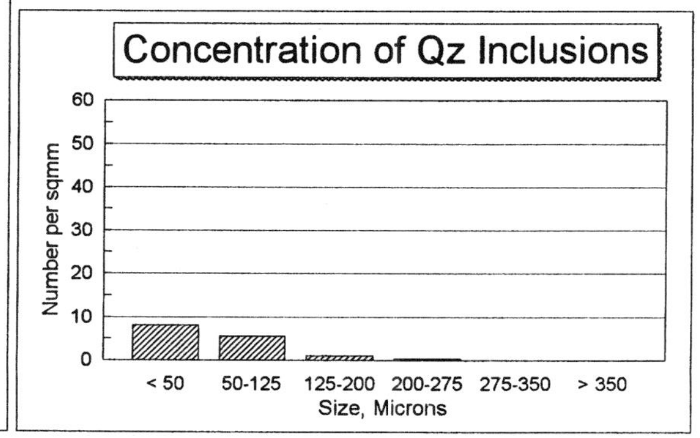

%matrix	86.49
%sand	7.01
%silt	6.49
conc<50	8.05
conc>50	6.88
%<50	53.93
%>50	46.07

RHD 54	period	level	morph type	rim dia	thick	sqmm
B IV/14	2	11	AII	10	7	12.5

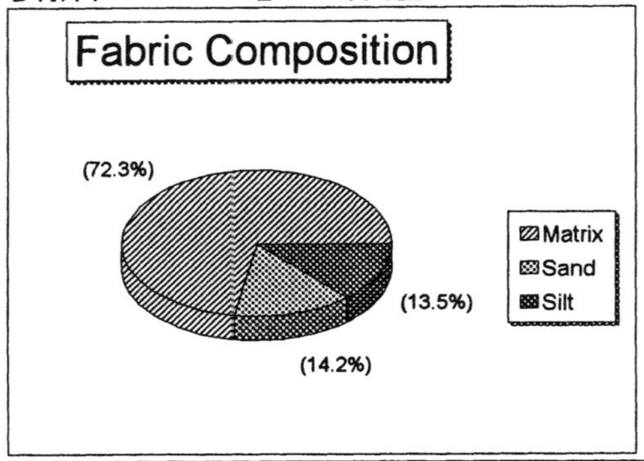

Qz Size	< 50	50-125	125-200	200-275	275-350	> 350	numpts
Number	410	84	32	18	3	3	550
%	74.55	15.27	5.82	3.27	0.55	0.55	
sigma	1.86	1.53	1.00	0.76	0.31	0.31	
2psigma	78.26	18.34	7.81	4.79	1.17	1.17	
2msigma	70.83	12.20	3.82	1.76	-0.08	-0.08	
Conc	32.8	6.72	2.56	1.44	0.24	0.24	
Fabric	Matrix	Sand	Silt	Micac	Biref	Colour	
	285	56	53	1	1	dkbrn	394
Minerals	Size	Ang	Sph	Wth	Conc	Type	CarOpa
PFs	3	4	3		4		
MFs					1		
Bio	3			3	3		
Mus	2			3	4		
Amp	3				3		
Car					2	2	undef
Quartz		4	3			m,p	
Pyrox					3		
Epidote					2		
Schist					2		

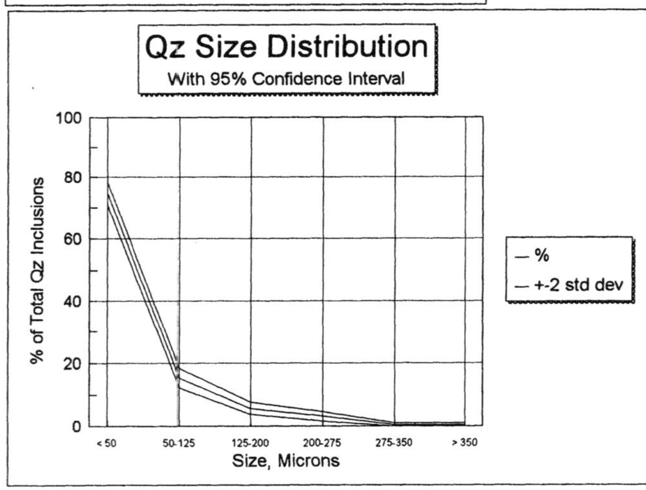

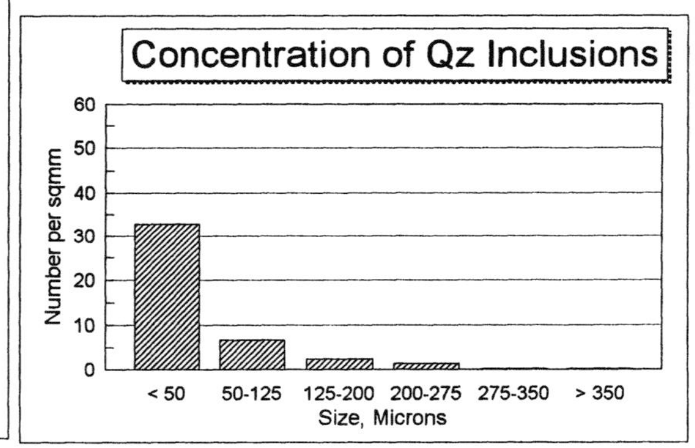

%matrix	72.34
%sand	14.21
%silt	13.45
conc<50	32.80
conc>50	11.20
%<50	74.55
%>50	25.45

RHD 55	period	level	morph type	rim dia	thick	sqmm
B IV/9	3	1	AII	13	7	18.75

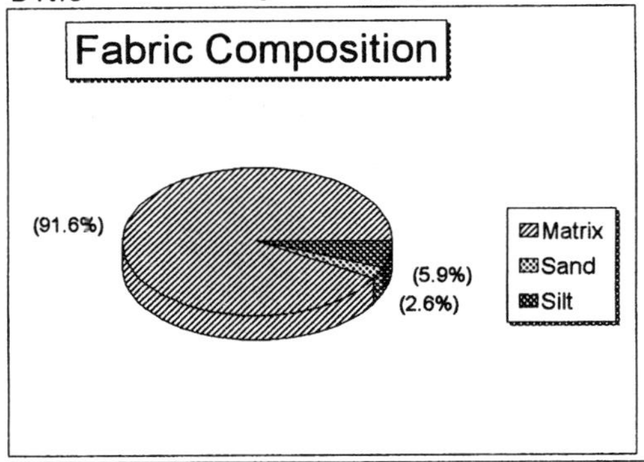

Qz Size	< 50	50-125	125-200	200-275	275-350	> 350	numpts
Number	270	28	17	10	0	0	325
%	83.08	8.62	5.23	3.08	0.00	0.00	
sigma	2.08	1.56	1.24	0.96	0.00	0.00	
2psigma	87.24	11.73	7.70	4.99	0.00	0.00	
2msigma	78.92	5.50	2.76	1.16	0.00	0.00	
Conc	14.40	1.49	0.91	0.53	0	0	
Fabric	Matrix	Sand	Silt	Micac	Biref	Colour	
	358	10	23	2	2	medbr	391

Minerals	Size	Ang	Sph	Wth	Conc	Type	CarOpa
PFs					1		
MFs					1		
Bio	2			3	3		
Mus	2			3	4		
Amp	1				2		
Car	2				2	2	undef
Quartz		3	2			m,p	
Pyrox					1		
Epidote					2		
Schist					1		

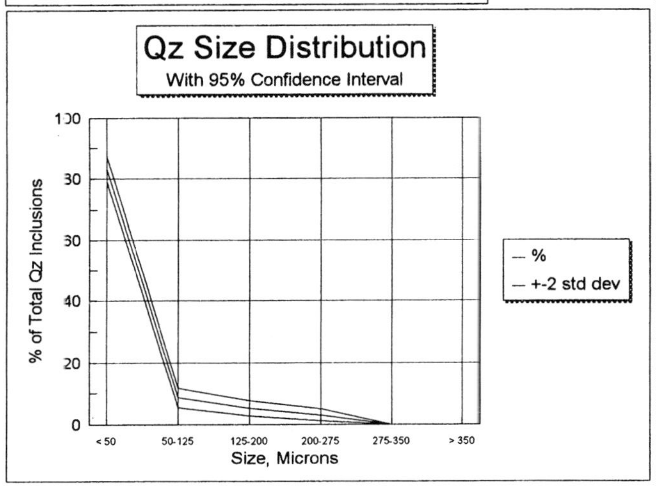

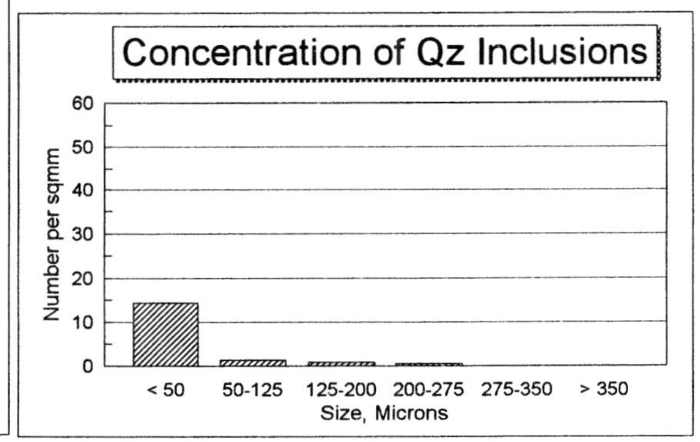

%matrix	91.56
%sand	2.56
%silt	5.88
conc<50	14.40
conc>50	2.93
%<50	83.08
%>50	16.92

APPENDIX A

RHD 56	period	level	morph type	rim dia	thick	sqmm
B IV/19	1	14	AII	10	7	31.25

Qz Size	< 50	50-125	125-200	200-275	275-350	> 350	numpts
Number	59	73	60	49	14	4	259
%	22.78	28.19	23.17	18.92	5.41	1.54	
sigma	2.61	2.80	2.62	2.43	1.41	0.77	
2psigma	27.99	33.78	28.41	23.79	8.22	3.08	
2msigma	17.57	22.59	17.92	14.05	2.60	0.01	
Conc	1.89	2.34	1.92	1.57	0.45	0.13	
Fabric	Matrix	Sand	Silt	Micac	Biref	Colour	
	290	91	9	3	3	dkbrn	390

Minerals	Size	Ang	Sph	Wth	Conc	Type	CarOpa
PFs	3	4	2		3		
MFs					1		
Bio	3			3	3		
Mus	3			3	4		
Amp	2						
Car	3				3	2	2,rare3
Quartz		4	2			m,p	
Pyrox					1		
Epidote					2		
Schist					3	epidote	

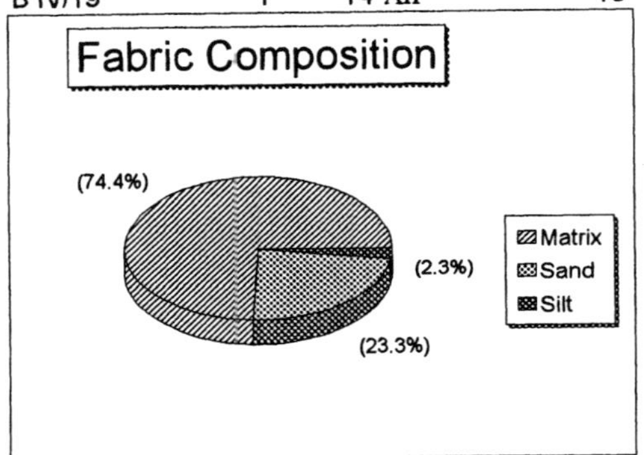

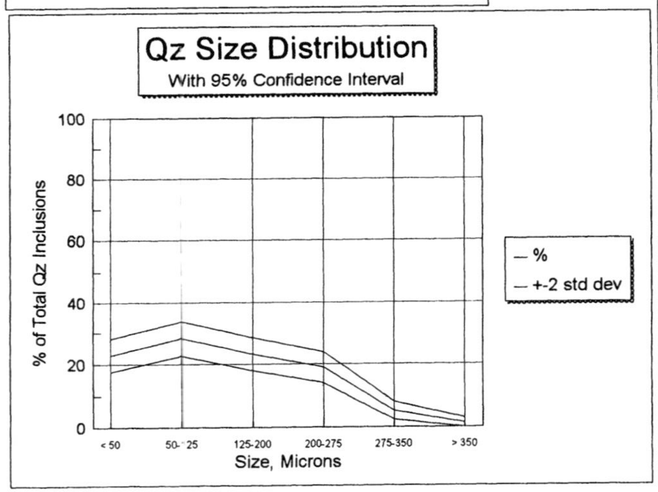

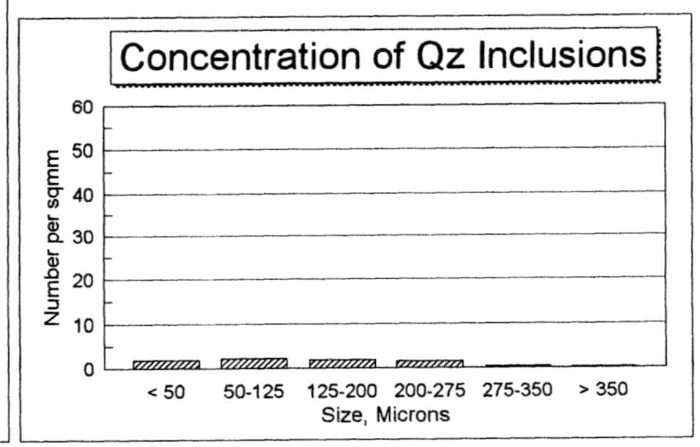

%matrix	74.36
%sand	23.33
%silt	2.31
conc<50	1.89
conc>50	6.40
%<50	22.78
%>50	77.22

RHD 60	period	level	morph type	rim dia	thick	sqmm
waster		surf	BIII		6	31.25

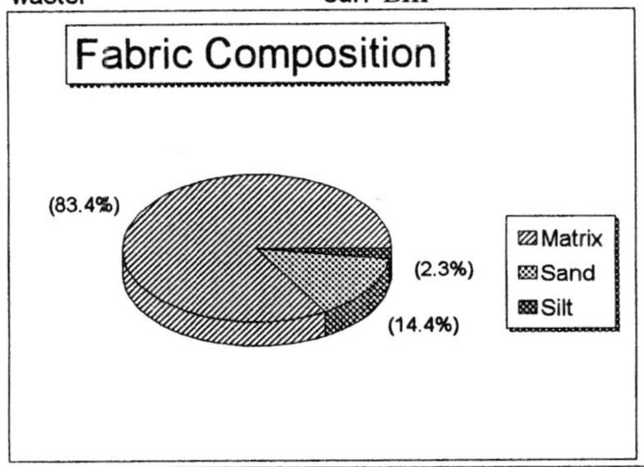

Qz Size	< 50	50-125	125-200	200-275	275-350	> 350	numpts
Number	122	53	45	37	6	4	267
%	45.69	19.85	16.85	13.86	2.25	1.50	
sigma	3.05	2.44	2.29	2.11	0.91	0.74	
2psigma	51.79	24.73	21.44	18.09	4.06	2.98	
2msigma	39.60	14.97	12.27	9.63	0.43	0.01	
Conc	3.90	1.70	1.44	1.18	0.19	0.13	
Fabric	Matrix	Sand	Silt	Micac	Biref	Colour	
	331	57	9	3	2	vdkbr	397

Minerals	Size	Ang	Sph	Wth	Conc	Type	CarOpa
PFs	3	4	2		3		
MFs					1		
Bio	3			3	3		
Mus	2			3	3		
Amp	3				3		
Car	3				3	2	2
Quartz		4	2			m,p	
Pyrox					1		
Epidote					2		
Schist					2	epidote	

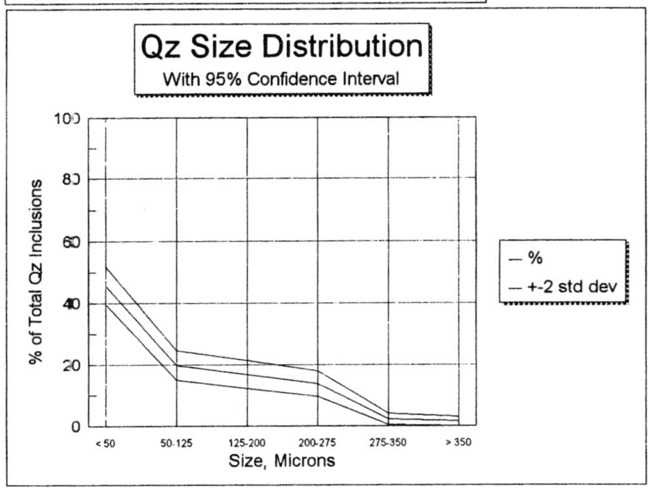

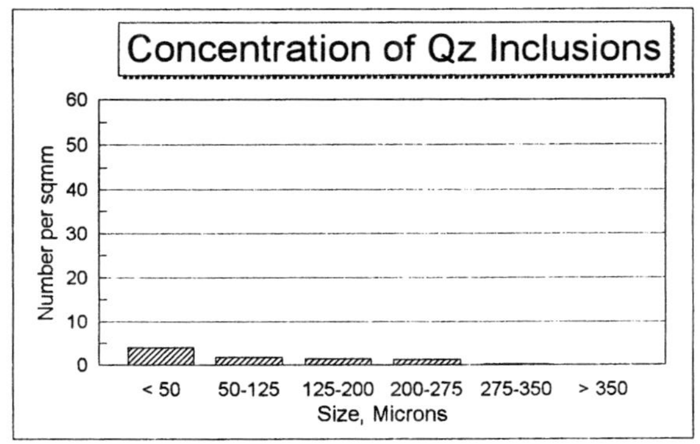

%matrix	83.38
%sand	14.36
%silt	2.27
conc<50	3.90
conc>50	4.64
%<50	45.69
%>50	54.31

APPENDIX A

RHD 61	period	level	morph type	rim dia	thick	sqmm
waster		surf			14	31.25

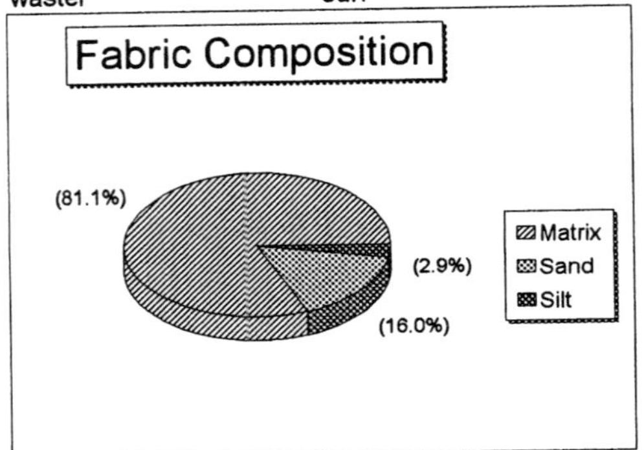

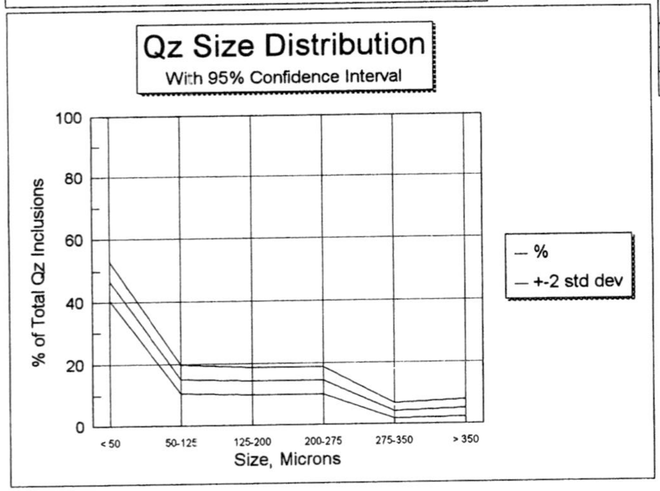

Qz Size	< 50	50-125	125-200	200-275	275-350	> 350	numpts
Number	117	38	36	36	11	13	251
%	46.61	15.14	14.34	14.34	4.38	5.18	
sigma	3.15	2.26	2.21	2.21	1.29	1.40	
2psigma	52.91	19.66	18.77	18.77	6.97	7.98	
2msigma	40.32	10.61	9.92	9.92	1.80	2.38	
Conc	3.74	1.22	1.15	1.15	0.35	0.42	
Fabric	Matrix	Sand	Silt	Micac	Biref	Colour	
	305	60	11	3	3	medbr	376

Minerals	Size	Ang	Sph	Wth	Conc	Type	CarOpa
PFs	3	3	2		3		
MFs					1		
Bio	3			2	3		
Mus	3			3	4		
Amp	3				4		
Car	3				2	2	undef
Quartz		3	2			m,p	
Pyrox					1		
Epidote					2		
Schist					3	epidote	

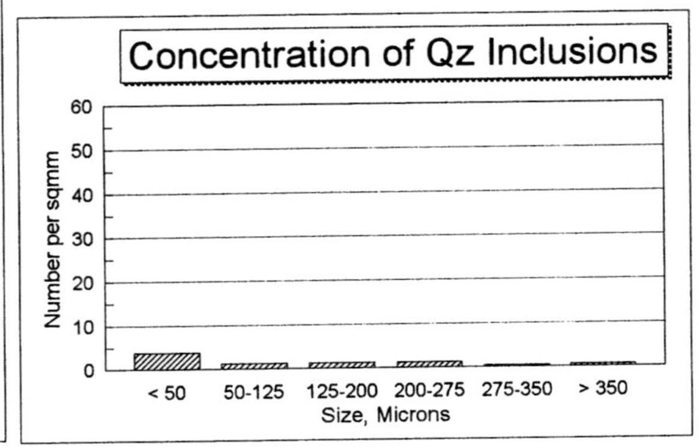

%matrix	81.12
%sand	15.96
%silt	2.93
conc<50	3.74
conc>50	4.29
%<50	46.61
%>50	53.39

APPENDIX A

RHD 62	period	level	morph type	rim dia	thick	sqmm
waster		surf	AII		6	18.75

Qz Size	< 50	50-125	125-200	200-275	275-350	> 350	numpts
Number	116	111	61	19	3	1	311
%	37.30	35.69	19.61	6.11	0.96	0.32	
sigma	2.74	2.72	2.25	1.36	0.55	0.32	
2psigma	42.78	41.12	24.12	8.83	2.07	0.96	
2msigma	31.81	30.26	15.11	3.39	-0.14	-0.32	
Conc	6.19	5.92	3.25	1.01	0.16	0.05	
Fabric	Matrix	Sand	Silt	Micac	Biref	Colour	
	302	59	15	1	1	vdkbr	376
Minerals	Size	Ang	Sph	Wth	Conc	Type	CarOpa
PFs	2	3	2		4		
MFs					1		
Bio					1		
Mus	3			3	2		
Amp					1		
Car					3	2	2,3
Quartz		3	2			m,p	
Pyrox					1		
Epidote					2		
Schist					2		

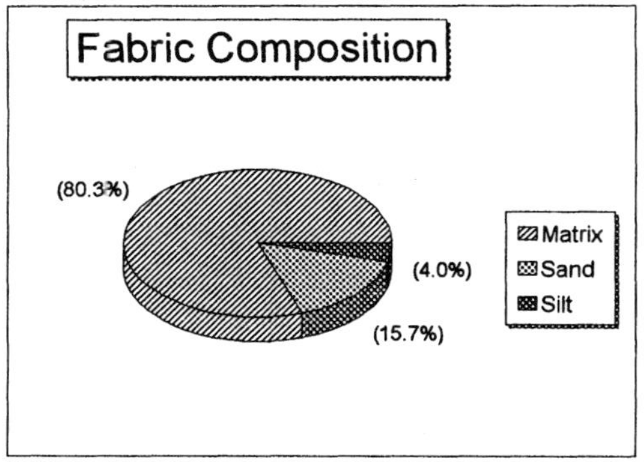

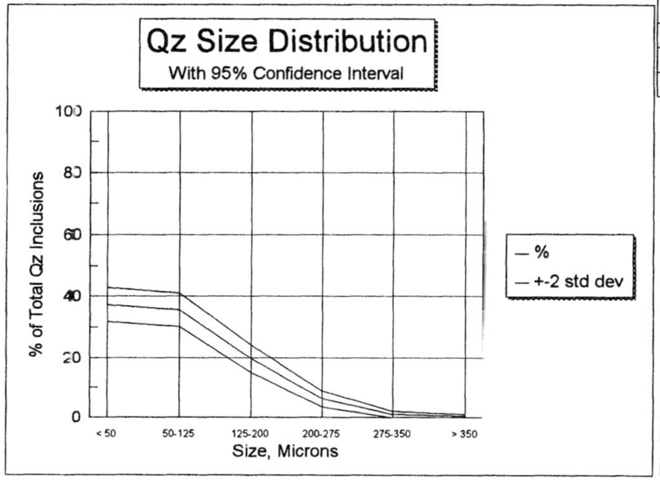

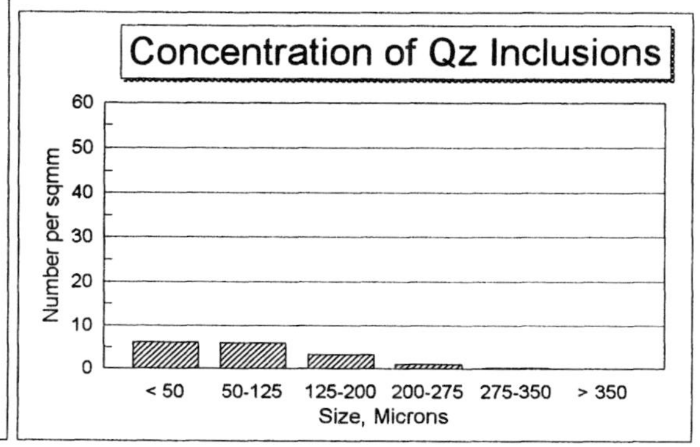

%matrix	80.32
%sand	15.69
%silt	3.99
conc<50	6.19
conc>50	10.40
%<50	37.30
%>50	62.70

RHD 63	period	level	morph type	rim dia	thick	sqmm
waster		surf	BIII/BIV		5	31.25

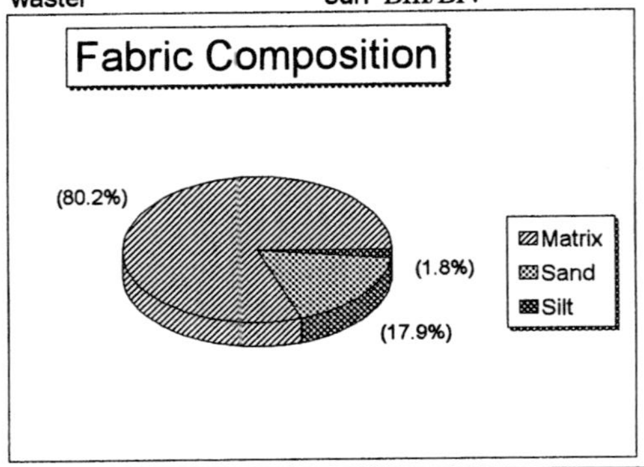

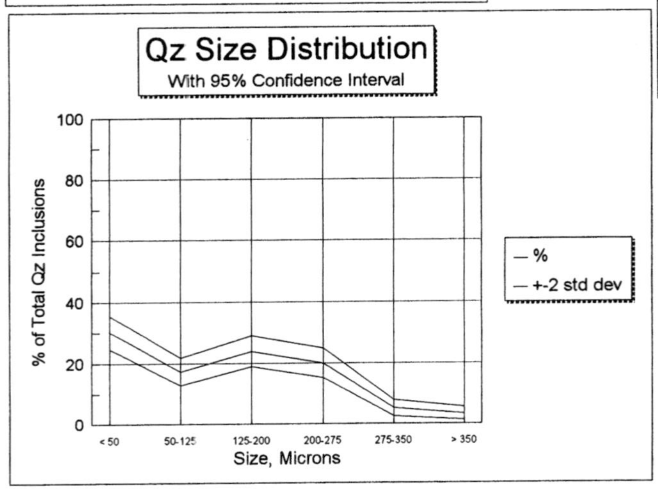

Qz Size	< 50	50-125	125-200	200-275	275-350	> 350	numpts
Number	87	50	69	58	15	10	289
%	30.10	17.30	23.88	20.07	5.19	3.46	
sigma	2.70	2.23	2.51	2.36	1.30	1.08	
2psigma	35.50	21.75	28.89	24.78	7.80	5.61	
2msigma	24.71	12.85	18.86	15.36	2.58	1.31	
Conc	2.78	1.60	2.21	1.86	0.48	0.32	
Fabric	Matrix	Sand	Silt	Micac	Biref	Colour	
	304	68	7	1	1	dkgray	379

Minerals	Size	Ang	Sph	Wth	Conc	Type	CarOpa
PFs	3	3	2		4		
MFs					1		
Bio	3			3	2		
Mus	3			3	3		
Amp					1		
Car	4				4	2	2,3
Quartz		3	2			m,p	
Pyrox					1		
Epidote					2		
Schist					3		

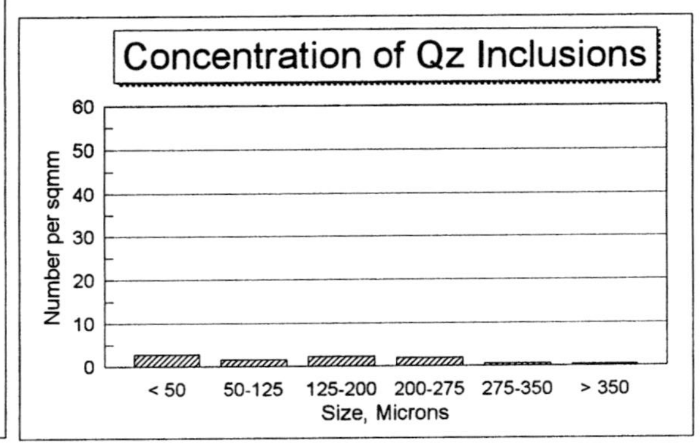

%matrix	80.21
%sand	17.94
%silt	1.85
conc<50	2.78
conc>50	6.46
%<50	30.10
%>50	69.90

RHD 64	period	level	morph type	rim dia	thick	sqmm
waster		surf			12	25

Qz Size	< 50	50-125	125-200	200-275	275-350	> 350	numpts
Number	113	72	83	41	5	1	315
%	35.87	22.86	26.35	13.02	1.59	0.32	
sigma	2.70	2.37	2.48	1.90	0.70	0.32	
2psigma	41.28	27.59	31.31	16.81	3.00	0.95	
2msigma	30.47	18.13	21.39	9.22	0.18	-0.32	
Conc	4.52	2.88	3.32	1.64	0.20	0.04	
Fabric	Matrix	Sand	Silt	Micac	Biref	Colour	
	293	84	8	2	2	dkgray	385
Minerals	Size	Ang	Sph	Wth	Conc	Type	CarOpa
PFs	4	4	2		4		
MFs					2		
Bio	3			3	3		
Mus	3			3	4		
Amp	2				3		
Car	3	4,3	2		3	2	2
Quartz						m,p	
Pyrox					2		
Epidote					2		
Schist					3		

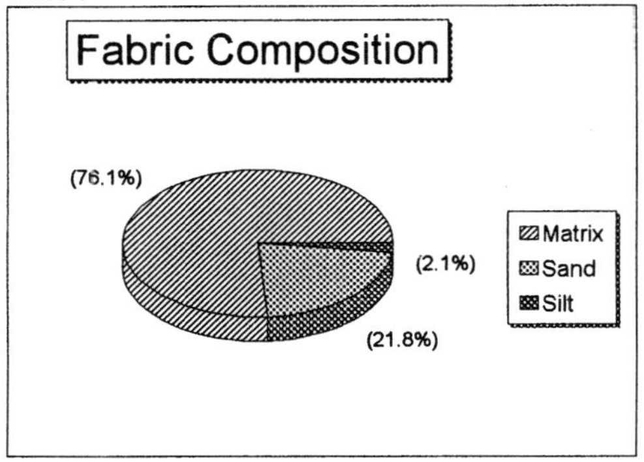

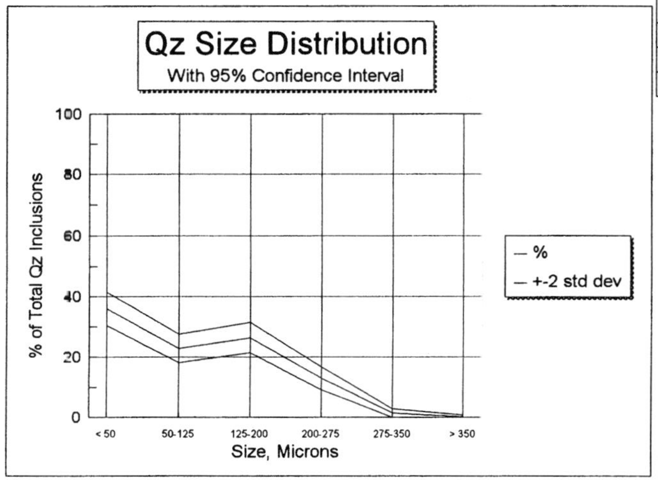

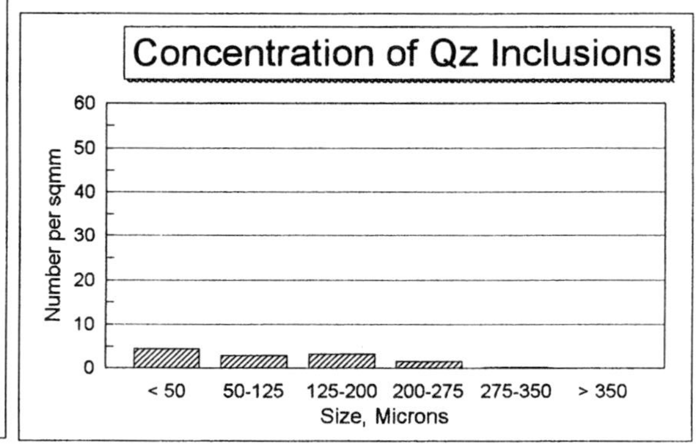

%matrix	76.10
%sand	21.82
%silt	2.08
conc<50	4.52
conc>50	8.08
%<50	35.87
%>50	64.13

RHD 65
Waster

Section damaged; not reliable

RHD 66	period	level	morph type	rim dia	thick	sqmm
waster		surf			8	31.25

Qz Size	< 50	50-125	125-200	200-275	275-350	> 350	numpts
Number	61	45	56	58	14	6	240
%	25.42	18.75	23.33	24.17	5.83	2.50	
sigma	2.81	2.52	2.73	2.76	1.51	1.01	
2psigma	31.04	23.79	28.79	29.69	8.86	4.52	
2msigma	19.80	13.71	17.87	18.64	2.81	0.48	
Conc	1.95	1.44	1.79	1.86	0.45	0.19	
Fabric	Matrix	Sand	Silt	Micac	Biref	Colour	
	285	104	6	1	1	black	395

Minerals	Size	Ang	Sph	Wth	Conc	Type	CarOpa
PFs	4	3	2		4		
MFs					1		
Bio					1		
Mus	3			3	3		
Amp					2		
Car	3				3	2	2,3
Quartz		4,3	2			m,p	
Pyrox					2		
Epidote					2		
Schist					3		

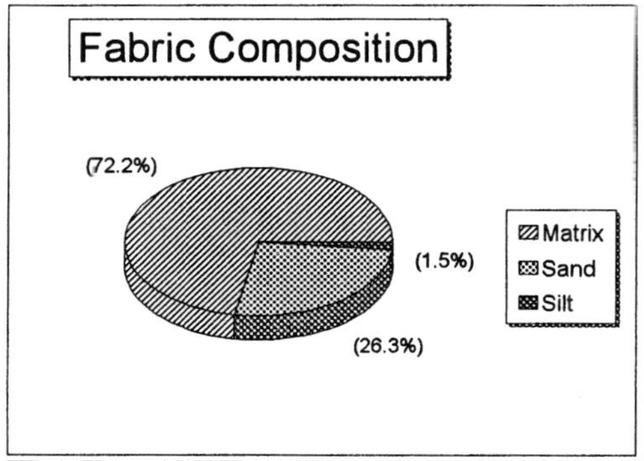

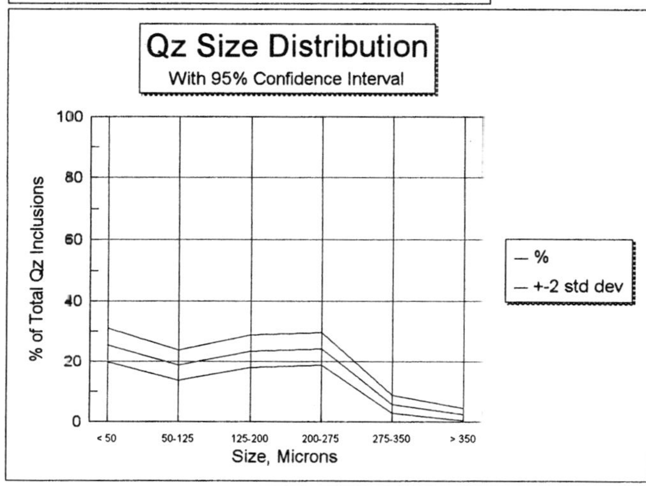

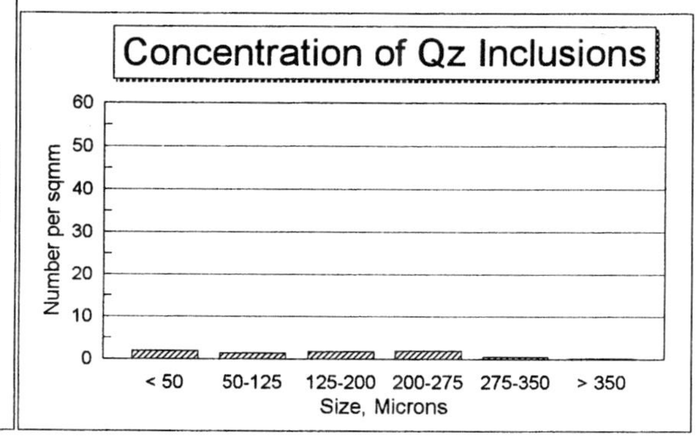

%matrix	72.15
%sand	26.33
%silt	1.52
conc<50	1.95
conc>50	5.73
%<50	25.42
%>50	74.58

RHD 67	period	level	morph type	rim dia	thick	sqmm
waster		surf			8	

Section damaged; not countable

Qz Size	< 50	50-125	125-200	200-275	275-350	> 350	numpts
Number							0
%							
sigma							
2psigma							
2msigma							
Conc							

Fabric

Minerals	Size	Ang	Sph	Wth	Conc	CarTyp	CarOpa
PFs					1		
MFs					1		
Bio	3			3	3		
Mus	3				3		
Amp					2		
Car					3	2	undef
Quartz		4	2				
Pyrox					1		
Epidote					2		
Schist					1		

RHD 69	period	level	morph type	rim dia	thick	sqmm
waster		surf			11	31.25

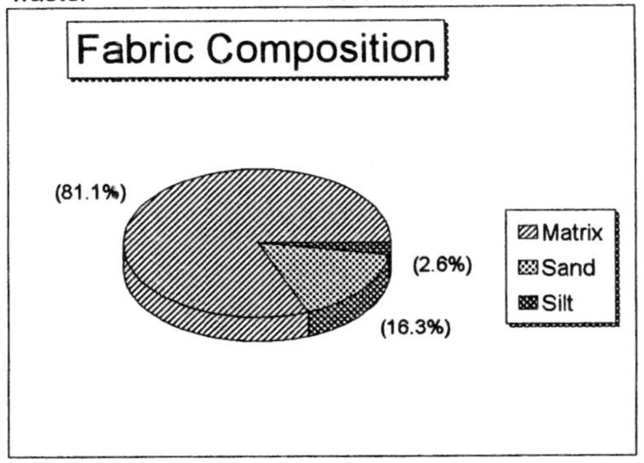

Qz Size	< 50	50-125	125-200	200-275	275-350	> 350	numpts
Number	74	54	73	50	8	3	262
%	28.24	20.61	27.86	19.08	3.05	1.15	
sigma	2.78	2.50	2.77	2.43	1.06	0.66	
2psigma	33.81	25.61	33.40	23.94	5.18	2.46	
2msigma	22.68	15.61	22.32	14.23	0.93	-0.17	
Conc	2.37	1.73	2.34	1.60	0.26	0.10	
Fabric	Matrix	Sand	Silt	Micac	Biref	Colour	
	318	64	10	1	1	black	392

Minerals	Size	Ang	Sph	Wth	Conc	Type	CarOpa
PFs	4	3	2		4		
MFs					1		
Bio	2			4	2		
Mus	3			4	3		
Amp					1		
Car	2			4	3	2	undef
Quartz		4,3	2			m,p	
Pyrox					1		
Epidote					2		
Schist					1		

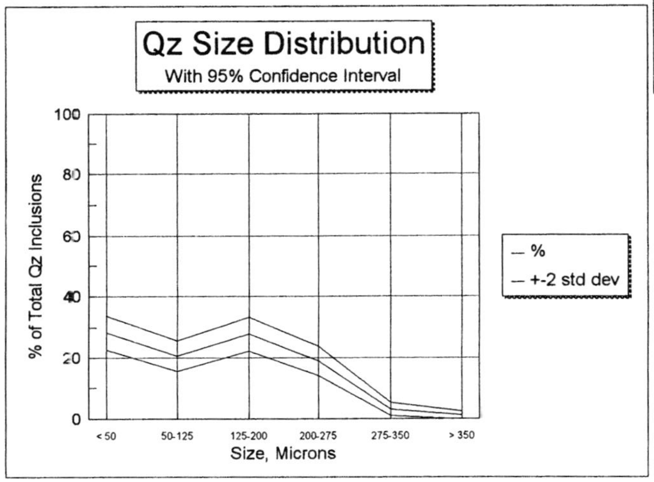

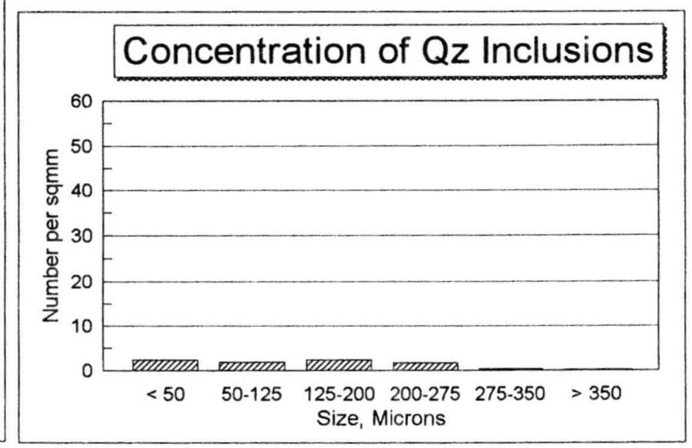

%matrix	81.12
%sand	16.33
%silt	2.55
conc<50	2.37
conc>50	6.02
%<50	28.24
%>50	71.76

RHD 70	period	level	morph type	rim dia	thick	sqmm
waster		surf				25

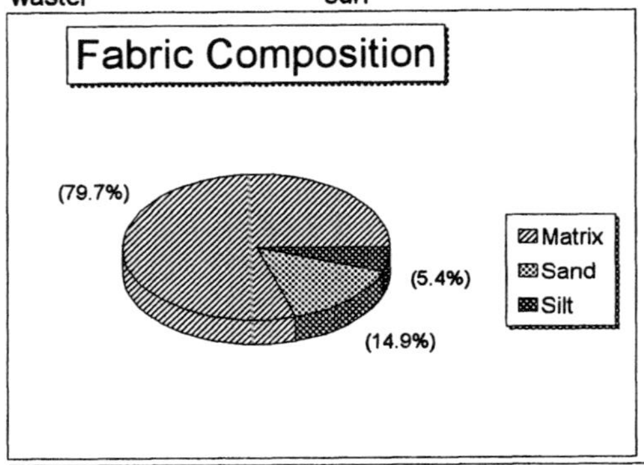

Qz Size	< 50	50-125	125-200	200-275	275-350	> 350	numpts
Number	100	67	73	41	11	5	297
%	33.67	22.56	24.58	13.80	3.70	1.68	
sigma	2.74	2.43	2.50	2.00	1.10	0.75	
2psigma	39.15	27.41	29.58	17.81	5.90	3.18	
2msigma	28.19	17.71	19.58	9.80	1.51	0.19	
Conc	4.00	2.68	2.92	1.64	0.44	0.20	
Fabric	Matrix	Sand	Silt	Micac	Biref	Colour	
	310	58	21	1	1	black	389

Minerals	Size	Ang	Sph	Wth	Conc	Type	CarOpa
PFs	4	3	2		4		
MFs	2	4	2	4	2		
Bio	3			3	4		
Mus	3			3	4		
Amp	2				4		
Car	4			3	4	2	2,3,4
Quartz		4	2			m,p	
Pyrox					1		
Epidote					2		
Schist					3		

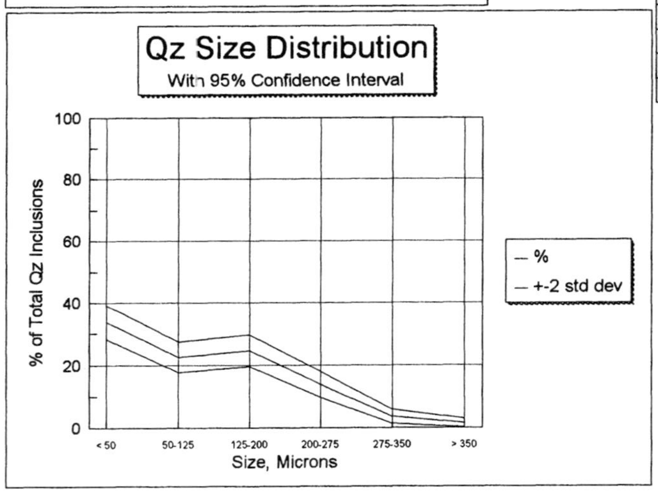

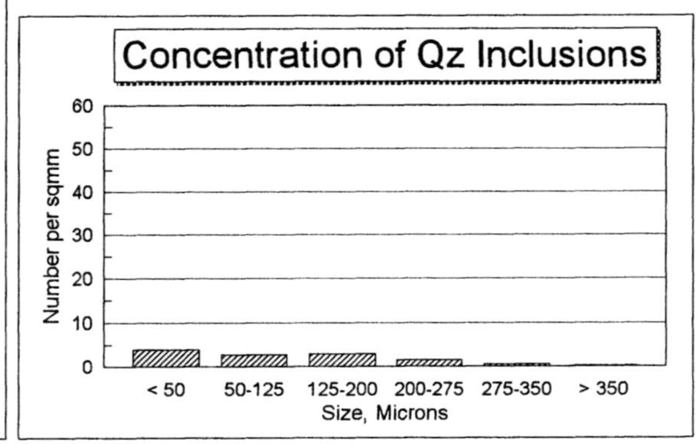

%matrix	79.69
%sand	14.91
%silt	5.40
conc<50	4.00
conc>50	7.88
%<50	33.67
%>50	66.33

RHD 71	period	level	morph type	rim dia	thick	sqmm
waster		surf			8	18.75

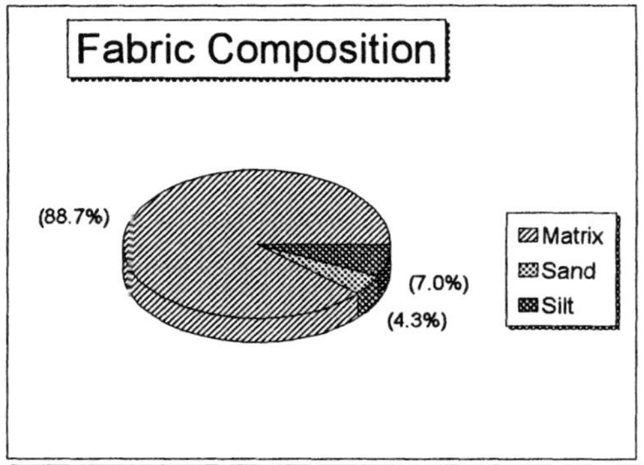

Qz Size	< 50	50-125	125-200	200-275	275-350	> 350	numpts
Number	183	69	24	7	0	1	284
%	64.44	24.30	8.45	2.46	0.00	0.35	
sigma	2.84	2.54	1.65	0.92	0.00	0.35	
2psigma	70.12	29.39	11.75	4.30	0.00	1.06	
2msigma	58.76	19.21	5.15	0.62	0.00	-0.35	
Conc	9.76	3.68	1.28	0.37	0.00	0.05	
Fabric	Matrix	Sand	Silt	Micac	Biref	Colour	
	353	17	28	2	2	black	398

Minerals	Size	Ang	Sph	Wth	Conc	Type	CarOpa
PFs					1		
MFs					1		
Bio					1		
Mus	1				4		
Amp					1		
Car	3				4	2	undef
Quartz		3	2			m,p	
Pyrox					1		
Epidote					1		
Schist					1		

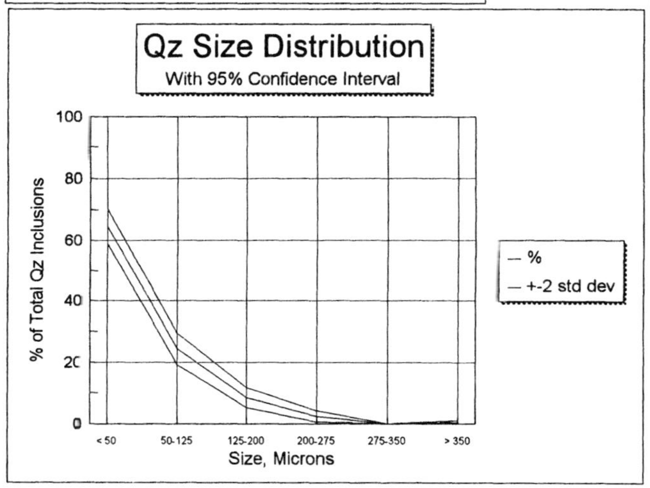

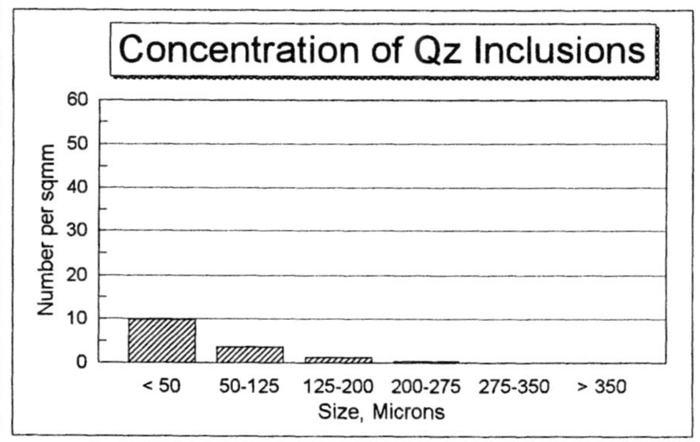

%matrix	88.69
%sand	4.27
%silt	7.04
conc<50	9.76
conc>50	5.39
%<50	64.44
%>50	35.56

APPENDIX B

BANNU SITES DATA

(LEWAN AND TARAKAI QILA)

See page 259 for key to abbreviations used

lwn200	period	level	morph type	r m dia	thick	sqmm
lwn-2 A2			BI	14	4.6	18.75

Qz Size	< 50	50-125	125-200	200-275	275-350	> 350	numpts
Number	385	15	7	5	1	0	413
%	93.22	3.63	1.69	1.21	0.24	0.00	
sigma	1.24	0.92	0.64	0.54	0.24	0.00	
p2sigma	95.69	5.47	2.97	2.29	0.73	0.00	
m2sigma	90.75	1.79	0.42	0.13	-0.24	0.00	
Conc	20.53	0.80	0.37	0.27	0.05	0.00	
Fabric	Matrix	Sand	Silt	Micac	Biref	Colour	
	366	13	18	3	3	medbr	397

Minerals	Size	Ang	Sph	Wth	Conc	Type	Opac
PFs	3	3	2		2		
MFs					1		
Biotite	2			2	2		
Mus	3			2	4		
Amph	2				2		
Carb	3				3	2,3	2
Quartz		3	2		4	m,p	
Pyrox					1		
Epidote					1		
Schist					2		
other							

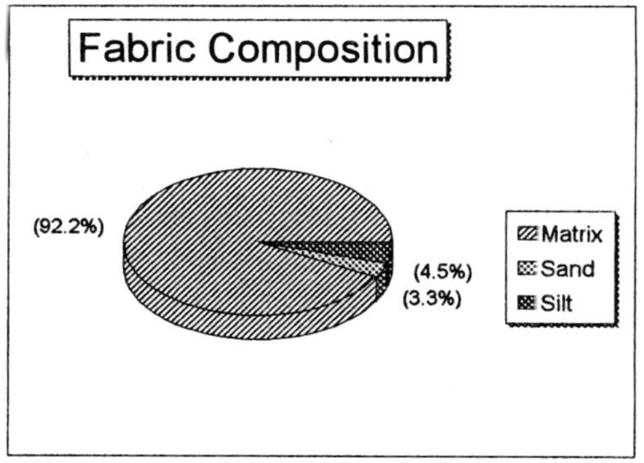

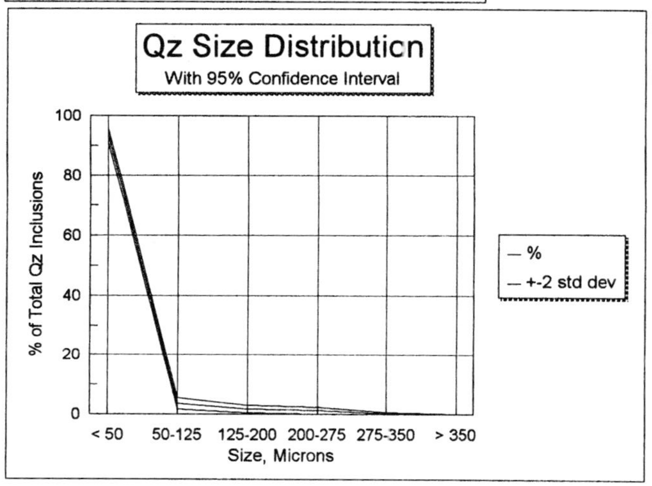

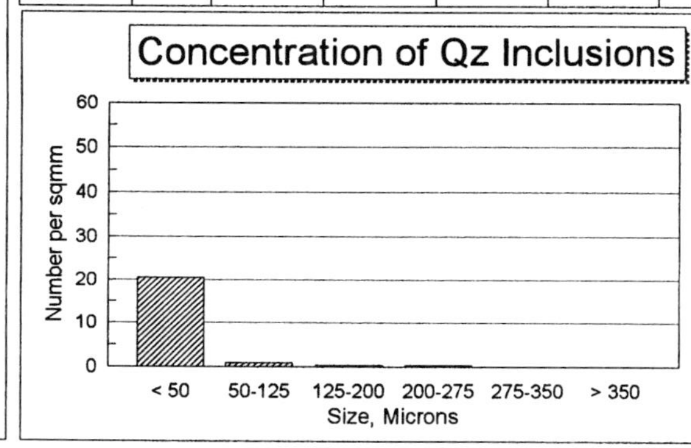

%matrix	92.19
%sand	3.27
%silt	4.53
conc<50	20.53
conc>50	1.49
%<50	93.22
%>50	6.78

APPENDIX B

lwn201	period	level	morph type	rim dia	thick	sqmm
lwn-2	A21st		BI	16	6	

Slide damaged; not countable

Qz Size	< 50	50-125	125-200	200-275	275-350	> 350	numpts
Number							0
%							
sigma							
p2sigma							
m2sigma							
Conc							
Fabric	Matrix	Sand	Silt	Micac	Biref	Colour	
				3	3	medbr	0

Minerals	Size	Ang	Sph	Wth	Conc	Type	Opac
PFs	2	3	2		2		
MFs					1		
Biotite				1	1		
Mus	1			2	4		
Amph					1		
Carb	4				3	2	2,3
Quartz		3	2		4		
Pyrox					2		
Epidote					1		
Schist					2		
other							

lwn202	period	level	morph type	rim dia	thick	sqmm
lwn-2 A2pit3			BIV or BII	29	7	12.5

Qz Size	< 50	50-125	125-200	200-275	275-350	> 350	numpts
Number	448	13	3	2	0	0	466
%	96.14	2.79	0.64	0.43	0.00	0.00	
sigma	0.89	0.76	0.37	0.30	0.00	0.00	
p2sigma	97.92	4.32	1.38	1.03	0.00	0.00	
m2sigma	94.35	1.26	-0.10	-0.18	0.00	0.00	
Conc	35.84	1.04	0.24	0.16	0.00	0.00	
Fabric	Matrix	Sand	Silt	Micac	Biref	Colour	
	369	6	24	3	3	medbr	399

Minerals	Size	Ang	Sph	Wth	Conc	Type	Opac
PFs	1	3	2		2		
MFs					1		
Biotite	2			2	2		
Mus	3			2	4		
Amph	1				2		
Carb	3				2	2,3	1
Quartz		3	2		4		
Pyrox					2		
Epidote					2		
Schist					2		
other					garnet	vrfrag	

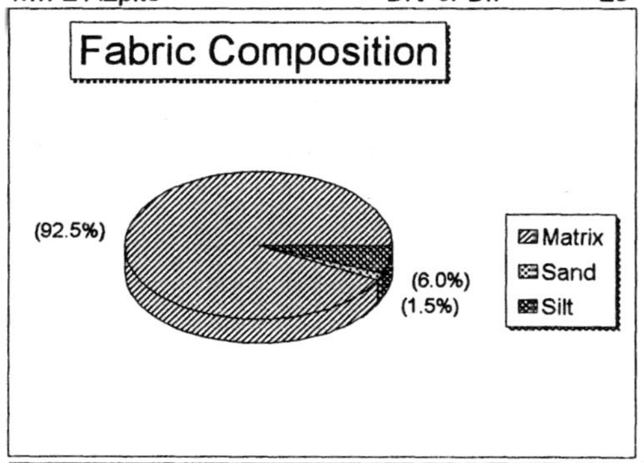

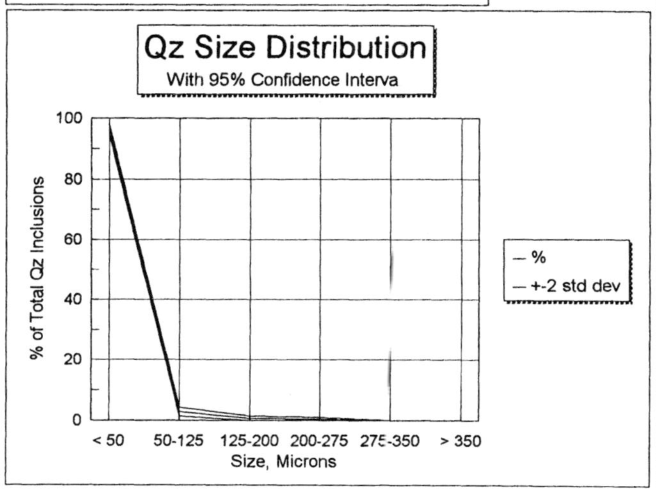

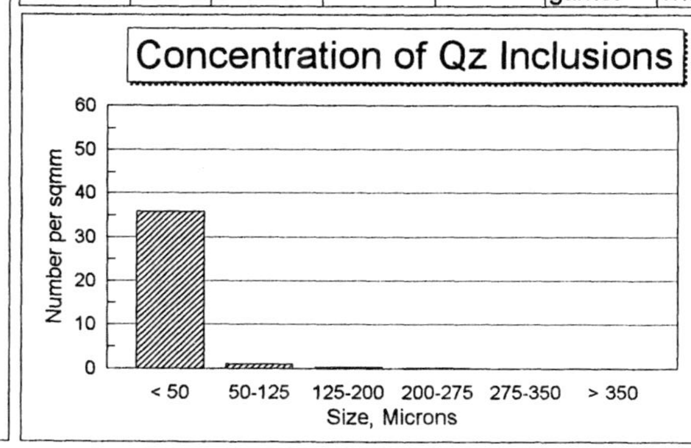

%matrix 92.48
%sand 1.50
%silt 6.02
conc<50 35.84
conc>50 1.44
%<50 96.14
%>50 3.86

APPENDIX B

lwn203	period	level	morph type	rim dia	thick	sqmm
lwn-2 A2pit3			BIV	27	7.8	12.5

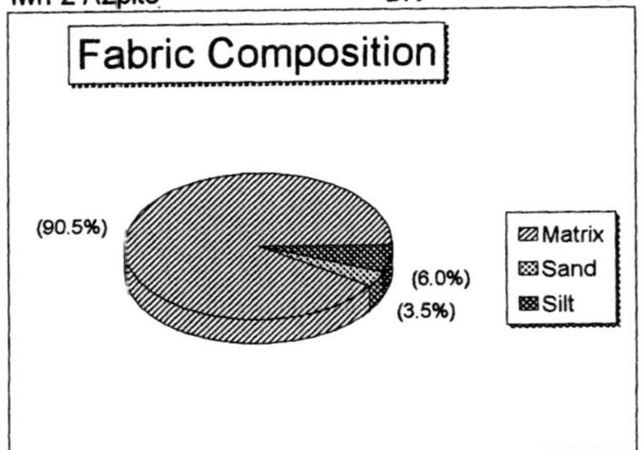

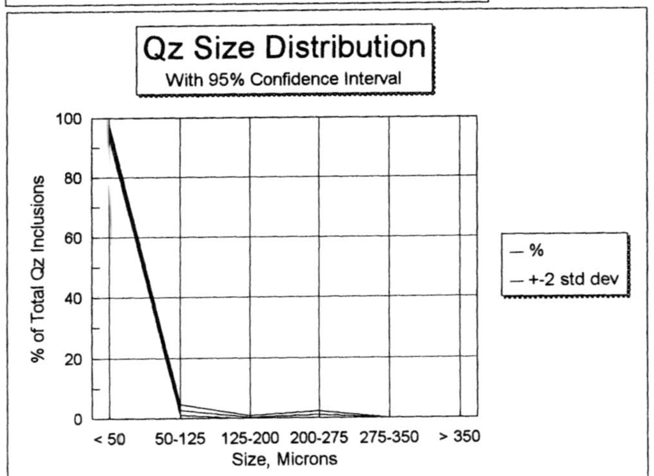

Qz Size	< 50	50-125	125-200	200-275	275-350	> 350	numpts
Number	337	10	1	4	0	0	352
%	95.74	2.84	0.28	1.14	0.00	0.00	
sigma	1.08	0.89	0.28	0.56	0.00	0.00	
p2sigma	97.89	4.61	0.85	2.27	0.00	0.00	
m2sigma	93.59	1.07	-0.28	0.01	0.00	0.00	
Conc	26.96	0.80	0.08	0.32	0.00	0.00	
Fabric	Matrix	Sand	Silt	Micac	Biref	Colour	
	362	14	24	2	2		400

Minerals	Size	Ang	Sph	Wth	Conc	Type	Opac
PFs	2	3	2		2		
MFs					1		
Biotite	1			2	2		
Mus	2			2	4		
Amph	2				2		
Carb	3				4	2,3	1
Quartz		3	2		4	m,p	
Pyrox					1		
Epidote					1		
Schist					1		
other						vrfrag	

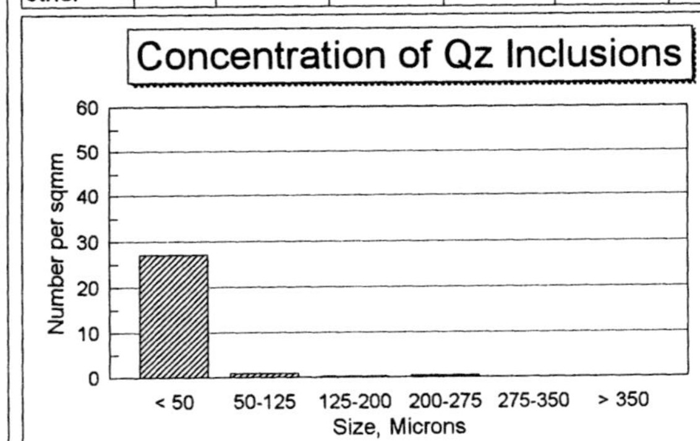

%matrix	90.50
%sand	3.50
%silt	6.00
conc<50	26.96
conc>50	1.20
%<50	95.74
%>50	4.26

wn204	period	level	morph type	rim dia	thick	sqmm
wn-1housb1			BIII	12	5.8	12.5

Qz Size	< 50	50-125	125-200	200-275	275-350	> 350	numpts
Number	422	13	2	0	0	0	437
%	96.57	2.97	0.46	0.00	0.00	0.00	
sigma	0.87	0.81	0.32	0.00	0.00	0.00	
p2sigma	98.31	4.60	1.10	0.00	0.00	0.00	
m2sigma	94.83	1.35	-0.19	0.00	0.00	0.00	
Conc	33.76	1.04	0.16	0.00	0.00	0.00	
Fabric	Matrix	Sand	Silt	Micac	Biref	Colour	
	371	7	22	3	3	medbr	400

Minerals	Size	Ang	Sph	Wth	Conc	Type	Opac
PFs	2	3	2		2		
MFs					1		
Biotite	2			2	2		
Mus	2			2	4		
Amph	2				1		
Carb	3				3	2,3	1
Quartz		3	2		4	m,p	
Pyrox					2		
Epidote					1		
Schist					2		
other						vrfrag	

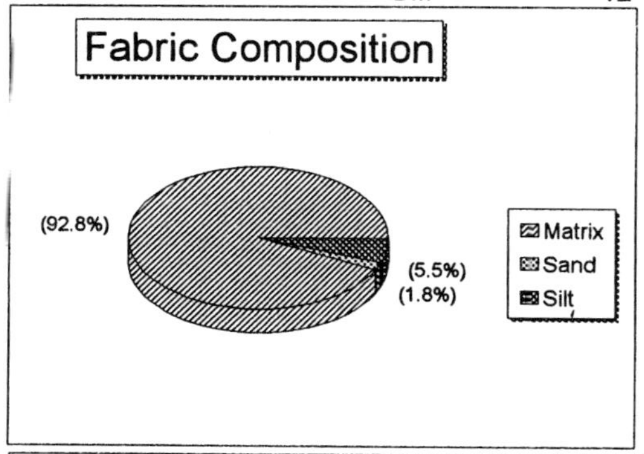

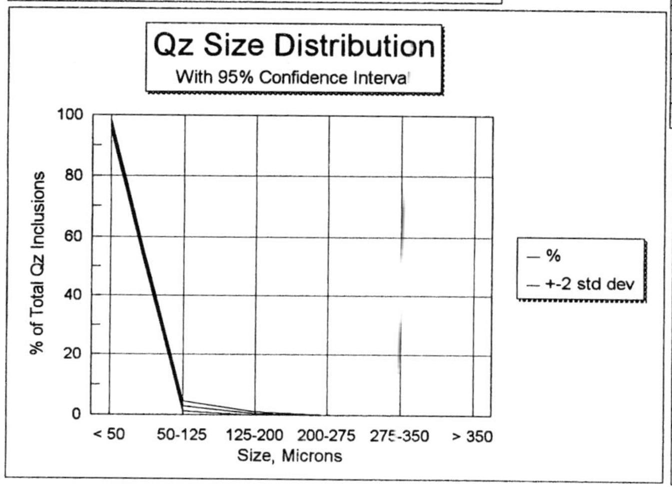

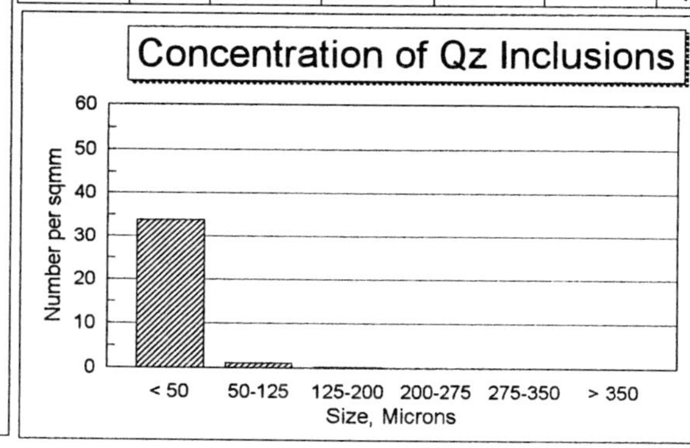

%matrix	92.75
%sand	1.75
%silt	5.50
conc<50	33.76
conc>50	1.20
%<50	96.57
%>50	3.43

APPENDIX B

lwn205	period	level	morph type	rim dia	thick	sqmm
lwn-1housb1			BIII	19	6.5	12.5

Qz Size	< 50	50-125	125-200	200-275	275-350	> 350	numpts
Number	485	34	8	0	0	0	527
%	92.03	6.45	1.52	0.00	0.00	0.00	
sigma	1.18	1.07	0.53	0.00	0.00	0.00	
p2sigma	94.39	8.59	2.58	0.00	0.00	0.00	
m2sigma	89.67	4.31	0.45	0.00	0.00	0.00	
Conc	38.80	2.72	0.64	0.00	0.00	0.00	
Fabric	Matrix	Sand	Silt	Micac	Biref	Colour	
	375	4	21	3	3	medbr	400

Minerals	Size	Ang	Sph	Wth	Conc	Type	Opac
PFs	2	3	2		2		
MFs					1		
Biotite	2			2	3		
Mus	2			2	4		
Amph	2				2		
Carb	3				4	2,3	1,2,3
Quartz		3	2		4	m,p	
Pyrox					1		
Epidote					1		
Schist					2		
other						vrfrag	

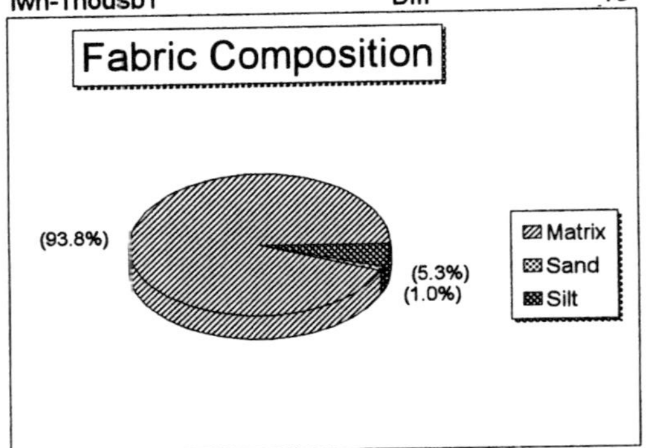

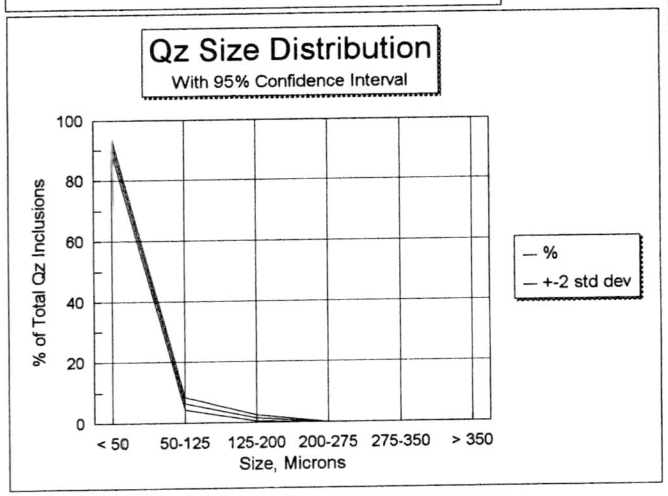

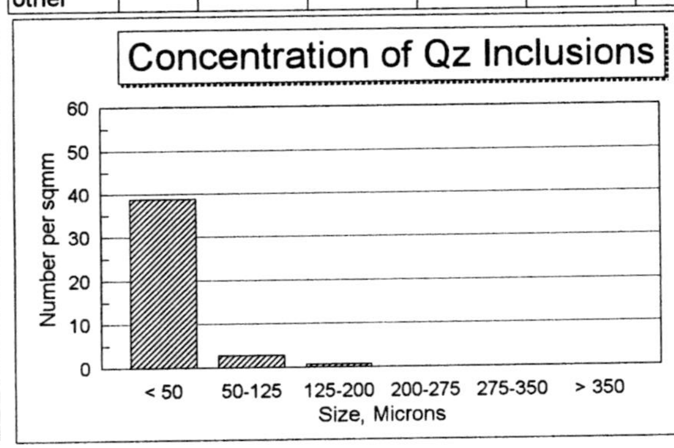

%matrix	93.75
%sand	1.00
%silt	5.25
conc<50	38.80
conc>50	3.36
%<50	92.03
%>50	7.97

lwn206	period	level	morph type	rim dia	thick	sqmm
lwn-2A2pit3			BIII or BII	36	9	12.5

Qz Size	< 50	50-125	125-200	200-275	275-350	> 350	numpts
Number	384	33	6	2	0	0	425
%	90.35	7.76	1.41	0.47	0.00	0.00	
sigma	1.43	1.30	0.57	0.33	0.00	0.00	
p2sigma	93.22	10.36	2.56	1.13	0.00	0.00	
m2sigma	87.49	5.17	0.27	-0.19	0.00	0.00	
Conc	30.72	2.64	0.48	0.16	0.00	0.00	
Fabric	Matrix	Sand	Silt	Micac	Biref	Colour	
	351	17	25	3	2	medbr	393

Minerals	Size	Ang	Sph	Wth	Conc	Type	Opac
PFs					1		
MFs					1		
Biotite	2			2	2		
Mus	3			2	4		
Amph	3				2		
Carb					3	2,3	1
Quartz		3	2		4	m,p	
Pyrox					1		
Epidote					1		
Schist					2		
other						vrfrag	

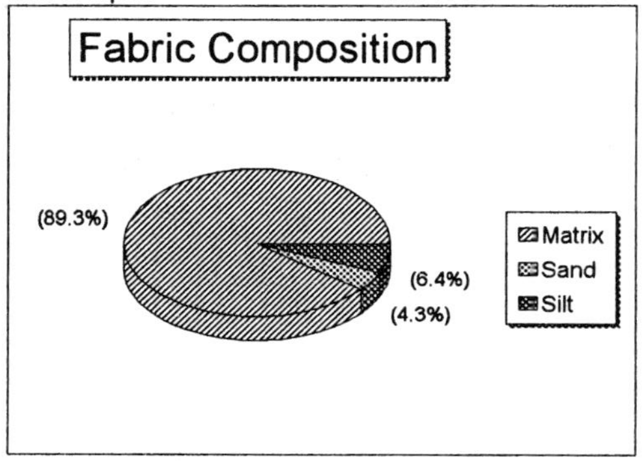

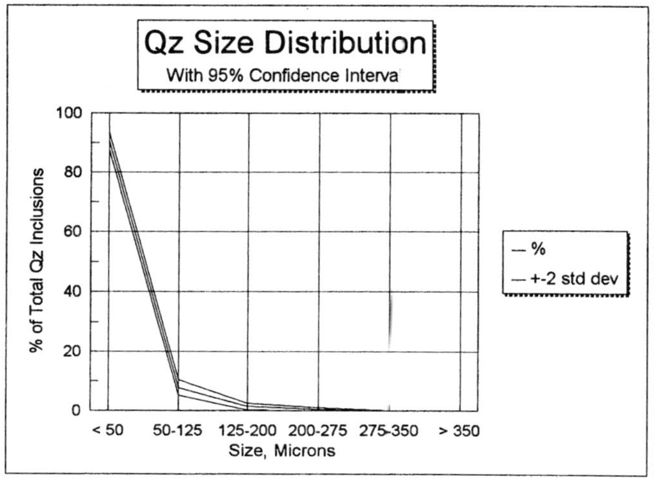

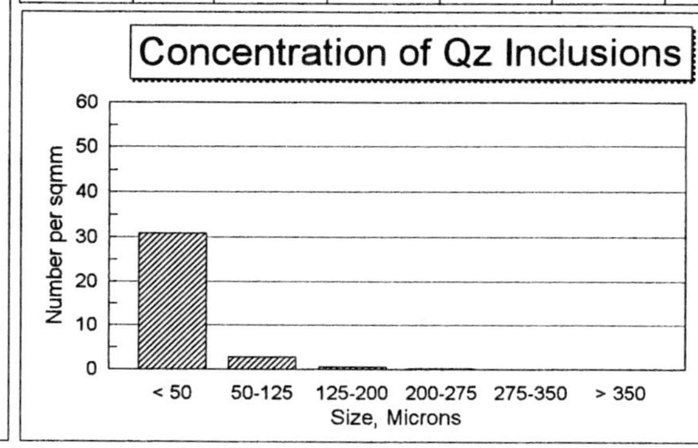

%matrix	89.31
%sand	4.33
%silt	6.36
conc<50	30.72
conc>50	3.28
%<50	90.35
%>50	9.65

lwn207	period	level	morph type	rim dia	thick	sqmm
lwn-2A2pit3			All	14	8.2	12.5

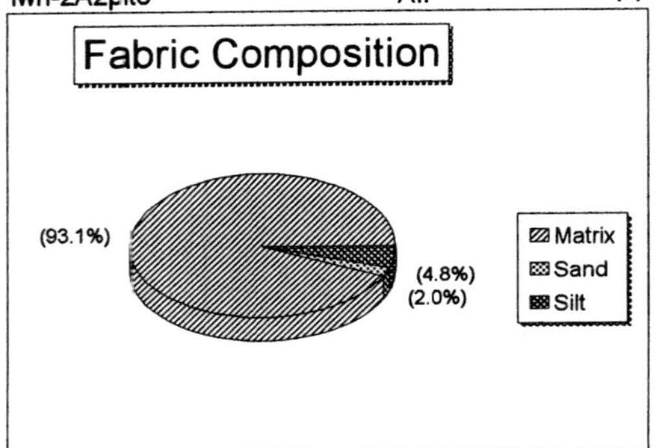

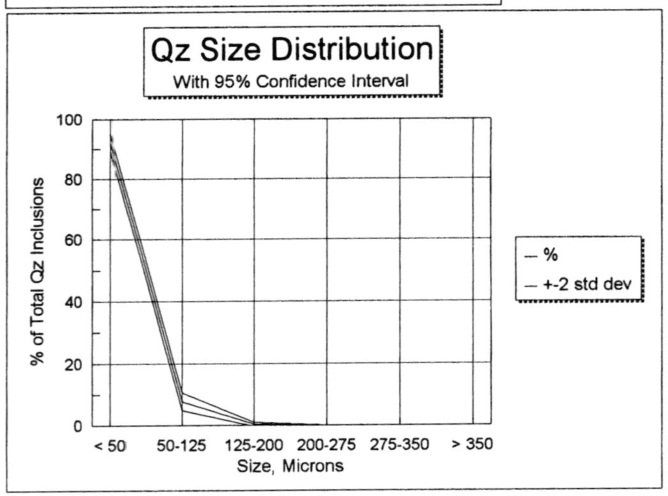

Qz Size	< 50	50-125	125-200	200-275	275-350	> 350	numpts
Number	309	26	1	0	0	0	336
%	91.96	7.74	0.30	0.00	0.00	0.00	
sigma	1.48	1.46	0.30	0.00	0.00	0.00	
p2sigma	94.93	10.65	0.89	0.00	0.00	0.00	
m2sigma	89.00	4.82	-0.30	0.00	0.00	0.00	
Conc	24.72	2.08	0.08	0.00	0.00	0.00	

Fabric	Matrix	Sand	Silt	Micac	Biref	Colour	
	366	8	19	2	2	medebr	393

Minerals	Size	Ang	Sph	Wth	Conc	Type	Opac
PFs	2	3	1		2		
MFs					1		
Biotite	2			2	2		
Mus	2			2	3		
Amph	2				2		
Carb					4	2,3	3
Quartz		3L 2S	2L 3S		4	m,p	
Pyrox					1		
Epidote					2		
Schist					1		
other						vrfrag	

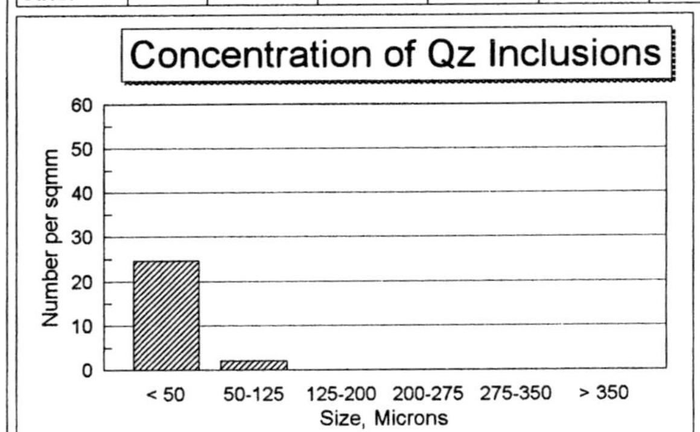

%matrix	93.13
%sand	2.04
%silt	4.83
conc<50	24.72
conc>50	2.16
%<50	91.96
%>50	8.04

lwn208	period	level	morph type	rim dia	thick	sqmm
lwn-1 housb			BIII	16	8	12.5

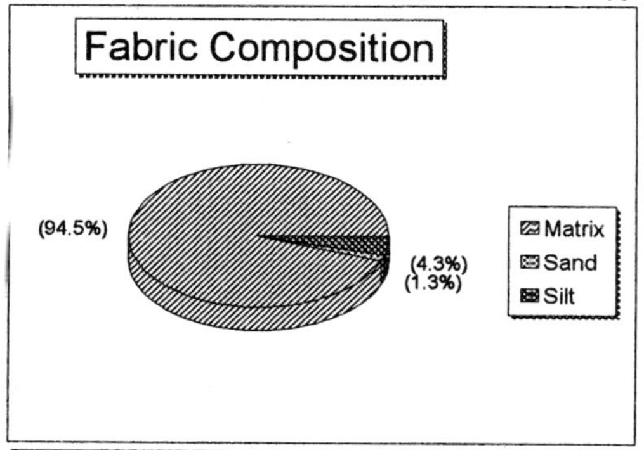

Qz Size	< 50	50-125	125-200	200-275	275-350	> 350	numpts
Number	257	17	4	0	0	0	278
%	92.45	6.12	1.44	0.00	0.00	0.00	
sigma	1.58	1.44	0.71	0.00	0.00	0.00	
p2sigma	95.62	8.99	2.87	0.00	0.00	0.00	
m2sigma	89.28	3.24	0.01	0.00	0.00	0.00	
Conc	20.56	1.36	0.32	0.00	0.00	0.00	
Fabric	Matrix	Sand	Silt	Micac	Biref	Colour	
	378	5	17	2	2	medbr	400

Minerals	Size	Ang	Sph	Wth	Conc	Type	Opac
PFs					1		
MFs					1		
Biotite					1		
Mus	2			2	3		
Amph					1		
Carb					2	2,3	1,3
Quartz		3L 2S	2L 3S		4	m,p	
Pyrox					2		
Epidote					1		
Schist					3		
other							

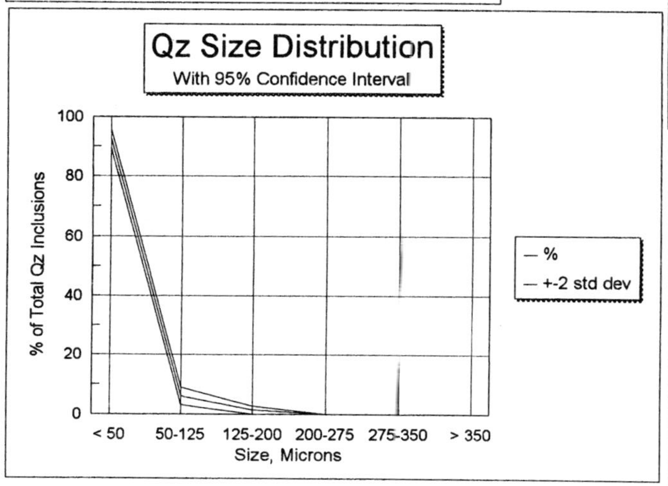

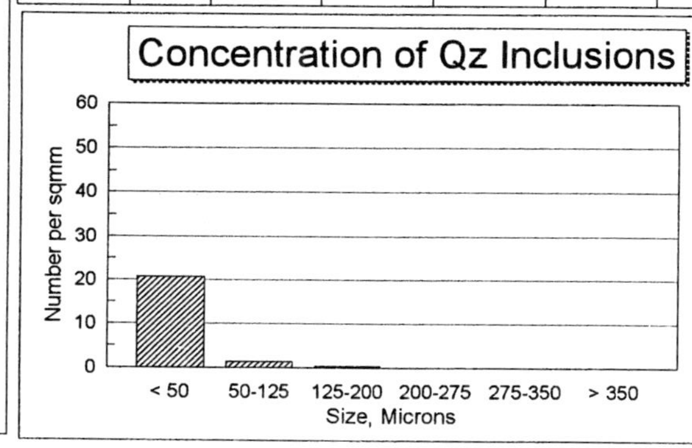

%matrix	94.50
%sand	1.25
%silt	4.25
conc<50	20.56
conc>50	1.68
%<50	92.45
%>50	7.55

APPENDIX B

lwn209	period	level	morph type	rim dia	thick	sqmm
lwn-1 housb1			AI	body	4.6	12.5

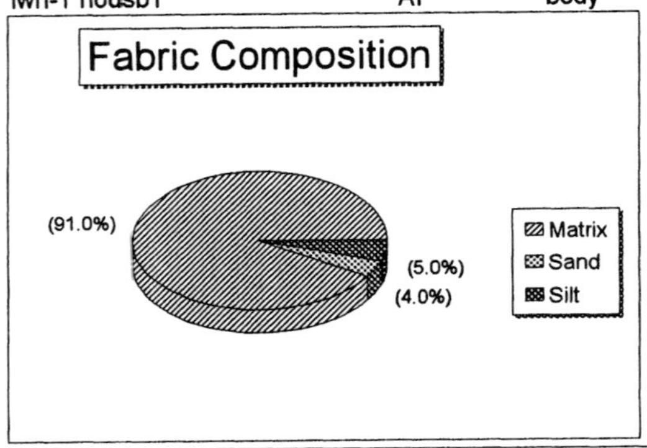

Qz Size	< 50	50-125	125-200	200-275	275-350	> 350	numpts
Number	334	24	8	5	0	0	371
%	90.03	6.47	2.16	1.35	0.00	0.00	
sigma	1.56	1.28	0.75	0.60	0.00	0.00	
p2sigma	93.14	9.02	3.66	2.54	0.00	0.00	
m2sigma	86.92	3.91	0.65	0.15	0.00	0.00	
Conc	26.72	1.92	0.64	0.40	0.00	0.00	
Fabric	Matrix	Sand	Silt	Micac	Biref	Colour	
	363	16	20	2	2	medbr	399

Minerals	Size	Ang	Sph	Wth	Conc	Type	Opac
PFs	3	4	2		2		
MFs					1		
Biotite					1		
Mus	2			2	3		
Amph					1		
Carb					1		
Quartz		3L 2S	2L 3S		4	m,p	
Pyrox					2		
Epidote					1		
Schist					2		
other					2	vrfrag	

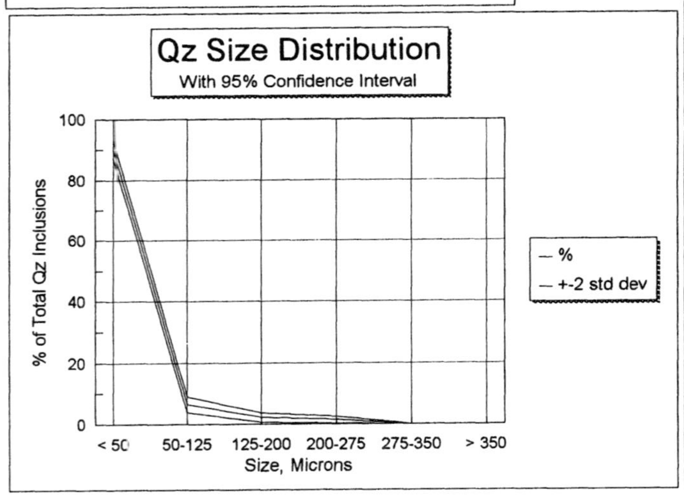

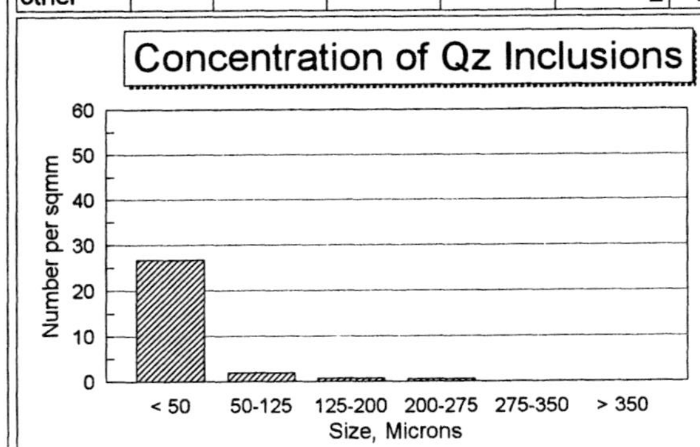

%matrix	90.98
%sand	4.01
%silt	5.01
conc<50	26.72
conc>50	2.96
%<50	90.03
%>50	9.97

APPENDIX B

lwn210	period	level	morph type	rim dia	thick	sqmm
lwn-1 housb1			All	body	4	12.5

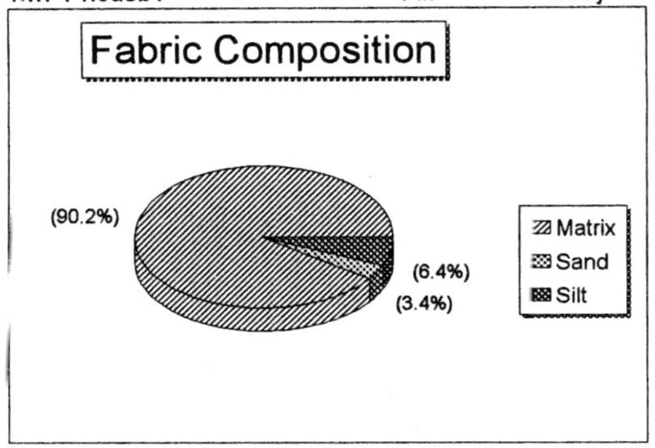

Qz Size	< 50	50-125	125-200	200-275	275-350	> 350	numpts
Number	339	13	2	2	1	1	358
%	94.69	3.63	0.56	0.56	0.28	0.28	
sigma	1.18	0.99	0.39	0.39	0.28	0.28	
p2sigma	97.06	5.61	1.35	1.35	0.84	0.84	
m2sigma	92.32	1.65	-0.23	-0.23	-0.28	-0.28	
Conc	27.12	1.04	0.16	0.16	0.08	0.08	
Fabric	Matrix	Sand	Silt	Micac	Biref	Colour	
	350	13	25	3	3	medbr	388

Minerals	Size	Ang	Sph	Wth	Conc	Type	Opac
PFs		4	2		2		
MFs					1		
Biotite	2			2	2		
Mus	2			2	4		
Amph					1		
Carb					3	2,3	1
Quartz		3L 2S	2L 3S		4	m,p	
Pyrox					1		
Epidote					1		
Schist					1		
other					2	vrfrag	

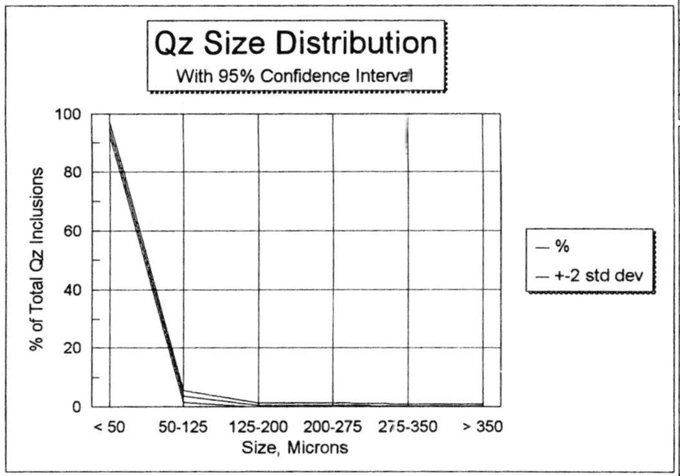

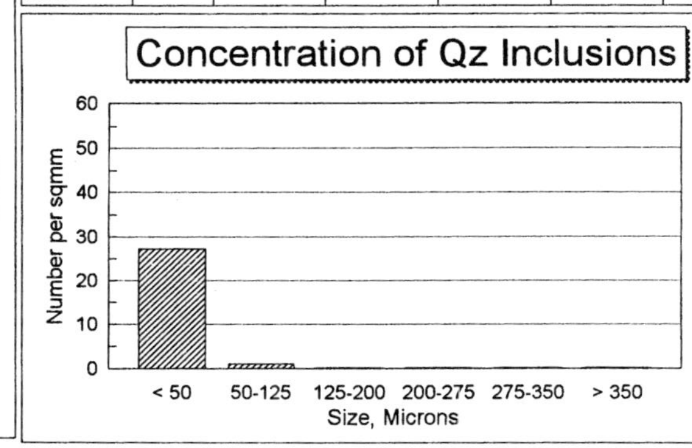

%matrix	90.21
%sand	3.35
%silt	6.44
conc<50	27.12
conc>50	1.52
%<50	94.69
%>50	5.31

lwn211	period	level	morph type	rim dia	thick	sqmm
lwn-2 A3			Bl	14	6.2	12.5

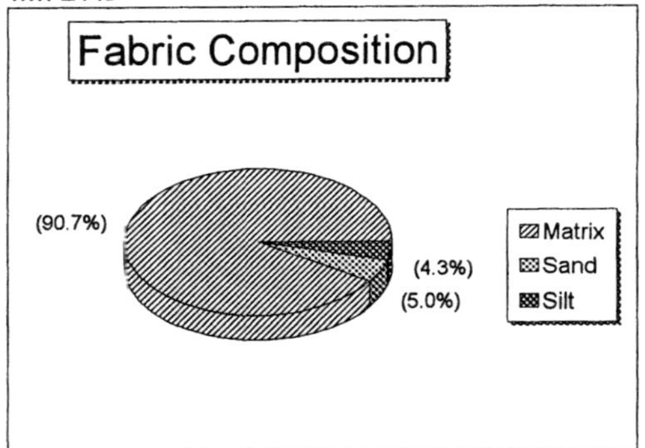

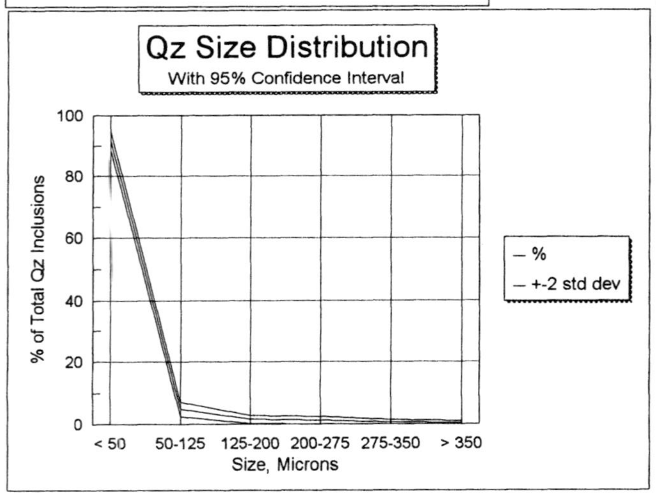

Qz Size	< 50	50-125	125-200	200-275	275-350	> 350	numpts
Number	308	16	5	4	2	1	336
%	91.67	4.76	1.49	1.19	0.60	0.30	
sigma	1.51	1.16	0.66	0.59	0.42	0.30	
p2sigma	94.68	7.09	2.81	2.37	1.43	0.89	
m2sigma	88.65	2.44	0.17	0.01	-0.24	-0.30	
Conc	24.64	1.28	0.40	0.32	0.16	0.08	
Fabric	Matrix	Sand	Silt	Micac	Biref	Colour	
	361	20	17	4	4	mica	398

Minerals	Size	Ang	Sph	Wth	Conc	Type	Opac
PFs	1	4	2		2		
MFs					1		
Biotite	2			2	2		
Mus	2			2	4		
Amph					2		
Carb					3	2,3	2
Quartz		3	2		4	m,p	
Pyrox					1		
Epidote					1		
Schist					1		
other					2	vrfrag	

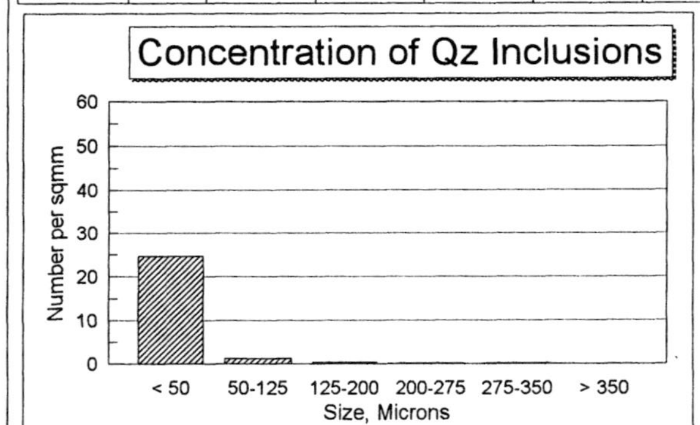

%matrix	90.70
%sand	5.03
%silt	4.27
conc<50	24.64
conc>50	2.24
%<50	91.67
%>50	8.33

lwn212	period	level	morph type	rim dia	thick	sqmm
lwn-1 housb			BIV	body	6.2	12.5

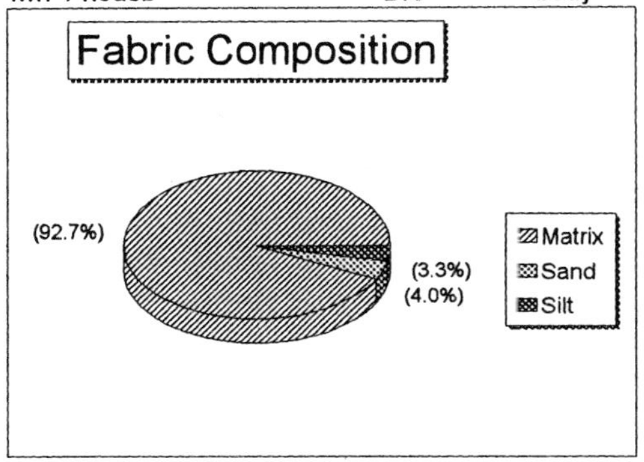

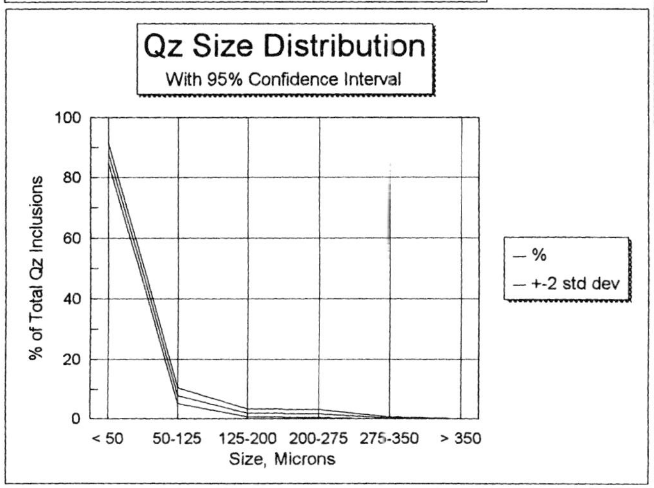

Qz Size	< 50	50-125	125-200	200-275	275-350	> 350	numpts
Number	345	31	8	7	1	0	392
%	88.01	7.91	2.04	1.79	0.26	0.00	
sigma	1.64	1.36	0.71	0.67	0.25	0.00	
p2sigma	91.29	10.63	3.47	3.12	0.76	0.00	
m2sigma	84.73	5.18	0.61	0.45	-0.25	0.00	
Conc	27.60	2.48	0.64	0.56	0.08	0.00	
Fabric	Matrix	Sand	Silt	Micac	Biref	Colour	
	367	16	13				396
Minerals	Size	Ang	Sph	Wth	Conc	Type	Opac
PFs	1	3	2		2		
MFs					1		
Biotite	2			2	2		
Mus	2			2	4		
Amph					2		
Carb					4	indet	
Quartz		3	2		4	m,p	
Pyrox					1		
Epidote					2		
Schist					2		
other					2	vrfrag	

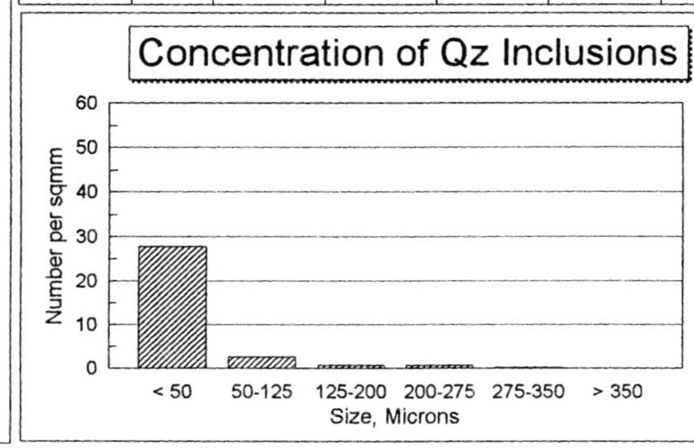

%matrix	92.68
%sand	4.04
%silt	3.28
conc<50	27.60
conc>50	3.76
%<50	88.01
%>50	11.99

lwn213	period	level	morph type	rim dia	thick	sqmm
lwn-1 housb1			BI	body	10.1	12.5

Qz Size	< 50	50-125	125-200	200-275	275-350	> 350	numpts
Number	330	36	5	0	0	0	371
%	88.95	9.70	1.35	0.00	0.00	0.00	
sigma	1.63	1.54	0.60	0.00	0.00	0.00	
p2sigma	92.20	12.78	2.54	0.00	0.00	0.00	
m2sigma	85.69	6.63	0.15	0.00	0.00	0.00	
Conc	26.40	2.88	0.40	0.00	0.00	0.00	
Fabric	Matrix	Sand	Silt	Micac	Biref	Colour	
				4	4	mica	0

Minerals	Size	Ang	Sph	Wth	Conc	Type	Opac
PFs	2	4	2		2		
MFs					1		
Biotite	2			2	2		
Mus	2			2	4		
Amph					1		
Carb	3				3	2,3	indet
Quartz		3	2		4	m,p	
Pyrox					1		
Epidote					1		
Schist					1		
other					3	grog	

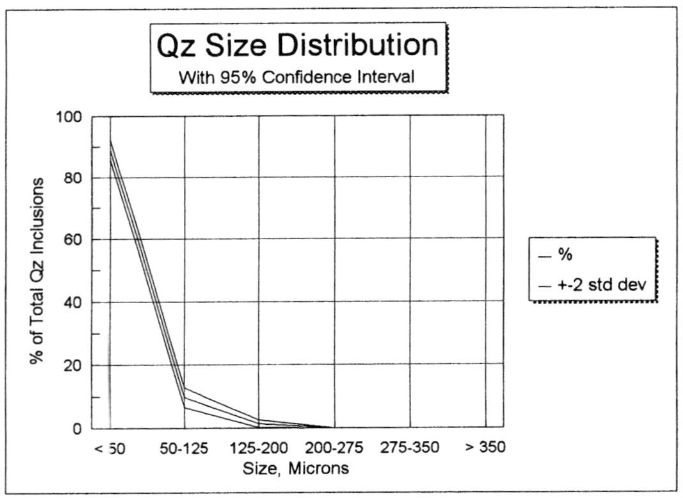

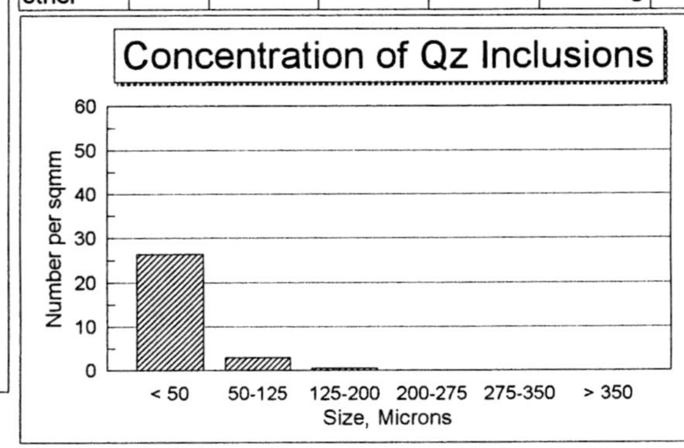

%matrix	ERR
%sand	ERR
%silt	ERR
conc<50	26.40
conc>50	3.28
%<50	88.95
%>50	11.05

lwn214	period	level	morph type	rim dia	thick	sqmm
lwn-1 housb1			BIII or BIV	base	4	12.5

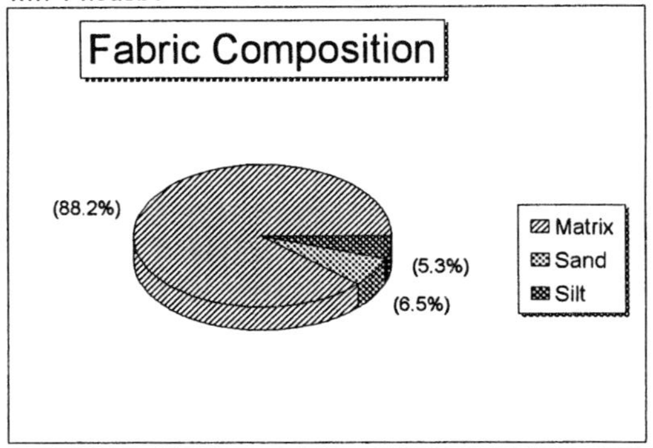

Qz Size	< 50	50-125	125-200	200-275	275-350	> 350	numpts
Number	218	27	6	4	0	0	255
%	85.49	10.59	2.35	1.57	0.00	0.00	
sigma	2.21	1.93	0.95	0.78	0.00	0.00	
p2sigma	89.90	14.44	4.25	3.12	0.00	0.00	
m2sigma	81.08	6.73	0.45	0.01	0.00	0.00	
Conc	17.44	2.16	0.48	0.32	0.00	0.00	
Fabric	Matrix	Sand	Silt	Micac	Biref	Colour	
	351	26	21	2	2	dkbrn	398
Minerals	Size	Ang	Sph	Wth	Conc	Type	Opac
PFs	2	3	2		2		
MFs					1		
Biotite	2			2	2		
Mus	2			2	4		
Amph					2		
Carb	3				4	2,3	2
Quartz		3	2		4	m,p	
Pyrox					1		
Epidote					1		
Schist					1		
other					3	vrfrag	

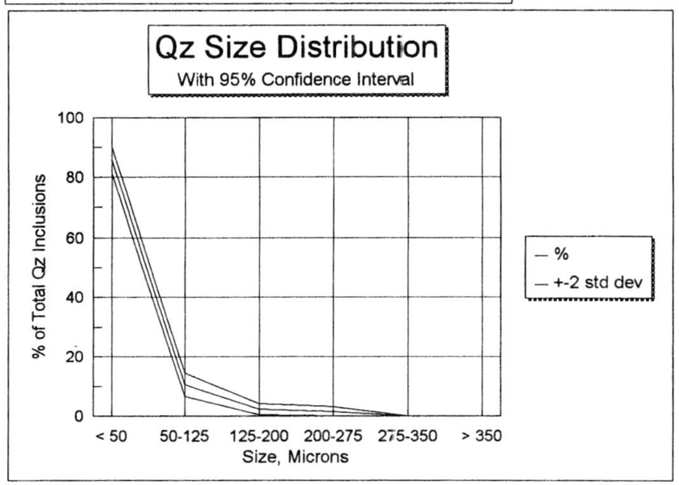

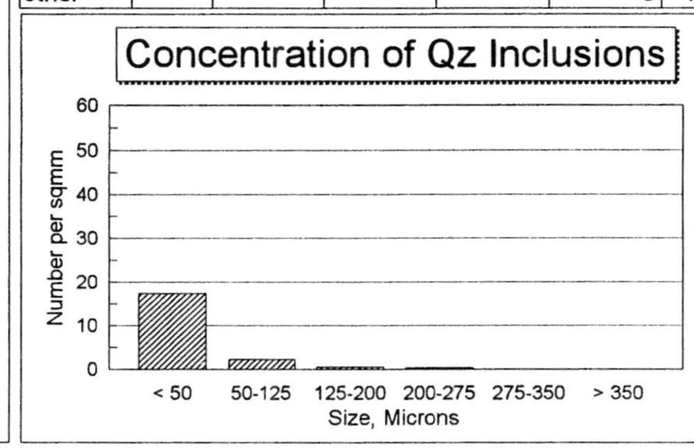

%matrix	88.19
%sand	6.53
%silt	5.28
conc<50	17.44
conc>50	2.96
%<50	85.49
%>50	14.51

APPENDIX B

lwn215	period	level	morph type	rim dia	thick	sqmm
lwn-1 housb1			Al or All	base	21.4	12.5

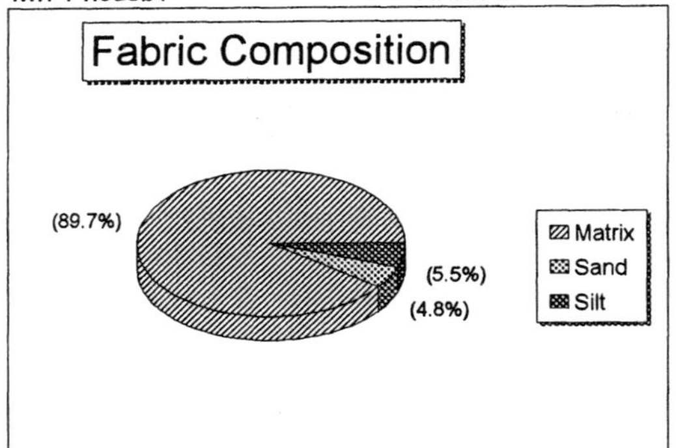

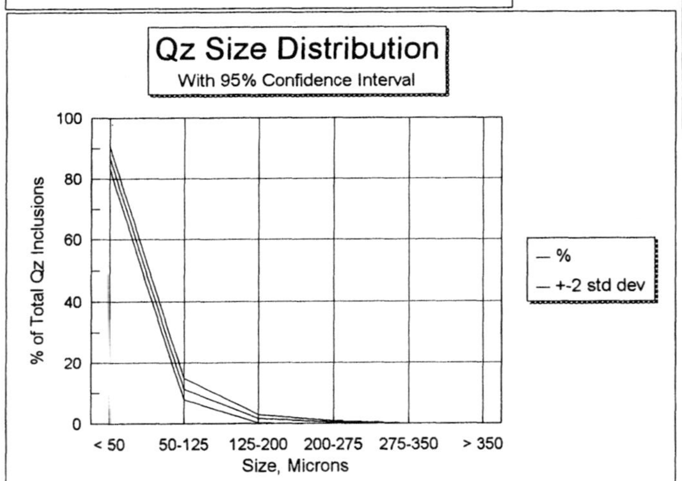

Qz Size	< 50	50-125	125-200	200-275	275-350	> 350	numpts
Number	284	37	5	1	0	0	327
%	86.85	11.31	1.53	0.31	0.00	0.00	
sigma	1.87	1.75	0.68	0.31	0.00	0.00	
p2sigma	90.59	14.82	2.89	0.92	0.00	0.00	
m2sigma	83.11	7.81	0.17	-0.30	0.00	0.00	
Conc	22.72	2.96	0.40	0.08	0.00	0.00	
Fabric	Matrix	Sand	Silt	Micac	Biref	Colour	
	356	19	22	3	3	medbr	397
Minerals	Size	Ang	Sph	Wth	Conc	Type	Opac
PFs	2	3	2		2		
MFs					1		
Biotite	2			3	2		
Mus	2			2	4		
Amph					1		
Carb	3				4	2,3	2,3
Quartz		3	2		4	m,p	
Pyrox					2		
Epidote					1		
Schist					1		
other					3	vrfrag	

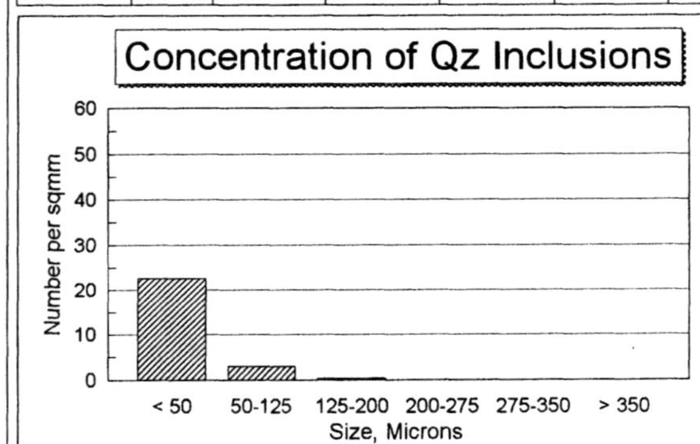

%matrix	89.67
%sand	4.79
%silt	5.54
conc<50	22.72
conc>50	3.44
%<50	86.85
%>50	13.15

lwn216	period	level	morph type	rim dia	thick	sqmm
lwn-1 housb1			Al or All	body	11.5	12.5

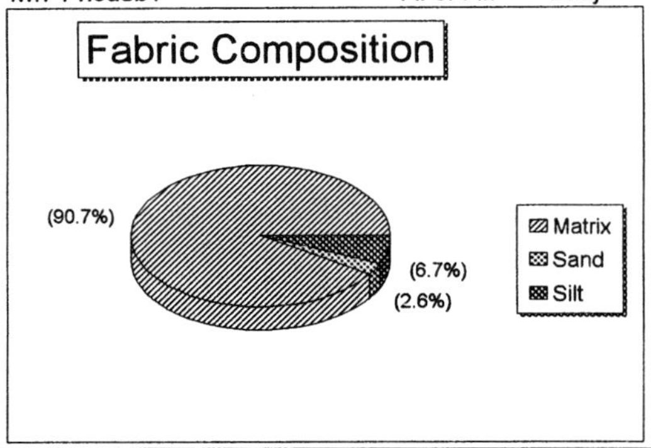

Qz Size	< 50	50-125	125-200	200-275	275-350	> 350	numpts
Number	264	12	7	6	1	0	290
%	91.03	4.14	2.41	2.07	0.34	0.00	
sigma	1.68	1.17	0.90	0.84	0.34	0.00	
p2sigma	94.39	6.48	4.22	3.74	1.03	0.00	
m2sigma	87.68	1.80	0.61	0.40	-0.34	0.00	
Conc	21.12	0.96	0.56	0.48	0.08	0.00	
Fabric	Matrix	Sand	Silt	Micac	Biref	Colour	
	353	10	26	4	4	ltbrn	389
Minerals	Size	Ang	Sph	Wth	Conc	Type	Opac
PFs	2	2	2		2		
MFs					1		
Biotite	2			3	2		
Mus	2			2	4		
Amph					1		
Carb	3				3	1,2	2
Quartz		2	2		4	m,p	
Pyrox					2		
Epidote					1		
Schist					1		
other					2	ag,grog	

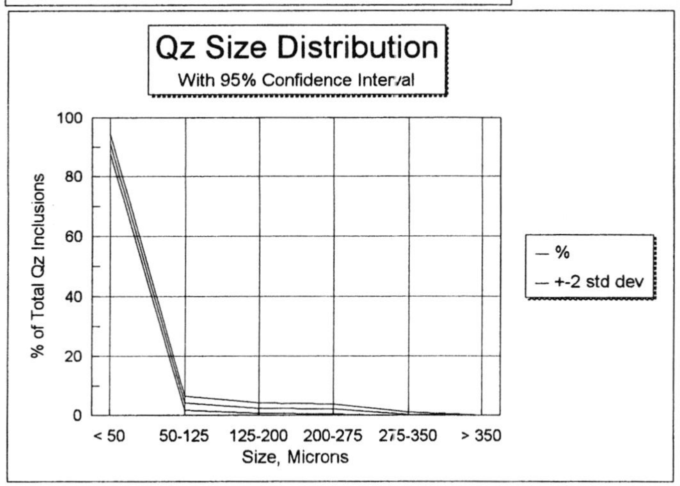

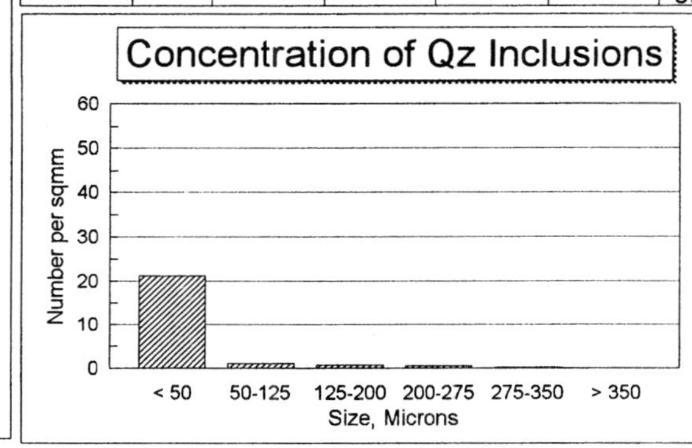

%matrix	90.75
%sand	2.57
%silt	6.68
conc<50	21.12
conc>50	2.08
%<50	91.03
%>50	8.97

APPENDIX B

lwn217	period	level	morph type	rim dia	thick	sqmm
lwn-IIA 2			BIII	12	3.2	

Poor section; not countable

Qz Size	< 50	50-125	125-200	200-275	275-350	> 350	numpts
Number							0
%							
sigma							
p2sigma							
m2sigma							
Conc							

Fabric	Matrix	Sand	Silt	Micac	Biref	Colour	
						medbr	0

Minerals	Size	Ang	Sph	Wth	Conc	Type	Opac
PFs					1		
MFs					1		
Biotite	2			3	2		
Mus				2	3		
Amph					1		
Carb	3				2	2,3	1,2,3
Quartz		2	2		4	m,p	
Pyrox					1		
Epidote					2		
Schist					1		
other							

lwn218	period	level	morph type	rim dia	thick	sqmm
lwn-IIB4			Al or All	base	8.2	12.5

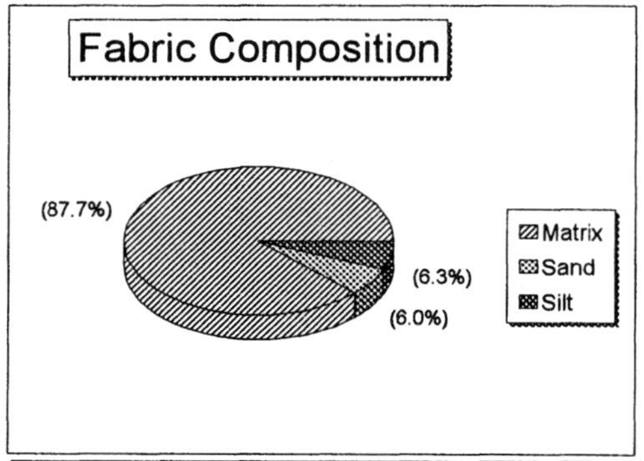

Qz Size	< 50	50-125	125-200	200-275	275-350	> 350	numpts
Number	318	15	11	2	1	0	347
%	91.64	4.32	3.17	0.58	0.29	0.00	
sigma	1.49	1.09	0.94	0.41	0.29	0.00	
p2sigma	94.61	6.51	5.05	1.39	0.86	0.00	
m2sigma	88.67	2.14	1.29	-0.24	-0.29	0.00	
Conc	25.44	1.20	0.88	0.16	0.08	0.00	
Fabric	Matrix	Sand	Silt	Micac	Biref	Colour	
	349	24	25	4	4	medbr	398

Minerals	Size	Ang	Sph	Wth	Conc	Type	Opac
PFs	3	3	1		2		
MFs					1		
Biotite	2			2	3		
Mus	2			2	4		
Amph					1		
Carb	2				3	2,3	1,2,3
Quartz		2	2		4	m,p	
Pyrox					2		
Epidote					2		
Schist					2	bi/amp	
other					2	grg,vrf	

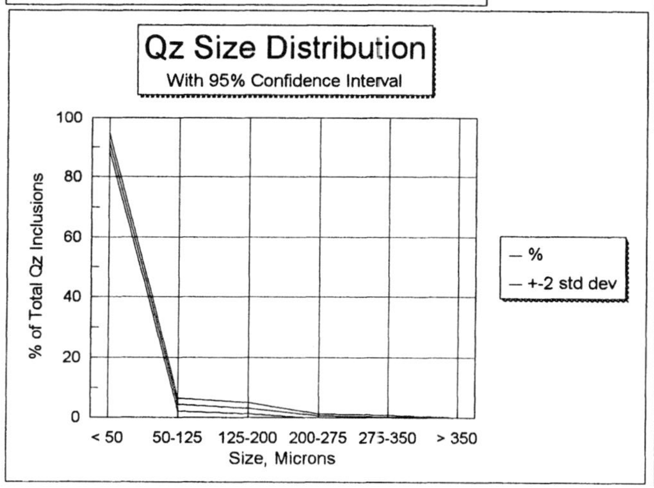

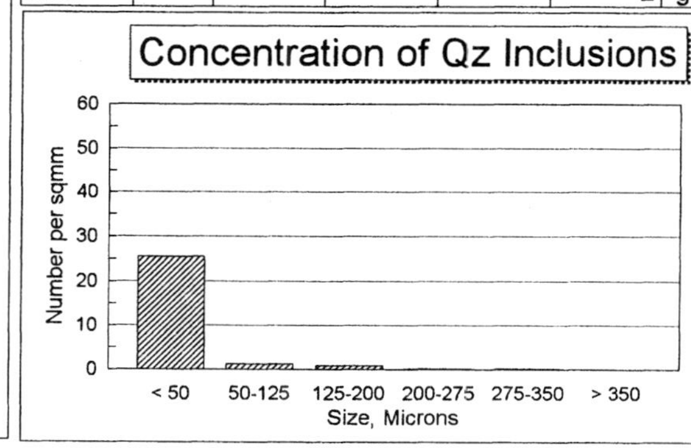

%matrix	87.69
%sand	6.03
%silt	6.28
conc<50	25.44
conc>50	2.32
%<50	91.64
%>50	8.36

lwn219	period	level	morph type	rim dia	thick	sqmm
lwn-1 housb1			Al or All	base	22.8	18.75

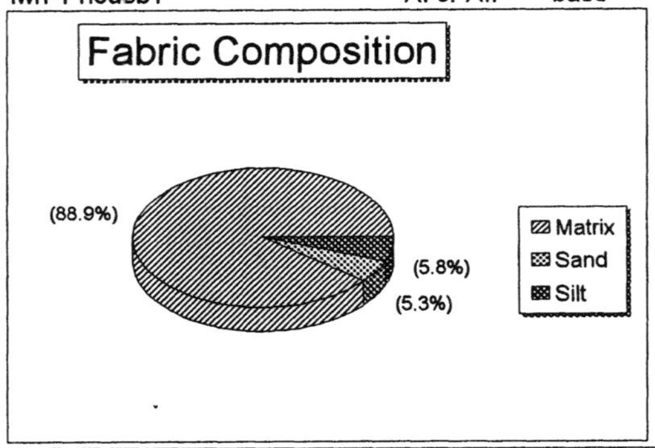

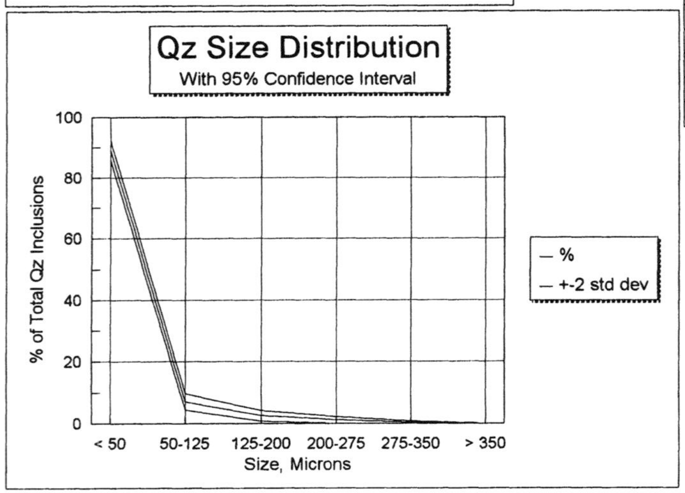

Qz Size	< 50	50-125	125-200	200-275	275-350	> 350	numpts
Number	313	25	9	4	1	0	352
%	88.92	7.10	2.56	1.14	0.28	0.00	
sigma	1.67	1.37	0.84	0.56	0.28	0.00	
p2sigma	92.27	9.84	4.24	2.27	0.85	0.00	
m2sigma	85.57	4.36	0.87	0.01	-0.28	0.00	
Conc	16.69	1.33	0.48	0.21	0.05	0.00	
Fabric	Matrix	Sand	Silt	Micac	Biref	Colour	
	354	21	23	4	4	grbrn	398
Minerals	Size	Ang	Sph	Wth	Conc	Type	Opac
PFs					1		
MFs					1		
Biotite	2			2	2		
Mus	2			2	4		
Amph					1		
Carb	2				3	2,3	4
Quartz		2	2		4	m,p	
Pyrox					2		
Epidote					2		
Schist					1		
other					2		

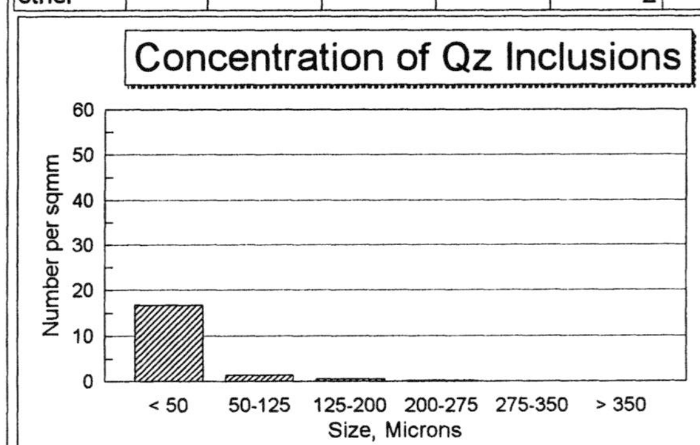

%matrix	88.94
%sand	5.28
%silt	5.78
conc<50	16.69
conc>50	2.08
%<50	88.92
%>50	11.08

lwn248	period	level	morph type	rim dia	thick	sqmm
lwn-1	waster	surf	Al or All			12.5

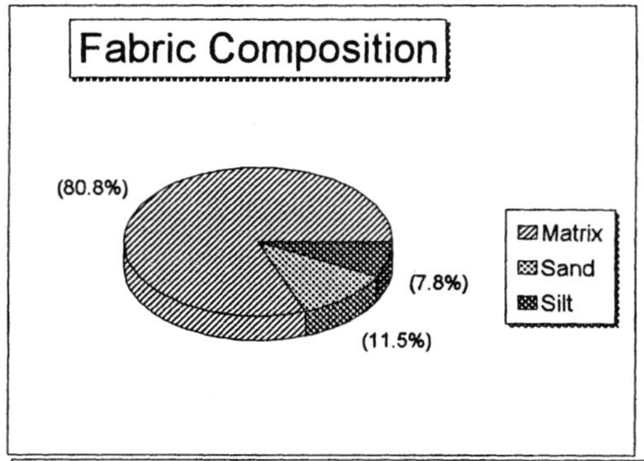

Qz Size	< 50	50-125	125-200	200-275	275-350	> 350	numpts
Number	272	16	10	13	1	0	312
%	87.18	5.13	3.21	4.17	0.32	0.00	
sigma	1.89	1.25	1.00	1.13	0.32	0.00	
p2sigma	90.96	7.63	5.20	6.43	0.96	0.00	
m2sigma	83.39	2.63	1.21	1.90	-0.32	0.00	
Conc	21.76	1.28	0.80	1.04	0.08	0.00	
Fabric	Matrix	Sand	Silt	Micac	Biref	Colour	
	323	46	31	2	2	grey	400
Minerals	Size	Ang	Sph	Wth	Conc	Type	Opac
PFs	3	3	2		2		
MFs					1		
Biotite					1		
Mus	2			2	3		
Amph					1		
Carb	2				4	1,2,3	2
Quartz		2	2		4	m,p	
Pyrox					2		
Epidote					2		
Schist					2		
other					2	vrfrag	

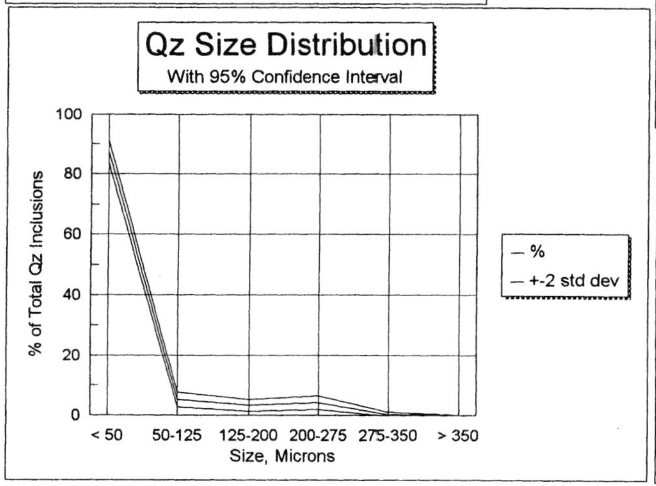

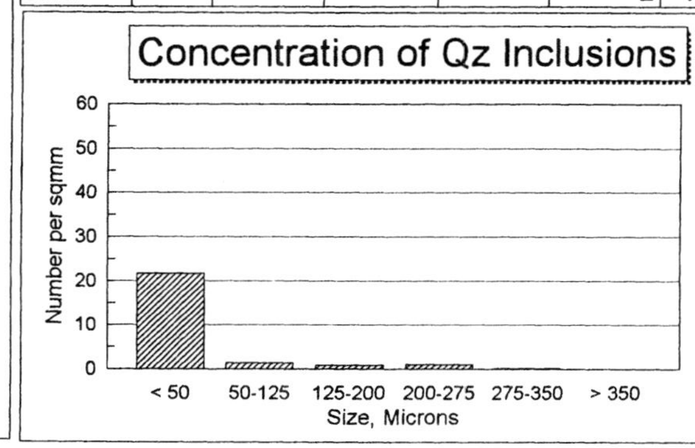

%matrix	80.75
%sand	11.50
%silt	7.75
conc<50	21.76
conc>50	3.20
%<50	87.18
%>50	12.82

lwn249	period	level	morph type	rim dia	thick	sqmm
lwn-2	waster	surf	Bl			12.5

Qz Size	< 50	50-125	125-200	200-275	275-350	> 350	numpts
Number	275	52	24	18	1	0	370
%	74.32	14.05	6.49	4.86	0.27	0.00	
sigma	2.27	1.81	1.28	1.12	0.27	0.00	
p2sigma	78.87	17.67	9.05	7.10	0.81	0.00	
m2sigma	69.78	10.44	3.93	2.63	-0.27	0.00	
Conc	22.00	4.16	1.92	1.44	0.08	0.00	
Fabric	Matrix	Sand	Silt	Micac	Biref	Colour	
	312	44	44	3	3	medbr	400

Minerals	Size	Ang	Sph	Wth	Conc	Type	Opac
PFs	2	3	2		2		
MFs					1		
Biotite					1		
Mus	2			2	3		
Amph					1		
Carb					4	indet	
Quartz		3	2		4	m,p	
Pyrox					2		
Epidote					2		
Schist					2		
other					1		

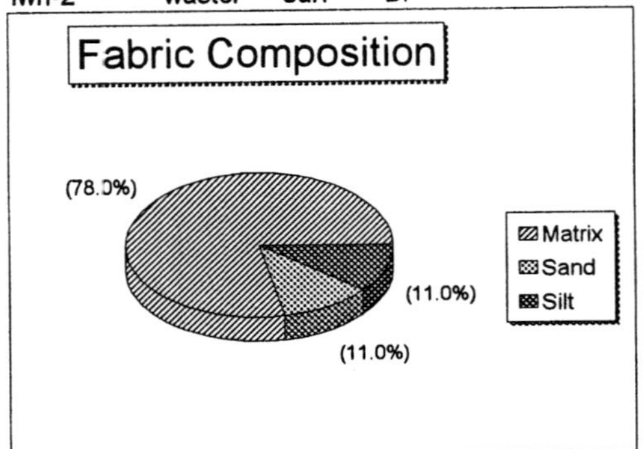

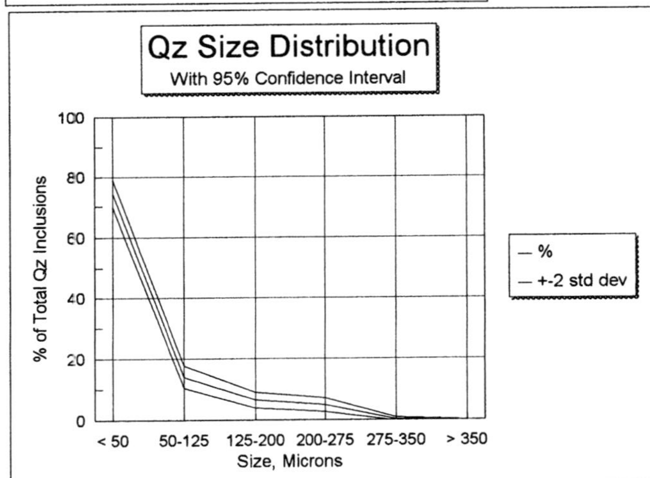

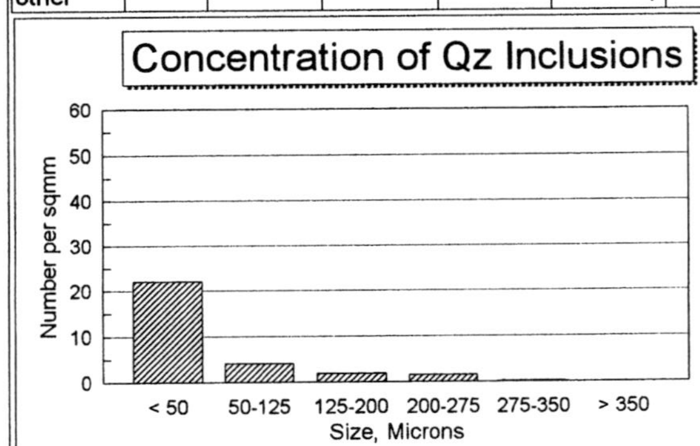

%matrix	78.00
%sand	11.00
%silt	11.00
conc<50	22.00
conc>50	7.60
%<50	74.32
%>50	25.68

lwn250	period	level	morph type	rim dia	thick	sqmm
lwn-3	waster	surf	indet	body		12.5

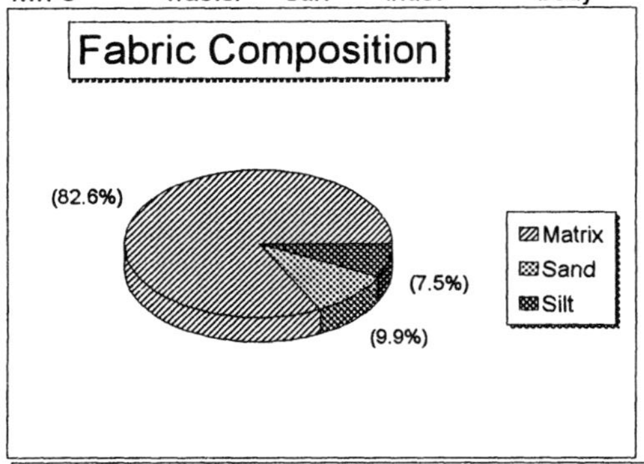

Qz Size	< 50	50-125	125-200	200-275	275-350	> 350	numpts
Number	234	20	7	10	4	5	280
%	83.57	7.14	2.50	3.57	1.43	1.79	
sigma	2.21	1.54	0.93	1.11	0.71	0.79	
p2sigma	88.00	10.22	4.37	5.79	2.85	3.37	
m2sigma	79.14	4.06	0.63	1.35	0.01	0.20	
Conc	18.72	1.60	0.56	0.80	0.32	0.40	
Fabric	Matrix	Sand	Silt	Micac	Biref	Colour	
	299	36	27	1	1	dkgrey	362
Minerals	Size	Ang	Sph	Wth	Conc	Type	Opac
PFs	2	3	2		2		
MFs					1		
Biotite					1		
Mus	2			3	3		
Amph					1		
Carb					4	2,3	1,2
Quartz		3	2		4	m,p	
Pyrox					2		
Epidote					2		
Schist					2		
other					2	vrfrag	

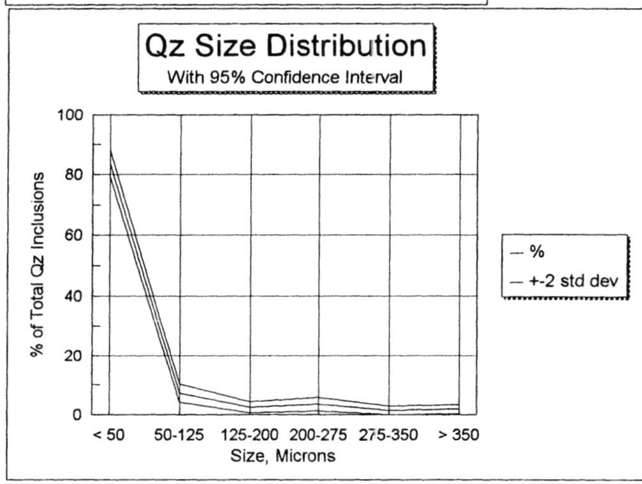

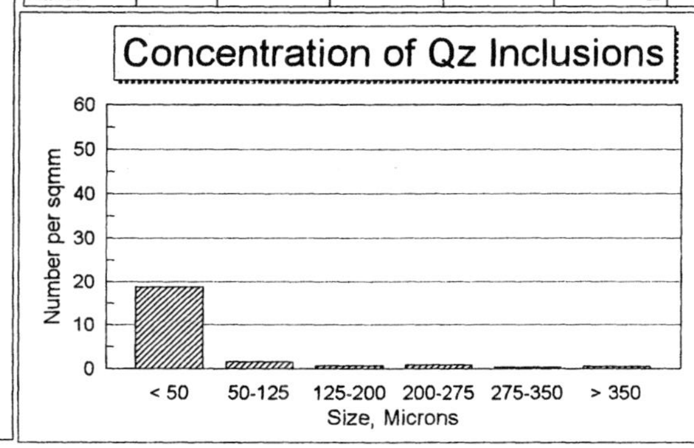

%matrix	82.60
%sand	9.94
%silt	7.46
conc<50	18.72
conc>50	3.68
%<50	83.57
%>50	16.43

lwn251	period	level	morph type	rim dia	thick	sqmm
lwn-4	waster	surf	indet	base		6.25

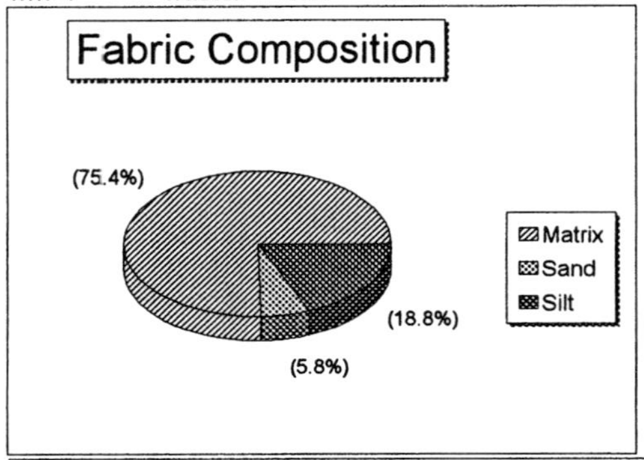

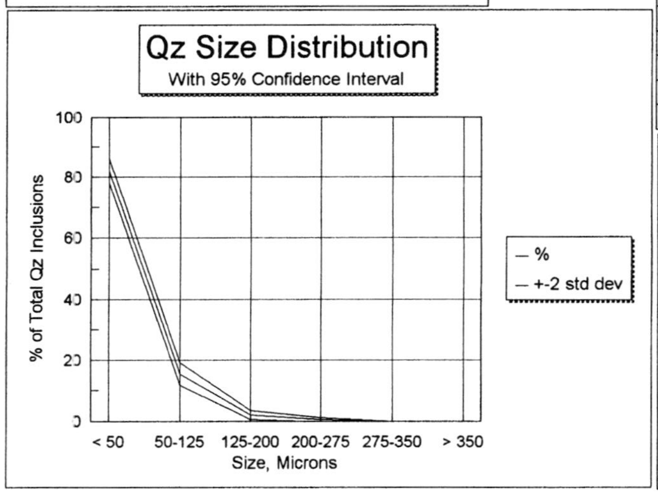

Qz Size	< 50	50-125	125-200	200-275	275-350	> 350	numpts
Number	313	59	8	2	0	0	382
%	81.94	15.45	2.09	0.52	0.00	0.00	
sigma	1.97	1.85	0.73	0.37	0.00	0.00	
p2sigma	85.87	19.14	3.56	1.26	0.00	0.00	
m2sigma	78.00	11.75	0.63	-0.21	0.00	0.00	
Conc	50.08	9.44	1.28	0.32	0.00	0.00	
Fabric	Matrix	Sand	Silt	Micac	Biref	Colour	
	297	23	74	4	4	medbr	394

Minerals	Size	Ang	Sph	Wth	Conc	Type	Opac
PFs	3	3	2		2		
MFs		4	1		2		
Biotite	2			3	4		
Mus	2			3	4		
Amph					1		
Carb					1		
Quartz		3	2		4	m,p	
Pyrox					1		
Epidote					2		
Schist					2		
other					2	vrfrag	

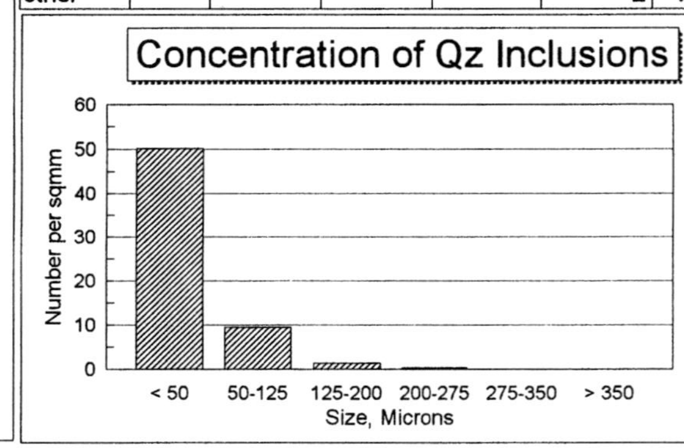

%matrix	75.38
%sand	5.84
%silt	18.78
conc<50	50.08
conc>50	11.04
%<50	81.94
%>50	18.06

lwn252	period	level	morph	type	rim dia	thick	sqmm
lwn-5	waster	surf	indet		indet		

Poor section; not countable

Qz Size	< 50	50-125	125-200	200-275	275-350	> 350	numpts
Number							0
%							
sigma							
p2sigma							
m2sigma							
Conc							
Fabric	Matrix	Sand	Silt	Micac	Biref	Colour	
				2	2	dkgrey	0
Minerals	Size	Ang	Sph	Wth	Conc	Type	Opac
PFs	2	3	2		2		
MFs					2		
Biotite				1	1		
Mus	2			3	2		
Amph					1		
Carb					4	2	2,3
Quartz		3	2		4	m,p	
Pyrox					2		
Epidote					2		
Schist					1		
other					2	vrfrag	

APPENDIX B

lwn253	period	level	morph type	rim dia	thick	sqmm
lwn-6	waster	surf	indet	indet		

Poor section; not countable

Qz Size	< 50	50-125	125-200	200-275	275-350	> 350	numpts
Number							0
%							
sigma							
p2sigma							
m2sigma							
Conc							
Fabric	Matrix	Sand	Silt	Micac	Biref	Colour	
				2	2	dkgrey	0

Minerals	Size	Ang	Sph	Wth	Conc	Type	Opac
PFs	2	3	2		2		
MFs					1		
Biotite					1		
Mus	2			3	3		
Amph					1		
Carb					4	indet	
Quartz		3	2		4	m,p	
Pyrox					2		
Epidote					2		
Schist					1		
other					1		

trq220	period	level	morph type	rim dia	thick	sqmm
trq II B2	EH	2	Al	13.6	8.7	18.75

Qz Size	< 50	50-125	125-200	200-275	275-350	> 350	numpts
Number	320	32	17	12	2	2	385
%	83.12	8.31	4.42	3.12	0.52	0.52	
sigma	1.91	1.41	1.05	0.89	0.37	0.37	
p2sigma	86.94	11.13	6.51	4.89	1.25	1.25	
m2sigma	79.30	5.50	2.32	1.35	-0.21	-0.21	
Conc	17.07	1.71	0.91	0.64	0.11	0.11	
Fabric	Matrix	Sand	Silt	Micac	Biref	Colour	
	345	16	24	2	2	medbr	385
Minerals	Size	Ang	Sph	Wth	Conc	Type	Opac
PFs	2	3	1		2		
MFs					1		
Biotite	2			2	2		
Mus	2			2	4		
Amph					1		
Carb					3	1,2	2
Quartz		3	2		4	m,p	
Pyrox					2		
Epidote					2		
Schist					1		
other						vrfrag	

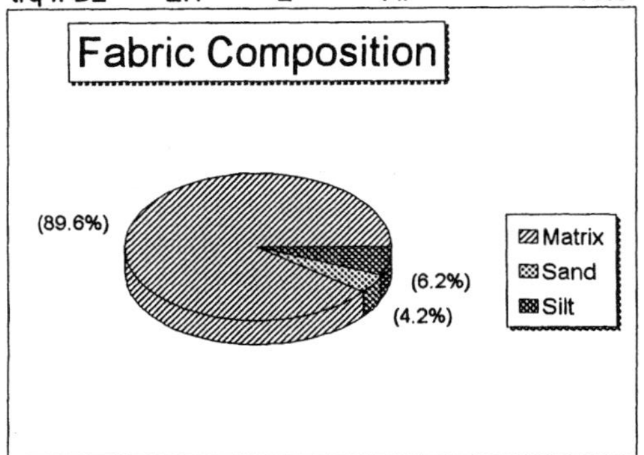

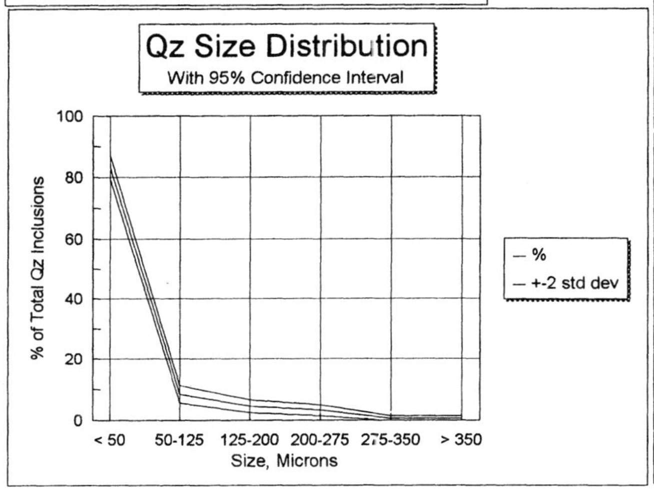

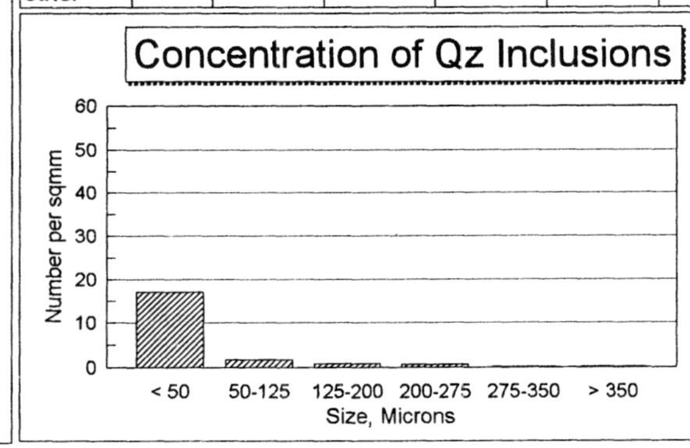

%matrix	89.61
%sand	4.16
%silt	6.23
conc<50	17.07
conc>50	3.47
%<50	83.12
%>50	16.88

trq221	period	level	morph type	rim dia	thick	sqmm
trq II B2	EH	2	Al	9	9.6	18.75

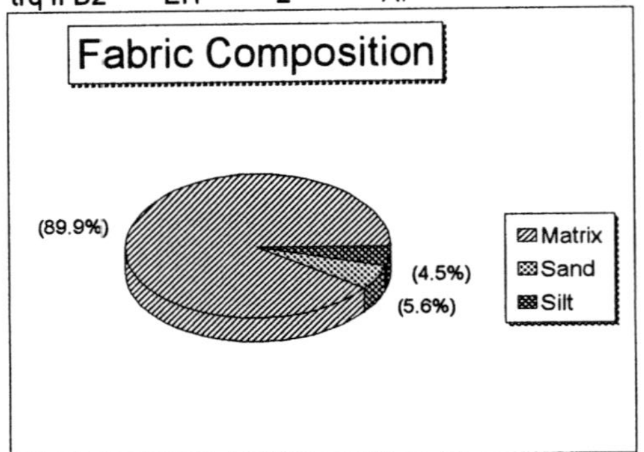

Qz Size	< 50	50-125	125-200	200-275	275-350	> 350	numpts
Number	389	23	[illegible]	9	0	0	450
%	86.44	5.11	6.44	2.00	0.00	0.00	
sigma	1.61	1.04	1.16	[illegible]	0.00	0.00	
p2sigma	89.67	7.19	8.76	3.32	[illegible]	0.00	
m2sigma	83.22	3.03	4.13	0.68	0.00	0.00	
Conc	20.75	1.23	1.55	0.48	0.00	0.00	
Fabric	Matrix	Sand	Silt	Micac	Biref	Colour	
	356	22	18	2	2	ltbrn	396

Minerals	Size	Ang	Sph	Wth	Conc	Type	Opac
PFs	1	3	1		2		
MFs					1		
Biotite	2			2	2		
Mus	2			2	4		
Amph					1		
Carb					2	1,2,3	1,2
Quartz		3	2		4	m,p	
Pyrox					1		
Epidote					1		
Schist					1		
other						vrfrag	

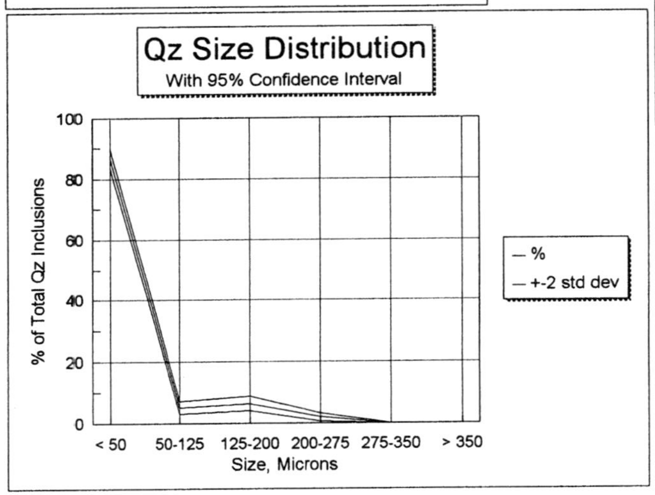

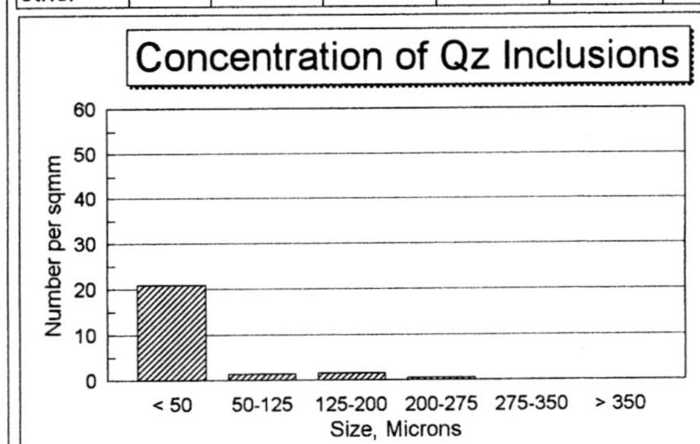

%matrix	89.90
%sand	5.56
%silt	4.55
conc<50	20.75
conc>50	3.25
%<50	86.44
%>50	13.56

trq222	period	level	morph type	rim dia	thick	sqmm
trq II B2	EH	2	AI	9.6	5.5	6.25

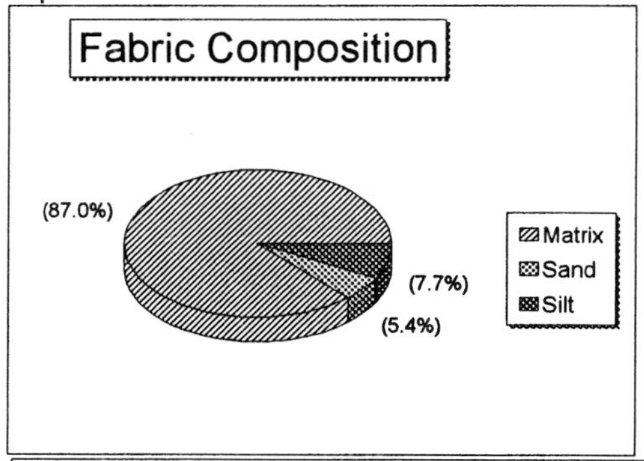

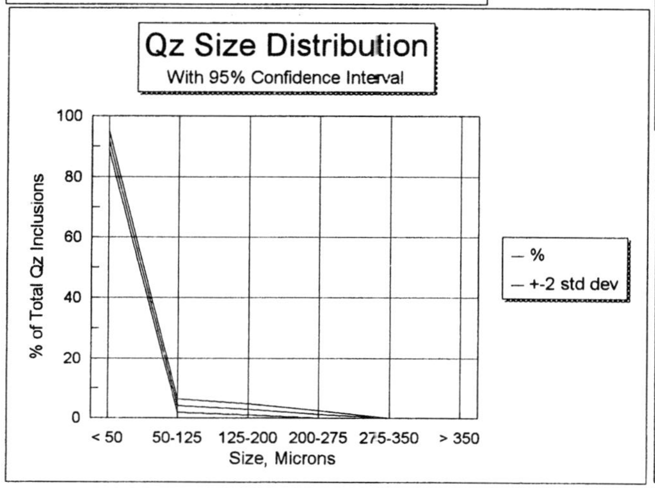

Qz Size	< 50	50-125	125-200	200-275	275-350	> 350	numpts
Number	291	13	9	4	0	0	317
%	91.80	4.10	2.84	1.26	0.00	0.00	
sigma	1.54	1.11	0.93	0.63	0.00	0.00	
p2sigma	94.88	6.33	4.70	2.52	0.00	0.00	
m2sigma	88.72	1.87	0.97	0.01	0.00	0.00	
Conc	46.56	2.08	1.44	0.64	0.00	0.00	
Fabric	Matrix	Sand	Silt	Micac	Biref	Colour	
	340	21	30	3	3	medbr	391

Minerals	Size	Ang	Sph	Wth	Conc	Type	Opac
PFs	2	3	1		2		
MFs					1		
Biotite	2			2	3		
Mus	1				4		
Amph					1		
Carb					3	2,3	1,3
Quartz		4L 2S	2L 3S		4	m,p	
Pyrox					1		
Epidote					2		
Schist					3	epidote	
other						vrfrag	

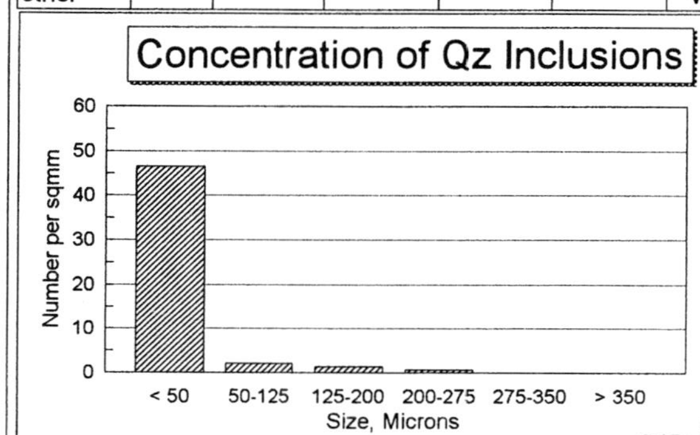

%matrix	86.96
%sand	5.37
%silt	7.67
conc<50	46.56
conc>50	4.16
%<50	91.80
%>50	8.20

trq223	period	level	morph type	rim dia	thick	sqmm
trq II B2	EH	2	Al	body	7.1	12.5

Qz Size	< 50	50-125	125-200	200-275	275-350	> 350	numpts
Number	321	21	5	6	1	0	354
%	90.68	5.93	1.41	1.69	0.28	0.00	
sigma	1.55	1.26	0.63	0.69	0.28	0.00	
p2sigma	93.77	8.44	2.67	3.07	0.85	0.00	
m2sigma	87.59	3.42	0.16	0.32	-0.28	0.00	
Conc	25.68	1.68	0.40	0.48	0.08	0.00	
Fabric	Matrix	Sand	Silt	Micac	Biref	Colour	
	343	19	35	2	2	medbr	397

Minerals	Size	Ang	Sph	Wth	Conc	Type	Opac
PFs	1	3	1		2		
MFs					1		
Biotite					1		
Mus	1			2	4		
Amph					1		
Carb					3	2,3	2
Quartz		3	2		4	m,p	
Pyrox					1		
Epidote					2		
Schist					1		
other						vrfrag	

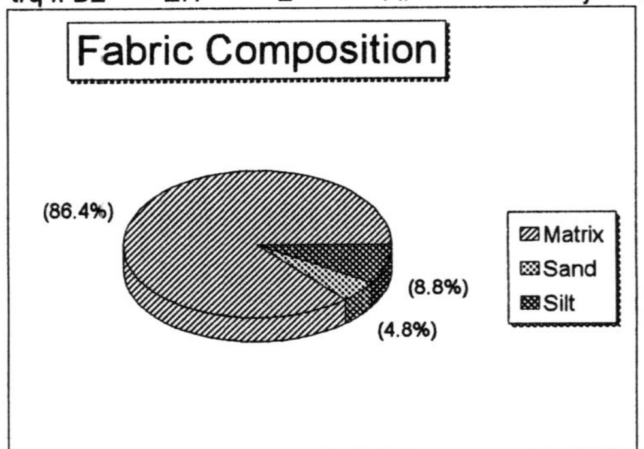

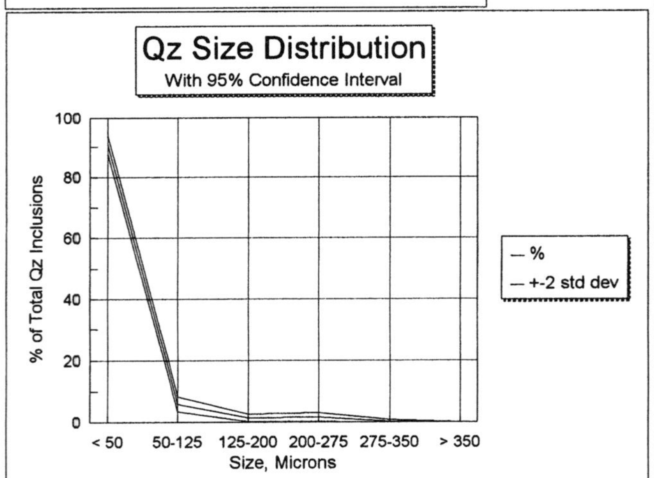

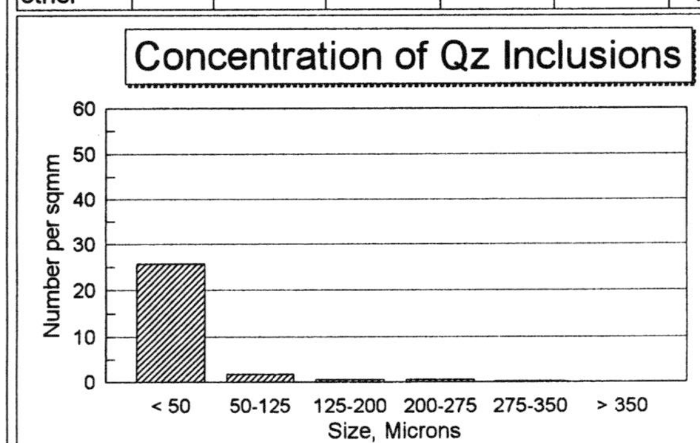

%matrix	86.40
%sand	4.79
%silt	8.82
conc<50	25.68
conc>50	2.64
%<50	90.68
%>50	9.32

trq224	period	level	morph type	rim dia	thick	sqmm
trq II B1	EH	2	Al	9	5.8	12.5

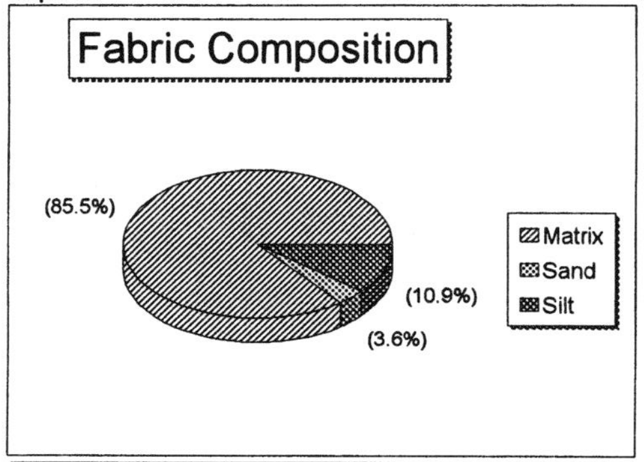

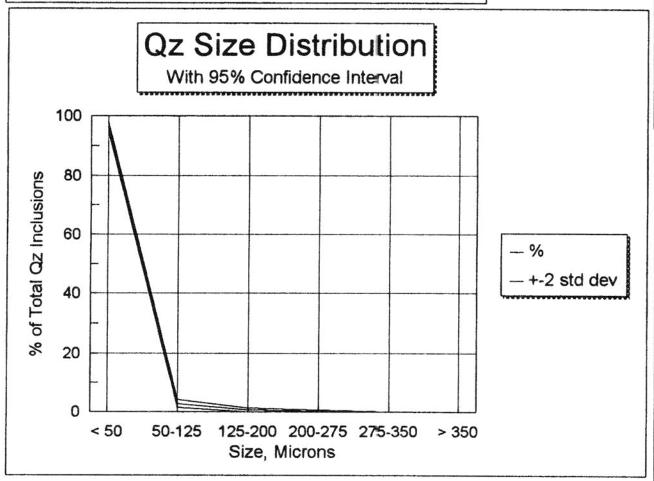

Qz Size	< 50	50-125	125-200	200-275	275-350	> 350	numpts
Number	517	15	3	1	0	0	536
%	96.46	2.80	0.56	0.19	0.00	0.00	
sigma	0.80	0.71	0.32	0.19	0.00	0.00	
p2sigma	98.05	4.22	1.20	0.56	0.00	0.00	
m2sigma	94.86	1.37	-0.08	-0.19	0.00	0.00	
Conc	41.36	1.20	0.24	0.08	0.00	0.00	
Fabric	Matrix	Sand	Silt	Micac	Biref	Colour	
	337	14	43	2	2	medbr	394
Minerals	Size	Ang	Sph	Wth	Conc	Type	Opac
PFs	2	3	1		2		
MFs					1		
Biotite	2			3	2		
Mus	2			2	4		
Amph					1		
Carb					3	2,3	1,2
Quartz		3	2		4	m,p	
Pyrox					1		
Epidote					1		
Schist					2	epidote	
other						vrfrag	

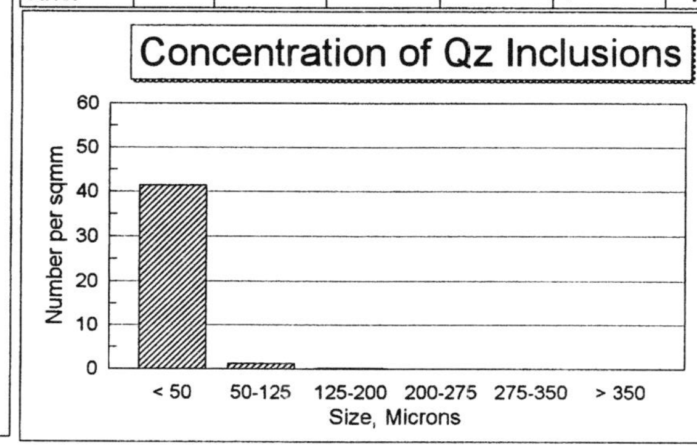

%matrix	85.53
%sand	3.55
%silt	10.91
conc<50	41.36
conc>50	1.52
%<50	96.46
%>50	3.54

trq225	period	level	morph type	rim dia	thick	sqmm
trq II B2	EH	2	All	body	7.8	6.25

Qz Size	< 50	50-125	125-200	200-275	275-350	> 350	numpts
Number	306	12	3	2	0	0	323
%	94.74	3.72	0.93	0.62	0.00	0.00	
sigma	1.24	1.05	0.53	0.44	0.00	0.00	
p2sigma	97.22	5.82	2.00	1.49	0.00	0.00	
m2sigma	92.25	1.61	-0.14	-0.25	0.00	0.00	
Conc	48.96	1.92	0.48	0.32	0.00	0.00	
Fabric	Matrix	Sand	Silt	Micac	Biref	Colour	
	351	14	30	2	2	ltbrn	395

Minerals	Size	Ang	Sph	Wth	Conc	Type	Opac
PFs	2	2	3		2		
MFs					1		
Biotite	2			2	2		
Mus	2			2	4		
Amph					1		
Carb					4	2,3	2,3
Quartz		3	2		4	m,p	
Pyrox					1		
Epidote					1		
Schist					1		
other							

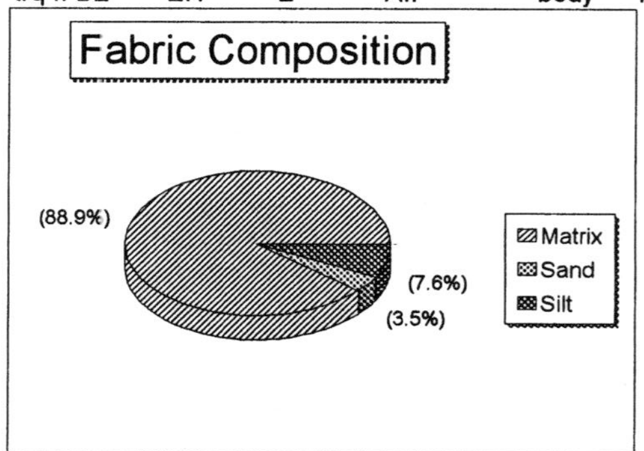

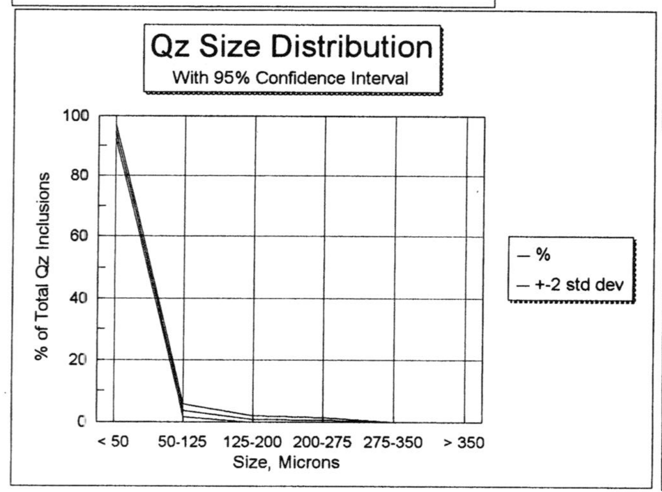

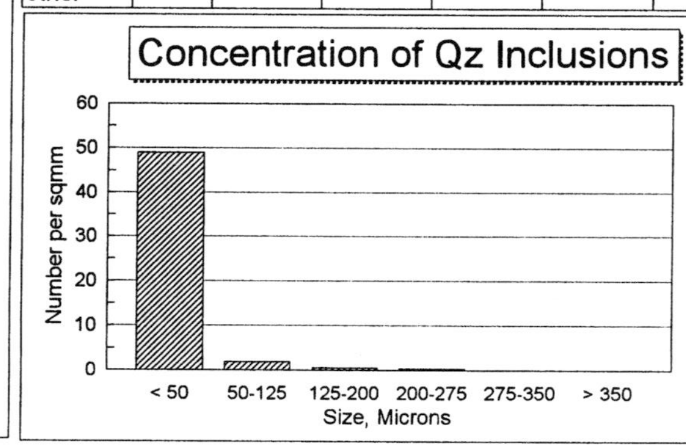

%matrix	88.86
%sand	3.54
%silt	7.59
conc<50	48.96
conc>50	2.72
%<50	94.74
%>50	5.26

trq226	period	level	morph type	rim dia	thick	sqmm
trq II B3	EH	2	All	9	5.6	6.25

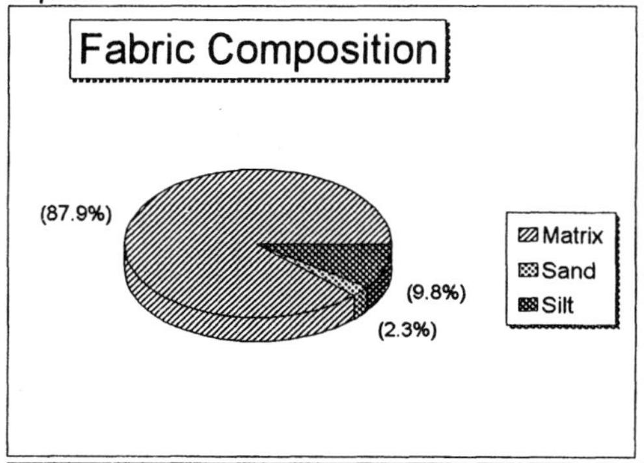

Qz Size	< 50	50-125	125-200	200-275	275-350	> 350	numpts
Number	317	19	2	2	0	0	340
%	93.24	5.59	0.59	0.59	0.00	0.00	
sigma	1.36	1.25	0.41	0.41	0.00	0.00	
p2sigma	95.96	8.08	1.42	1.42	0.00	0.00	
m2sigma	90.51	3.10	-0.24	-0.24	0.00	0.00	
Conc	50.72	3.04	0.32	0.32	0.00	0.00	
Fabric	Matrix	Sand	Silt	Micac	Biref	Colour	
	348	9	39	3	3	ltbrn	396

Minerals	Size	Ang	Sph	Wth	Conc	Type	Opac
PFs					1		
MFs					1		
Biotite	2			2	2		
Mus	2			2	4		
Amph					2		
Carb					3	2,3	1,2
Quartz		3	2		4	m,p	
Pyrox					1		
Epidote					1		
Schist					1		
other							

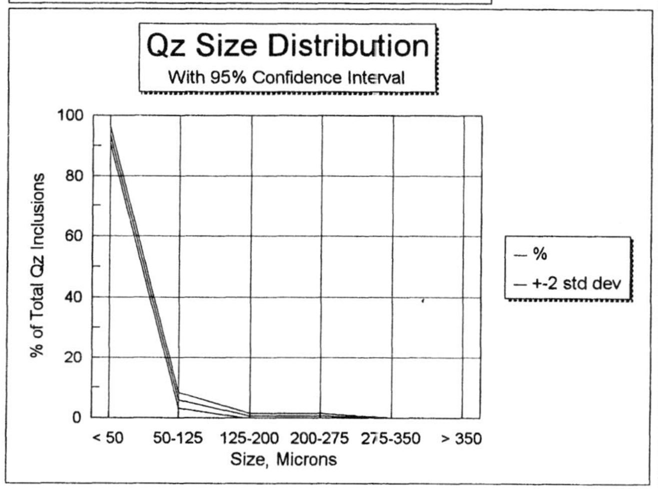

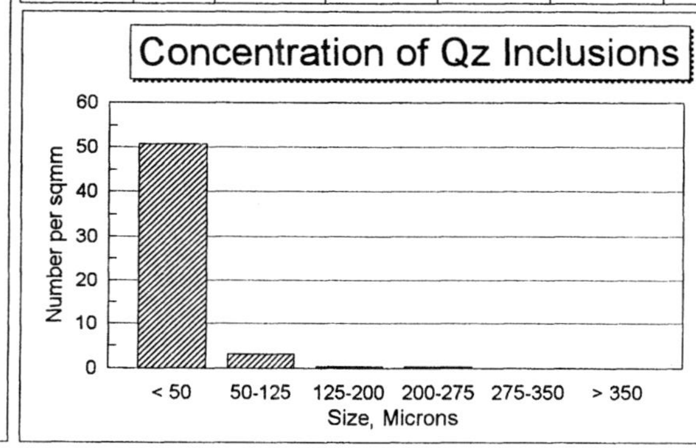

%matrix	87.88
%sand	2.27
%silt	9.85
conc<50	50.72
conc>50	3.68
%<50	93.24
%>50	6.76

trq227	period	level	morph type	rim dia	thick	sqmm
trq II B1	EH	2	All	15	9.5	12.5

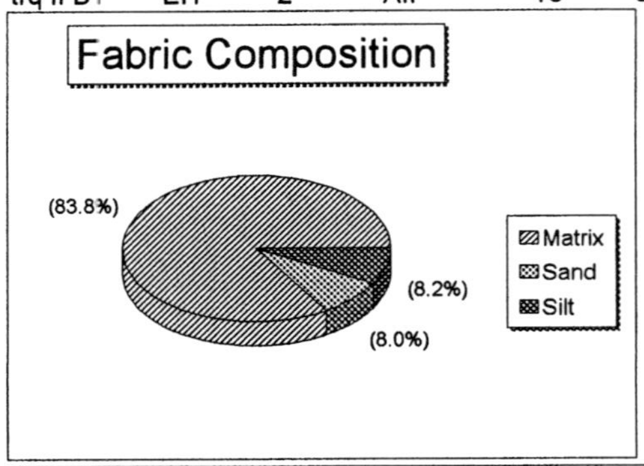

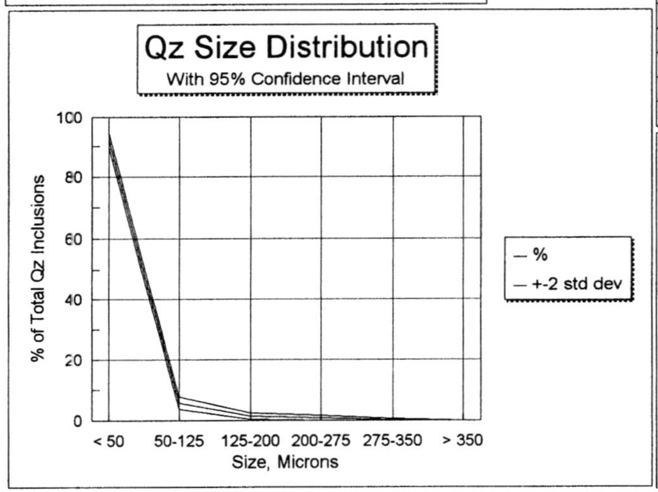

Qz Size	< 50	50-125	125-200	200-275	275-350	> 350	numpts
Number	451	28	7	4	1	0	491
%	91.85	5.70	1.43	0.81	0.20	0.00	
sigma	1.23	1.05	0.53	0.41	0.20	0.00	
p2sigma	94.32	7.80	2.50	1.63	0.61	0.00	
m2sigma	89.38	3.61	0.36	0.00	-0.20	0.00	
Conc	36.08	2.24	0.56	0.32	0.08	0.00	
Fabric	Matrix	Sand	Silt	Micac	Biref	Colour	
	325	31	32	3	3	medbr	388
Minerals	Size	Ang	Sph	Wth	Conc	Type	Opac
PFs	2	3	2		3		
MFs					1		
Biotite	1			3	2		
Mus	2			2	4		
Amph					1		
Carb					3	2	1,2
Quartz		3	2		4	m,p	
Pyrox					2		
Epidote					2		
Schist					1		
other						vrfrag	

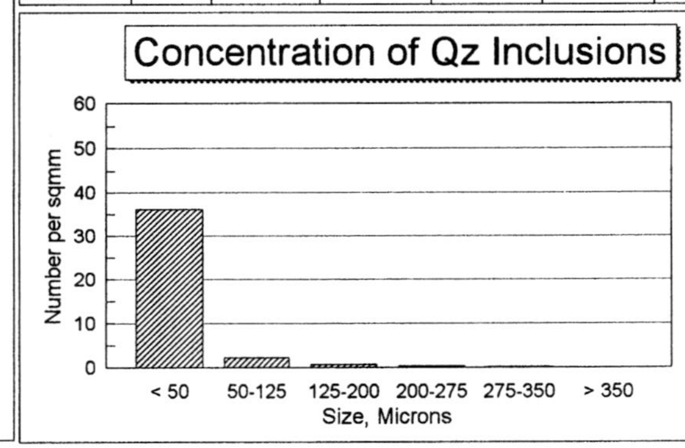

%matrix	83.76
%sand	7.99
%silt	8.25
conc<50	36.08
conc>50	3.20
%<50	91.85
%>50	8.15

trq228	period	level	morph type	rim dia	thick	sqmm
trq II B2	EH	2	Al or All	12	13.2	12.5

Qz Size	< 50	50-125	125-200	200-275	275-350	> 350	numpts
Number	407	13	1	5	2	3	431
%	94.43	3.02	0.23	1.16	0.46	0.70	
sigma	1.10	0.82	0.23	0.52	0.33	0.40	
p2sigma	96.64	4.66	0.70	2.19	1.12	1.50	
m2sigma	92.22	1.37	-0.23	0.13	-0.19	-0.10	
Conc	32.56	1.04	0.08	0.40	0.16	0.24	
Fabric	Matrix	Sand	Silt	Micac	Biref	Colour	
	354	12	24	3	3	medbr	390

Minerals	Size	Ang	Sph	Wth	Conc	Type	Opac
PFs	2	2	3		2		
MFs					1		
Biotite	1			2	2		
Mus	2			2	4		
Amph					1		
Carb	2				3	2,3	1,2
Quartz		3	2		4	m,p	
Pyrox					1		
Epidote					1		
Schist					2		
other							

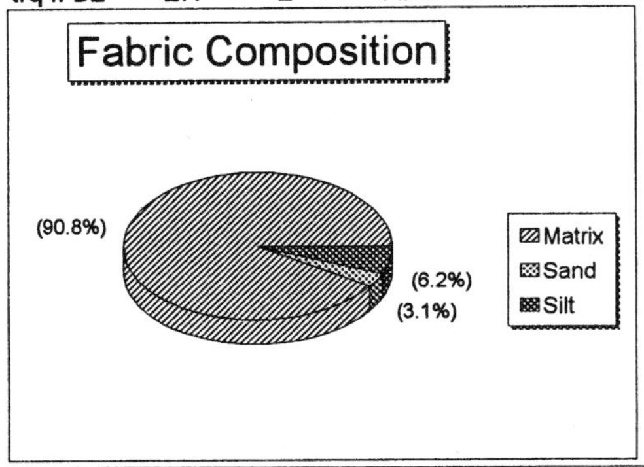

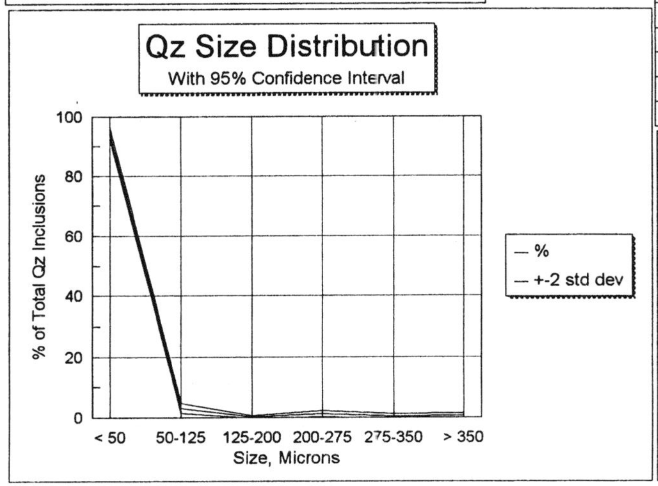

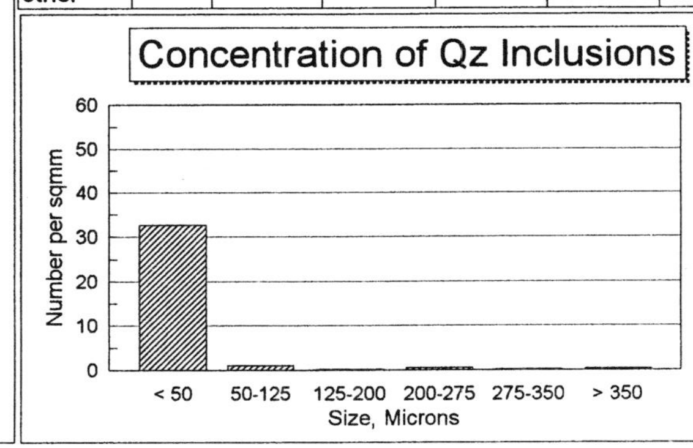

%matrix	90.77
%sand	3.08
%silt	6.15
conc<50	32.56
conc>50	1.92
%<50	94.43
%>50	5.57

trq229	period	level	morph type	rim dia	thick	sqmm
trq II B3	EH	2	Bl	10	6.2	6.25

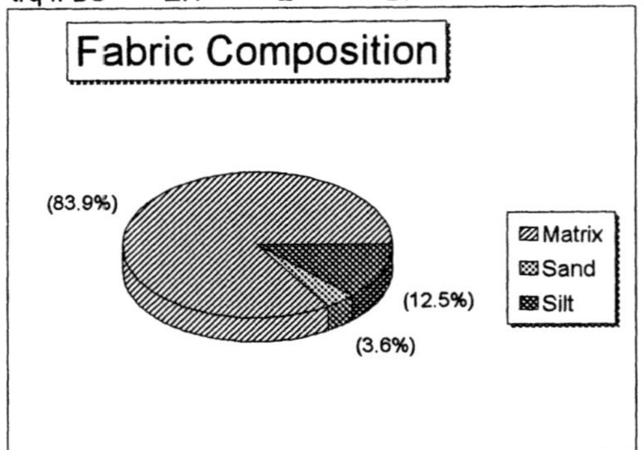

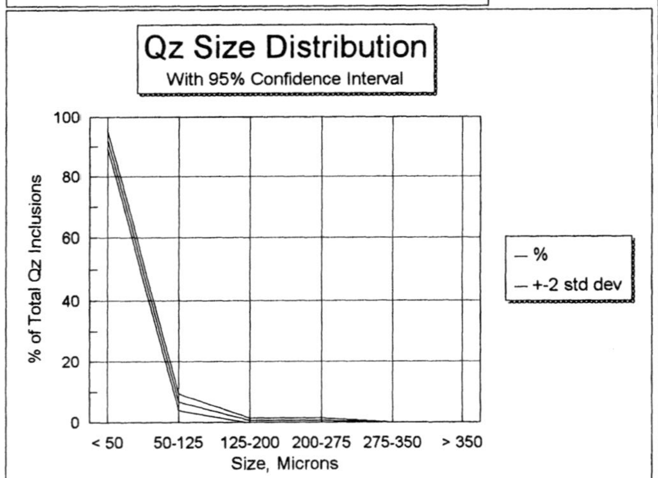

Qz Size	< 50	50-125	125-200	200-275	275-350	> 350	numpts
Number	301	22	2	2	0	0	327
%	92.05	6.73	0.61	0.61	0.00	0.00	
sigma	1.50	1.39	0.43	0.43	0.00	0.00	
p2sigma	95.04	9.50	1.47	1.47	0.00	0.00	
m2sigma	89.06	3.96	-0.25	-0.25	0.00	0.00	
Conc	48.16	3.52	0.32	0.32	0.00	0.00	
Fabric	Matrix	Sand	Silt	Micac	Biref	Colour	
	329	14	49	3	3	medbr	392

Minerals	Size	Ang	Sph	Wth	Conc	Type	Opac
PFs	2	2	2		2		
MFs	2	3	1		2		
Biotite	2			2	2		
Mus	3			2	4		
Amph					1		
Carb	2				3	2	1
Quartz		3	2		4	m,p	
Pyrox					2		
Epidote					1		
Schist					2		
other							

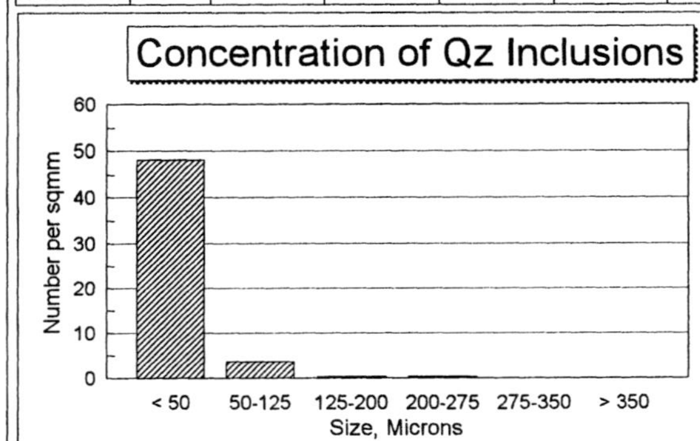

%matrix	83.93
%sand	3.57
%silt	12.50
conc<50	48.16
conc>50	4.16
%<50	92.05
%>50	7.95

trq230	period	level	morph type	rim dia	thick	sqmm
trq II B2	EH	2	BI	17	6.6	6.25

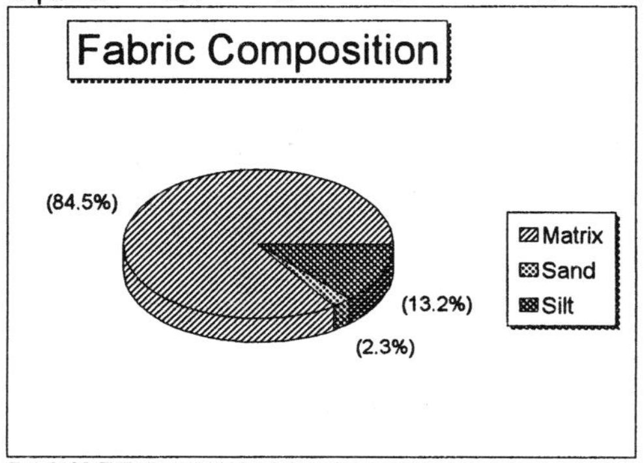

Qz Size	< 50	50-125	125-200	200-275	275-350	> 350	numpts
Number	285	17	3	3	0	0	308
%	92.53	5.52	0.97	0.97	0.00	0.00	
sigma	1.50	1.30	0.56	0.56	0.00	0.00	
p2sigma	95.53	8.12	2.09	2.09	0.00	0.00	
m2sigma	89.54	2.92	-0.15	-0.15	0.00	0.00	
Conc	45.60	2.72	0.48	0.48	0.00	0.00	
Fabric	Matrix	Sand	Silt	Micac	Biref	Colour	
	332	9	52	3	3	medbr	393
Minerals	Size	Ang	Sph	Wth	Conc	Type	Opac
PFs	2	3	2		2		
MFs					1		
Biotite	2			3	3		
Mus	2			2	4		
Amph					1		
Carb	2				3	2,3	1
Quartz		3	2		4	m,p	
Pyrox					2		
Epidote					1		
Schist					1		
other							

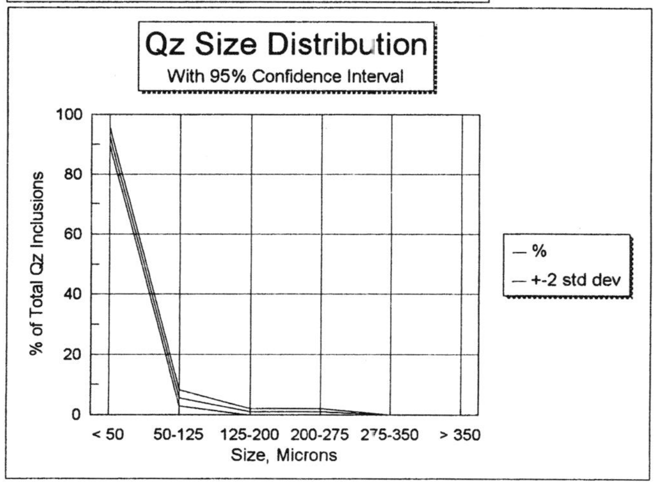

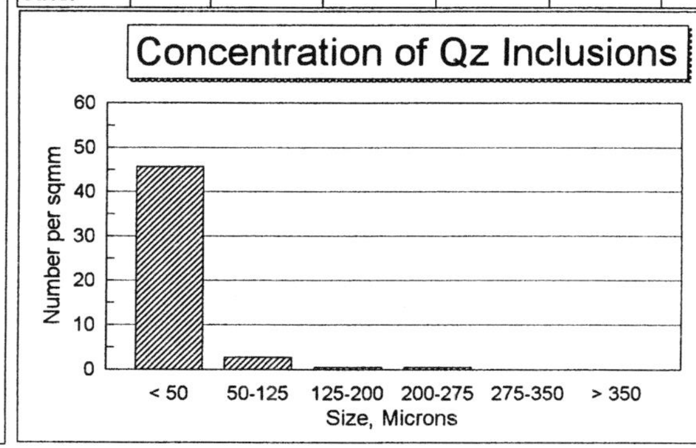

%matrix	84.48
%sand	2.29
%silt	13.23
conc<50	45.60
conc>50	3.68
%<50	92.53
%>50	7.47

trq231	period	level	morph type	rim dia	thick	sqmm
trq II B4	EH	2	BIII	16	6.2	12.5

Qz Size	< 50	50-125	125-200	200-275	275-350	> 350	numpts
Number	276	33	11	7	0	1	328
%	84.15	10.06	3.35	2.13	0.00	0.30	
sigma	2.02	1.66	0.99	0.80	0.00	0.30	
p2sigma	88.18	13.38	5.34	3.73	0.00	0.91	
m2sigma	80.11	6.74	1.37	0.54	0.00	-0.30	
Conc	22.08	2.64	0.88	0.56	0.00	0.08	
Fabric	Matrix	Sand	Silt	Micac	Biref	Colour	
	336	32	26	2	2	medbr	394

Minerals	Size	Ang	Sph	Wth	Conc	Type	Opac
PFs	3	3	2		2		
MFs					1		
Biotite	2			2	2		
Mus	2			2	4		
Amph					2		
Carb	2				3	2,3	2
Quartz		3	2		4	m,p	
Pyrox					2		
Epidote					1		
Schist					1		
other						vrfrag	

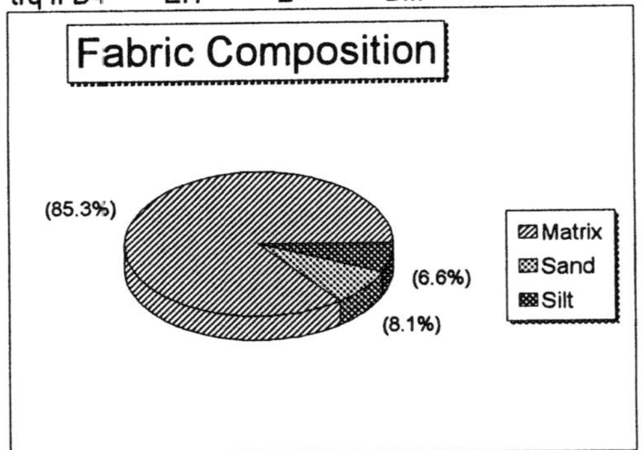

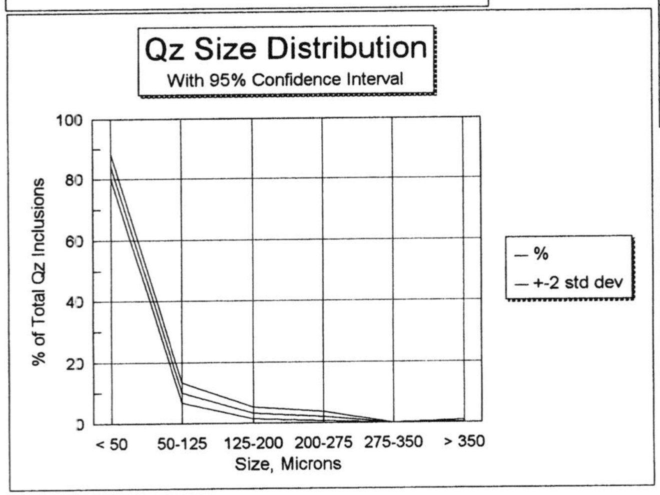

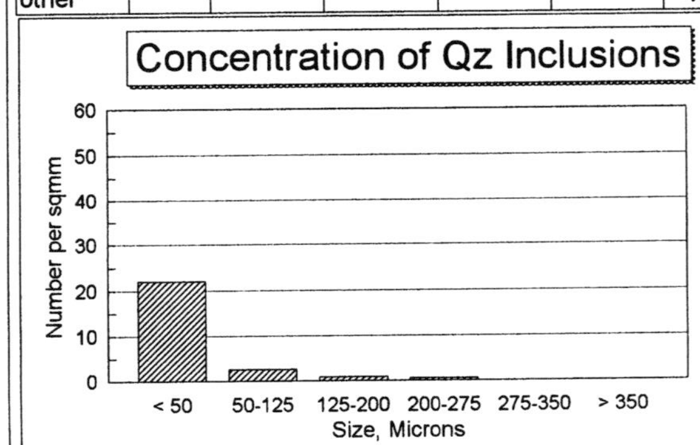

%matrix	85.28
%sand	8.12
%silt	6.60
conc<50	22.08
conc>50	4.16
%<50	84.15
%>50	15.85

trq232	period	level	morph type	rim dia	thick	sqmm
trq II B1	EH	2	BIII	23	5.4	12.5

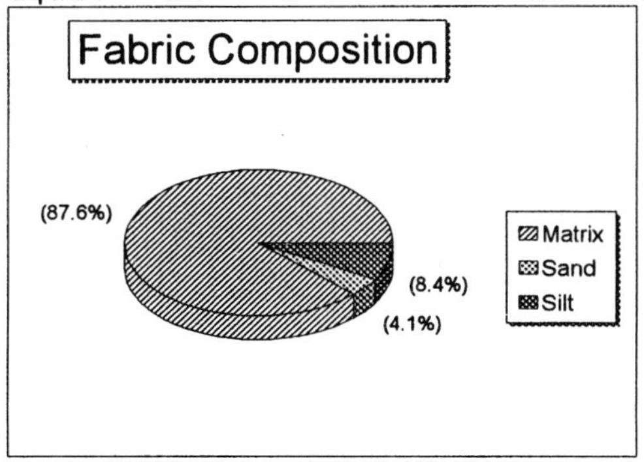

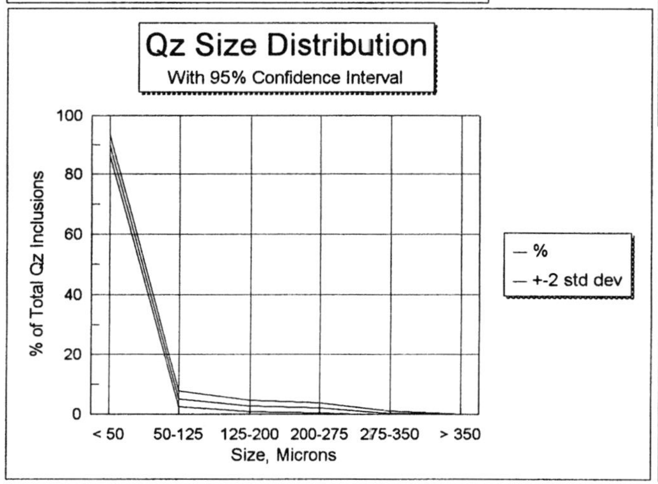

Qz Size	< 50	50-125	125-200	200-275	275-350	> 350	numpts
Number	266	15	8	6	1	0	296
%	89.86	5.07	2.70	2.03	0.34	0.00	
sigma	1.75	1.27	0.94	0.82	0.34	0.00	
p2sigma	93.37	7.62	4.59	3.67	1.01	0.00	
m2sigma	86.36	2.52	0.82	0.39	-0.34	0.00	
Conc	21.28	1.20	0.64	0.48	0.08	0.00	
Fabric	Matrix	Sand	Silt	Micac	Biref	Colour	
	346	16	33	3	3	medbr	395
Minerals	Size	Ang	Sph	Wth	Conc	Type	Opac
PFs	3	3	2		2		
MFs					1		
Biotite	2			2	2		
Mus	2			2	4		
Amph	2				2		
Carb	2				3	2	1,2
Quartz		3	2		4	m,p	
Pyrox					2		
Epidote					1		
Schist					2		
other						vrfrag	

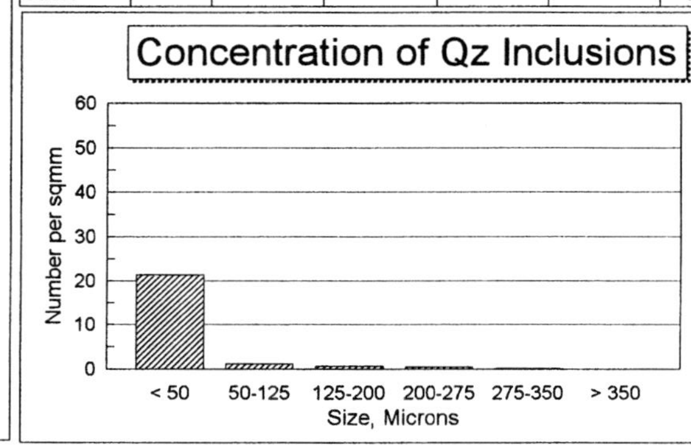

%matrix	87.59
%sand	4.05
%silt	8.35
conc<50	21.28
conc>50	2.40
%<50	89.86
%>50	10.14

trq233	period	level	morph type	rim dia	thick	sqmm
trq II B4	EH	2	BIII	25	7.5	12.5

Qz Size	< 50	50-125	125-200	200-275	275-350	> 350	numpts
Number	346	35	11	7	2	0	401
%	86.28	8.73	2.74	1.75	0.50	0.00	
sigma	1.72	1.41	0.82	0.65	0.35	0.00	
p2sigma	89.72	11.55	4.37	3.05	1.20	0.00	
m2sigma	82.85	5.91	1.11	0.44	-0.20	0.00	
Conc	27.68	2.80	0.88	0.56	0.16	0.00	
Fabric	Matrix	Sand	Silt	Micac	Biref	Colour	
	346	15	33	3	3	medbr	394

Minerals	Size	Ang	Sph	Wth	Conc	Type	Opac
PFs	1	3	2		2		
MFs					1		
Biotite	2			2	2		
Mus	3			2	4		
Amph	1				2		
Carb					3	2,3	1
Quartz		3	2		4	m,p	
Pyrox					1		
Epidote					1		
Schist					1		
other							

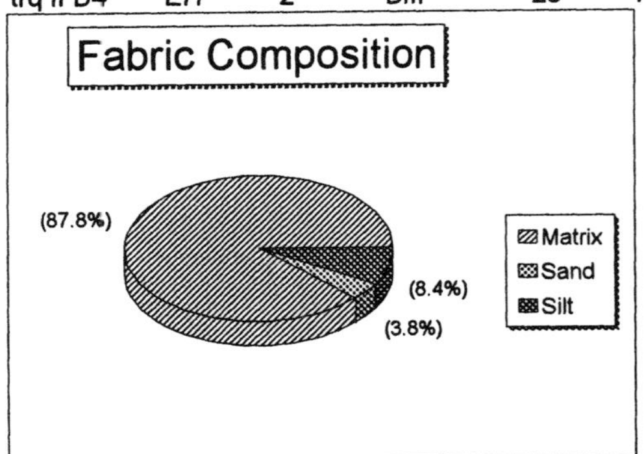

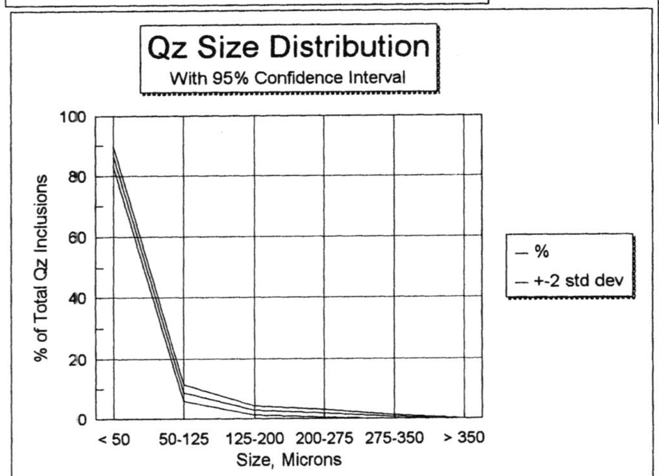

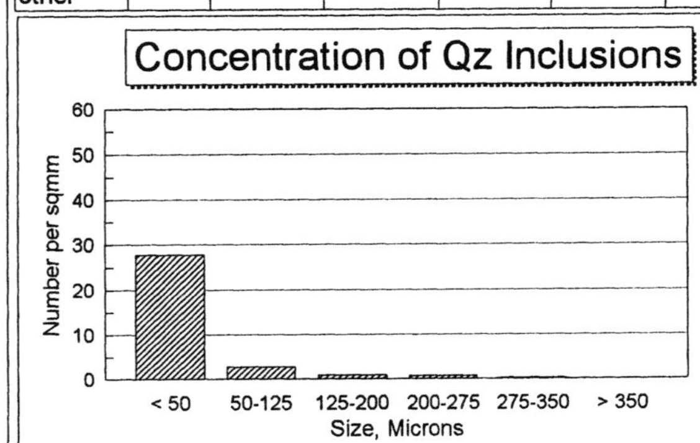

%matrix	87.82
%sand	3.81
%silt	8.38
conc<50	27.68
conc>50	4.40
%<50	86.28
%>50	13.72

trq234	period	level	morph type	rim dia	thick	sqmm
trq II B3	EH	2	BIII	33	12.2	12.5

Qz Size	< 50	50-125	125-200	200-275	275-350	> 350	numpts
Number	323	26	3	5	1	0	358
%	90.22	7.26	0.84	1.40	0.28	0.00	
sigma	1.57	1.37	0.48	0.62	0.28	0.00	
p2sigma	93.36	10.01	1.80	2.64	0.84	0.00	
m2sigma	87.08	4.52	-0.13	0.16	-0.28	0.00	
Conc	25.84	2.08	0.24	0.40	0.08	0.00	
Fabric	Matrix	Sand	Silt	Micac	Biref	Colour	
	352	17	26	3	3	medbr	395

Minerals	Size	Ang	Sph	Wth	Conc	Type	Opac
PFs	3	3	2		2		
MFs					1		
Biotite	2			2	2		
Mus	2			2	4		
Amph					2		
Carb					4	2	2
Quartz		3	2		4	m,p	
Pyrox					1		
Epidote					1		
Schist					2		
other						vrfrag	

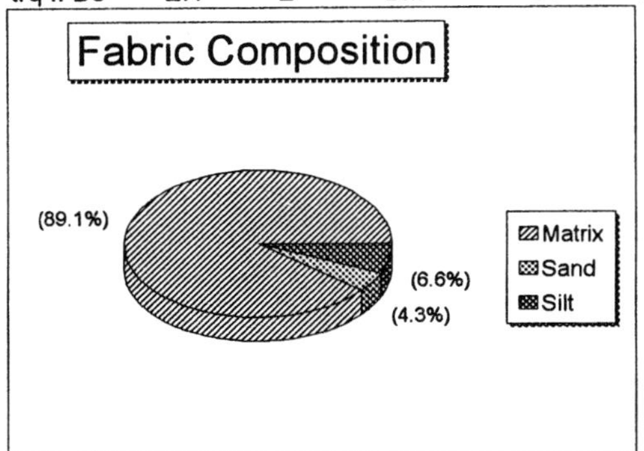

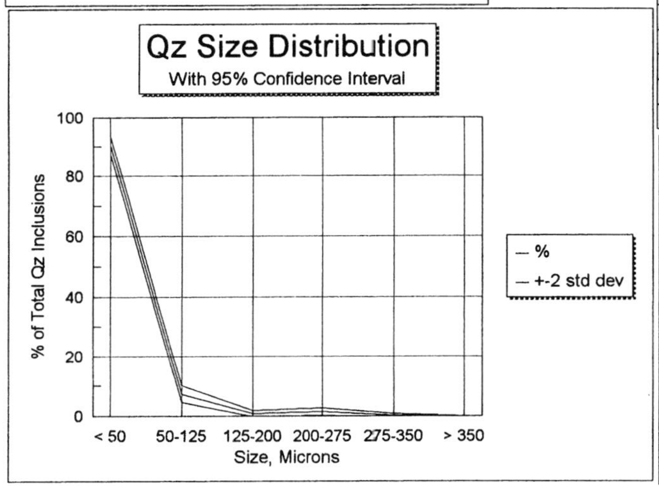

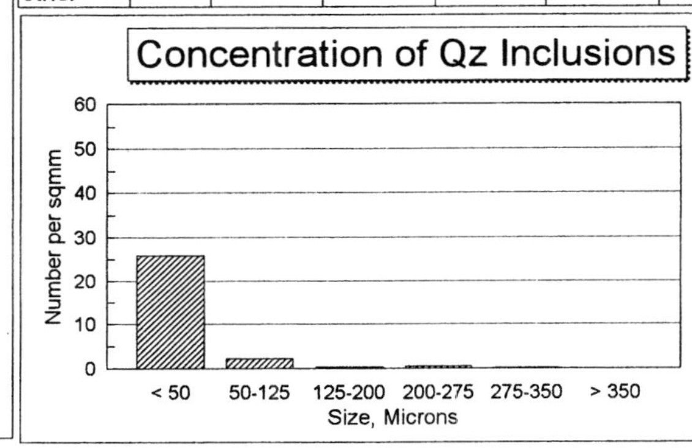

%matrix	89.11
%sand	4.30
%silt	6.58
conc<50	25.84
conc>50	2.80
%<50	90.22
%>50	9.78

trq235	period	level	morph type	rim dia	thick	sqmm
trq II B3	EH	2	BIII	28	12.1	

Poor section; not countable

Qz Size	< 50	50-125	125-200	200-275	275-350	> 350	numpts
Number							0
%							
sigma							
p2sigma							
m2sigma							
Conc							
Fabric	Matrix	Sand	Silt	Micac	Biref	Colour	
				1	1	medbr	0

Minerals	Size	Ang	Sph	Wth	Conc	Type	Opac
PFs	3	3	2		2		
MFs					1		
Biotite	2			2	2		
Mus	2			2	4		
Amph					1		
Carb					3	2,3	2
Quartz		3	2		4	m,p	
Pyrox					2		
Epidote					1		
Schist					1		
other						vrfrag	

trq236	period	level	morph type	rim dia	thick	sqmm
trq II B3	EH	2	BIV	24	13.8	

Not countable

Qz Size	< 50	50-125	125-200	200-275	275-350	> 350	numpts
Number							0
%							
sigma							
p2sigma							
m2sigma							
Conc							
Fabric	Matrix	Sand	Silt	Micac	Biref	Colour	
							0

Minerals	Size	Ang	Sph	Wth	Conc	Type	Opac
PFs	2	3	2		2		
MFs					1		
Biotite	2			2	2		
Mus				2	4		
Amph					1		
Carb					3	2	1
Quartz		3	2		4	m,p	
Pyrox					1		
Epidote					1		
Schist					1		
other						vrfrag	

trq237	period	level	morph type	rim dia	thick	sqmm
trq II B2	EH	2	BII	33	12.2	6.25

Qz Size	< 50	50-125	125-200	200-275	275-350	> 350	numpts
Number	207	12	1	3	2	0	225
%	92.00	5.33	0.44	1.33	0.89	0.00	
sigma	1.81	1.50	0.44	0.76	0.63	0.00	
p2sigma	95.62	8.33	1.33	2.86	2.14	0.00	
m2sigma	88.38	2.34	-0.44	-0.20	-0.36	0.00	
Conc	33.12	1.92	0.16	0.48	0.32	0.00	
Fabric	Matrix	Sand	Silt	Micac	Biref	Colour	
	339	10	51	2	2	medbr	400
Minerals	Size	Ang	Sph	Wth	Conc	Type	Opac
PFs					1		
MFs					1		
Biotite	1			2	2		
Mus	1			2	4		
Amph					1		
Carb					4	2,3	1,2,3
Quartz		3	2		4	m,p	
Pyrox					1		
Epidote					1		
Schist					2		
other							

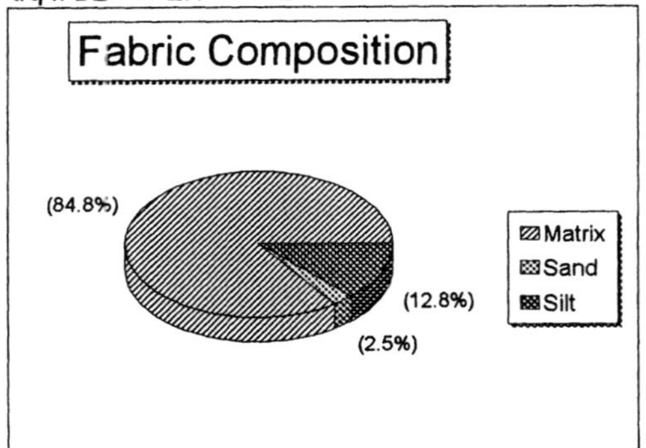

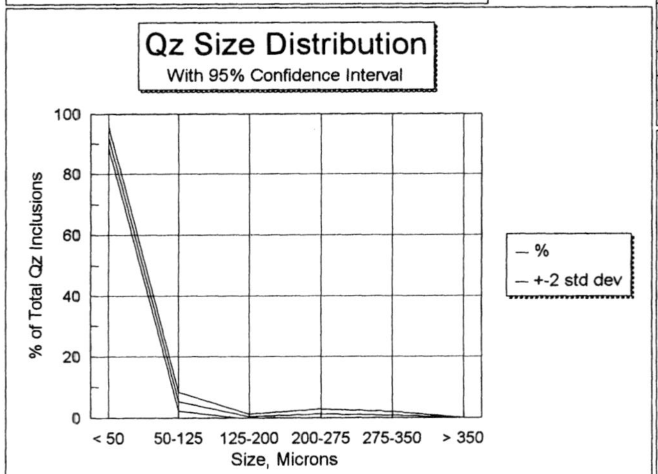

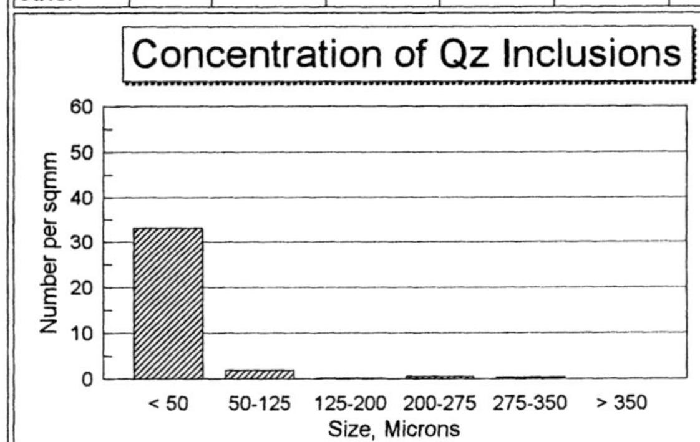

%matrix	84.75
%sand	2.50
%silt	12.75
conc<50	33.12
conc>50	2.88
%<50	92.00
%>50	8.00

trq238	period	level	morph type	rim dia	thick	sqmm
trq II B2	EH	2	BIV	36.6	4.6	

Not countable

Qz Size	< 50	50-125	125-200	200-275	275-350	> 350	numpts
Number							0
%							
sigma							
p2sigma							
m2sigma							
Conc							
Fabric	Matrix	Sand	Silt	Micac	Biref	Colour	
				1	1	dkbrn	0

Minerals	Size	Ang	Sph	Wth	Conc	Type	Opac
PFs					1		
MFs					1		
Biotite					1		
Mus	3			2	4		
Amph					1		
Carb					4	2,3	3
Quartz		3	2		4	m,p	
Pyrox					1		
Epidote					1		
Schist					1		
other						vrf,grg	

trq239	period	level	morph type	rim dia	thick	sqmm
trq II B1	EH	2	BIV or BV	25	7.3	12.5

Qz Size	< 50	50-125	125-200	200-275	275-350	> 350	numpts
Number	226	28	7	2	1	0	264
%	85.61	10.61	2.65	0.76	0.38	0.00	
sigma	2.16	1.90	0.99	0.53	0.38	0.00	
p2sigma	89.93	14.40	4.63	1.82	1.13	0.00	
m2sigma	81.29	6.82	0.67	-0.31	-0.38	0.00	
Conc	18.08	2.24	0.56	0.16	0.08	0.00	
Fabric	Matrix	Sand	Silt	Micac	Biref	Colour	
	348	16	36				400

Minerals	Size	Ang	Sph	Wth	Conc	Type	Opac
PFs					1		
MFs					1		
Biotite					1		
Mus	2			2	4		
Amph					1		
Carb					4	2,3	1
Quartz		3	2		4	m,p	
Pyrox					2		
Epidote					2		
Schist					1		
other							

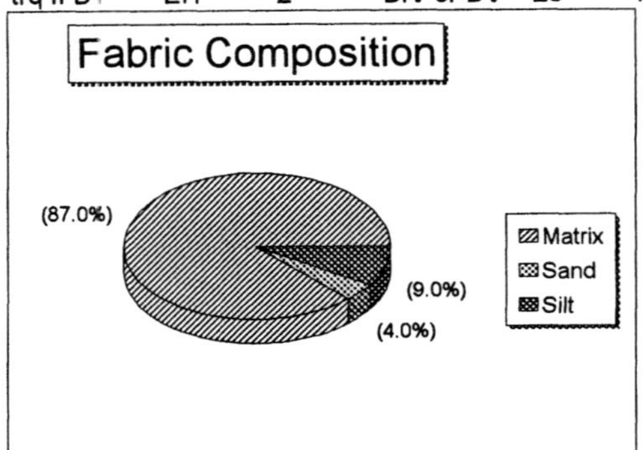

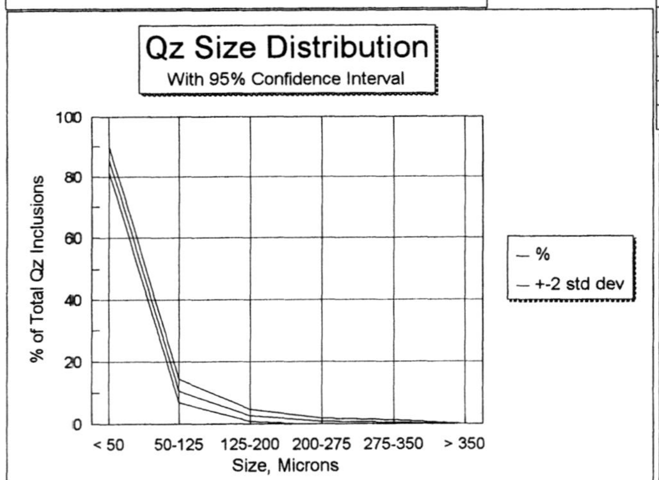

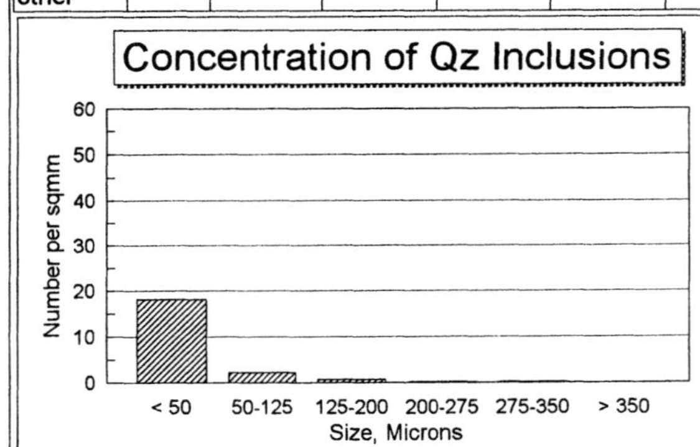

%matrix	87.00
%sand	4.00
%silt	9.00
conc<50	18.08
conc>50	3.04
%<50	85.61
%>50	14.39

APPENDIX B

trq240	period	level	morph type	rim dia	thick	sqmm
trq II B1	EH	2	BIV or BV	37	8.3	12.5

Qz Size	< 50	50-125	125-200	200-275	275-350	> 350	numpts
Number	248	28	15	7	3	0	301
%	82.39	9.30	4.98	2.33	1.00	0.00	
sigma	2.20	1.67	1.25	0.87	0.57	0.00	
p2sigma	86.78	12.65	7.49	4.06	2.14	0.00	
m2sigma	78.00	5.95	2.47	0.59	-0.15	0.00	
Conc	19.84	2.24	1.20	0.56	0.24	0.00	
Fabric	Matrix	Sand	Silt	Micac	Biref	Colour	
	329	36	30	3	3	medbr	395

Minerals	Size	Ang	Sph	Wth	Conc	Type	Opac
PFs					1		
MFs					1		
Biotite					1		
Mus	2			2	4		
Amph					2		
Carb					4	2,3	2
Quartz		3	2		4	m,p	
Pyrox					2		
Epidote					2		
Schist					2	epidote	
other						vrfrag	

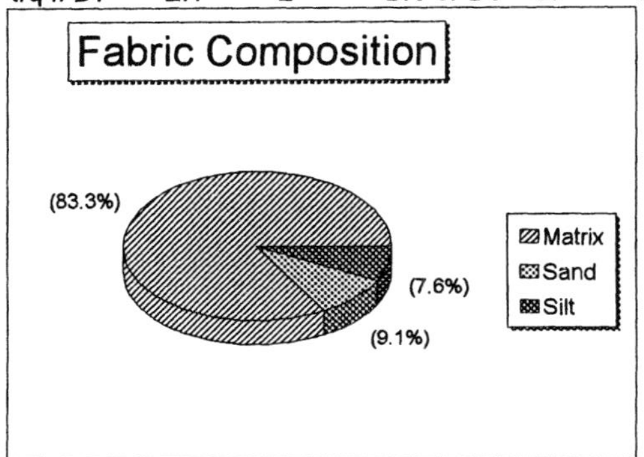

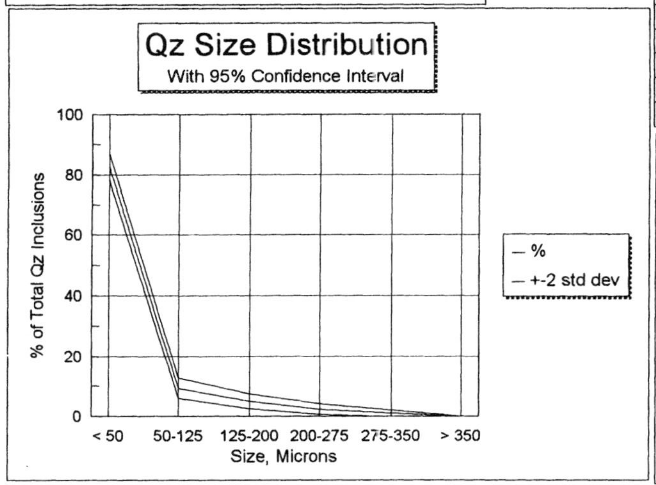

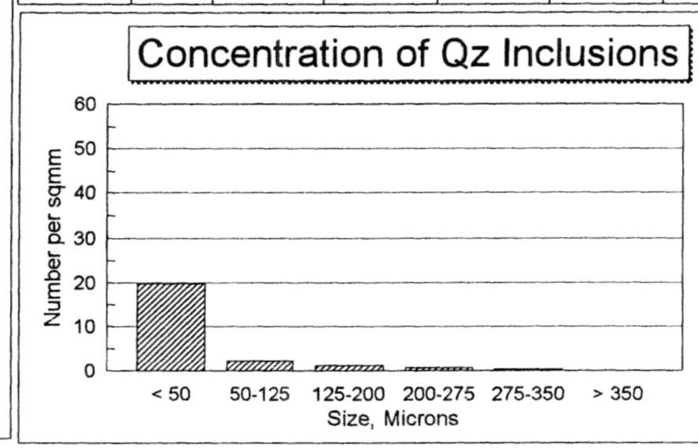

%matrix	83.29
%sand	9.11
%silt	7.59
conc<50	19.84
conc>50	4.24
%<50	82.39
%>50	17.61

trq241	period	level	morph type	rim dia	thick	sqmm
trq II B4	EH	2	BIV or BV	16	6.6	6.25

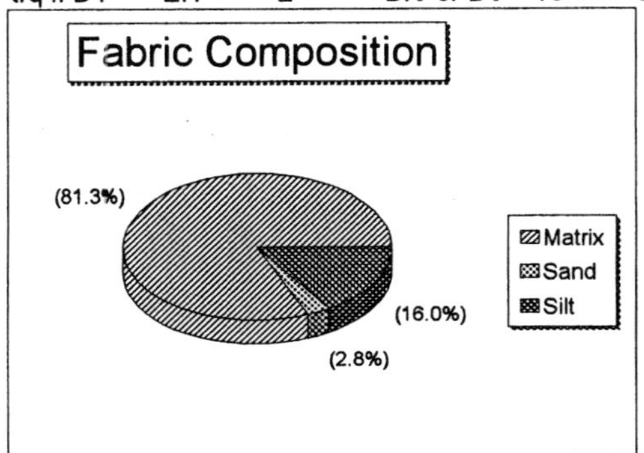

Qz Size	< 50	50-125	125-200	200-275	275-350	> 350	numpts
Number	419	34	2	3	0	0	458
%	91.48	7.42	0.44	0.66	0.00	0.00	
sigma	1.30	1.22	0.31	0.38	0.00	0.00	
p2sigma	94.09	9.87	1.05	1.41	0.00	0.00	
m2sigma	88.88	4.97	-0.18	-0.10	0.00	0.00	
Conc	67.04	5.44	0.32	0.48	0.00	0.00	
Fabric	Matrix	Sand	Silt	Micac	Biref	Colour	
	325	11	64	3	3	medbr	400

Minerals	Size	Ang	Sph	Wth	Conc	Type	Opac
PFs					1		
MFs					1		
Biotite	1			2	3		
Mus	2			2	4		
Amph					2		
Carb					4	2,3	2
Quartz		3	2		4	m,p	
Pyrox					1		
Epidote					2		
Schist					2		
other						vrfrag	

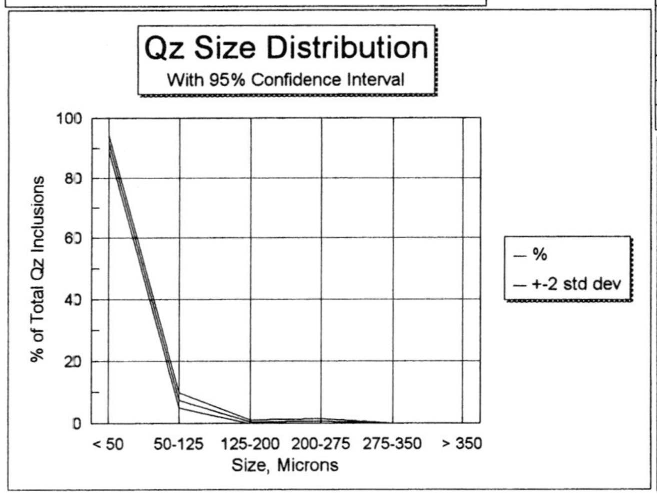

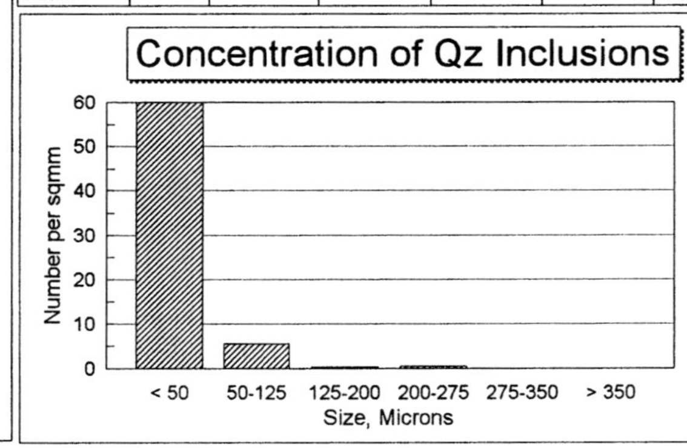

%matrix	81.25
%sand	2.75
%silt	16.00
conc<50	67.04
conc>50	6.24
%<50	91.48
%>50	8.52

APPENDIX B

trq242	period	level	morph type	rim dia	thick	sqmm
trq II B4	EH	2	BIV or BV	26	12.3	25

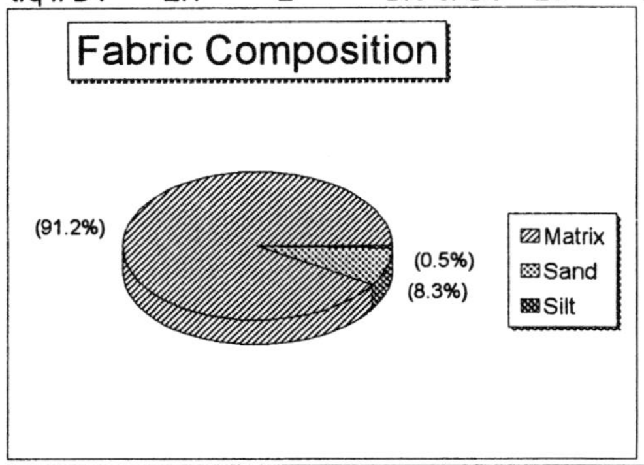

Qz Size	< 50	50-125	125-200	200-275	275-350	> 350	numpts
Number	213	20	8	6	1	0	248
%	85.89	8.06	3.23	2.42	0.40	0.00	
sigma	2.21	1.73	1.12	0.98	0.40	0.00	
p2sigma	90.31	11.52	5.47	4.37	1.21	0.00	
m2sigma	81.47	4.61	0.98	0.47	-0.40	0.00	
Conc	8.52	0.80	0.32	0.24	0.04	0.00	
Fabric	Matrix	Sand	Silt	Micac	Biref	Colour	
	364	33	2	1	1	dkbrn	399

Minerals	Size	Ang	Sph	Wth	Conc	Type	Opac
PFs					2		
MFs					1		
Biotite	1			2	2		
Mus	2			2	3		
Amph					2		
Carb					3	2,3	1
Quartz		3	2		4	m,p	
Pyrox					2		
Epidote					2		
Schist					2		
other						vrfrag	

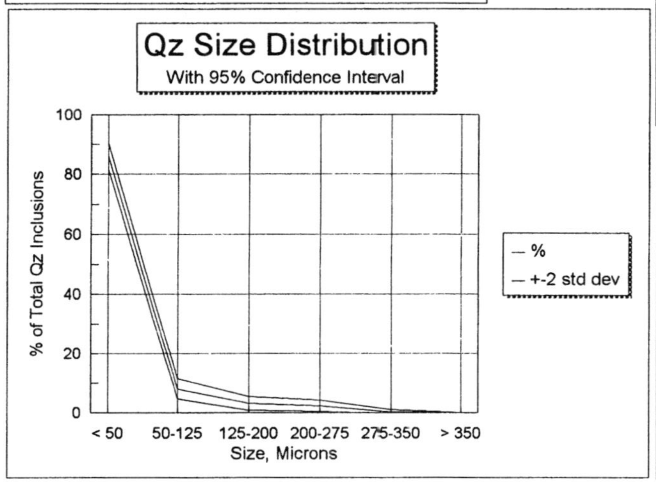

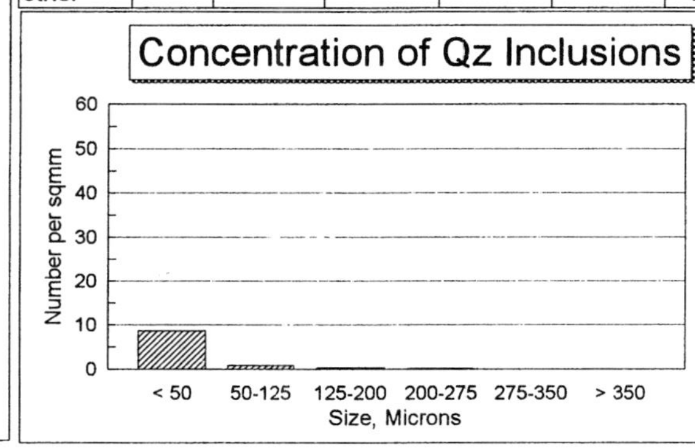

%matrix	91.23
%sand	8.27
%silt	0.50
conc<50	8.52
conc>50	1.40
%<50	85.89
%>50	14.11

APPENDIX B

trq243	period	level	morph type	rim dia	thick	sqmm
trq II B2	EH	2	BIV or BV	32	8.4	18.75

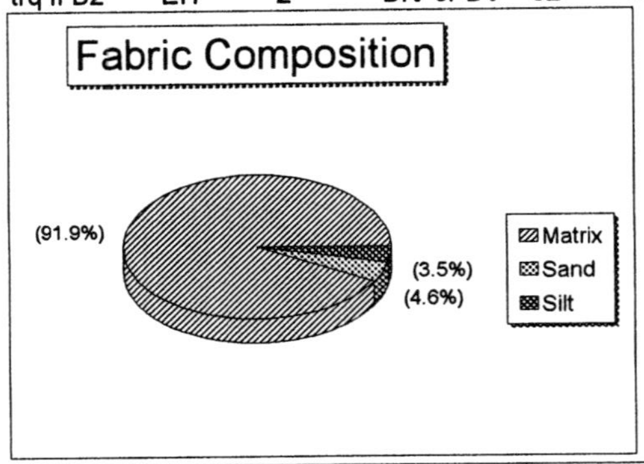

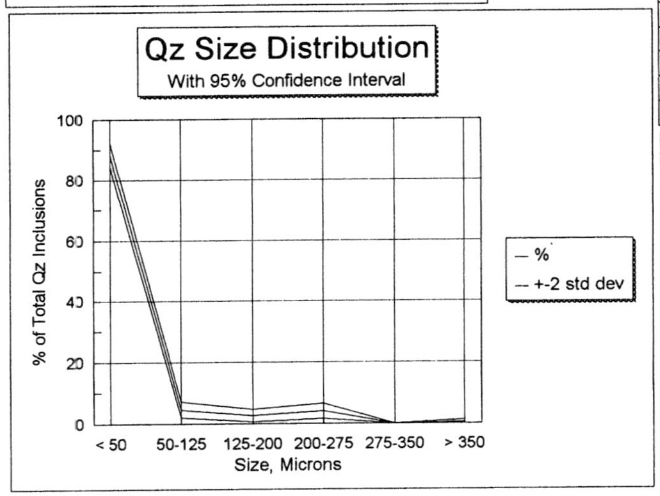

Qz Size	< 50	50-125	125-200	200-275	275-350	> 350	numpts
Number	232	12	7	11	0	1	263
%	88.21	4.56	2.66	4.18	0.00	0.38	
sigma	1.99	1.29	0.99	1.23	0.00	0.38	
p2sigma	92.19	7.14	4.65	6.65	0.00	1.14	
m2sigma	84.24	1.99	0.68	1.71	0.00	-0.38	
Conc	12.37	0.64	0.37	0.59	0.00	0.05	
Fabric	Matrix	Sand	Silt	Micac	Biref	Colour	
	363	18	14	1	1	dkbrn	395

Minerals	Size	Ang	Sph	Wth	Conc	Type	Opac
PFs	3	3	2		2		
MFs					1		
Biotite					1		
Mus	1			2	3		
Amph					1		
Carb					3	2,3	3
Quartz		3	2		4	m,p	
Pyrox					2		
Epidote					2		
Schist					2		
other							

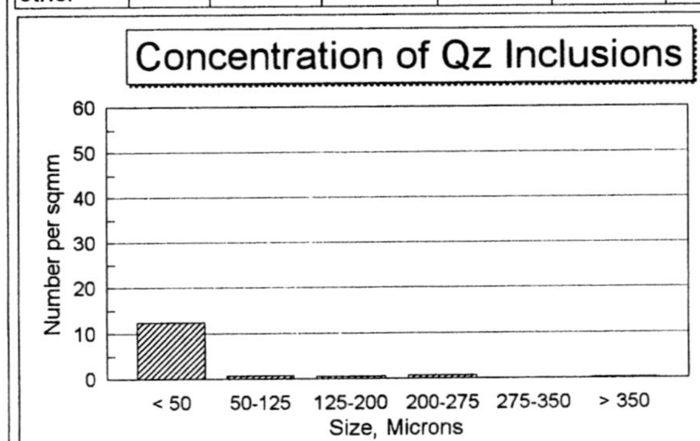

%matrix	91.90
%sand	4.56
%silt	3.54
conc<50	12.37
conc>50	1.65
%<50	88.21
%>50	11.79

trq244	period	level	morph type	rim dia	thick	sqmm
trq II B4	EH	2	BIV or BV	27	9	25

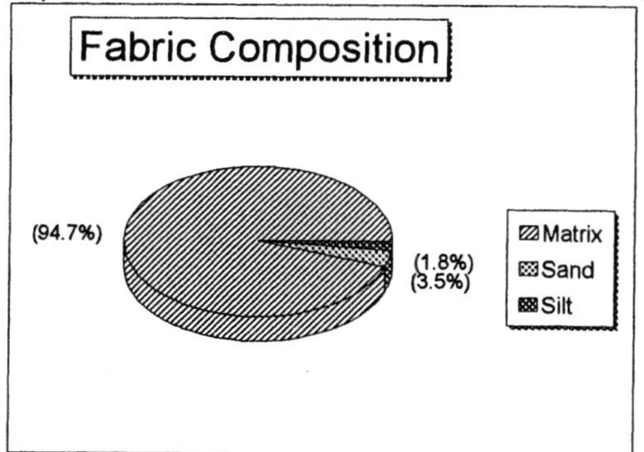

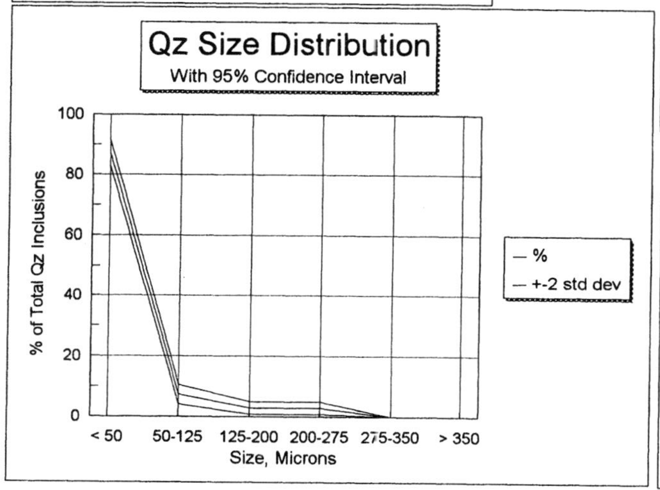

Qz Size	< 50	50-125	125-200	200-275	275-350	> 350	numpts
Number	215	18	7	7	0	0	247
%	87.04	7.29	2.83	2.83	0.00	0.00	
sigma	2.14	1.65	1.06	1.06	0.00	0.00	
p2sigma	91.32	10.60	4.95	4.95	0.00	0.00	
m2sigma	82.77	3.98	0.72	0.72	0.00	0.00	
Conc	8.60	0.72	0.28	0.28	0.00	0.00	

Fabric	Matrix	Sand	Silt	Micac	Biref	Colour	
	377	14	7	1	1	dkbrn	398

Minerals	Size	Ang	Sph	Wth	Conc	Type	Opac
PFs					1		
MFs					1		
Biotite				2	2		
Mus	1			2	3		
Amph					1		
Carb					3	2,3	2
Quartz		3	2		4	m,p	
Pyrox					2		
Epidote					1		
Schist					1		
other							

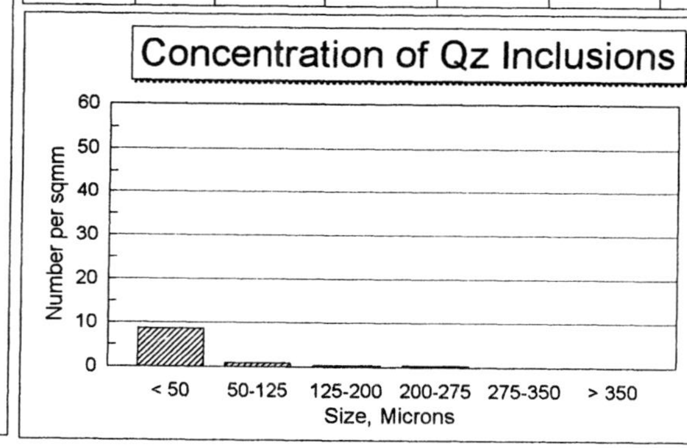

%matrix	94.72
%sand	3.52
%silt	1.76
conc<50	8.60
conc>50	1.28
%<50	87.04
%>50	12.96

APPENDIX B

trq245	period	level	morph type	rim dia	thick	sqmm
trq II B2	EH	2	BII	24	7.8	18.75

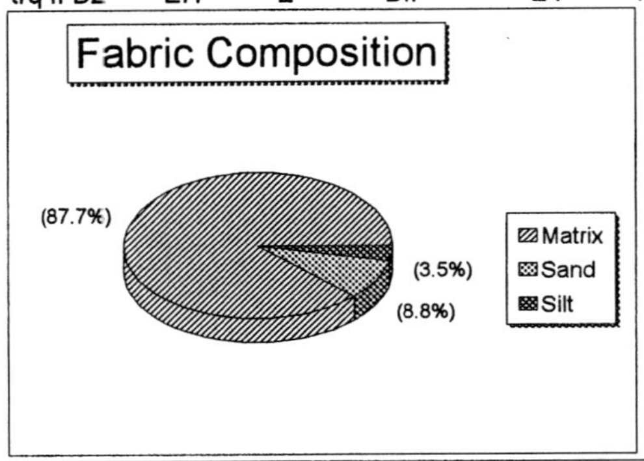

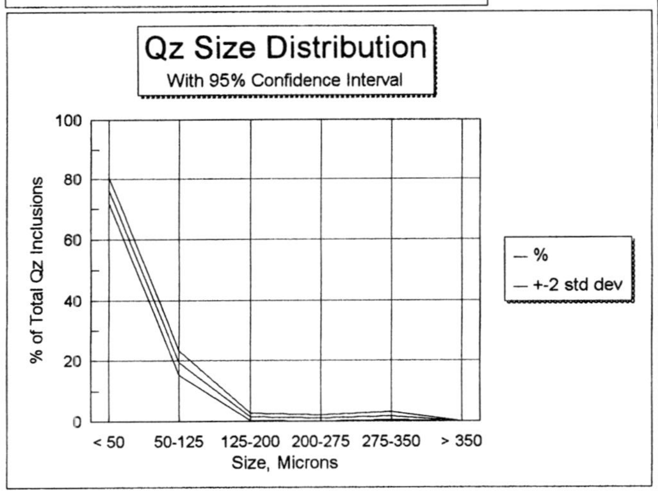

Qz Size	< 50	50-125	125-200	200-275	275-350	> 350	numpts
Number	293	75	6	4	7	0	385
%	76.10	19.48	1.56	1.04	1.82	0.00	
sigma	2.17	2.02	0.63	0.52	0.68	0.00	
p2sigma	80.45	23.52	2.82	2.07	3.18	0.00	
m2sigma	71.76	15.44	0.30	0.01	0.46	0.00	
Conc	15.63	4.00	0.32	0.21	0.37	0.00	
Fabric	Matrix	Sand	Silt	Micac	Biref	Colour	
	350	35	14	2	2	dkbrn	399

Minerals	Size	Ang	Sph	Wth	Conc	Type	Opac
PFs					1		
MFs					1		
Biotite	1			2	3		
Mus	1			2	4		
Amph					2		
Carb					4	2,3	2
Quartz		3	2		4	m,p	
Pyrox					2		
Epidote					2		
Schist					1		
other							

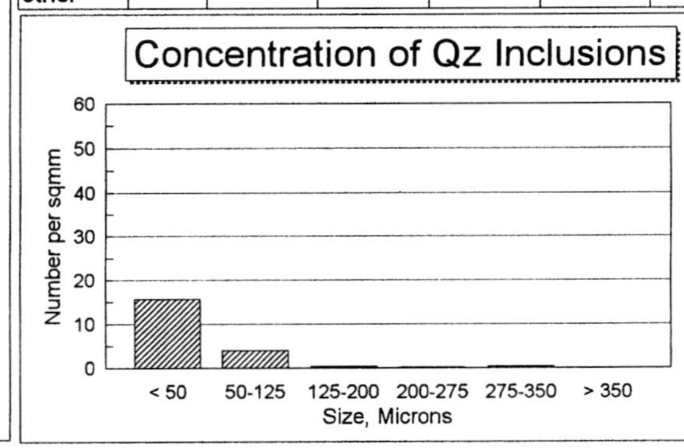

%matrix	87.72
%sand	8.77
%silt	3.51
conc<50	15.63
conc>50	4.91
%<50	76.10
%>50	23.90

trq246	period	level	morph type	rim dia	thick	sqmm
trq II B2	EH	2	BII	32.5	9.1	6.25

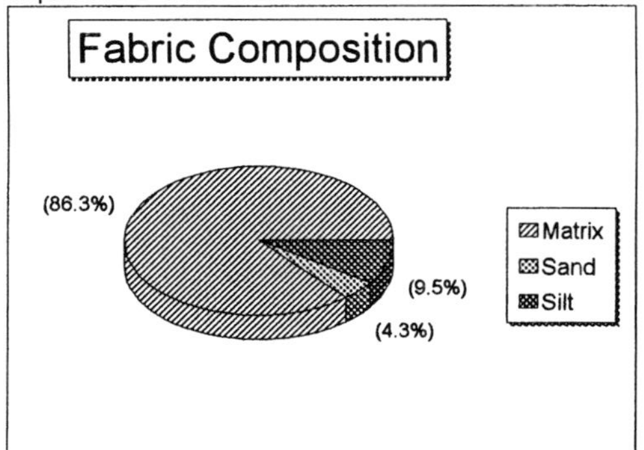

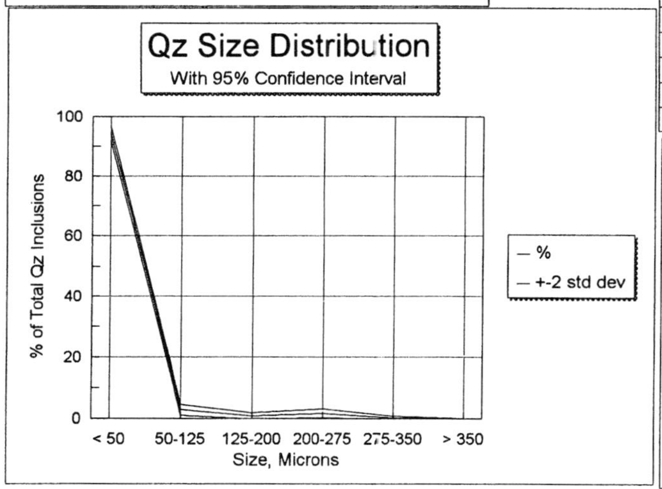

Qz Size	< 50	50-125	125-200	200-275	275-350	> 350	numpts
Number	334	10	3	6	1	0	354
%	94.35	2.82	0.85	1.69	0.28	0.00	
sigma	1.23	0.88	0.49	0.69	0.28	0.00	
p2sigma	96.80	4.59	1.82	3.07	0.85	0.00	
m2sigma	91.90	1.06	-0.13	0.32	-0.28	0.00	
Conc	53.44	1.60	0.48	0.96	0.16	0.00	
Fabric	Matrix	Sand	Silt	Micac	Biref	Colour	
	345	17	38	3	2	medbr	400

Minerals	Size	Ang	Sph	Wth	Conc	Type	Opac
PFs	3	4	1		2		
MFs					1		
Biotite	2			2	3		
Mus	2			2	4		
Amph					1		
Carb					4	2,3	1,2,3
Quartz		3	2		4	m,p	
Pyrox					2		
Epidote					1		
Schist					2		
other					2	vrf,arg	

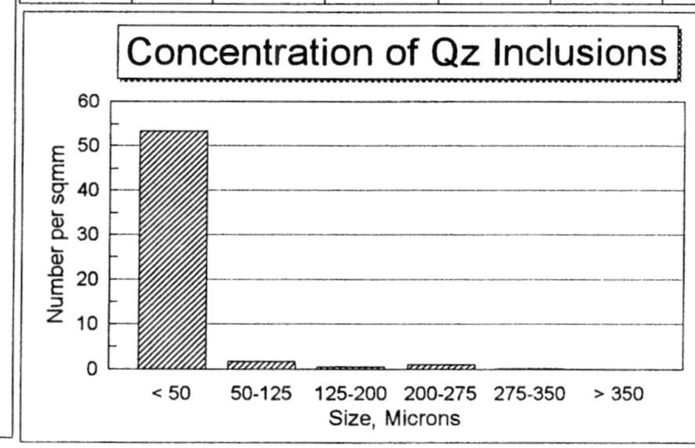

%matrix	86.25
%sand	4.25
%silt	9.50
conc<50	53.44
conc>50	3.20
%<50	94.35
%>50	5.65

APPENDIX B

trq247	period	level	morph type	rim dia	thick	sqmm
trq II B3	EH	2	BII	35	10.2	12.5

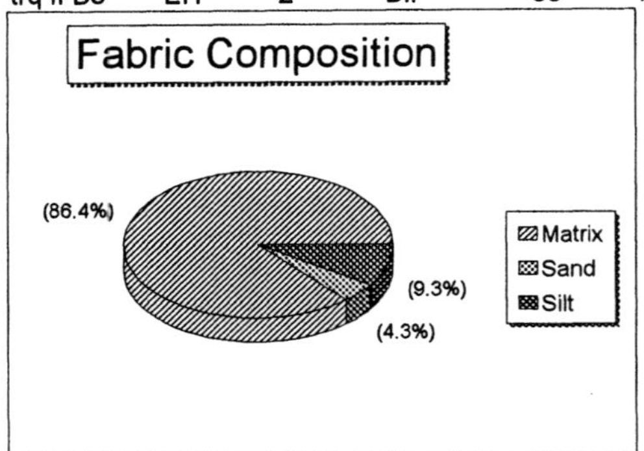

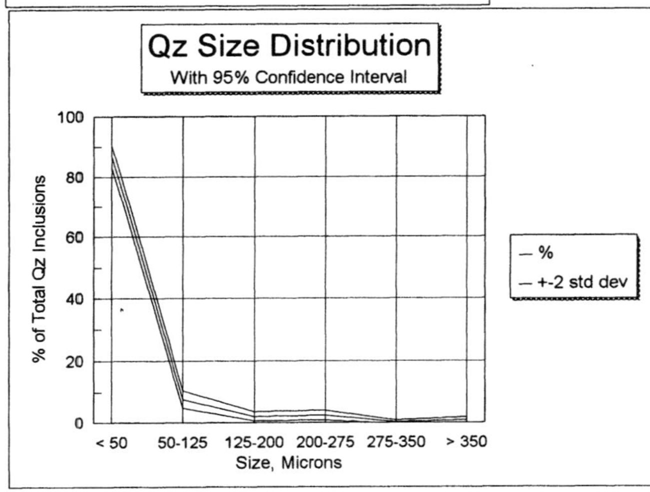

Qz Size	< 50	50-125	125-200	200-275	275-350	> 350	numpts
Number	293	26	7	8	1	3	338
%	86.69	7.69	2.07	2.37	0.30	0.89	
sigma	1.85	1.45	0.77	0.83	0.30	0.51	
p2sigma	90.38	10.59	3.62	4.02	0.89	1.91	
m2sigma	82.99	4.79	0.52	0.71	-0.29	-0.13	
Conc	23.44	2.08	0.56	0.64	0.08	0.24	
Fabric	Matrix	Sand	Silt	Micac	Biref	Colour	
	343	17	37	2	2	dkbrn	397

Minerals	Size	Ang	Sph	Wth	Conc	Type	Opac
PFs					1		
MFs					1		
Biotite	2			2	2		
Mus	2			2	4		
Amph					1		
Carb					4	2,3	2
Quartz		3	2		4	m,p	
Pyrox					1		
Epidote					2		
Schist					2		
other					2	vrf,arg	

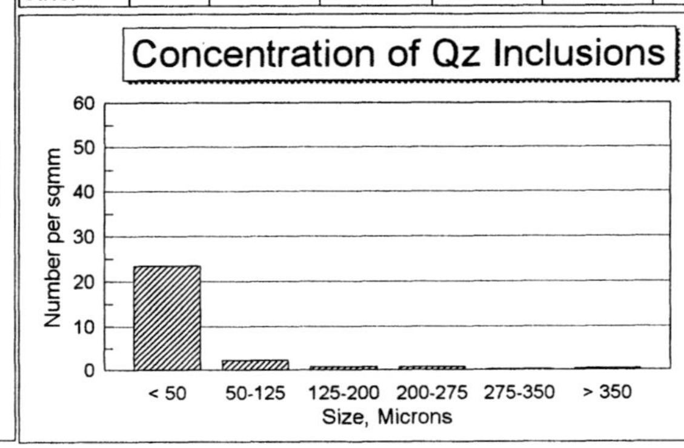

%matrix	86.40
%sand	4.28
%silt	9.32
conc<50	23.44
conc>50	3.60
%<50	86.69
%>50	13.31

APPENDIX C

HARAPPA DATA

See page 259 for key to abbreviations used

hrp100	period	level	morph type	rim dia	thick	sqmm
H90/1200-525		2 1/2	A1	7.2	4	12.5

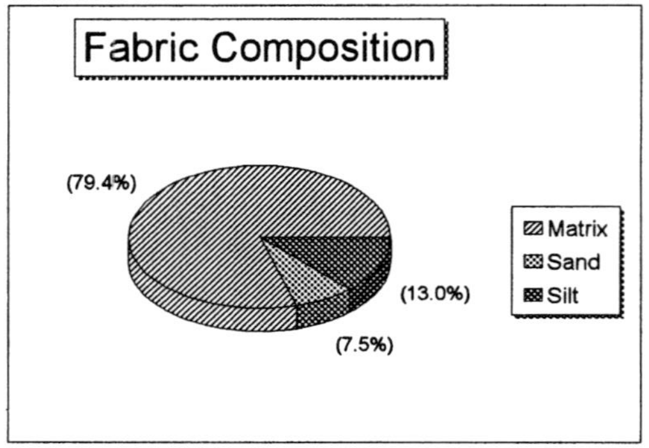

Qz Size	< 50	50-125	125-200	200-275	275-350	> 350	numpts
Number	342	46	3	0	0	0	391
%	87.47	11.76	0.77	0	0	0	
sigma	1.67	1.63	0.44	0	0	0	
p2sigma	90.82	15.02	1.65	0	0	0	
m2sigma	84.12	8.51	-0.12	0	0	0	
Conc	27.36	3.68	0.24	0	0	0	
Fabric	Matrix	Sand	Silt	Micac	Biref	Colour	
	317	30	52	3	3		399

Minerals	Size	Ang	Sph	Wth	Conc	Type	CarOpa
PFs	2				2		
MFs					1		
Biotite	2			1	4		
Mus	2			1	4		
Amph					2		
Carb					1		
Quartz		4L 2S	2L 3S			m,p	
Pyrox					1		
Epidote					1		
Schist					2		

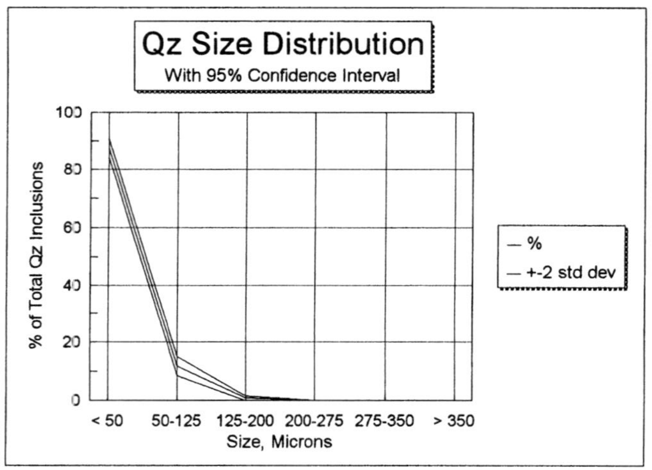

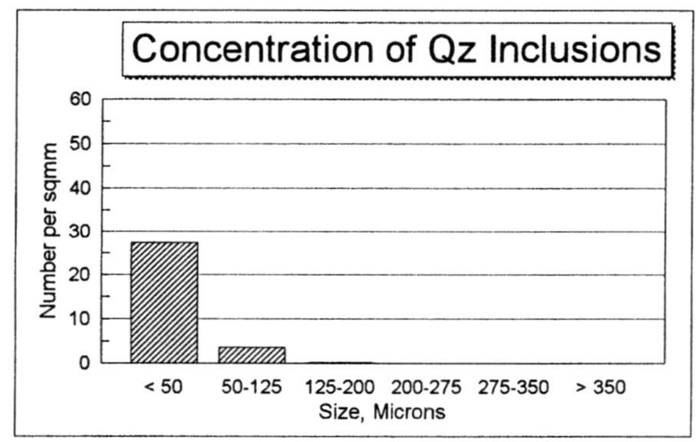

%matrix	79.45
%sand	7.52
%silt	13.03
conc<50	27.36
conc>50	3.92
%<50	87.47
%>50	12.53

APPENDIX C

hrp101	period level	morph type	rim dia	thick	sqmm
H89/1201-500	2 1/2	A1	7	4.8	18.75

Qz Size	< 50	50-125	125-200	200-275	275-350	> 350	numpts
Number	171	48	12	1	0	0	232
%	73.71	20.69	5.17	0.43	0	0	
sigma	2.89	2.66	1.45	0.43	0	0	
p2sigma	79.49	26.01	8.08	1.29	0	0	
m2sigma	67.93	15.37	2.26	-0.43	0	0	
Conc	9.12	2.56	0.64	0.05	0	0	
Fabric	Matrix	Sand	Silt	Micac	Biref	Colour	
	378	5	17	4	4		400
Minerals	Size	Ang	Sph	Wth	Conc	Type	CarOpa
PFs	2				2		
MFs					1		
Biotite	3			2	3		
Mus					4		
Amph					1		
Carb					1		
Quartz		3	2			m,p	
Pyrox					1		
Epidote					2		
Schist					2		

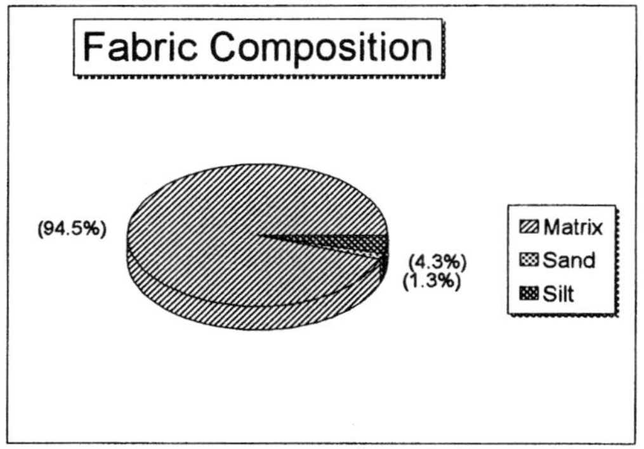

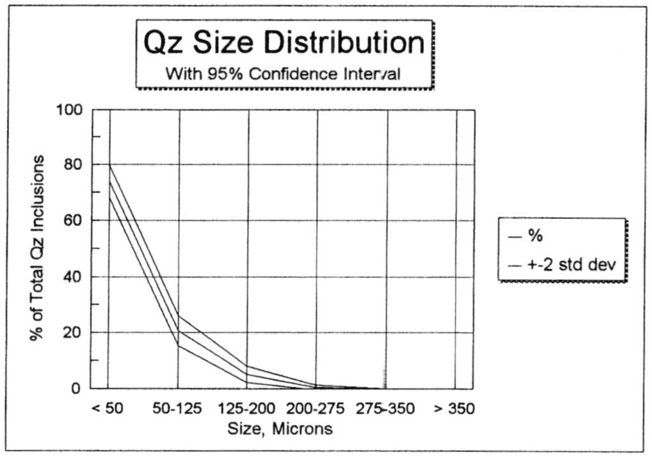

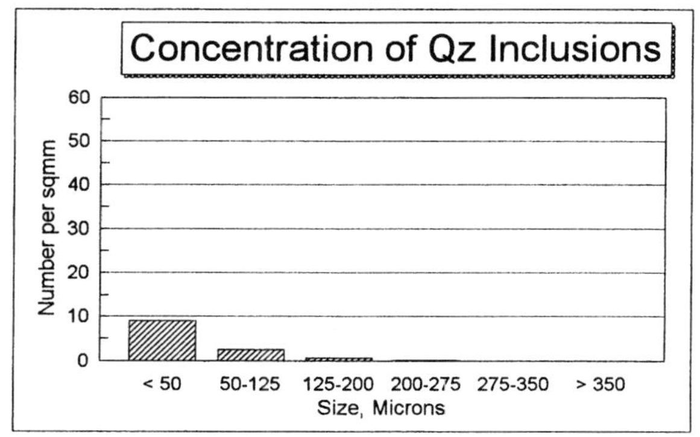

%matrix	94.50
%sand	1.25
%silt	4.25
conc<50	9.12
conc>50	3.25
%<50	73.71
%>50	26.29

hrp102	period level	morph type	rim dia	thick	sqmm
H90/1217-518	2 1/2	A1		6.8	12.5

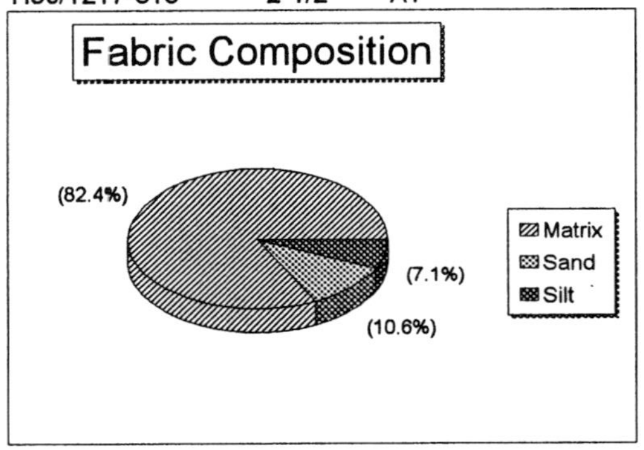

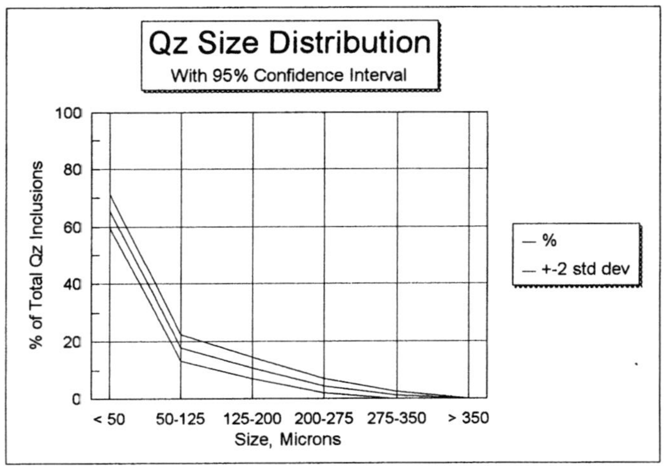

Qz Size	< 50	50-125	125-200	200-275	275-350	> 350	numpts
Number	176	48	29	12	3	1	269
%	65.43	17.84	10.78	4.46	1.12	0	
sigma	2.90	2.33	1.89	1.26	0.64	0	
p2sigma	71.23	22.51	14.56	6.98	2.40	0	
m2sigma	59.63	13.17	7.00	1.94	-0.17	0	
Conc	14.08	3.84	2.32	0.96	0.24	0.08	
Fabric	Matrix	Sand	Silt	Micac	Biref	Colour	
	327	42	28	3	2	dkbrn	397

Minerals	Size	Ang	Sph	Wth	Conc	Type	CarOpa
PFs	3	2	2		2		
MFs					1		
Biotite	3			2	4		
Mus	3			2	4		
Amph					1		
Carb					1		
Quartz		3	2			m,p	
Pyrox					1		
Epidote					2		
Schist					3		

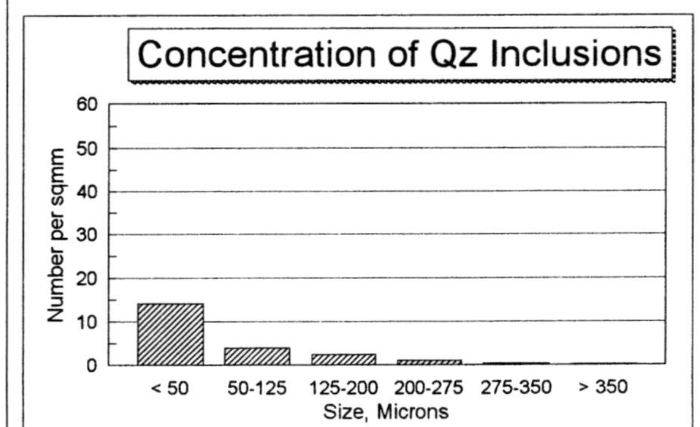

%matrix	82.37
%sand	10.58
%silt	7.05
conc<50	14.08
conc>50	7.44
%<50	65.43
%>50	34.57

APPENDIX C

hrp103	period	level	morph type	rim dia	thick	sqmm
H90/1222-540		2 1/2	A1		7.7	25

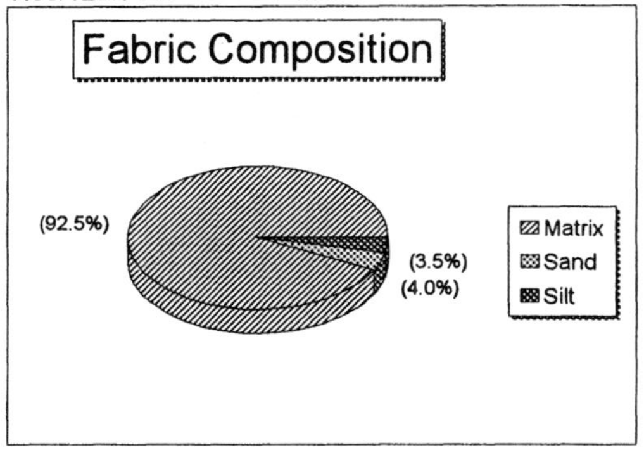

Qz Size	< 50	50-125	125-200	200-275	275-350	> 350	numpts
Number	192	91	42	6	4	0	335
%	57.31	27.16	12.54	1.79	1.19	0	
sigma	2.70	2.43	1.81	0.72	0.59	0	
p2sigma	62.72	32.02	16.16	3.24	2.38	0	
m2sigma	51.91	22.30	8.92	0.34	0.01	0	
Conc	7.68	3.64	1.68	0.24	0.16	0	
Fabric	Matrix	Sand	Silt	Micac	Biref	Colour	
	369	16	14	3	3	dkbrn	399
Minerals	Size	Ang	Sph	Wth	Conc	Type	CarOpa
PFs		2	2		3		
MFs					1		
Biotite	3			3	2		
Mus	3			2	4		
Amph					1		
Carb					1		
Quartz		3	3			m,p	
Pyrox					1		
Epidote					1		
Schist					2		

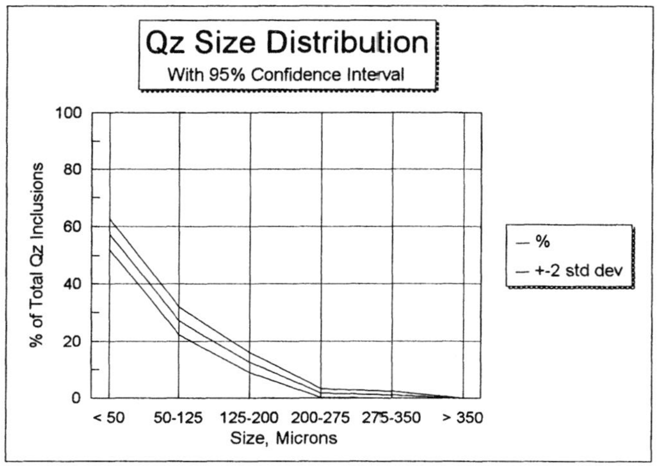

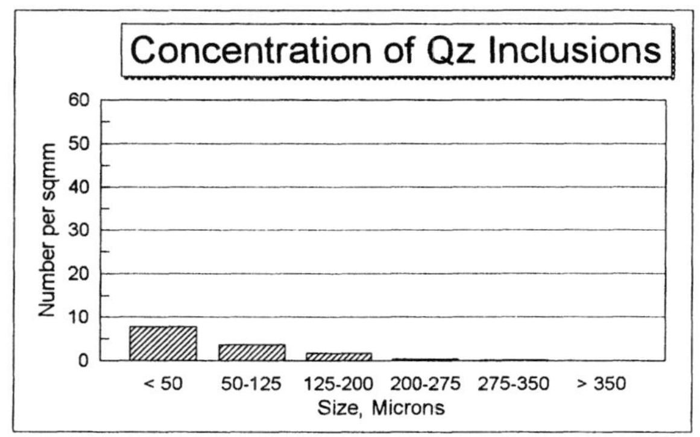

%matrix	92.48
%sand	4.01
%silt	3.51
conc<50	7.68
conc>50	5.72
%<50	57.31
%>50	42.69

hrp104	period	level	morph type	rim dia	thick	sqmm
H89/1200-504		2 1/2	All	16.1		25

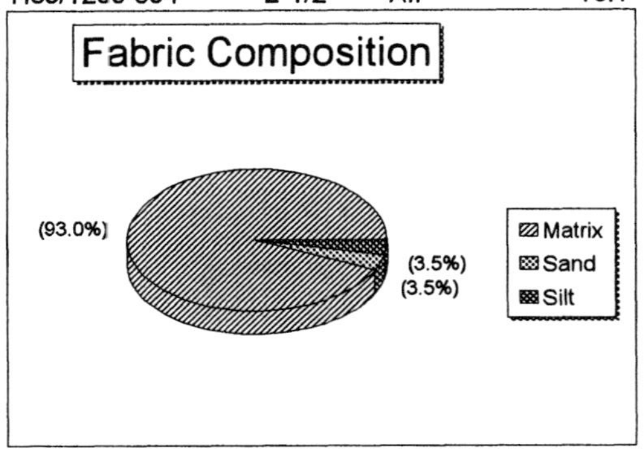

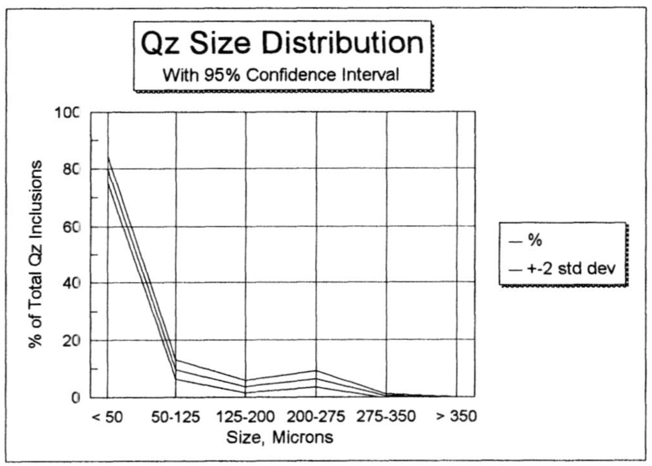

Qz Size	< 50	50-125	125-200	200-275	275-350	> 350	numpts
Number	238	29	11	19	1	1	299
%	79.60	9.70	3.68	6.35	0.33	0	
sigma	2.33	1.71	1.09	1.41	0.33	0	
p2sigma	84.26	13.12	5.86	9.18	1.00	0	
m2sigma	74.94	6.28	1.50	3.53	-0.33	0	
Conc	9.52	1.16	0.44	0.76	0.04	0.04	
Fabric	Matrix	Sand	Silt	Micac	Biref	Colour	
	371	14	14	3	2	dkbrn	399

Minerals	Size	Ang	Sph	Wth	Conc	Type	CarOpa
PFs	3	3	2		3		
MFs					1		
Biotite	2			2	2		
Mus	3			2	4		
Amph					1		
Carb					2	1	
Quartz		3	2			m,p	
Pyrox					1		
Epidote					2		
Schist					3		
Other						grog	

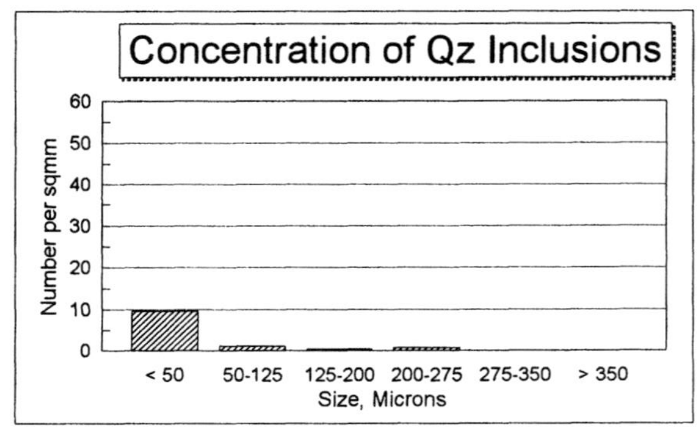

%matrix	92.98
%sand	3.51
%silt	3.51
conc<50	9.52
conc>50	2.44
%<50	79.60
%>50	20.40

hrp105	period level	morph type	rim dia	thick	sqmm
H89/1200-579	2 1/2	All	36	5.7	12.5

Qz Size	< 50	50-125	125-200	200-275	275-350	> 350	numpts
Number	357	55	16	10	0	1	439
%	81.32	12.53	3.64	2.28	0.00	0	
sigma	1.86	1.58	0.89	0.71	0.00	0	
p2sigma	85.04	15.69	5.43	3.70	0.00	0	
m2sigma	77.60	9.37	1.86	0.85	0.00	0	
Conc	28.56	4.40	1.28	0.80	0	0.08	
Fabric	Matrix	Sand	Silt	Micac	Biref	Colour	
	355	13	26	3	2	dkbrn	394
Minerals	Size	Ang	Sph	Wth	Conc	Type	CarOpa
PFs	1	3	2		2		
MFs					1		
Biotite	2			3	3		
Mus	3			3	4		
Amph					2		
Carb					2	2	1
Quartz		3	2			m,p	
Pyrox					1		
Epidote					2		
Schist					2		
Other						grog	

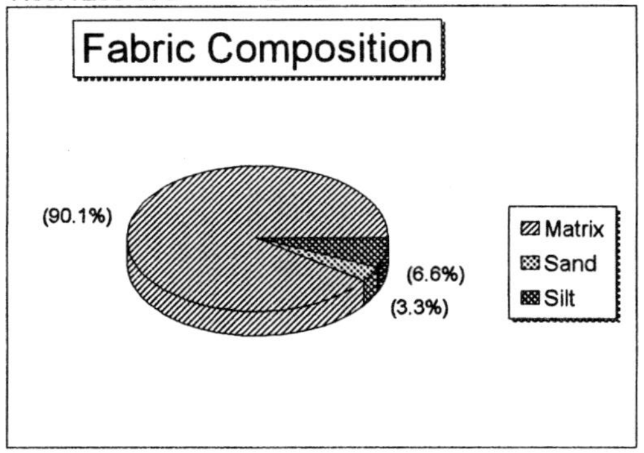

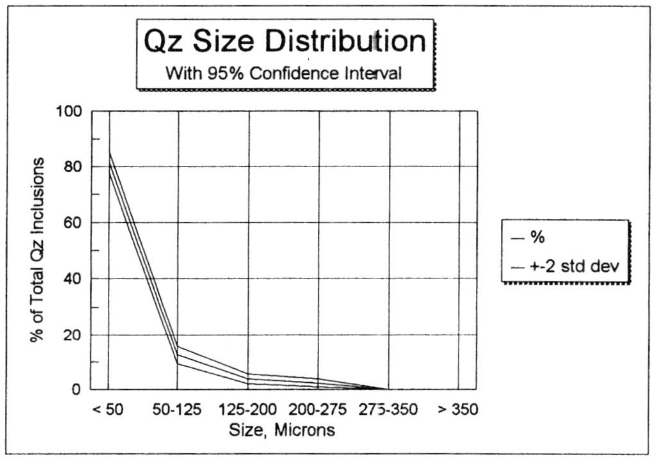

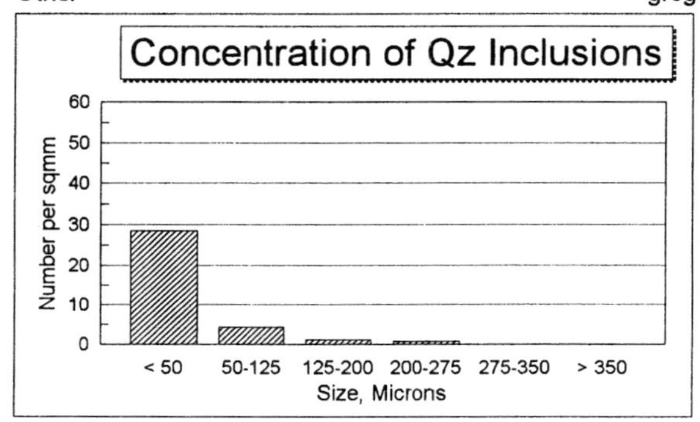

%matrix	90.10
%sand	3.30
%silt	6.60
conc<50	28.56
conc>50	6.56
%<50	81.32
%>50	18.68

APPENDIX C

hrp106	period	level	morph type	rim dia	thick	sqmm
H90/1215-526	2	1/2	All	12	5.1	25

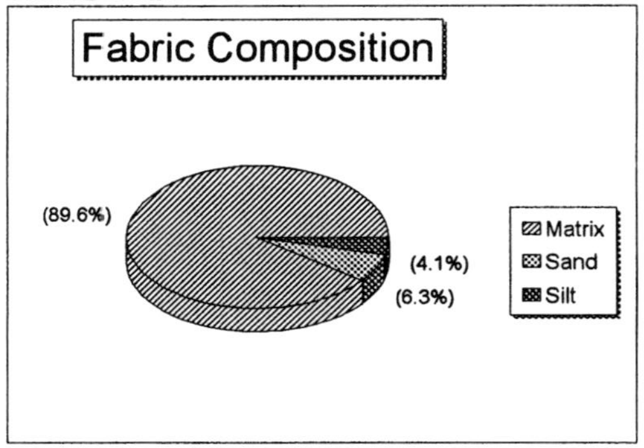

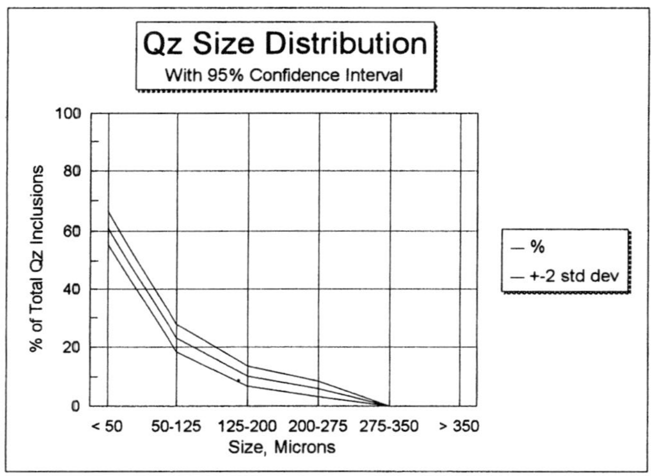

Qz Size	< 50	50-125	125-200	200-275	275-350	> 350	numpts
Number	198	75	33	19	0	0	325
%	60.92	23.08	10.15	5.85	0.00	0	
sigma	2.71	2.34	1.68	1.30	0.00	0	
p2sigma	66.34	27.75	13.50	8.45	0.00	0	
m2sigma	55.51	18.40	6.80	3.24	0.00	0	
Conc	7.92	3.00	1.32	0.76	0	0	
Fabric	Matrix	Sand	Silt	Micac	Biref	Colour	
	354	25	16	4	4	medbr	395
Minerals	Size	Ang	Sph	Wth	Conc	Type	CarOpa
PFs	3	3	2		3		
MFs					1		
Biotite	2			3	3		
Mus	2			3	4		
Amph					2		
Carb					1		
Quartz		3	2			m,p	
Pyrox					1		
Epidote					2		
Schist					2		

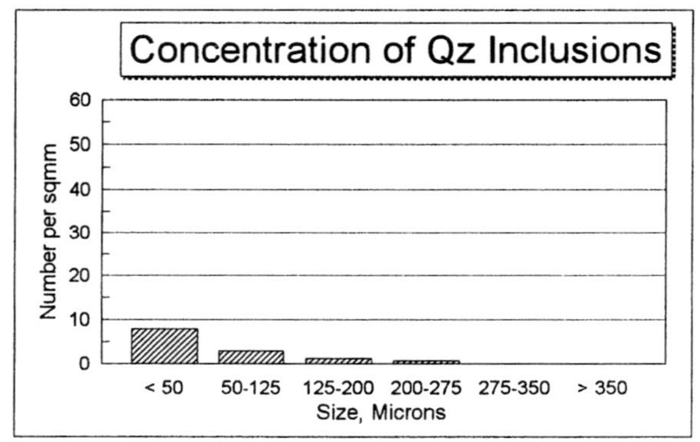

%matrix	89.62
%sand	6.33
%silt	4.05
conc<50	7.92
conc>50	5.08
%<50	60.92
%>50	39.08

hrp107	period	level	morph	type	rim dia	thick	sqmm
H90/1235-500		2 1/2	All		13	7.7	18.75

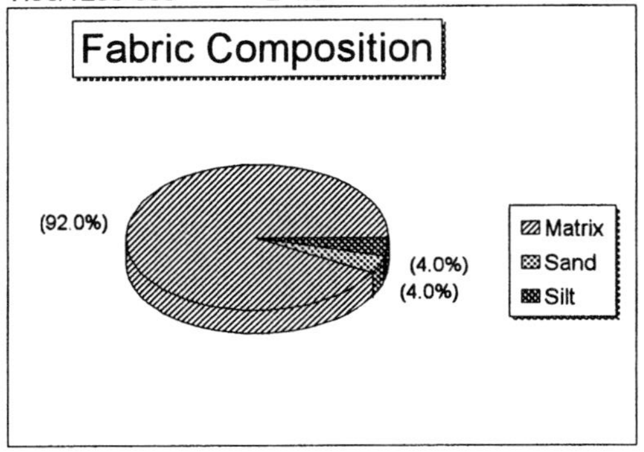

Qz Size	< 50	50-125	125-200	200-275	275-350	> 350	numpts
Number	179	27	8	5	7	2	228
%	78.51	11.84	3.51	2.19	3.07	0	
sigma	2.72	2.14	1.22	0.97	1.14	0	
p2sigma	83.95	16.12	5.95	4.13	5.36	0	
m2sigma	73.07	7.56	1.07	0.25	0.79	0	
Conc	9.55	1.44	0.43	0.27	0.3733	0.107	
Fabric	Matrix	Sand	Silt	Micac	Biref	Colour	
	367	16	16	3	3	dkbrn	399

Minerals	Size	Ang	Sph	Wth	Conc	Type	CarOpa
PFs		3	2		2		
MFs					1		
Biotite	3			3	3		
Mus	4				4		
Amph					1		
Carb					2	2	1
Quartz		3	2			m,p	
Pyrox					1		
Epidote					1		
Schist					2		

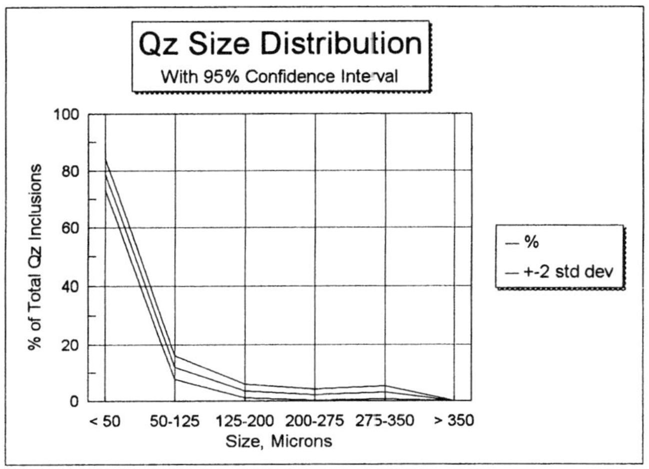

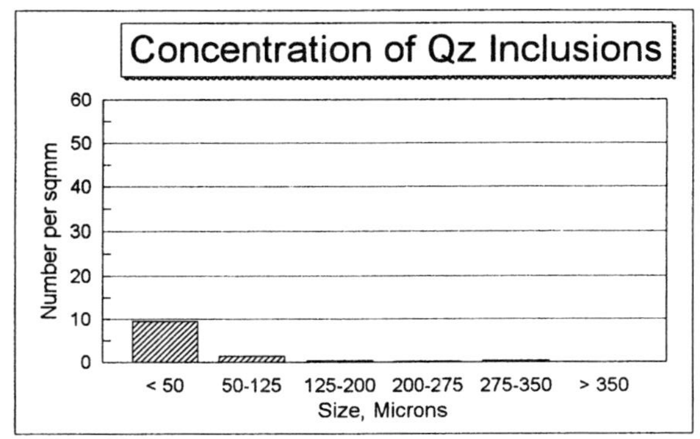

%matrix	91.98
%sand	4.01
%silt	4.01
conc<50	9.55
conc>50	2.61
%<50	78.51
%>50	21.49

APPENDIX C

hrp108	period level	morph type	rim dia	thick	sqmm
H90/1236-530	2 1/2	All	21	8	18.75

Qz Size	< 50	50-125	125-200	200-275	275-350	> 350	numpts
Number	222	55	28	24	2	1	332
%	66.87	16.57	8.43	7.23	0.60	0	
sigma	2.58	2.04	1.53	1.42	0.42	0	
p2sigma	72.03	20.65	11.48	10.07	1.45	0	
m2sigma	61.70	12.49	5.38	4.39	-0.25	0	
Conc	11.84	2.93	1.49	1.28	0.1067	0.053	
Fabric	Matrix	Sand	Silt	Micac	Biref	Colour	
	332	39	19	3	3	medbr	390

Minerals	Size	Ang	Sph	Wth	Conc	Type	CarOpa
PFs		3	2		3		
MFs					1		
Biotite	3			3	3		
Mus	3			3	4		
Amph					1		
Carb					2	2	1
Quartz		3	2			m,p	
Pyrox					1		
Epidote					1		
Schist					4		

Fabric Composition

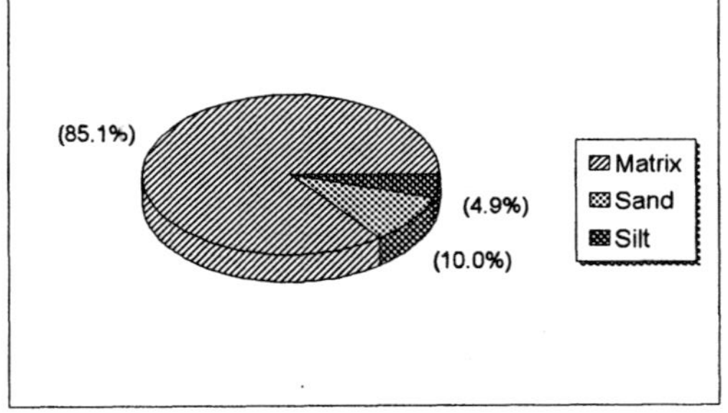

Qz Size Distribution

With 95% Confidence Interval

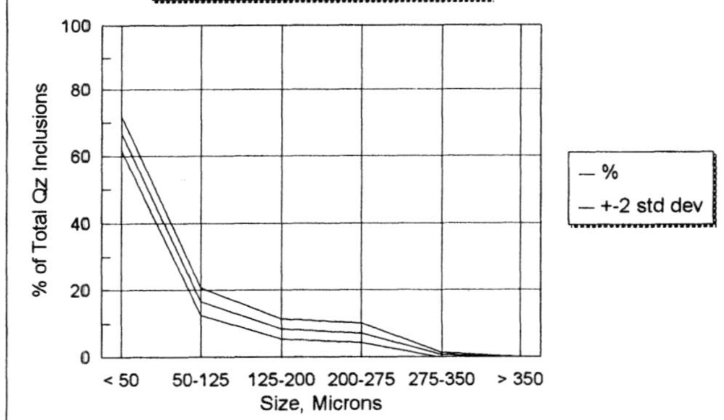

Concentration of Qz Inclusions

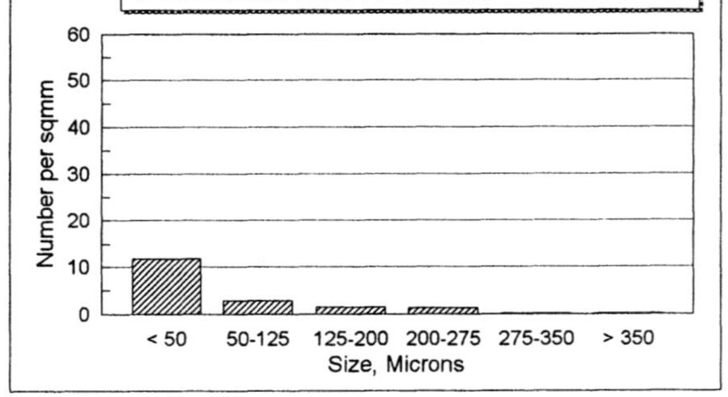

%matrix	85.13
%sand	10.00
%silt	4.87
conc<50	11.84
conc>50	5.87
%<50	66.87
%>50	33.13

hrp109	period	level	morph type	rim dia	thick	sqmm
H90/1217-505	2	1/2	Bl	7	4.8	

Not countable

Qz Size	< 50	50-125	125-200	200-275	275-350	> 350	numpts
Number							0
%							
sigma							
p2sigma							
m2sigma							
Conc							
Fabric	Matrix	Sand	Silt	Micac	Biref	Colour	
						dkbrn	0

Minerals	Size	Ang	Sph	Wth	Conc	Type	CarOpa
PFs					1		
MFs					1		
Biotite				3	3		
Mus	2			3	4		
Amph					1		
Carb							
Quartz		3	2			m,p	
Pyrox					1		
Epidote					1		
Schist					2		

hrp110	period	level	morph type	rim dia	thick	sqmm
H89/1200-628		2 1/2	BI	45	11	18.75

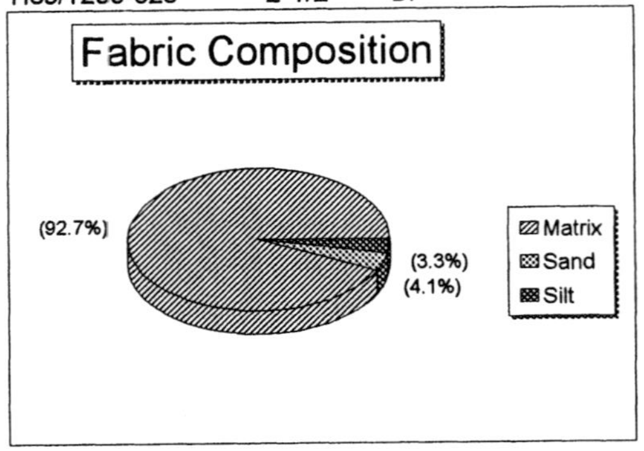

Qz Size	< 50	50-125	125-200	200-275	275-350	> 350	numpts
Number	244	105	15	0	0	0	364
%	67.03	28.85	4.12	0.00	0.00	0	
sigma	2.46	2.37	1.04	0.00	0.00	0	
p2sigma	71.96	33.60	6.20	0.00	0.00	0	
m2sigma	62.11	24.10	2.04	0.00	0.00	0	
Conc	13.01	5.60	0.80	0.00	0	0	
Fabric	Matrix	Sand	Silt	Micac	Biref	Colour	
	366	16	13	4	4	medbr	395

Minerals	Size	Ang	Sph	Wth	Conc	Type	CarOpa
PFs	2	3	2		2		
MFs					1		
Biotite	2			2	3		
Mus	3			3	4		
Amph					2		
Carb					1		
Quartz		3	2		4	m,p	
Pyrox					1		
Epidote					2		
Schist					3		

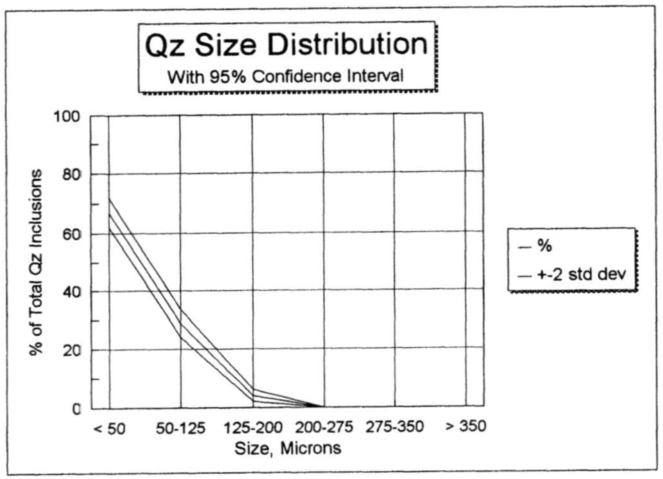

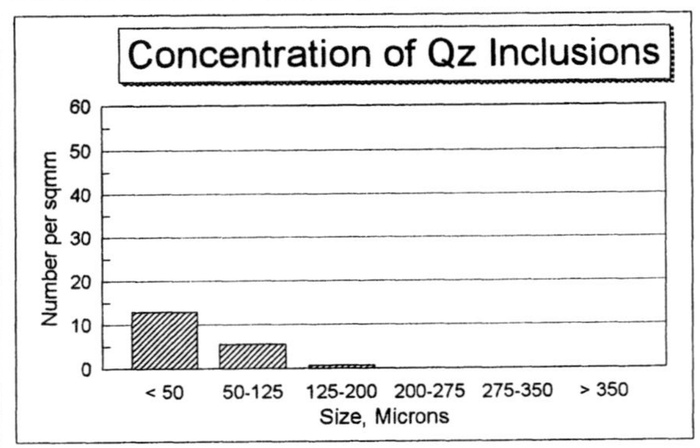

%matrix	92.66
%sand	4.05
%silt	3.29
conc<50	13.01
conc>50	6.40
%<50	67.03
%>50	32.97

hrp111	period level	morph type	rim dia	thick	sqmm
H89/1060-500	2 1/2	Bl	15	5.6	18.75

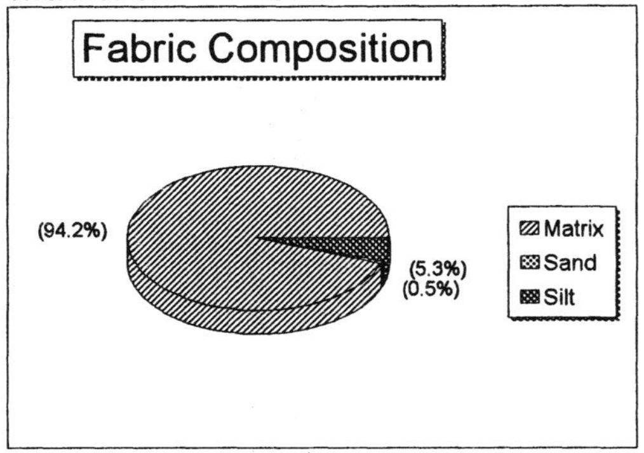

Qz Size	< 50	50-125	125-200	200-275	275-350	> 350	numpts
Number	303	33	3	4	0	0	343
%	88.34	9.62	0.87	1.17	0.00	0	
sigma	1.73	1.59	0.50	0.58	0.00	0	
p2sigma	91.80	12.81	1.88	2.33	0.00	0	
m2sigma	84.87	6.44	-0.13	0.01	0.00	0	
Conc	16.16	1.76	0.16	0.21	0	0	
Fabric	Matrix	Sand	Silt	Micac	Biref	Colour	
	373	2	21	4	3	redbr	396
Minerals	Size	Ang	Sph	Wth	Conc	Type	CarOpa
PFs	1	2	3		2		
MFs					1		
Biotite	2			3	3		
Mus	2			3	4		
Amph					1		
Carb					1		
Quartz		2	3		4	m,p	
Pyrox					1		
Epidote					1		
Schist					3		

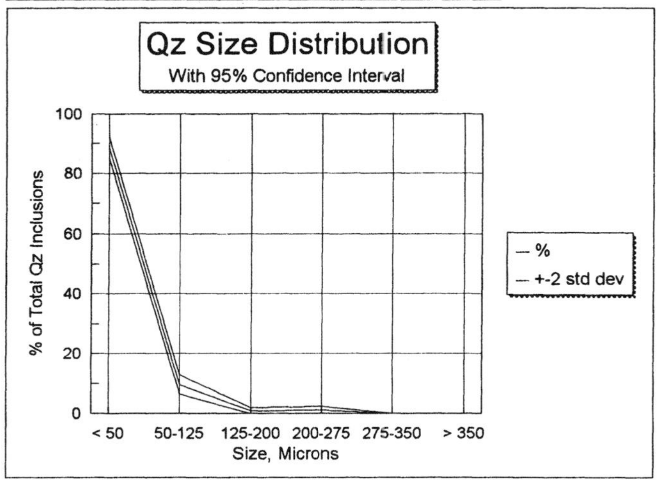

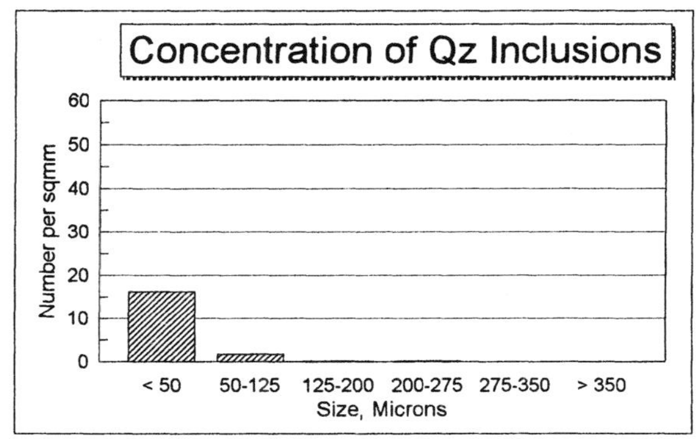

%matrix	94.19
%sand	0.51
%silt	5.30
conc<50	16.16
conc>50	2.13
%<50	88.34
%>50	11.66

APPENDIX C

hrp112	period	level	morph type	rim dia	thick	sqmm
H89/1095-13		2 1/2	Bl		5.6	12.5

Qz Size	< 50	50-125	125-200	200-275	275-350	> 350	numpts
Number	510	46	2	1	0	1	560
%	91.07	8.21	0.36	0.18	0.00	0	
sigma	1.21	1.16	0.25	0.18	0.00	0	
p2sigma	93.48	10.53	0.86	0.54	0.00	0	
m2sigma	88.66	5.89	-0.15	-0.18	0.00	0	
Conc	40.80	3.68	0.16	0.08	0	0.08	
Fabric	Matrix	Sand	Silt	Micac	Biref	Colour	
	339	5	53	4	4	medbr	397

Minerals	Size	Ang	Sph	Wth	Conc	Type	CarOpa
PFs	1	3	3		2		
MFs					1		
Biotite	2			2	3		
Mus	2			2	4		
Amph					1		
Carb					1		
Quartz		3	3		4	m,p	
Pyrox					1		
Epidote					1		
Schist					3		
Other						clay	

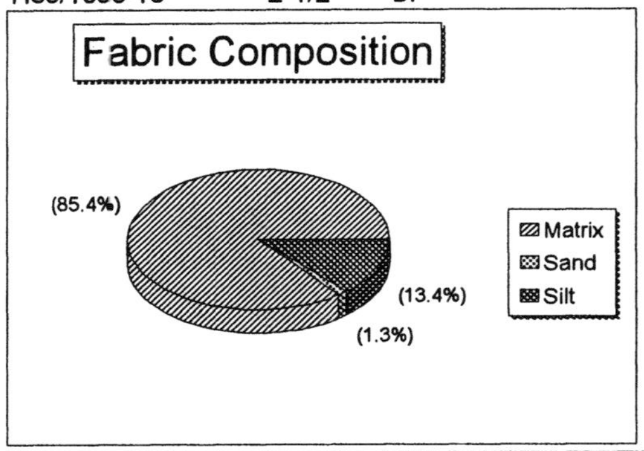

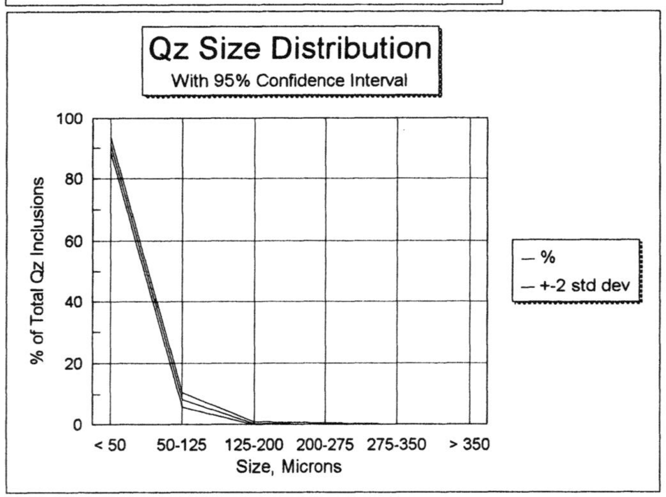

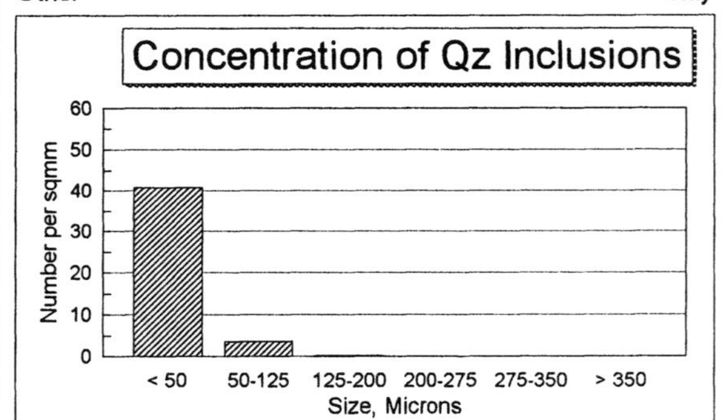

%matrix	85.39
%sand	1.26
%silt	13.35
conc<50	40.80
conc>50	4.00
%<50	91.07
%>50	8.93

hrp113	period level	morph type	rim dia	thick	sqmm
H90/1217-506	2 1/2	BI			18.75

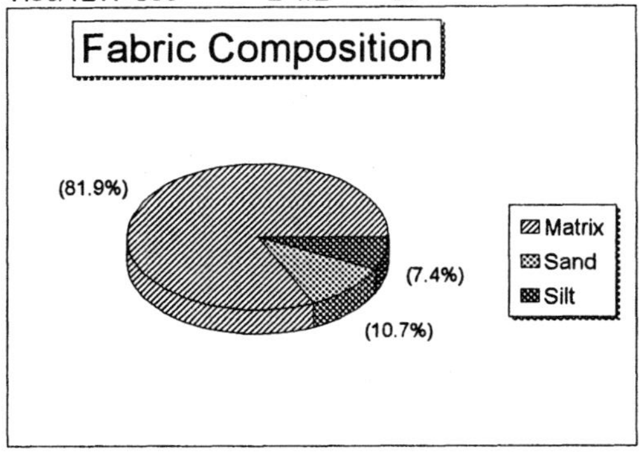

Qz Size	< 50	50-125	125-200	200-275	275-350	> 350	numpts
Number	165	72	46	37	7	1	328
%	50.30	21.95	14.02	11.28	2.13	0	
sigma	2.76	2.29	1.92	1.75	0.80	0	
p2sigma	55.83	26.52	17.86	14.77	3.73	0	
m2sigma	44.78	17.38	10.19	7.79	0.54	0	
Conc	8.80	3.84	2.45	1.97	0.3733	0.053	
Fabric	Matrix	Sand	Silt	Micac	Biref	Colour	
	322	42	29	4	4	medbr	393

Minerals	Size	Ang	Sph	Wth	Conc	Type	CarOpa
PFs	2	3	2		3		
MFs					1		
Biotite	3			2	3		
Mus	3			2	4		
Amph					2		
Carb					1		
Quartz		3	2		4	m,p	
Pyrox					1		
Epidote					1		
Schist					3	epidote	

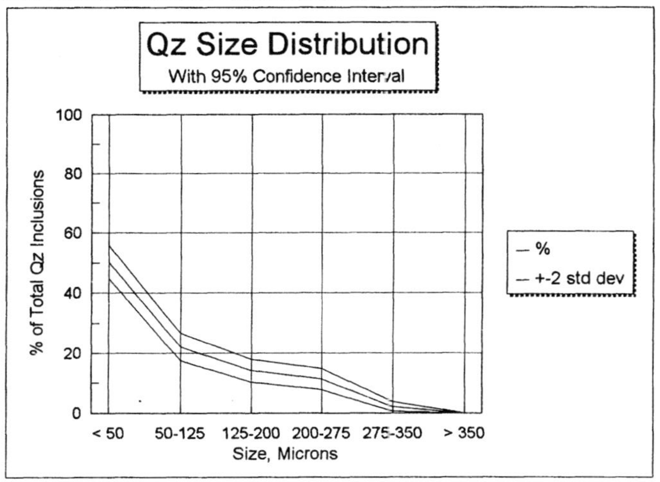

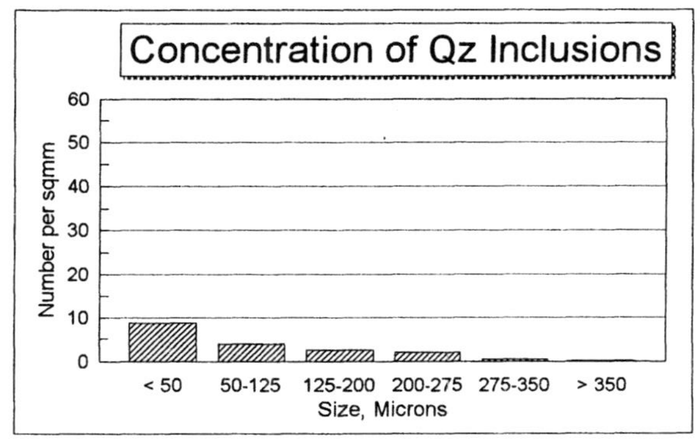

%matrix	81.93
%sand	10.69
%silt	7.38
conc<50	8.80
conc>50	8.69
%<50	50.30
%>50	49.70

hrp114	period	level	morph type	rim dia	thick	sqmm
H89/1200-625		2 1/2	BII	34	9.6	12.5

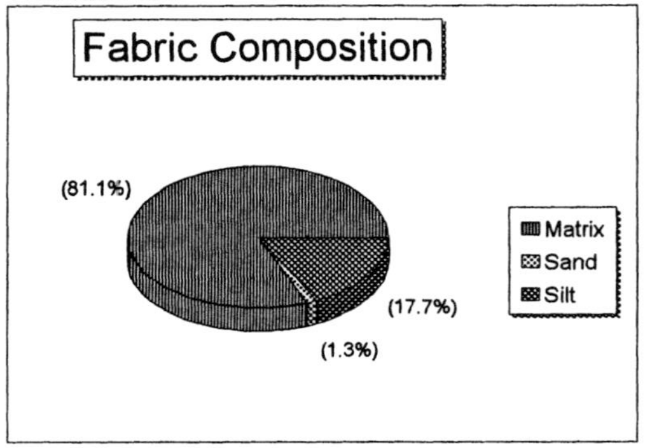

Qz Size	< 50	50-125	125-200	200-275	275-350	> 350	numpts
Number	527	75	11	2	0	0	615
%	85.69	12.20	1.79	0.33	0.00	0	
sigma	1.41	1.32	0.53	0.23	0.00	0	
p2sigma	88.52	14.83	2.86	0.78	0.00	0	
m2sigma	82.87	9.56	0.72	-0.13	0.00	0	
Conc	42.16	6.00	0.88	0.16	0	0	
Fabric	Matrix	Sand	Silt	Micac	Biref	Colour	
	321	5	70	4	3	medbr	396
Minerals	Size	Ang	Sph	Wth	Conc	Type	CarOpa
PFs	3	3	2		2		
MFs					1		
Biotite	2			2	3		
Mus	2			2	4		
Amph					1		
Carb					2	2	2
Quartz		3	3		4	m,p	
Pyrox					1		
Epidote					1		
Schist					3		

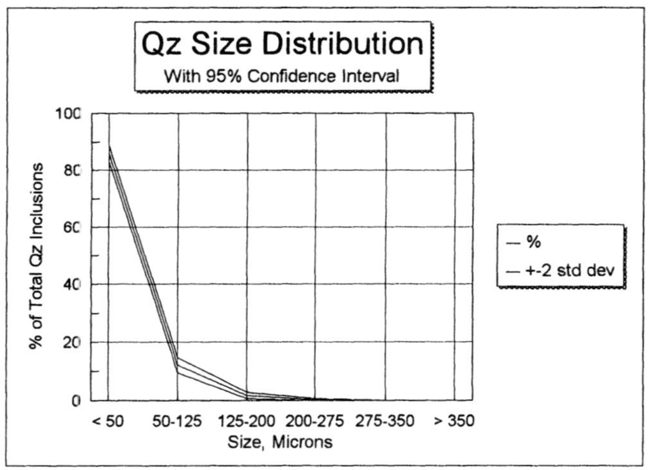

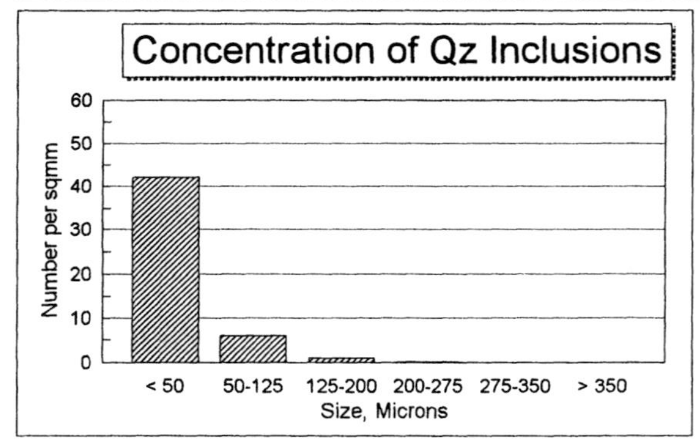

%matrix	81.06
%sand	1.26
%silt	17.68
conc<50	42.16
conc>50	7.04
%<50	85.69
%>50	14.31

hrp115	period	level	morph type	rim dia	thick	sqmm
H89/1200-519		2 1/2	BIII	24	6.8	25

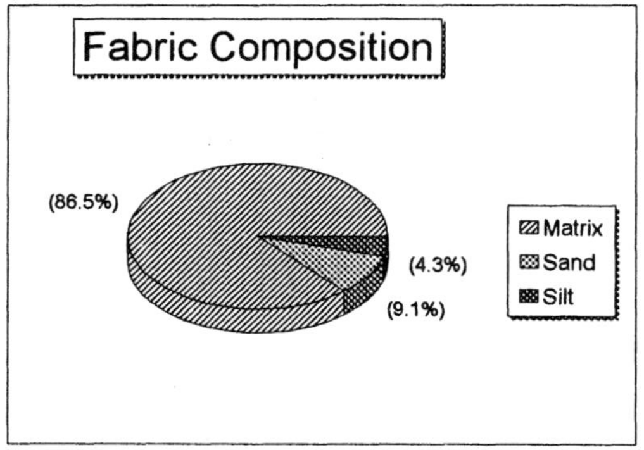

Qz Size	< 50	50-125	125-200	200-275	275-350	> 350	numpts
Number	201	120	45	11	2	0	379
%	53.03	31.66	11.87	2.90	0.53	0	
sigma	2.56	2.39	1.66	0.86	0.37	0	
p2sigma	58.16	36.44	15.20	4.63	1.27	0	
m2sigma	47.91	26.88	8.55	1.18	-0.22	0	
Conc	8.04	4.80	1.80	0.44	0.08	0	
Fabric	Matrix	Sand	Silt	Micac	Biref	Colour	
	341	36	17	4	4	medbr	394
Minerals	Size	Ang	Sph	Wth	Conc	Type	CarOpa
PFs	2	3	2		2		
MFs					1		
Biotite	2			2	2		
Mus	3			2	4		
Amph					1		
Carb							
Quartz		3	2		4	m,p	
Pyrox					1		
Epidote					2		
Schist					3		

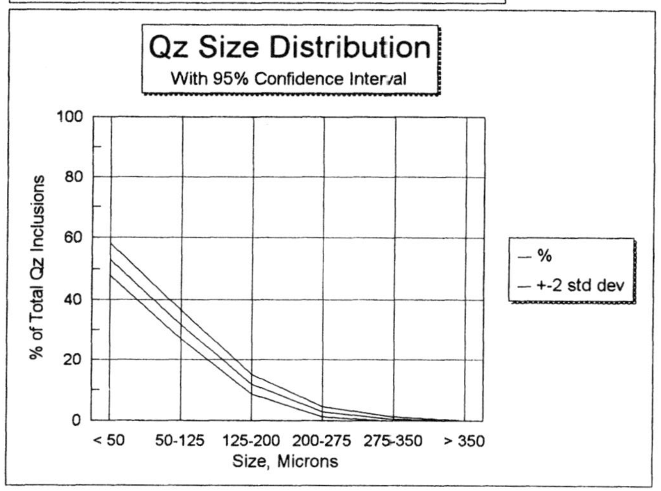

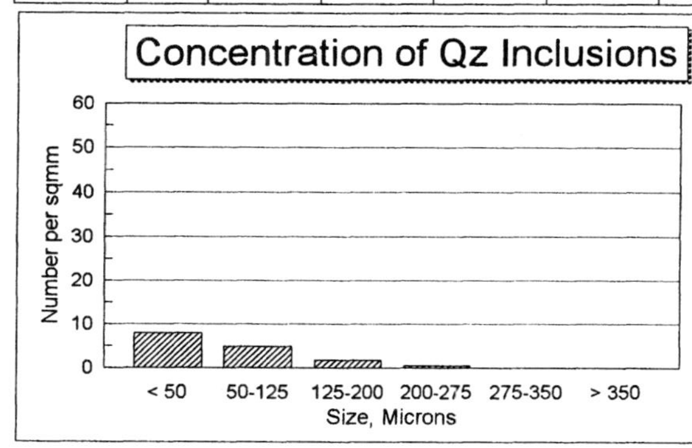

%matrix	86.55
%sand	9.14
%silt	4.31
conc<50	8.04
conc>50	7.12
%<50	53.03
%>50	46.97

hrp116	period	level	morph type	rim dia	thick	sqmm
H90/1217-545	2	1/2	BIII	18	8.1	12.5

Qz Size	< 50	50-125	125-200	200-275	275-350	> 350	numpts
Number	213	84	37	12	0	0	346
%	61.56	24.28	10.69	3.47	0.00	0	
sigma	2.62	2.31	1.66	0.98	0.00	0	
p2sigma	66.79	28.89	14.02	5.44	0.00	0	
m2sigma	56.33	19.67	7.37	1.50	0.00	0	
Conc	17.04	6.72	2.96	0.96	0	0	
Fabric	Matrix	Sand	Silt	Micac	Biref	Colour	
	344	38	15	4	4	medbr	397
Minerals	Size	Ang	Sph	Wth	Conc	Type	CarOpa
PFs	3	2	2		2		
MFs					1		
Biotite	2			2	2		
Mus	3			3	4		
Amph					1		
Carb					1		
Quartz		3	2		4	m,p	
Pyrox					1		
Epidote					2		
Schist					3		

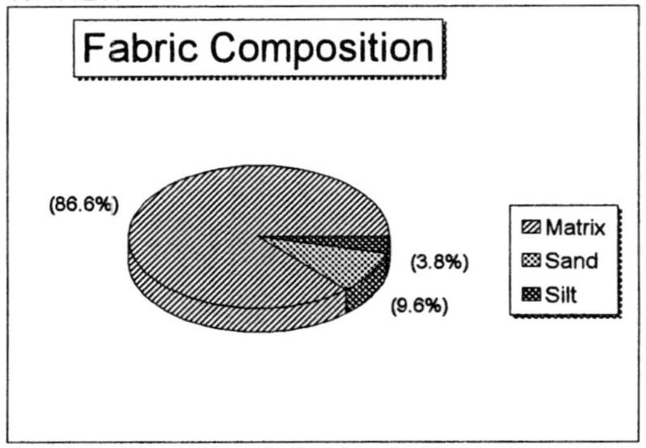

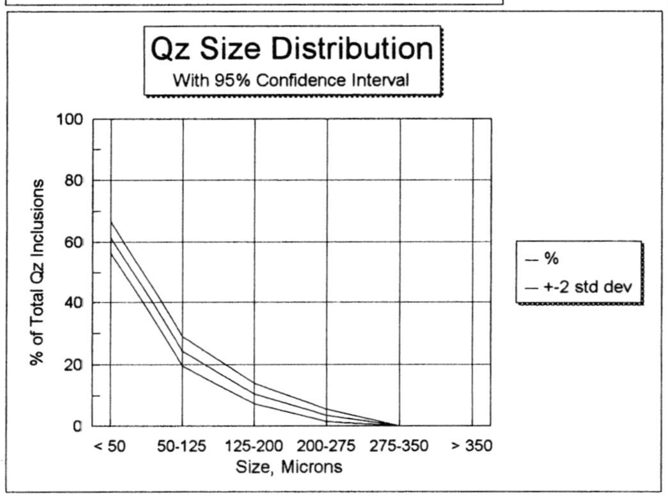

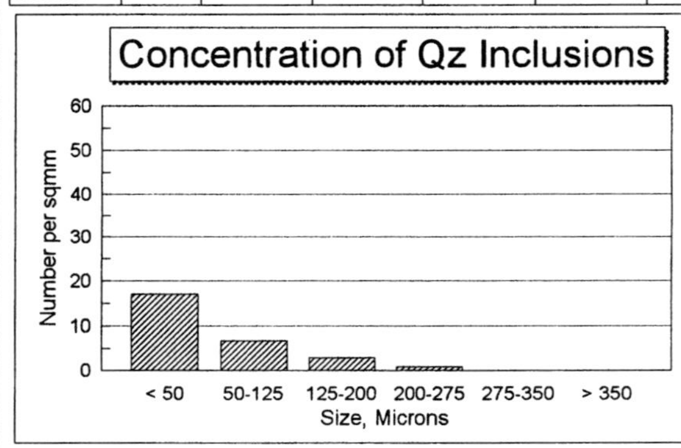

%matrix	86.65
%sand	9.57
%silt	3.78
conc<50	17.04
conc>50	10.64
%<50	61.56
%>50	38.44

hrp117	period	level	morph type	rim dia	thick	sqmm
H90/1235-507		2 1/2	BIII	22	8.1	18.75

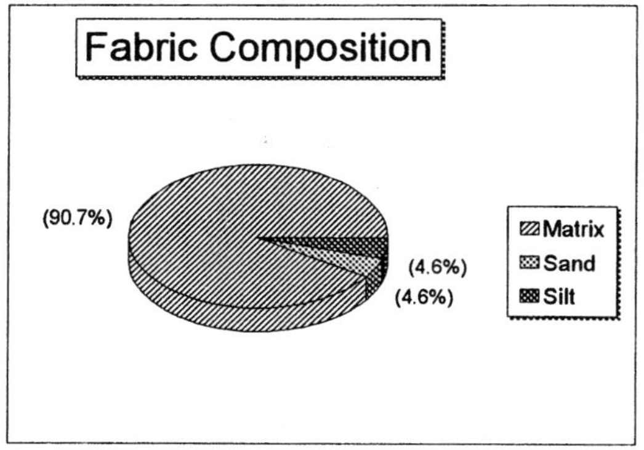

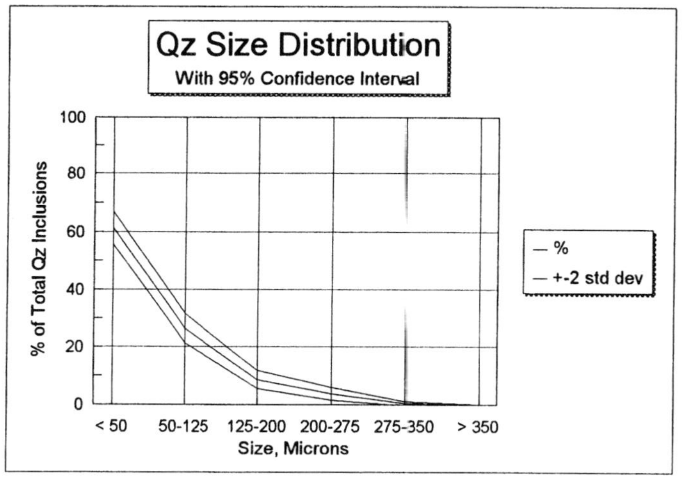

Qz Size	< 50	50-125	125-200	200-275	275-350	> 350	numpts
Number	188	81	26	11	1	0	307
%	61.24	26.38	8.47	3.58	0.33	0	
sigma	2.78	2.52	1.59	1.06	0.33	0	
p2sigma	66.80	31.41	11.65	5.70	0.98	0	
m2sigma	55.68	21.35	5.29	1.46	-0.32	0	
Conc	10.03	4.32	1.39	0.59	0.0533	0	
Fabric	Matrix	Sand	Silt	Micac	Biref	Colour	
	353	18	18	4	4	medbr	389

Minerals	Size	Ang	Sph	Wth	Conc	Type	CarOpa
PFs	2	3	2		2		
MFs					1		
Biotite	2			2	2		
Mus	2			2	4		
Amph					2		
Carb					1		
Quartz		3	2		4	m,p	
Pyrox					1		
Epidote					1		
Schist					3		

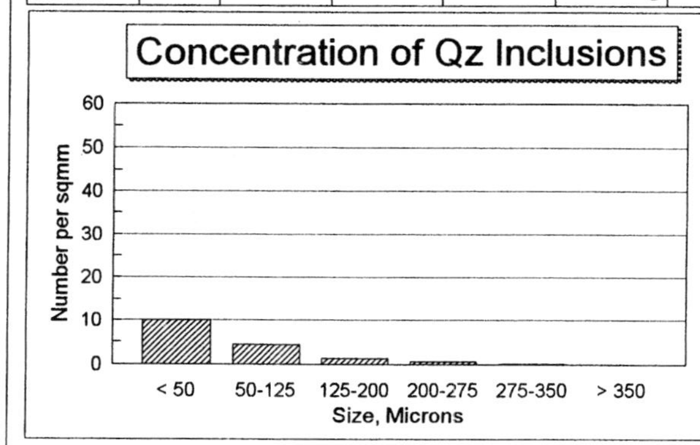

%matrix	90.75
%sand	4.63
%silt	4.63
conc<50	10.03
conc>50	6.35
%<50	61.24
%>50	38.76

APPENDIX C

hrp118	period	level	morph type	rim dia	thick	sqmm
H89/1200-627		2 1/2	BIII	41	12.2	12.5

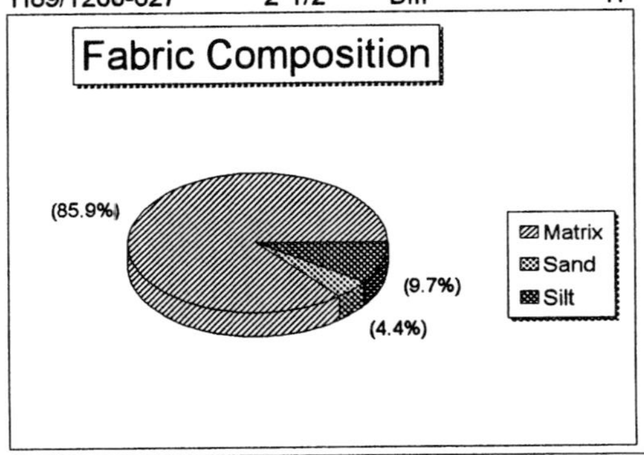

Qz Size	< 50	50-125	125-200	200-275	275-350	> 350	numpts
Number	438	66	4	4	1	0	513
%	85.38	12.87	0.78	0.78	0.19	0	
sigma	1.56	1.48	0.39	0.39	0.19	0	
p2sigma	88.50	15.82	1.56	1.56	0.58	0	
m2sigma	82.26	9.91	0.00	0.00	-0.19	0	
Conc	35.04	5.28	0.32	0.32	0.08	0	
Fabric	Matrix	Sand	Silt	Micac	Biref	Colour	
	335	17	38	4	4	medbr	390

Minerals	Size	Ang	Sph	Wth	Conc	Type	CarOpa
PFs	2	2	2		2		
MFs					1		
Biotite	2			3	3		
Mus	2			3	4		
Amph					2		
Carb					1		
Quartz		3	2		4	m,p	
Pyrox					1		
Epidote					2		
Schist					3		
Other						clay	

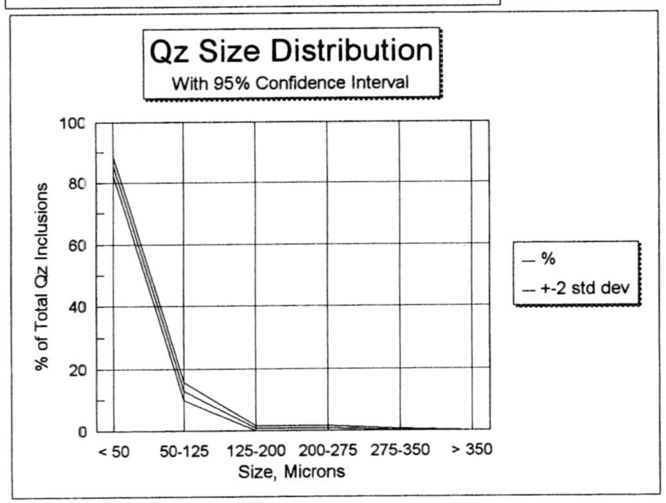

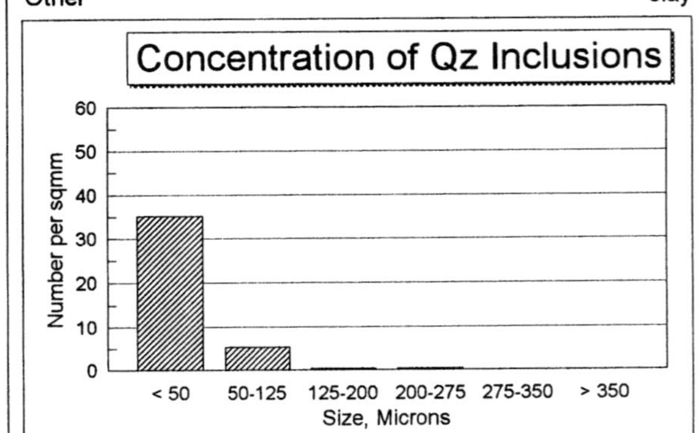

%matrix	85.90
%sand	4.36
%silt	9.74
conc<50	35.04
conc>50	6.00
%<50	85.38
%>50	14.62

hrp119	period level	morph type	rim dia	thick	sqmm
H89/1060-501	2 1/2	BIV	37	7to17.1	12.5

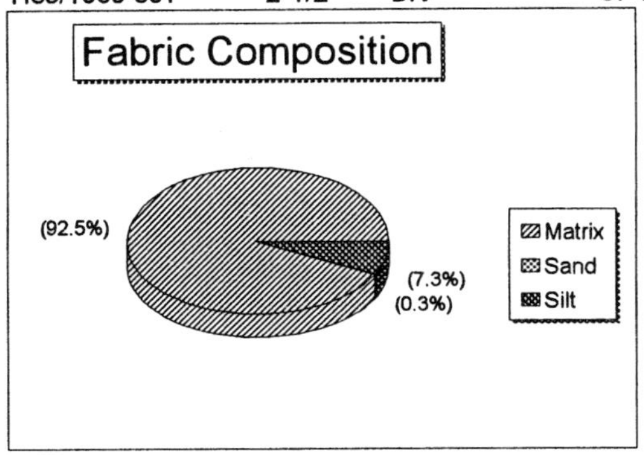

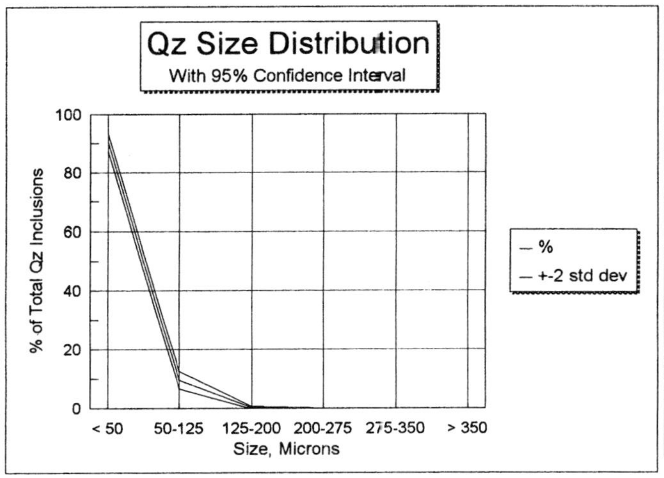

Qz Size	< 50	50-125	125-200	200-275	275-350	> 350	numpts
Number	349	37	1	0	0	0	387
%	90.18	9.56	0.26	0.00	0.00	0	
sigma	1.51	1.49	0.26	0.00	0.00	0	
p2sigma	93.21	12.55	0.77	0.00	0.00	0	
m2sigma	87.16	6.57	-0.26	0.00	0.00	0	
Conc	27.92	2.96	0.08	0.00	0	0	
Fabric	Matrix	Sand	Silt	Micac	Biref	Colour	
	368	1	29	3	3	dkrdbr	398

Minerals	Size	Ang	Sph	Wth	Conc	Type	CarOpa
PFs					1		
MFs					1		
Biotite	2			2	2		
Mus	2			2	4		
Amph					1		
Carb					1		
Quartz		3L 2S	2L 3S		4	m,p	
Pyrox					1		
Epidote					2		
Schist					3		

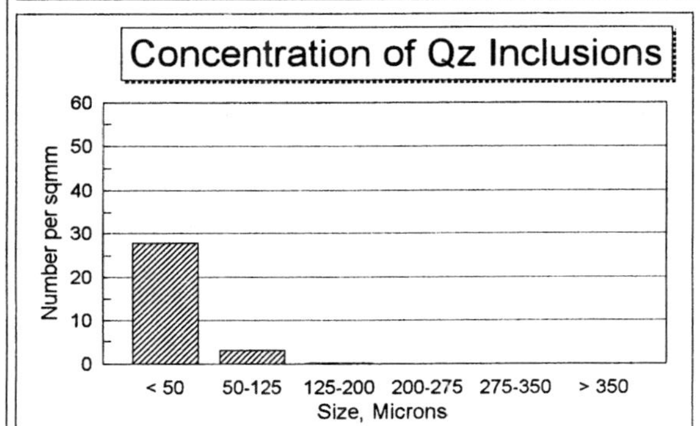

%matrix	92.46
%sand	0.25
%silt	7.29
conc<50	27.92
conc>50	3.04
%<50	90.18
%>50	9.82

hrp120	period level	morph type	rim dia	thick	sqmm
H89/1060-502	2 1/2	BIV	28	8.3	18.75

Qz Size	< 50	50-125	125-200	200-275	275-350	> 350	numpts
Number	360	102	17	5	0	0	484
%	74.38	21.07	3.51	1.03	0.00	0	
sigma	1.98	1.85	0.84	0.46	0.00	0	
p2sigma	78.35	24.78	5.19	1.95	0.00	0	
m2sigma	70.41	17.37	1.84	0.11	0.00	0	
Conc	19.20	5.44	0.91	0.27	0	0	
Fabric	Matrix	Sand	Silt	Micac	Biref	Colour	
	356	20	19	3	3	medbr	395

Minerals	Size	Ang	Sph	Wth	Conc	Type	CarOpa
PFs	2	3	2		2		
MFs					1		
Biotite	2			2	3		
Mus	2			2	4		
Amph					1		
Carb					2	2	2
Quartz		3L 2S	2L 3S		4	m,p	
Pyrox					1		
Epidote					1		
Schist					3		

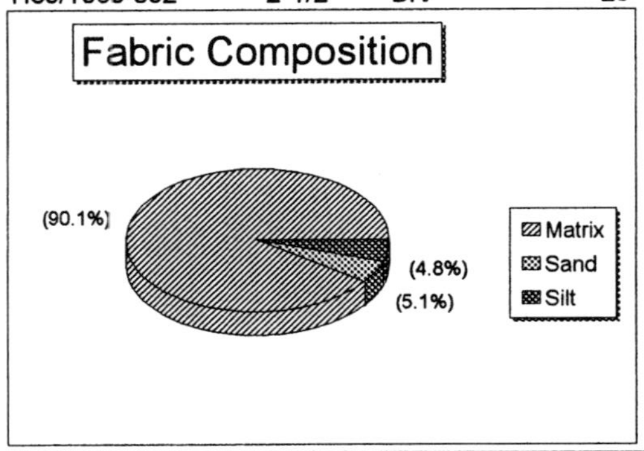

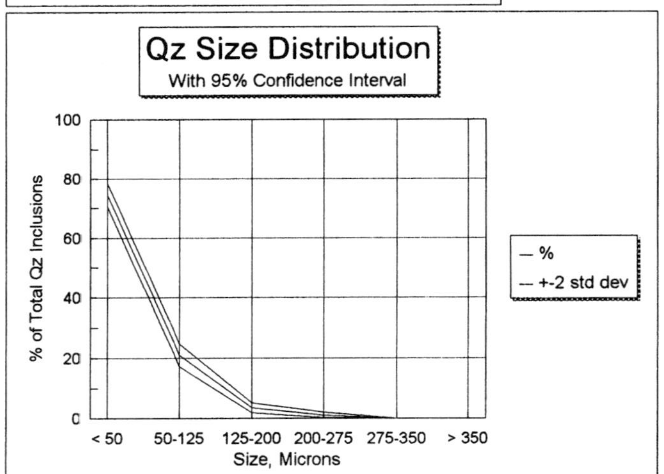

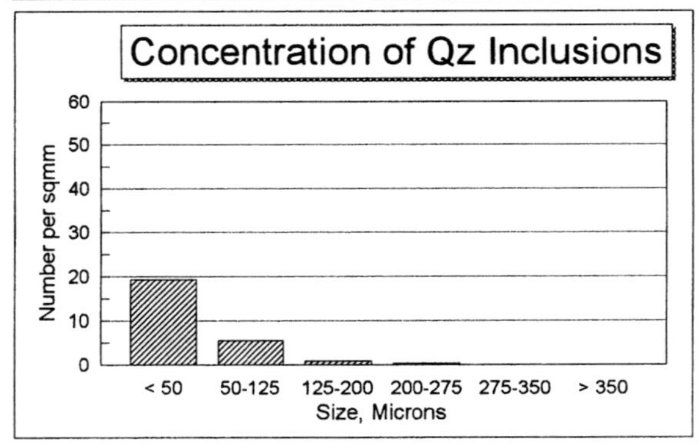

%matrix	90.13
%sand	5.06
%silt	4.81
conc<50	19.20
conc>50	6.61
%<50	74.38
%>50	25.62

hrp121	period level	morph type	rim dia	thick	sqmm
H89/1200-618	2 1/2	BIV	22	8.4to17.4	18.75

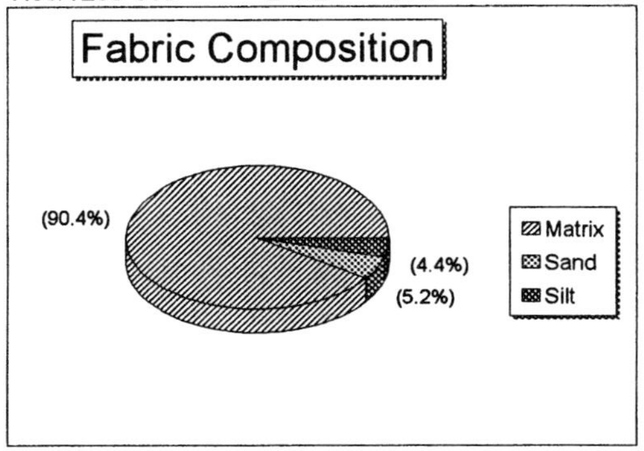

Qz Size	< 50	50-125	125-200	200-275	275-350	> 350	numpts
Number	312	115	11	3	0	0	441
%	70.75	26.08	2.49	0.68	0.00	0	
sigma	2.17	2.09	0.74	0.39	0.00	0	
p2sigma	75.08	30.26	3.98	1.46	0.00	0	
m2sigma	66.42	21.90	1.01	-0.10	0.00	0	
Conc	16.64	6.13	0.59	0.16	0	0	
Fabric	Matrix	Sand	Silt	Micac	Biref	Colour	
	349	20	17	3	3	medbr	386

Minerals	Size	Ang	Sph	Wth	Conc	Type	CarOpa
PFs	2	2	2		2		
MFs					1		
Biotite	2			3	3		
Mus	3			2	4		
Amph					2		
Carb					1		
Quartz		3L 2S	2L 3S		4	m,p	
Pyrox					2		
Epidote					1		
Schist					2		

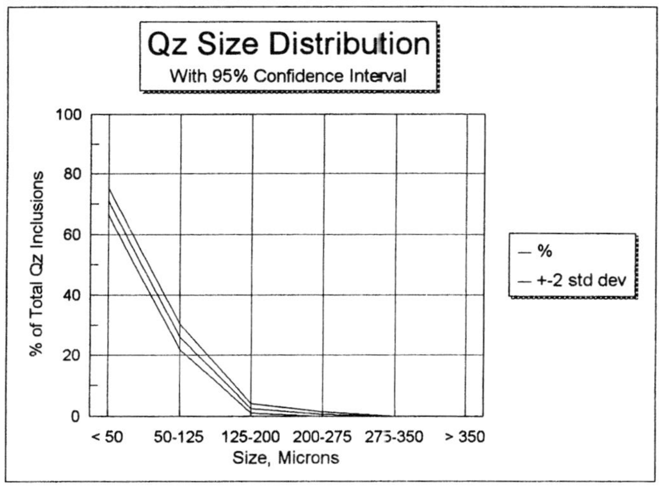

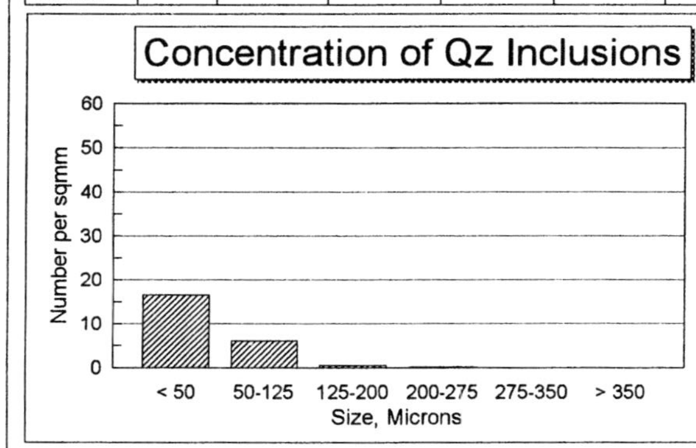

%matrix	90.41
%sand	5.18
%silt	4.40
conc<50	16.64
conc>50	6.88
%<50	70.75
%>50	29.25

APPENDIX C

hrp122	period	level	morph type	rim dia	thick	sqmm
H90/1217-507		2 1/2	BV	34	11.1	12.5

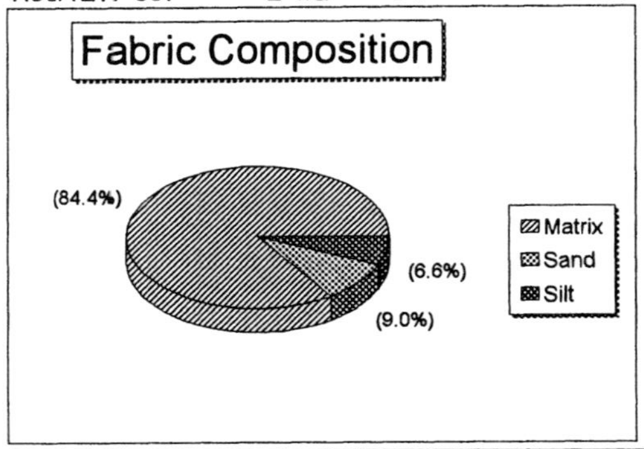

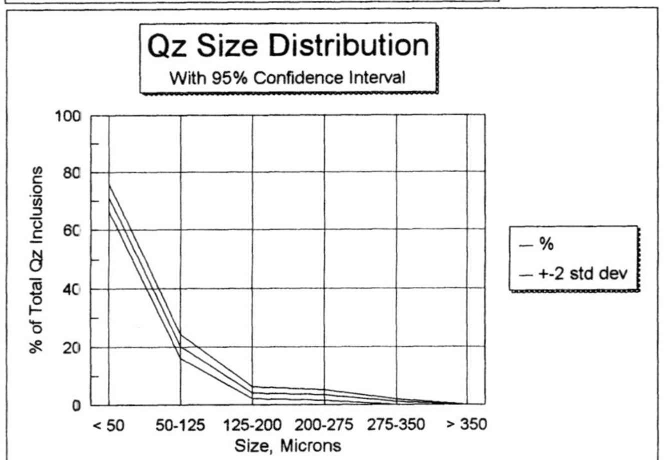

Qz Size	< 50	50-125	125-200	200-275	275-350	> 350	numpts
Number	268	76	16	13	4	0	377
%	71.09	20.16	4.24	3.45	1.06	0	
sigma	2.33	2.07	1.04	0.94	0.53	0	
p2sigma	75.76	24.29	6.32	5.33	2.12	0	
m2sigma	66.42	16.03	2.17	1.57	0.01	0	
Conc	21.44	6.08	1.28	1.04	0.32	0	
Fabric	Matrix	Sand	Silt	Micac	Biref	Colour	
	330	35	26	3	3	medbr	391

Minerals	Size	Ang	Sph	Wth	Conc	Type	CarOpa
PFs	2	2	2		2		
MFs					1		
Biotite	2			2	2		
Mus	3			2	4		
Amph					1		
Carb					2	2	2
Quartz		3L 2S	2L 3S		4	m,p	
Pyrox					1		
Epidote					1		
Schist					3		
Other						grog	

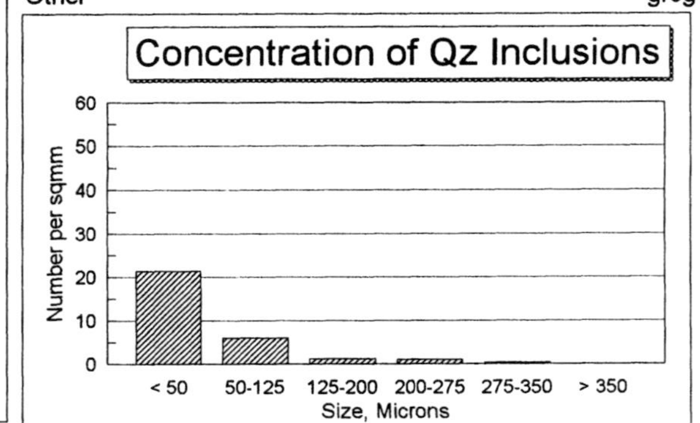

%matrix	84.40
%sand	8.95
%silt	6.65
conc<50	21.44
conc>50	8.72
%<50	71.09
%>50	28.91

hrp130	period	level	morph type	rim dia	thick	sqmm
H89/1049-3	2	1/2	bangle		4.5x4.5	

Not countable

Qz Size	< 50	50-125	125-200	200-275	275-350	> 350	numpts
Number						0	0
%							
sigma							
p2sigma							
m2sigma							
Conc							

Fabric	Matrix	Sand	Silt	Micac	Biref	Colour	
				4	4	medbr	0

Minerals	Size	Ang	Sph	Wth	Conc	Type	CarOpa
PFs					1		
MFs					1		
Biotite					1		
Mus	1				4		
Amph					1		
Carb					1		
Quartz		3	2		4	m,p	
Pyrox					1		
Epidote					1		
Schist					2		

APPENDIX C

hrp131	period level	morph type	rim dia	thick	sqmm
H89/1049-4	2 1/2	bangle		5.6x5.3	

Not countable

Qz Size	< 50	50-125	125-200	200-275	275-350	> 350	numpts
Number						0	0
%							
sigma							
p2sigma							
m2sigma							
Conc							
Fabric	Matrix	Sand	Silt	Micac	Biref	Colour	
				4	4	medbr	0

Minerals	Size	Ang	Sph	Wth	Conc	Type	CarOpa
PFs					1		
MFs					1		
Biotite					1		
Mus	1				4		
Amph					1		
Carb					1		
Quartz		3	2		4	m,p	
Pyrox					1		
Epidote					1		
Schist					1		

hrp132	period level	morph type	rim dia	thick	sqmm
H89/1049-5	2 1/2	bangle		5.5x4.4	18.75

Qz Size	< 50	50-125	125-200	200-275	275-350	> 350	numpts
Number	250	35	2	2	1	0	290
%	86.21	12.07	0.69	0.69	0.34	0	
sigma	2.02	1.91	0.49	0.49	0.34	0	
p2sigma	90.26	15.89	1.66	1.66	1.03	0	
m2sigma	82.16	8.24	-0.28	-0.28	-0.34	0	
Conc	13.33	1.87	0.11	0.11	0.0533	0	
Fabric	Matrix	Sand	Silt	Micac	Biref	Colour	
	369	3	19	4	4	grey	391

Minerals	Size	Ang	Sph	Wth	Conc	Type	CarOpa
PFs					1		
MFs					1		
Biotite	2			2	2		
Mus	2			2	4		
Amph					1		
Carb					1		
Quartz		2	3		4	m,p	
Pyrox					1		
Epidote					1		
Schist					2		

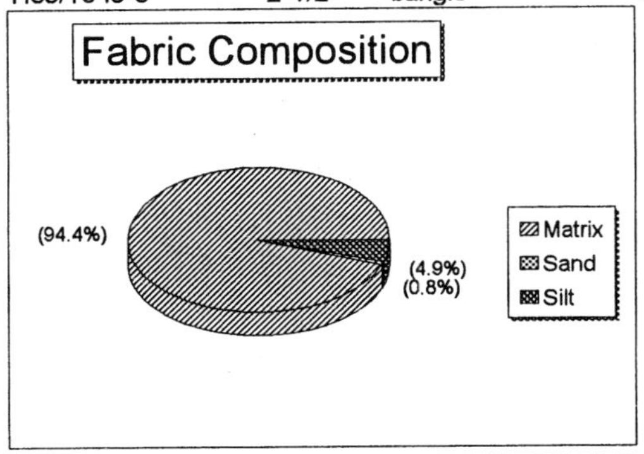

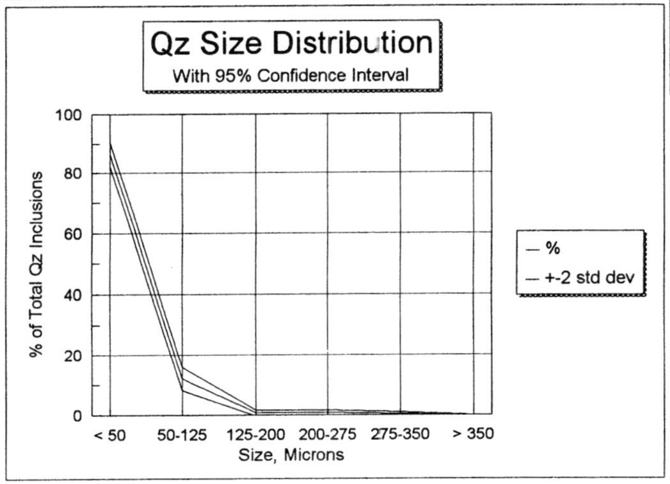

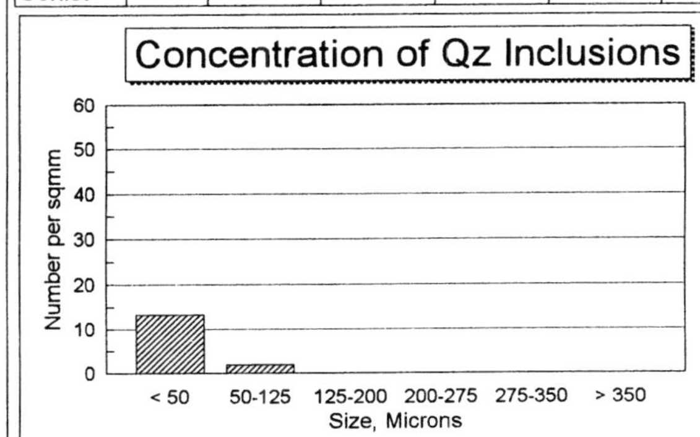

%matrix	94.37
%sand	0.77
%silt	4.86
conc<50	13.33
conc>50	2.13
%<50	86.21
%>50	13.79

hrp133	period level	morph type	rim dia	thick	sqmm
H89/1049-6	2 1/2	bangle		4.9x4.6	

Not countable

Qz Size	< 50	50-125	125-200	200-275	275-350	> 350	numpts
Number						0	0
%							
sigma							
p2sigma							
m2sigma							
Conc							
Fabric	Matrix	Sand	Silt	Micac	Biref	Colour	
							0

Minerals	Size	Ang	Sph	Wth	Conc	Type	CarOpa
PFs							
MFs							
Biotite							
Mus							
Amph							
Carb							
Quartz						m,p	
Pyrox							
Epidote							
Schist							

APPENDIX C

hrp134	period	level	morph type	rim dia	thick	sqmm
H89/1060-30	2	1/2	bangle		2.7x3.6	12.5

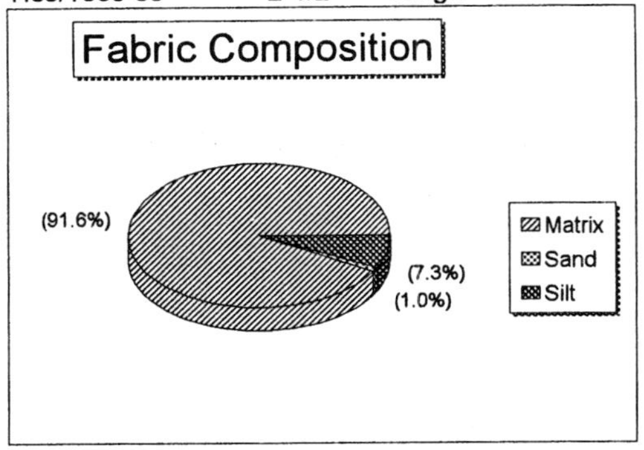

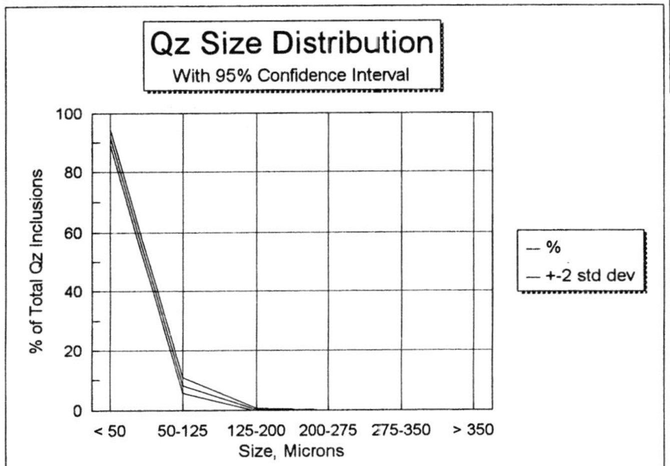

Qz Size	< 50	50-125	125-200	200-275	275-350	> 350	numpts
Number	405	36	1	0	0	0	442
%	91.63	8.14	0.23	0.00	0.00	0	
sigma	1.32	1.30	0.23	0.00	0.00	0	
p2sigma	94.26	10.75	0.68	0.00	0.00	0	
m2sigma	88.99	5.54	-0.23	0.00	0.00	0	
Conc	32.40	2.88	0.08	0.00	0	0	
Fabric	Matrix	Sand	Silt	Micac	Biref	Colour	
	362	4	29	3	3	dkgrey	395

Minerals	Size	Ang	Sph	Wth	Conc	Type	CarOpa
PFs					1		
MFs					1		
Biotite					1		
Mus	2			2	4		
Amph					1		
Carb					1		
Quartz					4	m,p	
Pyrox					1		
Epidote					1		
Schist					1		

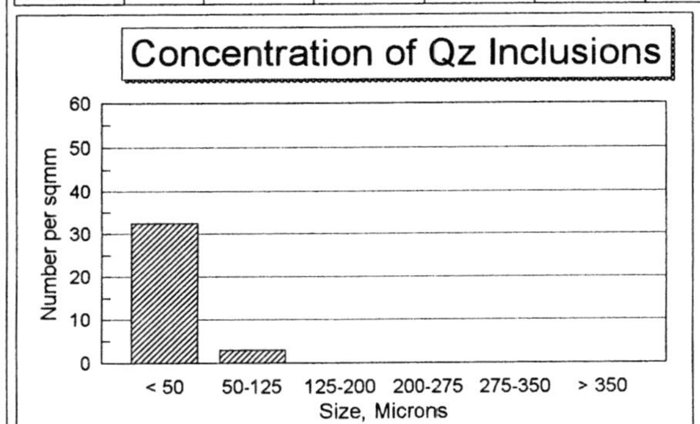

%matrix	91.65
%sand	1.01
%silt	7.34
conc<50	32.40
conc>50	2.96
%<50	91.63
%>50	8.37

APPENDIX C

hrp135	period	level	morph type	rim dia	thick	sqmm
H89/1060-42		2 1/2	bangle		7.8x7.5	12.5

Qz Size	< 50	50-125	125-200	200-275	275-350	> 350	numpts
Number	465	24	7	0	0	0	496
%	93.75	4.84	1.41	0.00	0.00	0	
sigma	1.09	0.96	0.53	0.00	0.00	0	
p2sigma	95.92	6.77	2.47	0.00	0.00	0	
m2sigma	91.58	2.91	0.35	0.00	0.00	0	
Conc	37.20	1.92	0.56	0.00	0	0	
Fabric	Matrix	Sand	Silt	Micac	Biref	Colour	
	379	3	18	3	2		400

Minerals	Size	Ang	Sph	Wth	Conc	Type	CarOpa
PFs	2	3	2		2		
MFs					1		
Biotite					2		
Mus	3			2	4		
Amph					1		
Carb					2	1	
Quartz		3	2		4	m,p	
Pyrox					1		
Epidote					1		
Schist					2		

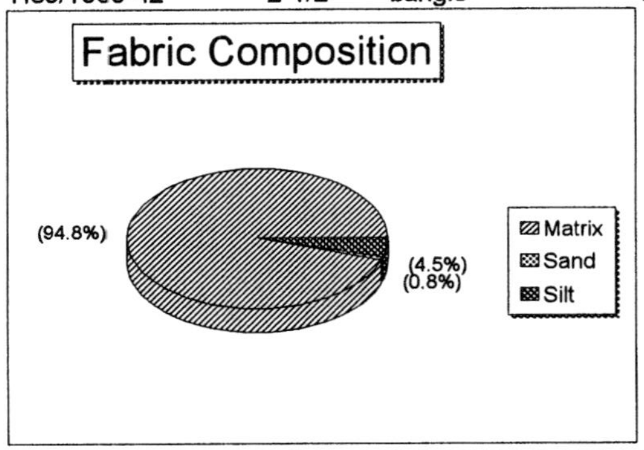

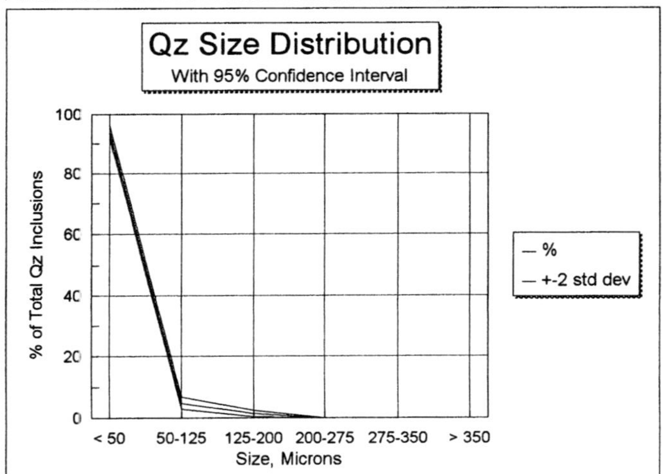

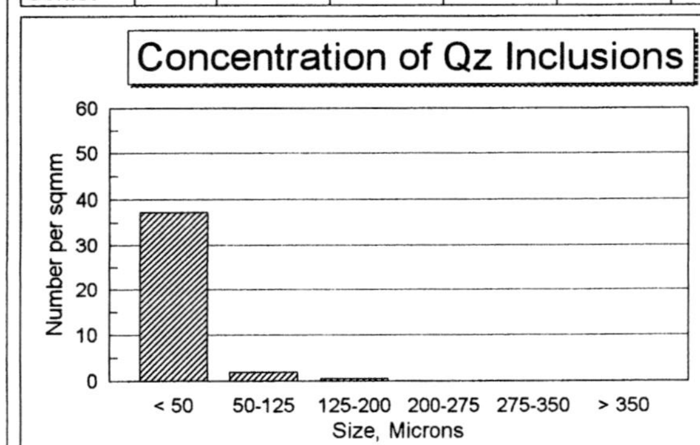

%matrix	94.75
%sand	0.75
%silt	4.50
conc<50	37.20
conc>50	2.48
%<50	93.75
%>50	6.25

hrp136	period	level	morph type	rim dia	thick	sqmm
H89/1060-73	2	1/2	bangle		12.4x21.0	6.25

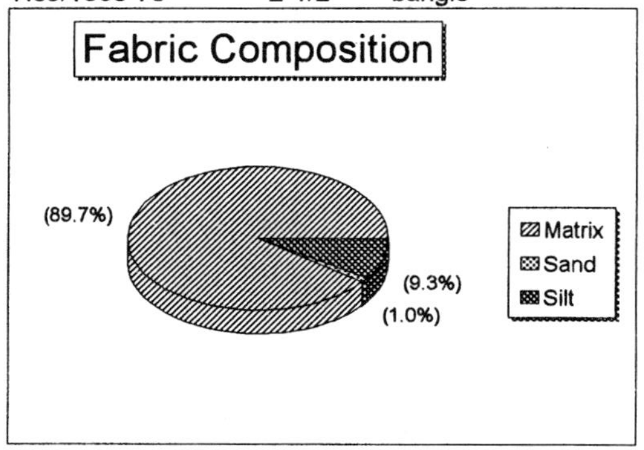

Qz Size	< 50	50-125	125-200	200-275	275-350	> 350	numpts
Number	348	10	0	0	0	0	358
%	97.21	2.79	0.00	0.00	0.00	0	
sigma	0.87	0.87	0.00	0.00	0.00	0	
p2sigma	98.95	4.54	0.00	0.00	0.00	0	
m2sigma	95.46	1.05	0.00	0.00	0.00	0	
Conc	55.68	1.60	0.00	0.00	0	0	

Fabric	Matrix	Sand	Silt	Micac	Biref	Colour	
	357	4	37	4	3	ltbrn	398

Minerals	Size	Ang	Sph	Wth	Conc	Type	CarOpa
PFs	1	3	2		2		
MFs					1		
Biotite	2			2	3		
Mus	3			2	4		
Amph					1		
Carb					1		
Quartz		3	2		4	m,p	
Pyrox					1		
Epidote					1		
Schist					2		

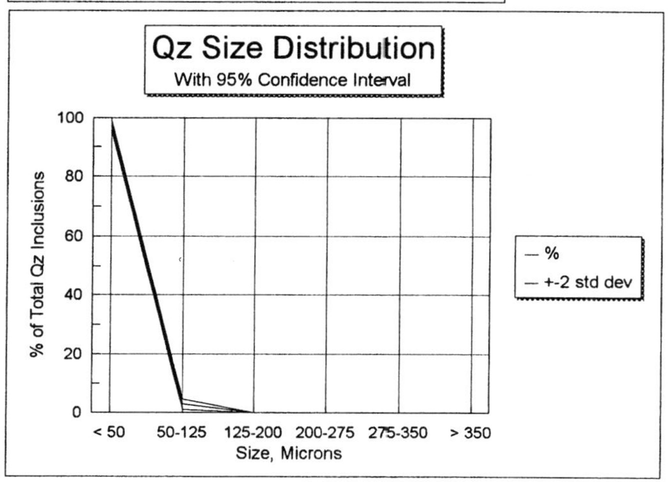

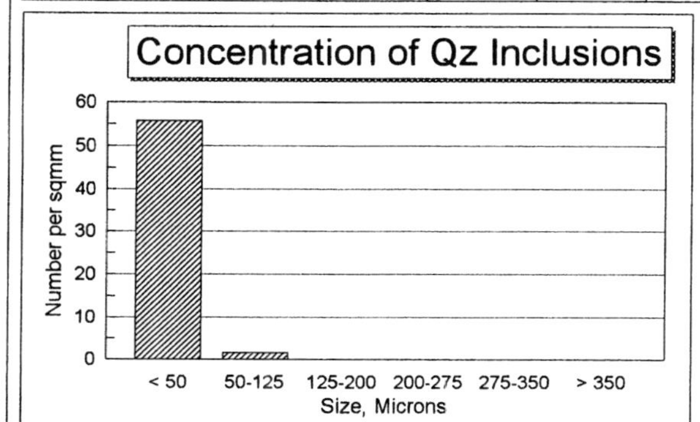

%matrix	89.70
%sand	1.01
%silt	9.30
conc<50	55.68
conc>50	1.60
%<50	97.21
%>50	2.79

hrp137	period level	morph type	rim dia	thick	sqmm
H89/1060-74	2 1/2	bangle		8.2x8.0	6.25

Qz Size	< 50	50-125	125-200	200-275	275-350	> 350	numpts
Number	303	7	2	0	0	0	312
%	97.12	2.24	0.64	0.00	0.00	0	
sigma	0.95	0.84	0.45	0.00	0.00	0	
p2sigma	99.01	3.92	1.54	0.00	0.00	0	
m2sigma	95.22	0.57	-0.26	0.00	0.00	0	
Conc	48.48	1.12	0.32	0.00	0	0	
Fabric	Matrix	Sand	Silt	Micac	Biref	Colour	
	363	3	31	3	2	dkbrn	397

Minerals	Size	Ang	Sph	Wth	Conc	Type	CarOpa
PFs	3	3	2		2		
MFs					1		
Biotite	2			2	2		
Mus	2			2	4		
Amph					1		
Carb					1		
Quartz		3	2		4	m,p	
Pyrox					2		
Epidote					1		
Schist					2	musc	

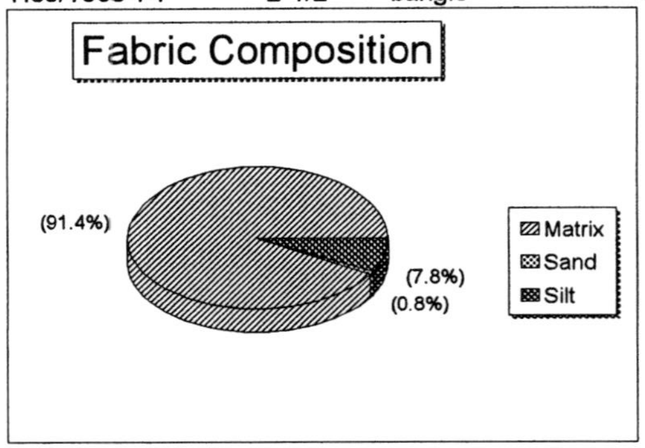

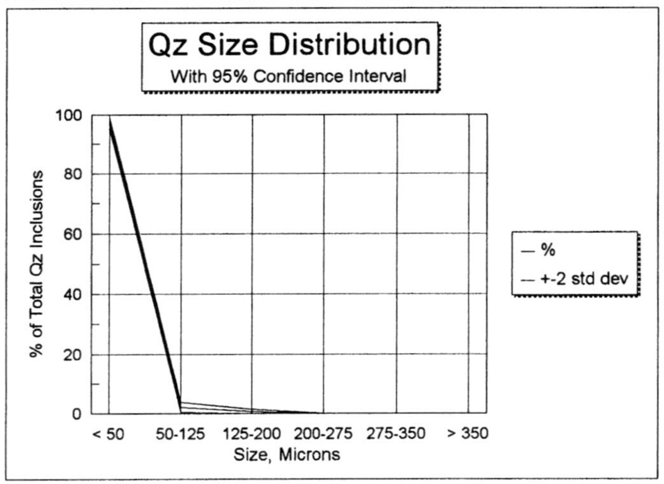

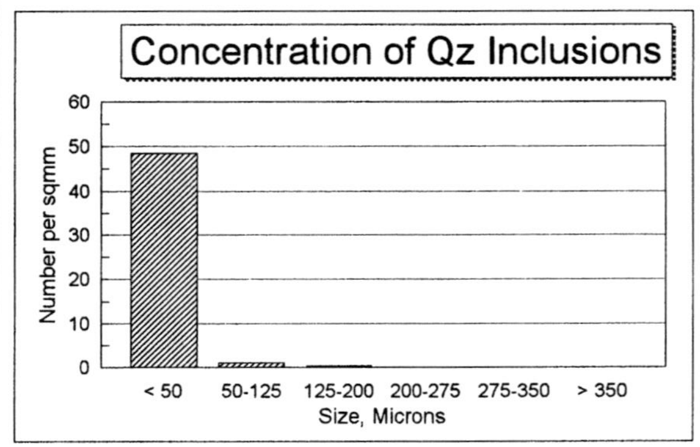

%matrix	91.44
%sand	0.76
%silt	7.81
conc<50	48.48
conc>50	1.44
%<50	97.12
%>50	2.88

hrp138	period level	morph type	rim dia	thick	sqmm
H89/1104-24	2 1/2	bangle		9.9x4.9	12.5

Qz Size	< 50	50-125	125-200	200-275	275-350	> 350	numpts
Number	396	12	5	0	0	0	413
%	95.88	2.91	1.21	0.00	0.00	0	
sigma	0.98	0.83	0.54	0.00	0.00	0	
p2sigma	97.84	4.56	2.29	0.00	0.00	0	
m2sigma	93.93	1.25	0.13	0.00	0.00	0	
Conc	31.68	0.96	0.40	0.00	0	0	
Fabric	Matrix	Sand	Silt	Micac	Biref	Colour	
	353	6	27	4	4	medbr	386
Minerals	Size	Ang	Sph	Wth	Conc	Type	CarOpa
PFs	3	3	2		2		
MFs					1		
Biotite	2			2	2		
Mus	3			2	4		
Amph					1		
Carb					1		
Quartz		3	2		4	m,p	
Pyrox					1		
Epidote					1		
Schist					2		

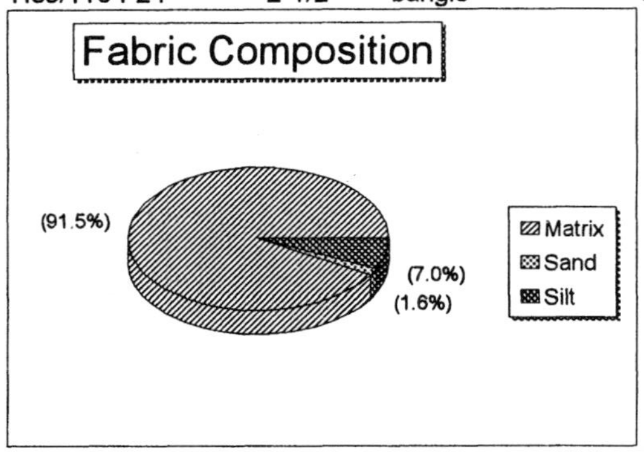

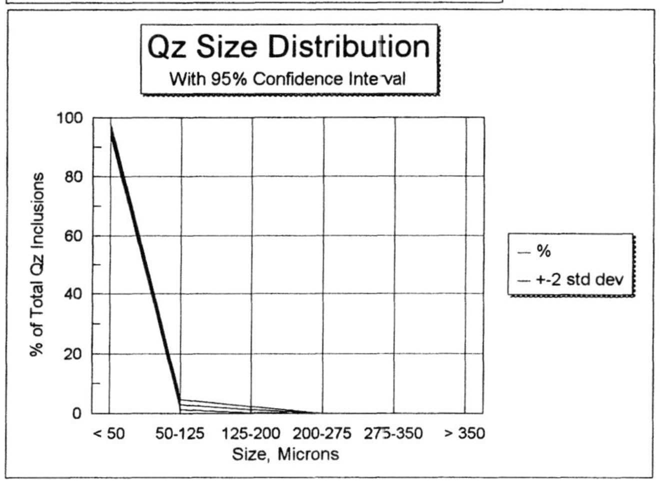

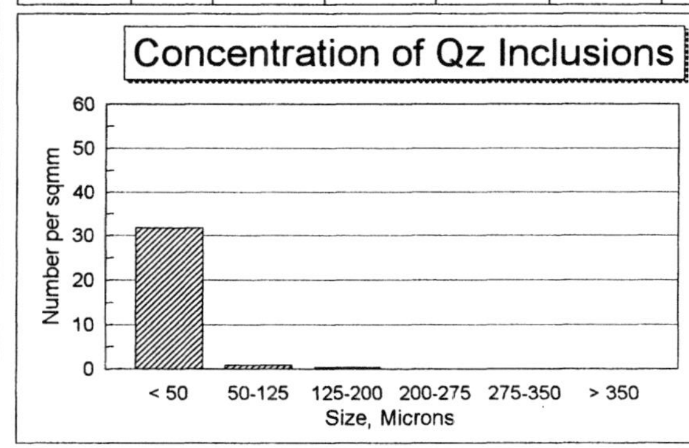

%matrix	91.45
%sand	1.55
%silt	6.99
conc<50	31.68
conc>50	1.36
%<50	95.88
%>50	4.12

APPENDIX C

hrp139	period level	morph type	rim dia	thick	sqmm
H89/1104-26	2 1/2	bangle		5.7x5.2	18.75

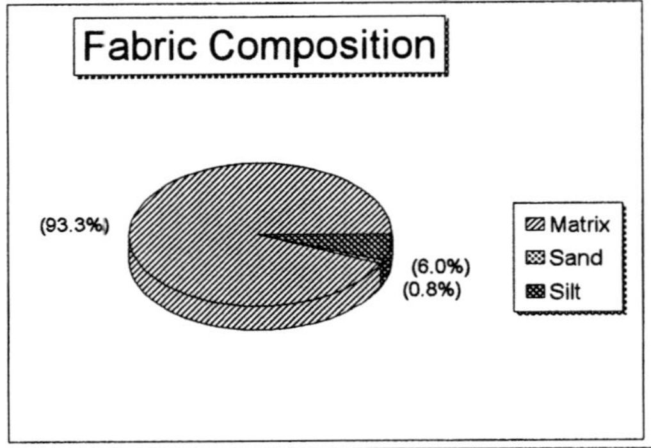

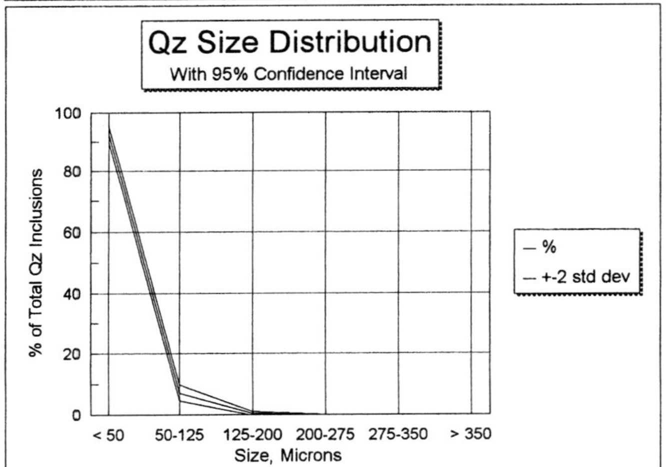

Qz Size	< 50	50-125	125-200	200-275	275-350	> 350	numpts
Number	372	29	2	0	0	0	403
%	92.31	7.20	0.50	0.00	0.00	0	
sigma	1.33	1.29	0.35	0.00	0.00	0	
p2sigma	94.96	9.77	1.20	0.00	0.00	0	
m2sigma	89.65	4.62	-0.20	0.00	0.00	0	
Conc	19.84	1.55	0.11	0.00	0	0	
Fabric	Matrix	Sand	Silt	Micac	Biref	Colour	
	373	3	24	4	4	ltgrey	400

Minerals	Size	Ang	Sph	Wth	Conc	Type	CarOpa
PFs					1		
MFs					1		
Biotite	2			2	2		
Mus	2			2	4		
Amph					1		
Carb					1		
Quartz		2	2		4	m,p	
Pyrox					1		
Epidote					2		
Schist					1		

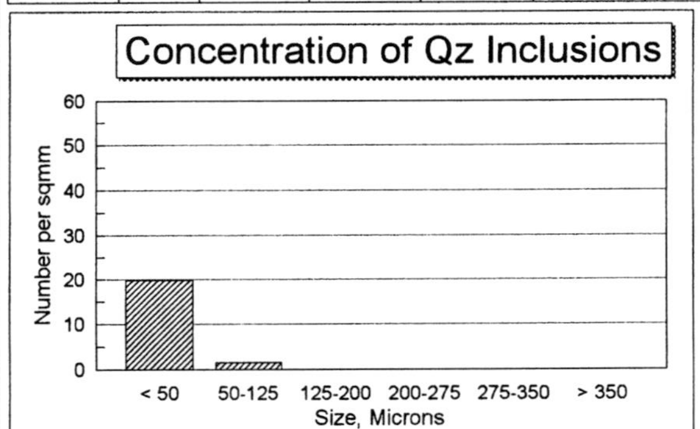

%matrix	93.25
%sand	0.75
%silt	6.00
conc<50	19.84
conc>50	1.65
%<50	92.31
%>50	7.69

hrp140	period	level	morph type	rim dia	thick	sqmm
H89/1104-27	2	1/2	bangle		7.8x4.9	12.5

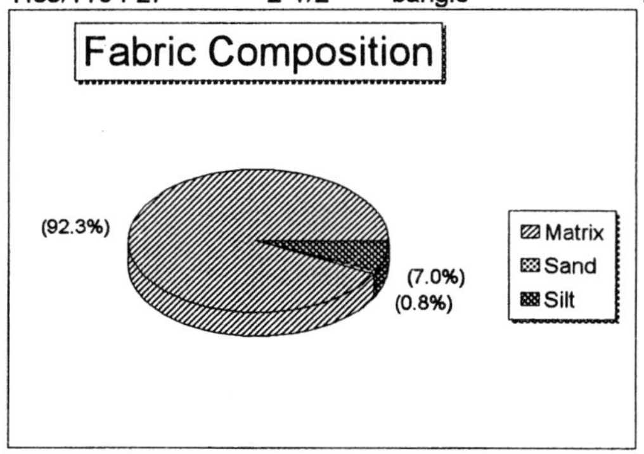

Qz Size	< 50	50-125	125-200	200-275	275-350	> 350	numpts
Number	354	41	6	0	0	0	401
%	88.28	10.22	1.50	0.00	0.00	0	
sigma	1.61	1.51	0.61	0.00	0.00	0	
p2sigma	91.49	13.25	2.71	0.00	0.00	0	
m2sigma	85.07	7.20	0.28	0.00	0.00	0	
Conc	28.32	3.28	0.48	0.00	0	0	
Fabric	Matrix	Sand	Silt	Micac	Biref	Colour	
	369	3	28	4	4	medbr	400

Minerals	Size	Ang	Sph	Wth	Conc	Type	CarOpa
PFs					1		
MFs					1		
Biotite	3			2	3		
Mus	3			2	4		
Amph					1		
Carb					1		
Quartz		3	2		4	m,p	
Pyrox					1		
Epidote					1		
Schist					1		

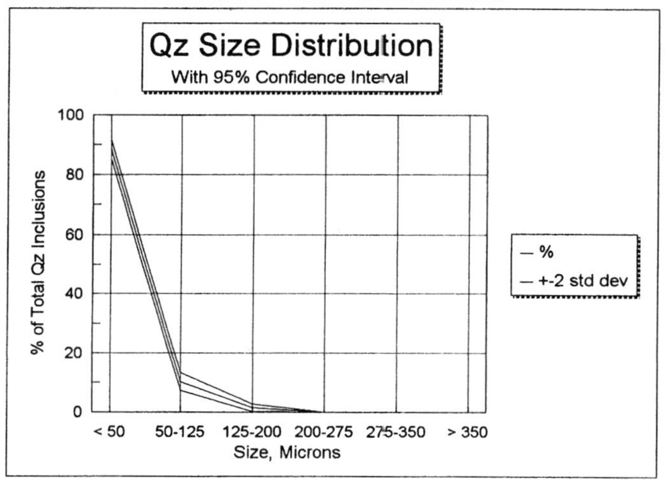

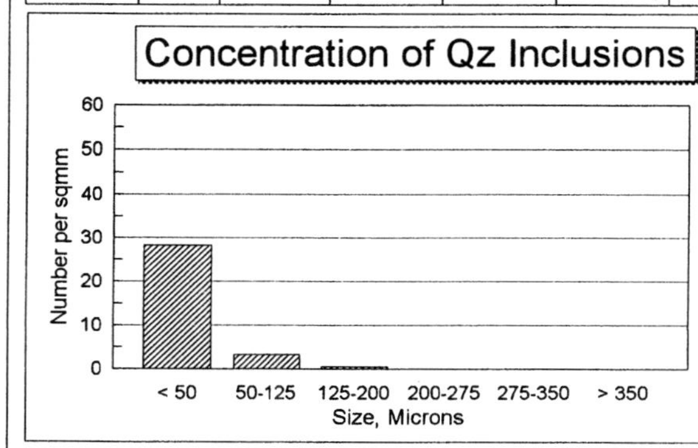

%matrix	92.25
%sand	0.75
%silt	7.00
conc<50	28.32
conc>50	3.76
%<50	88.28
%>50	11.72

APPENDIX C

hrp141	period	level	morph type	rim dia	thick	sqmm
H89/1104-29		2 1/2	bangle		6.3x5.3	12.5

Qz Size	< 50	50-125	125-200	200-275	275-350	> 350	numpts
Number	348	67	6	0	0	0	421
%	82.66	15.91	1.43	0.00	0.00	0	
sigma	1.85	1.78	0.58	0.00	0.00	0	
p2sigma	86.35	19.48	2.58	0.00	0.00	0	
m2sigma	78.97	12.35	0.27	0.00	0.00	0	
Conc	27.84	5.36	0.48	0.00	0	0	
Fabric	Matrix	Sand	Silt	Micac	Biref	Colour	
	351	19	29	3	3	brgrey	399

Minerals	Size	Ang	Sph	Wth	Conc	Type	CarOpa
PFs	2	3	2		2		
MFs					1		
Biotite	2				2		
Mus	3				4		
Amph					1		
Carb					1		
Quartz		3	2		4	m,p	
Pyrox					1		
Epidote					2		
Schist					1		

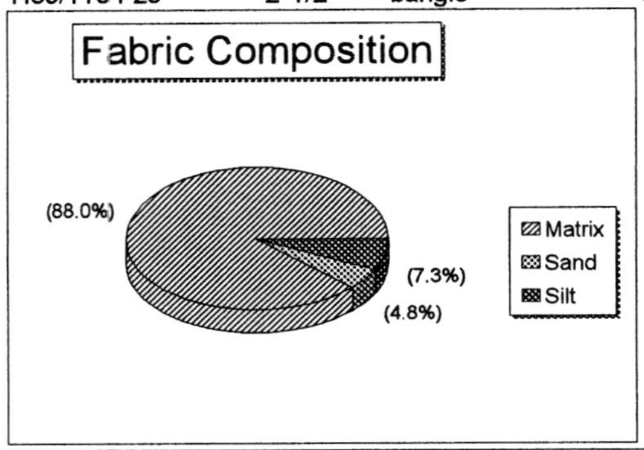

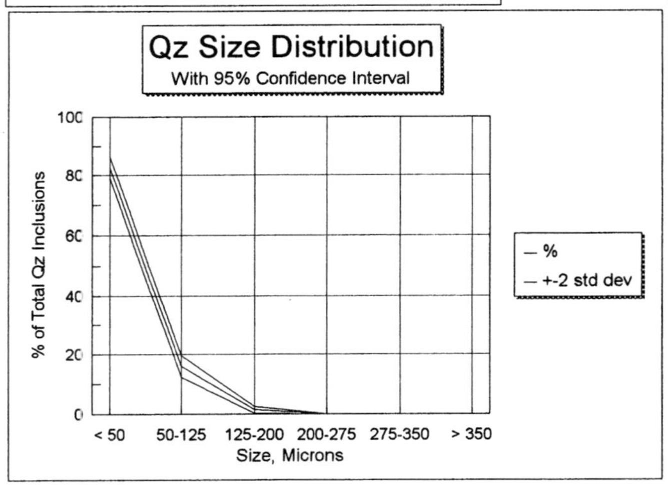

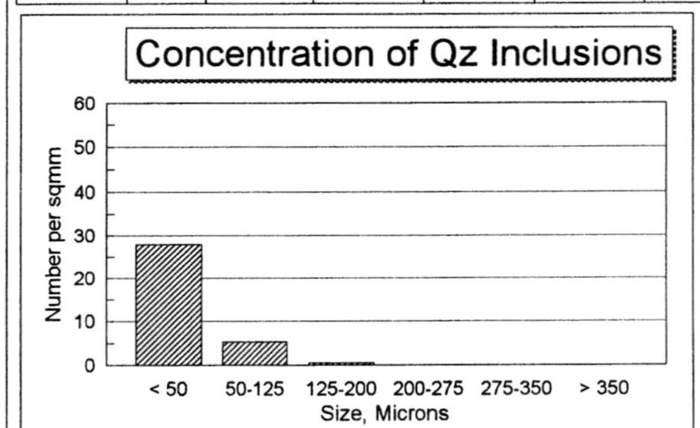

%matrix	87.97
%sand	4.76
%silt	7.27
conc<50	27.84
conc>50	5.84
%<50	82.66
%>50	17.34

hrp142	period	level	morph type	rim dia	thick	sqmm
H89/1106-7		2 1/2	bangle		5.8x3.9	18.75

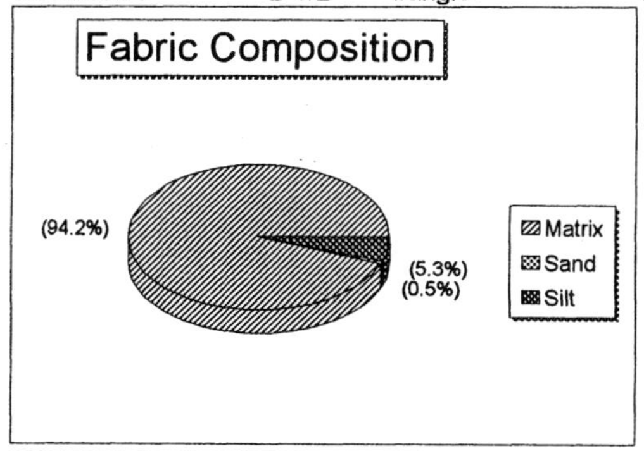

Qz Size	< 50	50-125	125-200	200-275	275-350	> 350	numpts
Number	366	27	1	0	0	0	394
%	92.89	6.85	0.25	0.00	0.00	0	
sigma	1.29	1.27	0.25	0.00	0.00	0	
p2sigma	95.48	9.40	0.76	0.00	0.00	0	
m2sigma	90.30	4.31	-0.25	0.00	0.00	0	
Conc	19.52	1.44	0.05	0.00	0	0	
Fabric	Matrix	Sand	Silt	Micac	Biref	Colour	
	372	2	21	4	4	biref	395

Minerals	Size	Ang	Sph	Wth	Conc	Type	CarOpa
PFs					1		
MFs					1		
Biotite	2			2	2		
Mus	3			2	4		
Amph					1		
Carb					1		
Quartz		3	2		4	m,p	
Pyrox					1		
Epidote					1		
Schist					1		
Other						arf	

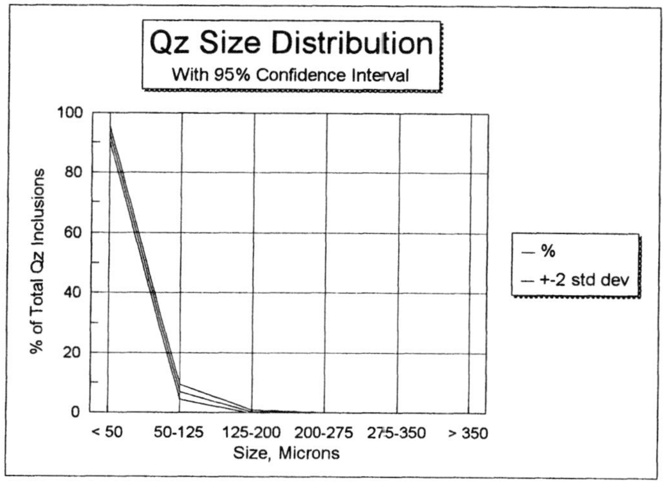

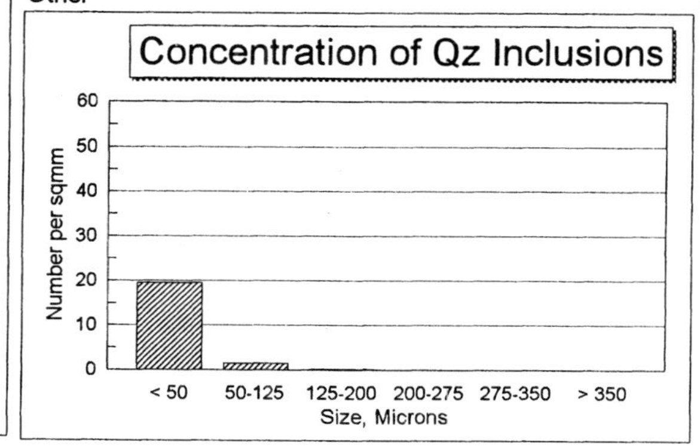

%matrix	94.18
%sand	0.51
%silt	5.32
conc<50	19.52
conc>50	1.49
%<50	92.89
%>50	7.11

APPENDIX C

hrp143	period	level	morph type	rim dia	thick	sqmm
H89/1106-?		2 1/2	bangle		6.4x3.6	18.75

Qz Size	< 50	50-125	125-200	200-275	275-350	> 350	numpts
Number	365	38	7	0	0	0	410
%	89.02	9.27	1.71	0.00	0.00	0	
sigma	1.54	1.43	0.64	0.00	0.00	0	
p2sigma	92.11	12.13	2.99	0.00	0.00	0	
m2sigma	85.94	6.40	0.43	0.00	0.00	0	
Conc	19.47	2.03	0.37	0.00	0	0	
Fabric	Matrix	Sand	Silt	Micac	Biref	Colour	
	363	10	27	4	4	ltbrn	400

Minerals	Size	Ang	Sph	Wth	Conc	Type	CarOpa
PFs	3	3	2		2		
MFs					1		
Biotite	2			2	2		
Mus	2			2	4		
Amph					1		
Carb					2		
Quartz		3	2		4	m,p	
Pyrox					1		
Epidote					1		
Schist					2		
Other						arf	

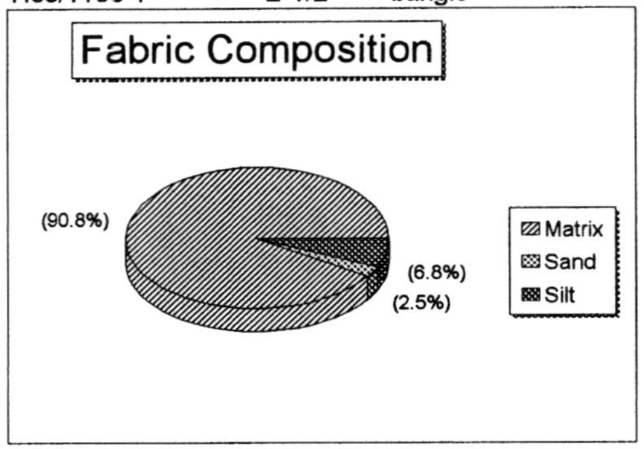

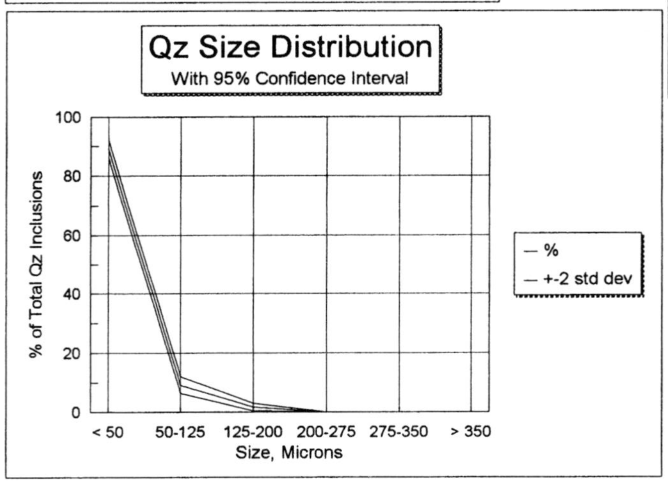

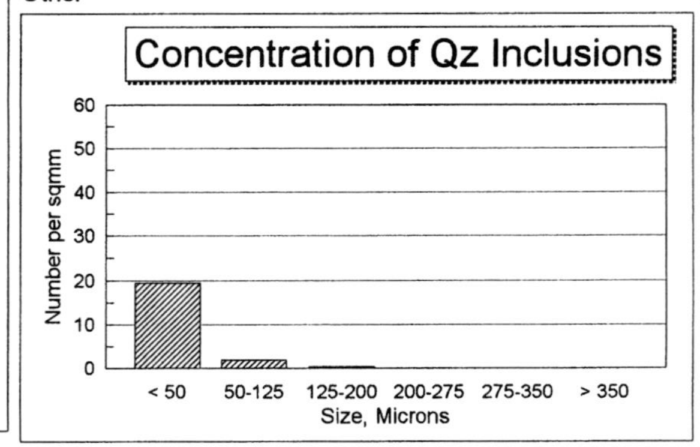

%matrix	90.75
%sand	2.50
%silt	6.75
conc<50	19.47
conc>50	2.40
%<50	89.02
%>50	10.98

APPENDIX C

hrp144	period	level	morph type	rim dia	thick	sqmm
H89/1106-?	2	1/2	bangle		6.7x4.0	12.5

Qz Size	< 50	50-125	125-200	200-275	275-350	> 350	numpts
Number	441	33	5	0	0	0	479
%	92.07	6.89	1.04	0.00	0.00	0	
sigma	1.23	1.16	0.46	0.00	0.00	0	
p2sigma	94.54	9.20	1.97	0.00	0.00	0	
m2sigma	89.60	4.57	0.12	0.00	0.00	0	
Conc	35.28	2.64	0.40	0.00	0	0	
Fabric	Matrix	Sand	Silt	Micac	Biref	Colour	
	369	5	26	4	4	medbr	400
Minerals	Size	Ang	Sph	Wth	Conc	Type	CarOpa
PFs	2	3	2		2		
MFs					1		
Biotite	2			2	2		
Mus	2			2	4		
Amph					1		
Carb					2		
Quartz		3	2		4	m,p	
Pyrox					1		
Epidote					1		
Schist					1		

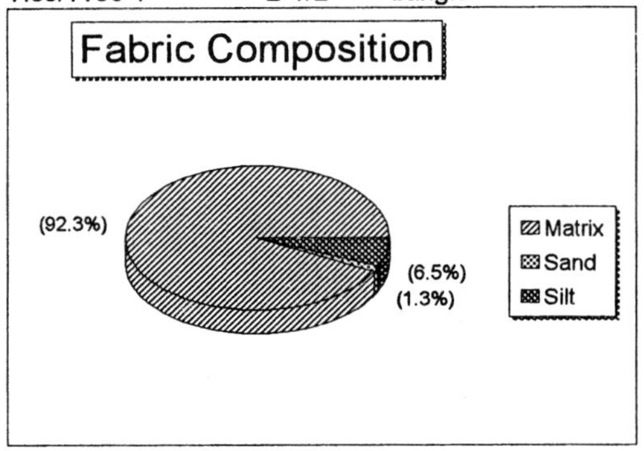

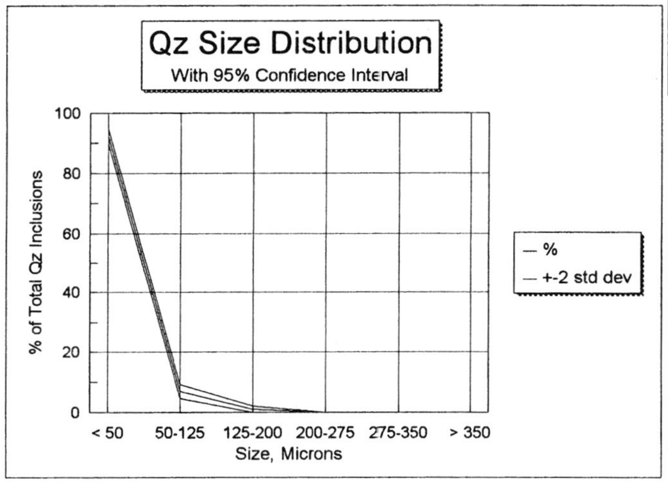

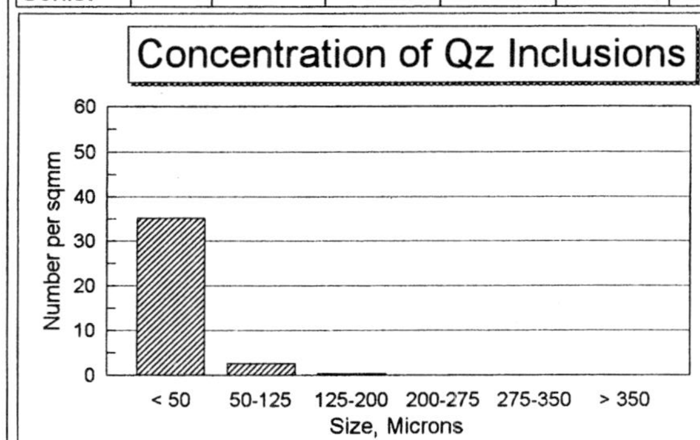

%matrix	92.25
%sand	1.25
%silt	6.50
conc<50	35.28
conc>50	3.04
%<50	92.07
%>50	7.93

hrp145	period	level	morph type	rim dia	thick	sqmm
H89/1106-16		2 1/2	bangle		8.3x3.9	6.25

Qz Size	< 50	50-125	125-200	200-275	275-350	> 350	numpts
Number	294	60	6	2	0	0	362
%	81.22	16.57	1.66	0.55	0.00	0	
sigma	2.05	1.95	0.67	0.39	0.00	0	
p2sigma	85.32	20.48	3.00	1.33	0.00	0	
m2sigma	77.11	12.67	0.32	-0.23	0.00	0	
Conc	47.04	9.60	0.96	0.32	0	0	
Fabric	Matrix	Sand	Silt	Micac	Biref	Colour	
	327	16	55	4	4	medbr	398

Minerals	Size	Ang	Sph	Wth	Conc	Type	CarOpa
PFs	2	3	2		2		
MFs					1		
Biotite	2			2	3		
Mus	3			2	4		
Amph					1		
Carb					1		
Quartz		3	2		4	m,p	
Pyrox					1		
Epidote					2		
Schist					1		

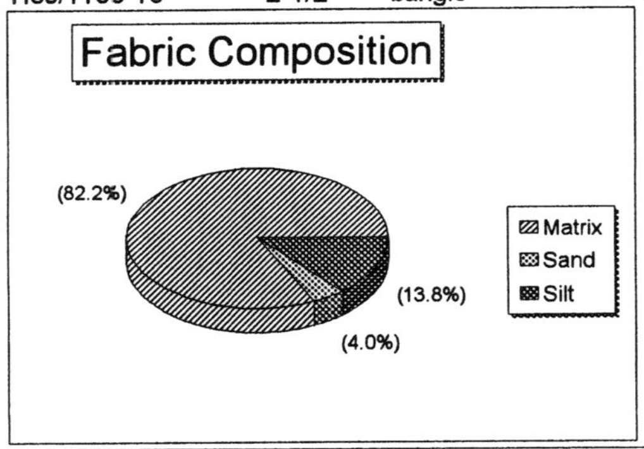

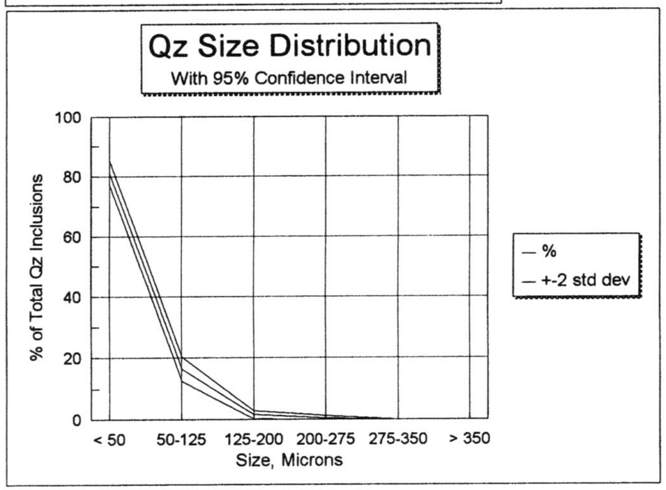

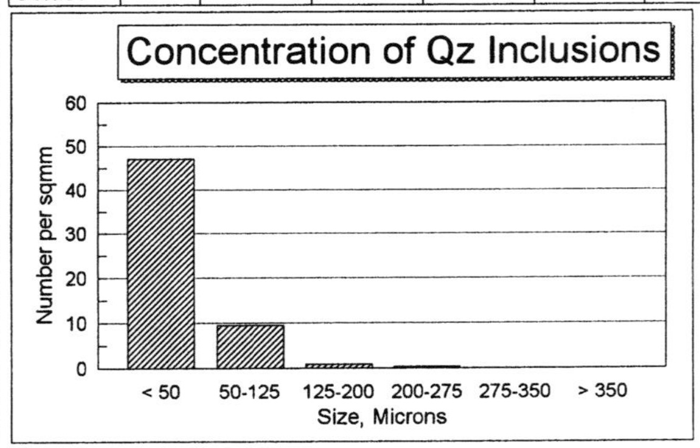

%matrix	82.16
%sand	4.02
%silt	13.82
conc<50	47.04
conc>50	10.88
%<50	81.22
%>50	18.78

hrp146	period	level	morph type	rim dia	thick	sqmm
H89/1114-5		2 1/2	bangle		4.4x3.7	6.25

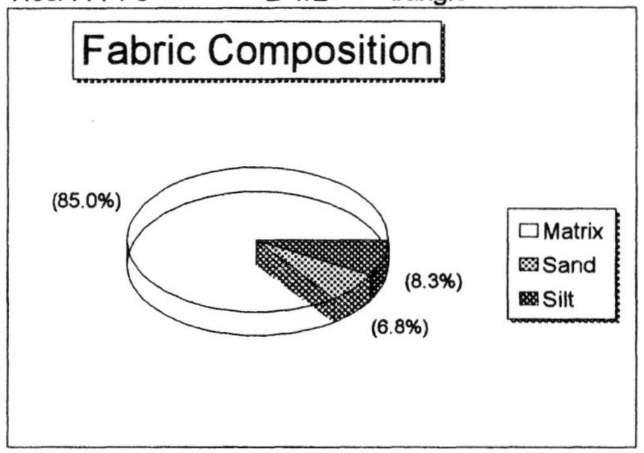

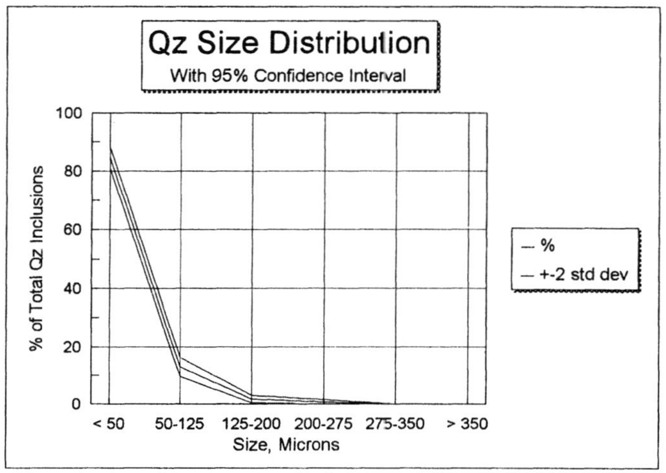

Qz Size	< 50	50-125	125-200	200-275	275-350	> 350	numpts
Number	338	52	7	3	0	0	400
%	84.50	13.00	1.75	0.75	0.00	0	
sigma	1.81	1.68	0.66	0.43	0.00	0	
p2sigma	88.12	16.36	3.06	1.61	0.00	0	
m2sigma	80.88	9.64	0.44	-0.11	0.00	0	
Conc	54.08	8.32	1.12	0.48	0	0	
Fabric	Matrix	Sand	Silt	Micac	Biref	Colour	
	339	27	33	3	3	dkbrn	399
Minerals	Size	Ang	Sph	Wth	Conc	Type	CarOpa
PFs	2	3	2		2		
MFs					1		
Biotite	2			2	2		
Mus	3			2	4		
Amph					1		
Carb					1		
Quartz		3	2		4	m,p	
Pyrox					2		
Epidote					1		
Schist					2		

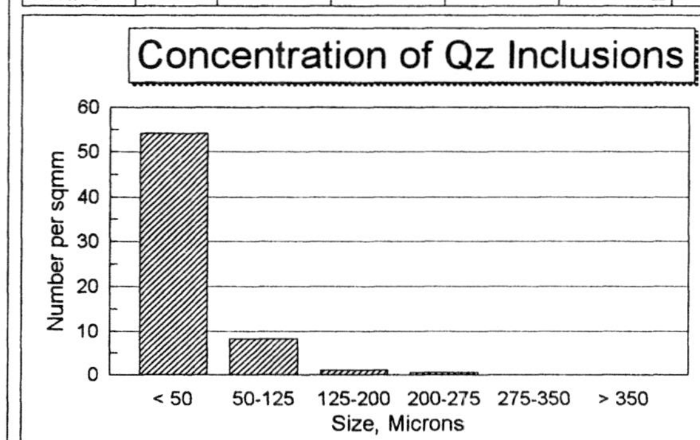

%matrix	84.96
%sand	6.77
%silt	8.27
conc<50	54.08
conc>50	9.92
%<50	84.50
%>50	15.50

APPENDIX C

hrp147	period	level	morph type	rim dia	thick	sqmm
H89/1114-8	2	1/2	bangle		2.1x3.9	

Not countable

Qz Size	< 50	50-125	125-200	200-275	275-350	> 350	numpts
Number							0
%							
sigma							
p2sigma							
m2sigma							
Conc							
Fabric	Matrix	Sand	Silt	Micac	Biref	Colour	
				4	4	mica	0

Minerals	Size	Ang	Sph	Wth	Conc	Type	CarOpa
PFs					1		
MFs					1		
Biotite	2			2	2		
Mus	3			2	4		
Amph					1		
Carb					1		
Quartz		2	3		4	m,p	
Pyrox					1		
Epidote					2		
Schist					1		

hrp148	period	level	morph type	rim dia	thick	sqmm
local pot	n/a	n/a	bangle			12.5

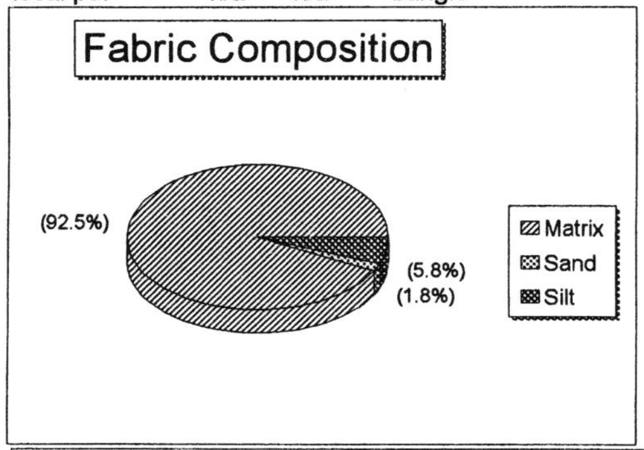

Qz Size	< 50	50-125	125-200	200-275	275-350	> 350	numpts
Number	427	41	3	0	0	0	471
%	90.66	8.70	0.64	0.00	0.00	0	
sigma	1.34	1.30	0.37	0.00	0.00	0	
p2sigma	93.34	11.30	1.37	0.00	0.00	0	
m2sigma	87.98	6.11	-0.10	0.00	0.00	0	
Conc	34.16	3.28	0.24	0.00	0	0	
Fabric.	Matrix	Sand	Silt	Micac	Biref	Colour	
	368	7	23	4	4	medbr	398

Minerals	Size	Ang	Sph	Wth	Conc	Type	CarOpa
PFs	2	3	2		2		
MFs					1		
Biotite	2			2	2		
Mus	3			2	4		
Amph					2		
Carb					1		
Quartz		3	2		4	m,p	
Pyrox					1		
Epidote					1		
Schist					2		

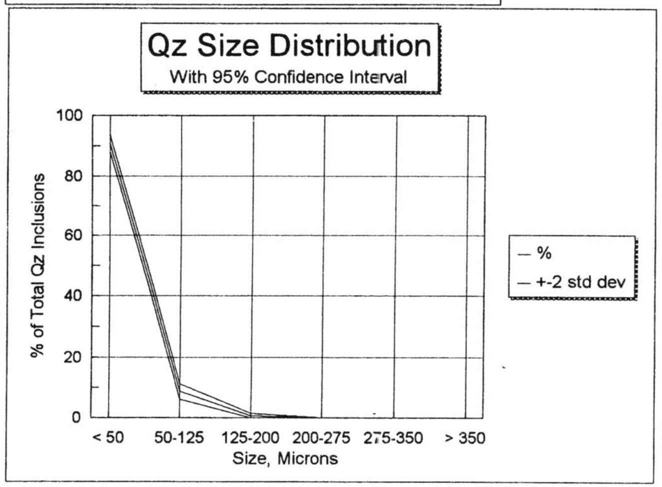

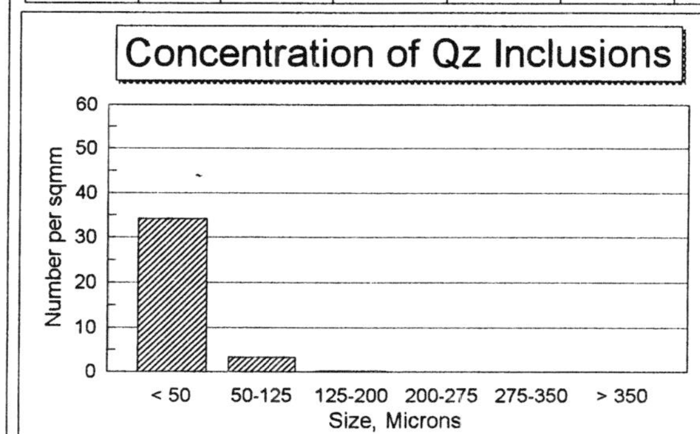

%matrix	92.46
%sand	1.76
%silt	5.78
conc<50	34.16
conc>50	3.52
%<50	90.66
%>50	9.34

APPENDIX C